THE ROUTLEDGE HANDBOOK OF WASTE STUDIES

The Routledge Handbook of Waste Studies offers a comprehensive survey of the new field of waste studies, critically interrogating the cultural, social, economic, and political systems within which waste is created, managed, and circulated.

While scholars have not settled on a definitive categorization of what waste studies is, more and more researchers claim that there is a distinct cluster of inquiries, concepts, theories, and key themes that constitute this field. In this handbook the editors and contributors explore the research questions, methods, and case studies preoccupying academics working in this field, in an attempt to develop a set of criteria by which to define and understand waste studies as an interdisciplinary field of study.

This handbook will be invaluable to those wishing to broaden their understanding of waste studies and to students and practitioners of geography, sociology, anthropology, history, environment, and sustainability studies.

Zsuzsa Gille is Professor of Sociology and Director of Global Studies at the University of Illinois at Urbana-Champaign. She is the author of *Paprika, Foie Gras, and Red Mud: The Politics of Materiality in the European Union* (2016) and *From the Cult of Waste to the Trash Heap of History: The Politics of Waste in Socialist and Postsocialist Hungary* (2007—recipient of honorable mention of the AAASS Davis Prize).

Josh Lepawsky is Professor of Geography at Memorial University, Canada. He is author of *Reassembling Rubbish: Worlding Electronic Waste* and "Planet of fixers? Mapping the middle grounds of independent and do-it-yourself information and communication technology maintenance and repair".

THE ROUTLEDGE HANDBOOK OF WASTE STUDIES

Edited by Zsuzsa Gille and Josh Lepawsky

Routledge
Taylor & Francis Group

LONDON AND NEW YORK

First published 2022
by Routledge
2 Park Square, Milton Park, Abingdon, Oxon OX14 4RN

and by Routledge
605 Third Avenue, New York, NY 10158

Routledge is an imprint of the Taylor & Francis Group, an informa business

British Library Cataloguing-in-Publication Data
A catalogue record for this book is available from the British Library

Library of Congress Cataloging-in-Publication Data
Names: Gille, Zsuzsa, editor. | Lepawsky, Josh, 1972- editor.
Title: The Routledge handbook of waste studies / edited by Zsuzsa Gille and Josh Lepawsky.
Description: Abingdon, Oxon ; New York, NY : Routledge, 2022. |
Series: Routledge international handbooks | Includes bibliographical references and index. |
Identifiers: LCCN 2021036980 (print) | LCCN 2021036981 (ebook) | ISBN 9780367894207 (hardback) |
ISBN 9781032188959 (paperback) | ISBN 9781003019077 (ebook)
Subjects: LCSH: Refuse and refuse disposal--Study and teaching. |
Recycling (Waste, etc.)--Study and teaching.
Classification: LCC TD793.2 .R68 2022 (print) |
LCC TD793.2 (ebook) | DDC 363.72/8--dc23
LC record available at https://lccn.loc.gov/2021036980
LC ebook record available at https://lccn.loc.gov/2021036981

ISBN: 978-0-367-89420-7 (hbk)
ISBN: 978-1-032-18895-9 (pbk)
ISBN: 978-1-003-01907-7 (ebk)

DOI: 10.4324/9781003019077

Typeset in Bembo
by MPS Limited, Dehradun

CONTENTS

List of figures and tables   *viii*

List of contributors   *x*

PART I

Introducing the field of waste studies   **1**

1   Introduction: waste studies as a field   3
    Zsuzsa Gille and Josh Lepawsky

2   At home with the waste scholar   20
    Zsuzsa Gille, Josh Lepawsky, Catherine Alexander, and Nicky Gregson

PART II

Questions waste scholars ask   **29**

3   Matter out of place   31
    Max Liboiron

4   Waste and whiteness   41
    Joshua O. Reno and Britt Halvorson

5   Landfill life and the many lives of landfills   55
    Patrick O'Hare

6   Reading the signs: some ways waste is framed in Tunisia   68
    Jamie Furniss

7   Unmaking the made: the troubled temporalities of waste   88
    Heike Weber

8   Commodification and respect: indigenous contributions to the so-
    ciology of waste   103
    Michelle Schmidt

PART III
Methods waste scholars use   **119**

 9   Comparative methods for the study of waste   121
     Raul Pacheco-Vega

10   Teaching critical waste studies in higher education   139
     Kate Parizeau

11   Hunting for hidden treasures: a research methodology on China's
     informal recycling sector   154
     Benjamin Steuer

12   Waste metrics from the ground up   169
     Samantha MacBride

13   The potential role of gamification: an innovative intervention method
     in waste studies   196
     Tammara Soma, Belinda Li, and Virginia Maclaren

PART IV
Cases waste scholars investigate   **209**

14   The experience of nuclear waste   211
     Romain J. Garcier

15   Uranium legacies and settler-colonial imaginaries: nuclear waste as
     history, proximity, and colonial matter   224
     Emily Potter

16   Brownfields as waste/race governance: U.S. contaminated property
     redevelopment and racial capitalism   238
     Shiloh Krupar

Contents

17 Of ships of doom and icebergs: early perspectives on the global hazardous waste trade   254
   Kate O'Neill

18 Oil-wasting: the necroaesthetics of energy expenditure   267
   Amanda Boetzkes

19 Waste-picker organizations and urban sustainability   275
   Jutta Gutberlet

20 Waste, labor, and livelihoods in South Africa   291
   Mary Lawhon, Nate Millington, and Kathleen Stokes

21 Prepping for the [insert here] apocalypse and wasting the future   305
   Myra J. Hird and Jacob Riha

Index   322

FIGURES AND TABLES

Figures

5.1 Description: damaged waste containers lie beyond the fenced-in perimeter of Montevideo's Felipe Cardoso landfill. Image: Patrick O'Hare 60

5.2 Description: Waste-pickers recovering metals from a truck at Montevideo's Felipe Cardoso landfill as birds fly overhead. Image: Patrick O'Hare 61

6.1 Tunisian Transit Authority: Let's collaborate together, so that we can work in a space that is clean. Photo by the author 75

6.2 La Soukra (Tunis): The cleanliness of the city is a matter of awareness and responsibility. Photograph by the author 77

6.3 Bab al-Khadra (Tunis): Those who put garbage (here) will not receive (God's) reward or blessing. Photograph by the author 80

6.4 La Marsa: "Our environment is our life. So let's all cooperate together in protecting it." Photograph by the author 81

6.5 Enfidha, on the highway between Tunis and Sousse: Labib, the Mascot of the Environment. Photograph by the author 83

6.6 Mutuelleville (Tunis): Do not throw waste and dirty things here. Thank you. – the residents of the house. Photograph by the author 84

7.1 Debris which has recently been recovered from a site on Berlin's periphery where waste from Berlin households had once been used to meliorate the grounds. Copyright: Oswin Nikolaus, 2021 92

7.2 Dismantling mainframes in Switzerland, 1968 by courtesy of SRF 95

7.3 Dismantling mainframes in Switzerland, 1968 by courtesy of SRF 96

7.4 The Ouroboros icon on the cover of an EPA study on recycling markets 97

7.5 Nineteenth-century visualization of the "life stair", illustrating the stages of human life from cradle to grave in ten-year stages. Source: Museum Europäischer Kulturen der Staatlichen Museen zu Berlin—Preußischer Kulturbesitz, D (33 R 497) 488/1977 99

8.1 Author's diagram depicting respect-based relationships with nature 109

11.1 Composition of household waste in urban China (adapted from Steuer, Ramusch, and Salhofer 2017b) 156

11.2 Waste collectors on tricycles in Jingmen (Hubei Province) (left) and Changchun (Jilin Province) (right) (© the author) 162
11.3 A middleman's truck in Changchun (Jilin Province) (© the author) 162
12.1 Boundary of operational jurisdiction 170
12.2 The flow of generation to a first disposition point 170
12.3 Generation 171
12.4 Calculating generation 172
12.5 Ranges of average national waste generation by region 172
12.6 Disposition 173
12.7 Disposal 173
12.8 Diversion 174
12.9 The diversion rate 175
12.10 Materials and data flows 176
12.11 Composition 177
12.12 Waste composition percentages. Totals may not add to 100 due to rounding 178
12.13 Recycling plant diagram 179
12.14 The capture rate 180
12.15 Calculating the capture rate 181
12.16 Contamination 181
13.1 Example of multiple-choice question 203
13.2 Participant engagement with online game 203
14.1 Storage room for high-level waste at Cap La Hague, France. February 2018. Credits: ©PHOTOPQR/OUEST FRANCE/Arnaud Le Gall 212
17.1 Keith Haring artwork © Keith Haring Foundation, depicted on the cover of Vallette and Spalding (1990) 257
18.1 Monira Al Qadiri, behind the sun, single-channel video, 2013 268
19.1 Avemare, a waste-picker cooperative in Santana de Parnaiba. Photo by Jutta Gutberlet 277
19.2 Some of the brands that use aseptic PET for the packaging of milk drinks. Photo by Jutta Gutberlet 283
19.3 Bales of milk bottles at Cooper Viva Bem, in São Paulo that have no customer because of the composite materials. Photo by Jutta Gutberlet 284

Tables

9.1 Five questions to consider for robust research design of comparative studies 124
9.2 A proposed framework to think about how to conduct a comparative study 128
9.3 Comparing two different research projects on waste and the decisions made in the research design of comparative analyses 133
11.1 Survey questionnaire design (based on Steuer et al. 2017a and Steuer, Ramusch, and Salhofer 2018b) 164
13.1 Characteristics of the Gamification Case Studies 200

CONTRIBUTORS

Dr. Catherine Alexander is a professor of anthropology at Durham University. Her key publications include *Indeterminacy: Waste, Value and The Imagination* (with Andrew Sanchez, 2018, Berghahn Books) and *Economies of Recycling: The Global Transformation of Materials, Values and Social Relations* (With Joshua O. Reno, 2012, Zed Books.)

Dr. Amanda Boetzkes is a professor of contemporary art history and theory at the University of Guelph. She is the author of numerous books and articles including *Plastic Capitalism: Contemporary Art and the Drive to Waste* (MIT Press, 2019) and *The Ethics of Earth Art* (University of Minnesota Press, 2010). amandaboetzkes.com.

Dr. Jamie Furniss is a lecturer of social anthropology at the University of Edinburgh, on secondment to the *Institut de recherche sur le Maghreb contemporain* in Tunis, where he directs the "Environment, waste and recycling economies in the contemporary Maghreb" research program. He has authored and edited a diverse range of publications on environment, the global waste trade, and informal sector recycling economies in the disciplines of anthropology, international development, and human geography, based principally on fieldwork in Egypt and Tunisia. Among these are, in 2017, "What Type of Problem is Waste in Egypt?" *Social Anthropology/Anthropologie Sociale* (2017) 25(3): 301–317, and in 2015 "Alternative framings of transnational waste flows: reflections based on the Egypt–China PET plastic trade" *AREA* 41(24–30).

Dr. Romain J. Garcier is an associate professor of geography at Ecole normale supérieure de Lyon, in France. His two recent publications are Romain Garcier. "Déchets radioactifs" in *Dictionnaire critique de l'Anthropocène*, CNRS Editions, 2020. Garcier Romain J, "Disperse, Confine or Recycle? A geo-legal approach to the management and spatial circulations of low-level radioactive waste in France", *L'Espace géographique*, 2014/3 (Volume 43), pp. 265–283. doi:10.3917/eg.433.0265. http://garcier.net

Dr. Zsuzsa Gille is a professor of sociology and director of global studies at the University of Illinois at Urbana-Champaign. She is the author of *Paprika, Foie Gras, and Red Mud: The Politics of Materiality in the European Union* (2016 Indiana University Press) and *From the Cult of Waste to* *the Trash Heap of History: The Politics of Waste in Socialist and Postsocialist Hungary* (Indiana University Press 2007—recipient of honorable mention of the AAASS Davis Prize).

Dr. Nicky Gregson is an emerita professor of geography at Durham University. She was a principal investigator of the ESRC-funded Waste of the World project. Her key publications include *Living with Things: Ridding, Accommodation, Dwelling* (2007, Sean Kingston Publishing) and *Second-Hand Cultures* (co-authored with Louise Crewe, 2003, Berg).

Dr. Jutta Gutberlet is a professor in the geography department at the University of Victoria (UVic). She is also the director of the Community-Based Research Laboratory at UVic. In 2016, she has published the book *Urban Recycling Cooperatives: Building Resilient Communities*, with Taylor & Francis and numerous articles on waste studies in diverse journals engaging in participatory action research with a focus on social and environmental justice, social grassroots innovations, and governance issues. Two recent publications are Gutberlet, J. (2021) Grassroots waste picker organizations addressing the UN sustainable development goals. World Development, 138 (2021) 105195 and Gutberlet, J. Bramryd, T. & Johansson, M. (2020) Expansion of the Waste-Based Commodity Frontier: Insights from Sweden and Brazil. Sustainability, 12, 2628. www.JuttaGutberlet.com and www.cbrl.uvic.ca

Dr. Britt Halvorson is an assistant professor of anthropology at Colby College. She has published essays on religion and waste, such as one in *Discard Studies*, and her first book, *Conversionary Sites: Transforming Medical Aid and Global Christianity from Madagascar to Minnesota* (University of Chicago Press, 2018), explores the role of medical waste economies in Christian aid programs. www.britthalvorson.com

Dr. Gabrielle Hecht is a professor of history and anthropology at Stanford University. Her key publications include *The Radiance of France: Nuclear Power and National Identity after World War II* (MIT Press 1998, 2009) and *Being Nuclear: Africans and the Global Uranium Trade* (MIT Press and Wits University Press, 2012), both of which received awards in several disciplines.

Dr. Myra J. Hird is a professor in the School of Environmental Studies at Queen's University, Canada. Hird is the author of *Canada's Waste Flows* (2021, McGill-Queen's University Press) and "Waste, Environmental Politics and Dis/Engaged Publics" (2015, *Theory, Culture and Society*). More information is available at myrahird.com.

Dr. Shiloh Krupar is a geographer and Provost's Distinguished Associate Professor in the School of Foreign Service at Georgetown University, teaching in the Culture and Politics Program. Her research examines the administration of land and asymmetrical life, geographies of military waste, cities, biomedicine, race, and the politics of geodata. Among her waste-oriented projects, she is author of *Hot Spotter's Report: Military Fables of Toxic Waste* (University of Minnesota Press, 2013), co-editor of the multimedia research platform *A People's Atlas of Nuclear Colorado*, and co-director of the National Toxic Land/Labor Conservation Service. Two recent articles are "Sustainable World Expo? The Governing Function of Spectacle in Shanghai and Beyond", *Theory, Culture & Society* 35, no. 2 (2018): 91–113 and "Green Death: Sustainability and the Administration of the Dead", *cultural geographies* 25, no. 2 (2018): 267–284. https://www.shilohkrupar.com/

Dr. Mary Lawhon is senior lecturer in the Geography and the Lived Environment Institute at the University of Edinburgh. Her research examines political ecologies of infrastructure, primarily in African cities. Her co-authored publications include "Claiming value in a heterogeneous solid waste configuration in Kampala" (with Hakimu Sseviiri, Schuaib Lwasa, Henrik Ernstson, and Revocatus Twinomuhangi in *Urban Geography*) and "A labour question for the 21st century: Perpetuating the work ethic in the absence of jobs in South Africa's waste sector" (with Nate Millington and Kathleen Stokes in the *Journal of Southern African Studies*). Her website is www.marylawhon.com.

Dr. Josh Lepawsky is a professor of geography at Memorial University, Canada. He is author of *Reassembling Rubbish: Worlding Electronic Waste* and "Planet of fixers? Mapping the middle grounds of independent and do-it-yourself information and communication technology maintenance and repair". https://electronicplanet.xyz/

Belinda Li is a research associate in the School of Resource and Environmental Management at Simon Fraser University. Her two main publications are "Urban household food waste: drivers and practices in Toronto, Canada" and "An evaluation of a consumer food waste awareness campaign using the motivation opportunity ability framework." www.foodsystemslab.ca

Dr. Max Liboiron is an associate professor of geography at Memorial University, Canada. They are author of *Pollution is Colonialism* (2021) and "Redefining pollution and action: The matter of plastics" (2018), and is the managing editor of *Discard Studies*. www.maxliboiron.com

Samantha MacBride is Adjunct Assistant Professor, Marxe School of Public and International Affairs, Baruch College, City University of New York. Samantha teaches urban environmentalism, and is a professional in public works management of garbage and sewage. Samantha's interests center on discards measurement and characterization methods, biological materials flows, plastics pollution, composting and land use, corporate hegemony, environmental justice, and waste colonialism. Samantha also studies the contemporary UFO phenomenon in the Western hemisphere, and its relations to questions of technology, religion, knowledge, and ecology. She is the author of 2021. "Composting and Garbage in New York City: a Twentieth-Century History." In *Coastal Metropolis: Environmental Histories of Modern New York City*, edited by Carl Zimring and Steven H. Corey. Pittsburgh, PA: University of Pittsburgh Press "The History and Future of Municipal Solid Waste Characterization: New York City and the Study of Fortunes in Refuse." In *Handbook on Waste Management*, edited by Thomas C. Kinnaman and Kenji Takeuchi, Northampton, MA: Edward Elgar Publishing, 2014 and *Recycling Reconsidered: The Present Failure and Future Promise of Environmental Action in the United States*. MIT Press, 2011.

Dr. Virginia Maclaren is Associate Professor and Chair of the Department of Geography and Program in Planning. Professor Maclaren and fellow professor Joseph Whitney were recently recognized by the Government of Vietnam for their joint contribution to science and technology in Vietnam. Professor Maclaren received her B.A. from Bishop's University, an M.Pl. from the University of Ottawa, and an M.S. and Ph.D. from Cornell University. She has published on food waste and urban waste management.

Dr. Nate Millington is a presidential fellow in urban studies in the geography department and the Manchester Urban Institute at the University of Manchester. His research is focused on the politics of the urban environment in an era of climate crisis, with particular interests in the governance of water and waste. He was previously a postdoctoral fellow with the African Centre for Cities at the University of Cape Town and a visiting researcher at the University of São Paulo. His previous publications include "Geographies of waste: Conceptual vectors from the Global South" (with Mary Lawhon in *Progress in Human Geography*) and "A labour question for the 21st century: Perpetuating the work ethic in the absence of jobs in South Africa's waste sector" (with Kathleen Stokes and Mary Lawhon in the *Journal of Southern African Studies*). More information can be found on his website: https://www.research.manchester.ac.uk/portal/nate.millington.html.

Dr. Patrick O'Hare is a senior researcher in social anthropology at the University of St Andrews. He is the author of several articles on waste including "Creating waste and resisting recovery: contested practices and metaphors in post-neoliberal Argentina" (Ethnos 2020). His monograph "'Rubbish Belongs to the Poor': Hygienic Enclosure and the Waste Commons, will be published by Pluto Press in February 2022. More information can be found on his website: https://patrickohareanthropology.com/.

Dr. Kate O'Neill is a professor in the Department of Environmental Science, Policy and Management at the University of California at Berkeley. She has written widely on the subject of global waste, including two books: *Waste Trading Among Rich Nations: Building a New Theory of Environmental Regulation* (MIT Press 2000) and *Waste* (Polity Press 2019). https://ourenvironment.berkeley.edu/people/kate-o039neill

Dr. Raul Pacheco-Vega is an associate professor with the Methods Lab at the Facultad Latinoamericana de Ciencias Sociales (FLACSO) in Mexico. Two recent publications are Pacheco-Vega, R. (2021). *Policy transfer of environmental policy: where are we now and where are we going? Examples from water, climate, energy, and waste sectors.* In: Porto de Oliveira, Osmany (ed). **Handbook of Policy Transfer, Diffusion and Circulation**. Edward Elgar Publishing. pp. 384–403. Pacheco-Vega, R. (2019) *Policy Styles in Mexico: Still Muddling Through Centralized Bureaucracy, not yet Through the Democratic Transition* In: Howlett, Michael and Tosun, Jale (Eds.) **Policy Styles and Policy-Making: Exploring the Linkages**. Routledge. pp. 89–112. http://www.raulpacheco.org

Dr. Kate Parizeau is an associate professor in the Department of Geography, Environment & Geomatics at the University of Guelph. Waste studies publications include research on informal recycling, food waste, and waste management planning; Parizeau, K. (2015). "Urban political ecologies of informal recyclers' health in Buenos Aires, Argentina." Health & Place, 33, pp. 67–74. Parizeau, K., von Massow, M., & Martin, R. (2015). "Household-level dynamics of food waste production and related beliefs, attitudes, and behaviors in a municipality in Southwestern Ontario," Waste Management, 35, pp. 207–217. https://geg.uoguelph.ca/faculty/parizeau-kate

Dr. Emily Potter is an associate professor in the School of Communication and Creative Arts, Deakin University. She is the author of *Writing Belonging at the Millennium: Notes from the Field of Settler Colonial Place* (Intellect) and co-author of *Plastic Water: The Social and Material Life of Bottled Water* (MIT Press). https://www.deakin.edu.au/about-deakin/people/emily-potter

Dr. Joshua O. Reno is a professor of anthropology at Binghamton University. He is the author of *Waste Away: Working and Living with a North American Landfill* (2016) and *Military Waste: The Unexpected Consequences of Permanent War Readiness* (2019), both published by University of California Press.

Jacob Riha is a Master's student in the School of Environmental Studies at Queen's University where he is completing his dissertation on the relationship between plastics and fossil fuels and its implications for the Canadian economy.

Dr. Michelle Schmidt is an assistant professor of sociology at Eastern New Mexico University. She completed her Ph.D. at the University of Illinois Urbana-Champaign in 2018. Her ethnographic research focuses on community and environmental health in the context of transnational economic development. She has an article in *Agriculture and Human Values* entitled "Cultivating Health: Diabetes resilience through neo-traditional farming in Mopan Maya communities of Belize". https://enmu.academia.edu/MichelleSchmidt

Dr. Tammara Soma MCIP RPP is an assistant professor at the School of Resource and Environmental Management at Simon Fraser University. She is also the research director of the Food Systems Lab, Canada's first social innovation lab tackling the issue of food waste. She is the co-founder of the International Food Loss and Food Waste Studies group and is the co-editor of the *Routledge Handbook of Food Waste*. She was recently appointed as a committee member of the National Academy of Science of "A Systems Approach to Consumer Food Waste". You can find her projects and learn more at foodsystemslab.ca. Two recent publications include Reynolds, C., Soma, T., Spring, C., & Lazell, J. (Eds.). (2020). *Routledge Handbook of Food Waste*. Routledge. Taylor Francis, UK. And Soma, T., Kozhikode, R., Krishnan, R. (2021). Tilling food under: Barriers and opportunities to address the loss of edible food at the farm-level in British Columbia, Canada. *Resources, Conservation and Recycling*.

Dr. Benjamin Steuer earned his Ph.D. (Chinese studies) an MAs (Chinese studies/East Asian economy and society) at the University of Vienna, Austria. He currently holds the position of assistant professor at the Division of Environment and Sustainability within the Hong Kong University of Science and Technology, where he dedicates his research to institutional (rule and routine related) evolution in the context of the circular economy, waste recycling, and sustainable development. Two recent articles are Steuer, B., Staudner, M., Ramusch, R. (2021) Role and potential of the circular economy in managing end-of-life ships in China. *Resources, Conservation and Recycling*, Volume 164 and Steuer, B. (2020). Identifying effective institutions for China's circular economy: Bottom-up evidence from waste management. *Waste Management & Research*. https://envr.ust.hk/our-division/people/faculty-staff/bst.html

Dr. Kathleen Stokes is a postdoctoral research fellow in the geography department at Trinity College Dublin. Her research focuses on infrastructures, waste, labor, and citizenship(s). Her co-authored publications include "A labour question for the 21st century: Perpetuating the work ethic in the absence of jobs in South Africa's waste sector" (with Mary Lawhon and Nate Millington in the Journal of Southern African Studies) and "Turning livelihood to rubbish? The politics of value and valuation in South Africa's urban waste sector" (with Henrik Ernstson, Mary Lawhon, Anesu Makina, Nate Millington, and Erik Swyngedouw in *African Cities and Collaborative Futures*, eds. Michael Keith & Andreza Aruska de Souza Santos).

Dr. Heike Weber is a professor for history of technology at the Technische Universität Berlin (Institute for Philosophy, Literature, and History of Science and Technology). Her main research lies at the intersection of consumption history, environmental history, and history of technology. Recent publications include a special issue on *Rethinking Waste within Business History: A Transnational Perspective on Waste Recycling in World War II* (edited together with Chad Denton, Business History, 2021) and the open access publication *Histories of Technology's Persistence: Repair, Reuse and Disposal* (2021, with Stefan Krebs).

PART I

Introducing the field of waste studies

1

INTRODUCTION: WASTE STUDIES AS A FIELD

Zsuzsa Gille and Josh Lepawsky

Waste studies is a new and vibrant academic field. In the last couple decades, not only has the number of publications whose central concern is waste from a not strictly technical perspective exploded but there are now two open-source journals dedicated to this scholarship. The older one is *Discard Studies*, an online collection of mostly shorter pieces written from a critical, activist perspective, and the more recent and more traditionally scholarly journal is the *Worldwide Waste: Journal of Interdisciplinary Studies*. Various disciplinary encyclopedias and annual reviews have also published ambitious summary and programmatic articles on waste research. Anthropology, sociology, geography, and interdisciplinary journals have all dedicated special issues to various waste-related topics in the last decade. A handbook on waste studies could be said to be overdue.

Our goal in this introduction is not to define the field, certainly not in a way that would canonize it, and thus draw exclusionary boundaries around it. Rather we aim to give an account of the many areas of inquiry that arguably constitute the field. Three principles have guided our efforts to create this volume and in writing this introduction. First, we think of these areas as in a dialogic relationship with each another, informing and building on one another. Therefore, while it is certainly possible and helpful to describe origin points, the transformations and the disciplinary connections of the field, our way of doing so is guided by this relational under-standing of waste studies. Another principle guiding our introduction is a critical theoretical perspective, that is, we aim to focus on the implicit and explicit potential of the field to provide a critical lens on society. Finally, we understand this field in pragmatic terms. This means that rather than imposing what we think waste studies should do, we are presenting it as a dynamic assemblage of strategies and practices that waste studies scholars are employing. This latter point is what informs the structure of the *Handbook* and which gathers the chapters into three sections. Each section responds to a different question:

1. What questions do scholars in the field ask?
2. What methods do scholars in the field use?
3. What cases do scholars in the field investigate?

In what follows, we will provide an overview of the history of the field, noting in particular the empirical themes, theoretical frameworks, and methodologies utilized. In the process, we will

DOI: 10.4324/9781003019077-1

point out distinctions from other related fields. Third, we will summarize how the contributions in the volume answer the three guiding questions above. And finally, we will suggest research directions for the future.

A history of waste studies

Etymologically the word "waste" has origins in the Latin *vastus*, which denotes unoccupied, uncultivated, void, and immense, and Sanskrit for wanting or deficient. Yet, waste defies easy definition. We may know it when we see it (or smell it), but defining what waste is universally is not possible. Waste is always situated. It is waste for some individual or group—not necessarily only human ones (Reno 2014)—in some place at some time. One way to illustrate the situated character of waste is to recall that all aerobic life is sustained by oxygen, including human life, and depends for its existence on the continual renewal of that gas by photosynthesizing plants as a metabolic excreta crucial to generating the atmosphere on which aerobic lives depend (Volk 2004). Humans and other animal life in a certain practical sense dwell at the bottom of an atmospheric ocean of gas excreted by other life without which they would perish. At the same time, this example of the situatedness of waste is somewhat misleading since waste is also infused with a vast range of values that are at once moral, ethical, even spiritual, not to mention political and economic as well. Waste can be as cosmological as it can be mundane. Waste provokes art, philosophical thought, and concrete measures to avoid, disperse, dilute, rework—in short, to manage—it.

Arguably, the emergence of the field that sometimes goes by the name "waste studies" has no single point of origin in time or place. Somewhat like the materials that often form the empirical foci of its practitioners—garbage, trash, litter, and other forms of cast-offs—waste studies is a mixed assemblage that nucleates around questions, methods, and cases.

The history of the field we sketch out below is partial and situated. It is inflected by our own training, upbringing, and social location. We restrict ourselves to mostly Anglophone texts, which as such tend to be by authors from the Global North. Our review then is open in some ways but closed in others. A further limitation is our immersion in the social sciences. This means that our reading of the waste studies field will overemphasize the contribution of anthropology, geography, sociology, political economy, and environmental studies; and the humanities here will mostly be represented by history, only briefly mentioning philosophy, and cultural studies. While this is in part the result of our positionality, it is also informed by the more influential role the social sciences and history have played in the formation and development of waste studies. Neither we, nor any of the contributors to this volume, are practicing engineers or policy makers. This is not a text about the professional concerns of waste management, although this handbook may also be of interest (we hope) to professional waste managers with those kinds of backgrounds.

Today, the growing collection of scholars who identify with a field recognizable as "waste studies" tend to reference a few key early texts as informing their inquiry and as providing concepts and arguments with which they intend to be in conversation with. No other classical text has received more attention than anthropologist Mary Douglas' definition of dirt. Some directly import her understanding of dirt as matter out of place into their inquiries of waste issues, while some others note that dirt is not waste, and worry that equating the two unnecessarily limits our view of waste, trash and garbage as something polluting. While, as Liboiron in this volume argues, Douglas' goal wasn't so much to theorize dirt or waste, but rather demonstrating how power operates through creating clear boundaries around cleanliness and order, Douglas' conceptualization has exerted a strong influence on waste studies. In what is virtually a defining axiom, waste scholars all accept that waste is socially constructed or, to put it another way, that waste *is not*, but it *is* made. What this making implies for different authors however varies greatly. For some, as in another classical text by Thompson (1979), waste is the opposite of value, and since value is certainly suspended in a cultural matrix, and different registers of valuation, arguably waste too is in the eye of the beholder. The proverb "one man's waste is another's treasure" might serve as an approximate vernacular equivalent (and one doesn't have to endorse its gendering while acknowledging its prevalence as a frequently invoked saying).

For others, the social construction of waste extends beyond the realm of value, most notably to the materiality of waste. These authors demonstrate not only that waste can slide along a wide spectrum of utility/value and uselessness/lack of value, and in multiple directions to boot, but its material composition is also socially determined with significant implications for what materials are treated as waste by different social actors. Indeed, the evaluation of the matter constituting discarded objects changes historically (Strasser 1999).

As is clear from this brief overview of the most-cited origins of waste studies, the key animating question for the classics has been "what is waste?" It is not so much that later studies ignored this question (for useful conceptual frameworks see Sarah Moore (2012) and Reno (2018)) but started demonstrating what this definition matters for. What is at stake in defining waste in one way or another? K. A. Gourlay (1992) started this line of inquiry by critiquing common policy definitions of waste as discarded material, because as he argued, it immediately allowed the shift to technocratic concerns about waste management. Not only do these definitions give the impression that waste is already managed, thus, not really a problem, but shortcut analysis that would allow social scrutiny into who decides, and based on what considerations, the material composition of wastes that pose so much environmental and social trouble. Following Gourlay, as several scholars did in defining waste not as material discarded but as material we failed to use (for whatever reason) provided the first major critical impulse for waste studies, in as much as asking about the making of waste or the practices of wasting, allowed for scrutinizing social agency and social inequalities around waste issues.

From here, the scholarship proceeded in two distinct directions. In the United States, much informed by a new social movement, environmental justice became the main lens through which waste became a subject for the social sciences. The movement took off in the 1980s by protesting the siting of hazardous waste facilities (incinerators and landfills) near minority neighborhoods. Key authors in this tradition are Szasz and Meuser (1997), Bullard (1990), Brook (1998), Pellow and Brulle (2005).

In Western Europe, where the social sciences had already been much more attuned to consumption, practices of wasting in households and thrift shopping became the next influential area of study. Consumer waste studies tend to reach back to a different origin, not social movements and political economy, as in the United States, but one that has been influential in the consumption scholarship, namely the classical work by Thorstein Veblen (2007 (1899)). His concepts of conspicuous consumption and conspicuous waste in the service of demonstrating one's distinction (Varul 2006) are to this day, the most influential. Another influential classic is Vance Packard's (1960) *Waste Makers*. Packard demonstrated how obsolescence is built in the design of consumer goods, which serves as a major factor in ringing in a throw-away society. European scholarship has been especially critical of "blaming the consumer" (see on this Evans 2012, 2014); O'Brien (2007) but has contributed relatively little to an analysis of corporate practices that limit individuals' choices in what and how much to discard. That task has mostly been undertaken in some studies of food waste

(Stuart 2009; Bloom 2011; Alexander, Gregson and Gille 2013) and to a lesser extent of repair (Graham and Thrift 2007; Graziano and Trogal 2017; Persaud, Lepawsky and Liboiron 2019). Nicky Gregson's (2007) study of UK households, and Peter Evans' (2014) work on food waste carefully dissected the practices and social and familial relationships that inform what and how households throw away. Using research subjects' journals and other qualitative methods, they demonstrate that there are more social, cultural, and economic determinants of the composition and quantity of household wastes than corporate-dictated choices, whether in materiality or portion sizing.

Simultaneously, the field science, technology and society (STS) on both continents started focusing on a specific waste materials and waste treatment technologies. Myra Hird's work on landfills (2013, 2016, 2017) and more recently on waste and Canadian settler colonialism (Hird 2021); Alexander and Reno's (2014) research on anaerobic digesters; Liborion's studies of marine plastics (2021), pollution, and colonialism (Liboiron 2021a); and Gabrielle Hecht's (2014, 2018) historical studies of mining waste and nuclear waste are the most influential. Waste has become a central concern for scholars around the same time as STS shifted from science-as-culture paradigm (a wonderful example of which in waste studies is Kuletz 1998) to one that engages with wastes not only in the cultural register but also in the material one, interrogating particular waste matters for their nonhuman agency (Hird 2012; Muecke 2009). While materiality has never been absent from STS, with the appearance of various novel theories in the late 20th century, such as postructuralism, including feminist/performative theories of the body, Actor Network Theory (which is probably better thought of as a method of inquiry), new materialism, including feminist materialism, it was now possible not only to draw out in concrete terms how nonhumans constrained, tweaked, or enabled human action, but also, in more theoretical terms, how modernist conceptions of human and social agency need to be modified. This new materialist approach also did away with our old assumptions about a static biological surface on which society inscribes itself and thus has been particularly useful for demonstrating the indeterminacy of material transformations and waste technologies most notably in the work of Myra Hird and Joshua Reno. Here, to the extent that politics or power relations are part of the analysis, they tend to be manifest and prove most consequential at the micro-level, while also hinting at policy implications, for example, for how to treat organic wastes. Social institutions that have a capacity to act at a distance are usually captured as dangerously simplifying and hiding the uncertainties and fluidities of these material metamorphoses. Gabrielle Hecht's research, while also documenting unexpected geological and chemical transformations, in contrast, foregrounds the ways in which global, colonial, and racial inequalities are exploited and reinforced by the governance of nuclear technology at multiple social scales. Max Liboiron (2017, 2021) is also most concerned with the practices and structures of governing the science of pollution. Their many collaborative projects aim at developing open-access and affordable technological tools to assist citizen science in demonstrating their levels of exposure to toxic waste and in countering colonialist and sexist knowledge production and environmental policies.

By the late 1990s, studies concentrating on waste issues in the Global South, though initially mostly penned by scholars from the Global North, also started to appear. This scholarship contributed the most to our understanding of precarious waste labor in collecting and salvaging household wastes in poorly serviced megacities and on urban landfills. In addition to demonstrating the many ways in which material and human disposability depend on one another (Gidwani 2013; O'Hare 2019; Knowles 2017; Millar 2008, 2018; Fredericks 2009, 2018), scholars also document the struggles of informal waste collectors in organizing themselves, carving out sources of livelihood in the face of the professionalization and privatization of waste collection (Dinler 2016; Furniss 2017; Miraftab 2004), and gaining various health and legal protections from municipal governments (Gutberlet 2009; O'Hare 2020; Van Horen 2004).

Waste labor has continued to remain one of the strongest themes in waste studies both in the Global South and North. In the latter, environmental sociologists and geographers documented the intersectionality of the dirtiest aspects of waste labor (Weinberg, Pellow, and Schnaiberg 2000; Gregson et al. 2016, and a special issue of the *International Journal of Labor and Working-Class History,* No 95. Spring 2019). A different waste labor of sorts is the subject of an increasing number of publications focusing on scavenging either as a social movement or alternative lifestyle. Freeganism or dumpster diving has been particularly central in these studies and has gone the furthest in teasing out the theoretical and political implications of wasting and reusing practices (Giles 2014; Barnard 2016).

Waste emerges in the matrix of not only materiality, value, culture, and society, but of space as well. It is, however, not just that where an object is that defines its waste status, but that by placing it in a particular place, usually out of sight, or on the margins of mainstream social activity, society can be compelled to ignore it, as the proverb "out of sight, out of mind" suggests. While the waste-space nexus is certainly not the exclusive focus of geography, this discipline has contributed the most to our understanding of the paths materials deemed useless or too risky travel globally. Commodity chains necessarily entail waste chains, that is a series of links in discarding, salvaging and recycling materials that are cast off by some people somewhere. Hazardous wastes, plastics, and electronic wastes have received much attention in this subfield.

Early studies on electronic waste assumed that the primary direction of this stream was from the Global North to the Global South. Today however, a new consensus is emerging that suggests a more complex set of relations between waste producers, collectors, distributors, disassemblers and resellers (Lepawsky 2019; Crang et al. 2013). Studies of these kinds challenged thinking in fields beyond waste studies by taking waste as a critical case with which to investigate broader concepts such as knowledge and governance. Wynne (1987), for instance, documents how both hazardousness and waste exist in an indeterminate twilight zone that indiscriminately mingles physical properties of materials with ultimately arbitrary classification criteria developed by people in wildly diverse institutional settings. Hazardous waste as such will always elide fixed, universal attempts to govern it because "no clear, 'natural' definition of [hazard and waste] can be given within wide margins of uncertainty and variation" (Wynne, 1987, 1). While seemingly abstract, the consequences for governance of waste are both practical and intractable. Two jurisdictions, such as the United Kingdom and United States, may both recognize a given substance, such as polychlorinatedbiphenyls (PCBs) as a hazardous waste, yet quantify the threshold for meeting that designation at quite different levels. There is no neutral third party where such competing classifications can be taken to decide once and for all which one is right and which one is wrong. They are "both and".

A key theme that Wynne's work signals is the issue of waste governance, or the inter-organizational coordination of action to manage waste at a variety of scales, be they local households, municipalities, sub-national regions, the nation state or, indeed, internationally. A considerable literature exists in waste studies that examines the international governance of waste flows across borders. Wynne (1989) and Clapp (1994) for example, were especially concerned with the hazardous or toxic waste trade and its impact on countries of the Global South. Both Wynne's and Clapp's attention were directed at international agreements, particularly the Basel Convention, that represented a shift away from voluntary initiatives of transnational corporations, to binding international law. Governance of such waste is, Wynne showed, bound to fail if it is premised solely on downstream disclosure and description of materials as regulated wastes in large part because of "the insuperable indeterminacies in defining hazardous wastes" his earlier work had identified as a key governance challenge (Wynne, 1989, 124). To be effective, solutions would have to be upstream long before export was an issue. Clapp (2001), too, was concerned with the governance of the transboundary waste trade. Her attention was especially on the complex inter-scalar negotiations between state and non-state actors involved in the making of the Basel Convention. Like Wynne, Clapp argued that it was upstream where most governance work needed to be done. She argued for, among other things, mandating shifts in manufacturing toward clean production and the establishment publicly accessible firm level pollution release and transfer registries (PRTRs).

O'Neill (2000) took a different look at waste governance. Instead of concerns about rich nations shipping waste to poor ones, O'Neill shifted the attention toward waste trading among wealthy nations. Why would such nations accept hazardous waste sourced from similarly wealthy countries across their own borders? Waste governance in this respect came down to differences in regulatory structure and style between sending and receiving national jurisdictions. Other authors began to examine conflicts over waste governance within wealthier regions of the world especially around community opposition to new incineration and waste-to-energy plants being built (e.g., Fagan 2003; Leonard et al. 2009).

A key tension in the waste governance literature exists around the issue of scale. If Wynne, Clapp, and O'Neill emphasized the multiplicity of actors (e.g., state representatives, NGOs, trade associations) influencing international scales of waste governance, others have been keen to examine how waste governance works from the "bottom up" within national and sub-national jurisdictions. Authors such as Bulkeley et al. (2007) demonstrated that waste governance within the United Kingdom, for example, entails a multitude of both state and non-state agents operating in a plurality of coordinating roles with respect to household and municipal waste (see also, Bulkeley and Gregson 2010; Davies 2008). Much more recently, Gregson and Foreman (2021) working from municipal level waste flow data across the United Kingdom build up a geohistorical picture of changing waste governance regimes that are now dominated by European transnationals and the financialization of those flows. The governance of waste in the United Kingdom is increasingly bound up with the financial economy, with multi-decadal contracts that guarantee certain tonnages of residual material and, thus, ultimately work against waste minimization.

So far, we have mostly focused on the social sciences contribution to waste studies. However, the foundational axiom of waste studies, the socially constructed nature of waste, in the two senses we described previously, is also an attitude or a scholarly posture that resonates all throughout the humanities. We focus on two fields in which waste scholars with a humanities orientation have been the most active. One is history, the other is cultural studies.

Histories of garbage and sanitation have provided much of the initial empirical impetus for waste studies. A lot of this history was initially focused on Europe and the United States. Perhaps the most influential of these studies has been Susan Strasser's *Waste and Want: A* Social History of Trash (1999). Focusing on U.S. households in the 19th and 20th centuries, she demonstrates how changing technologies of production, the decline of handwork, and the spread of mass consumption led not only to increasing quantity and a radically different material composition of wastes but also to new cultural meanings of disposal. Her narrative arc is echoed in many other histories of waste that sees the birth of our contemporary meaning of the word waste as intricately involved in modernization, and its associated social processes, urbanization and industrialization. The connection between waste and modernity, including the Enlightenment, has been studied by John Scanlan and Tim Cooper who also attribute a certain emancipatory potential to waste in as much as it resists the easy assumption of human mastery of nature, but also in a more philosophical sense, questioning and occasionally reversing the hierarchy assumed in the dichotomies Western Enlightenment has been founded on. Others extended this connection to humans as disposable (Bauman 2004; Oloko 2018).

Histories of "waste" in its contemporary—one could say modernist meaning—are, in some sense, also histories of an emerging entity now called "Europe"…Paris' walled defenses were overshadowed by mounds of the city's solid waste outside its gates by 1400. The agglomeration of more and more people in concentrated areas brought with it attendant needs to deal with their wastes of various kinds, solid and liquid, individual and household, industrial and otherwise. City sanitation likely has its origins in street cleaning (Melosi 2008). This is in part because streets formed a commons of public and private interactions in their role as conveyances for goods, people, animals, and of course everything that is rid by them.

Readers who already have some knowledge of waste studies will likely be familiar with the sanitary idea—the notion that the physical environment has a profound effect on individual well-being. Articulated in urban and industrializing, England of the 19th century entrained with it new state formations, too. The miasmatic theory of disease predominated and solutions it suggested to the various epidemics sweeping Europe's cities included the organization of new administrative structures to govern hygiene and disease prevention. In England, Edwin Chadwick, was an advocate of Jeremy Bentham's utilitarian philosophy (and eventually Bentham's secretary). Utilitarianism in this form provided a philosophical basis for a variety of state-based interventions into public health. Reformers like Chadwick were embroiled in decades long struggles between centralizing tendencies of national governments and desires for autonomy of municipal and regional authorities as well as business interests. Waste and state formation go hand in hand. Indeed, the utilitarian approaches to hygiene management, particularly of the urban poor, by figures such as Bentham and Chadwick speak to the genealogy of what philosopher Michel Foucault called biopower (Foucault 2004). In this sense, histories of waste and its management are also the *geo*histories of the emergence of modern capital, state, and civic power along the colonial modernization front.

Well before the Benthamite and Chadwickian reforms of 19th-century England, the long history of enclosure turned idle land not being used for cultivation—that is, waste lands—into enclosed private property. Wastes in the form of common land entered the long history of legal modernity with the Magna Carta (Gidwani 2013). The Great Charter of the Forest of 1225 defended common rights to common land—wastes—but soon after with the Statute of Merton of 1235 those rights were patchily and unevenly eroded. That erosion was a lengthy process, covering at least 200 years. In the drama of monarchic attempts to regulate enclosure of waste lands by the manorial elite so as to mitigate its ill effects on those who were displaced by it are also the emergence of institutions of modernist state apparati and capital. What emerged was a shift from wastes as commons to commons as wastes (Goldstein 2013). Enabling that shift were all the techniques of regulation and enforcement that depopulated the commons as wastes, often violently. In 1700, most land in England was still open waste land, but by 1840 at least 20% of it had been enclosed by acts of Parliament between 1750 and 1820. These enclosures chart a history of waste in its multiple, ramifying yet distinctly emplaced forms: initially with regard to the English commons waste had a largely benign or even positive connotation as land available for common use, but it later came to have strongly negative connotations when waste became the constitutive outside of the political society imagined in the philosophy of liberalism from the 17th-century natural rights doctrine of John Locke through to Adam Smith's liberal political economy.

Yet, recognizing waste in these terms also means recognizing that its concrete and symbolic force go hand in hand with the making of colonial frontiers. A fuller tracing of these entanglements is beyond the scope of this introduction or this handbook, but it is quite possible that "waste" as it is understood today emerges in the ramifying fronts of colonial encounter. Contemporaneous with the long process of enclosure of the English commons, "waste" was a similarly tactically useful idea and practice to justify and enact a nexus of colonizer, colonized, and land, albeit patchily and unevenly along colonial frontiers. In a certain sense, the "waste" of waste studies was born colonial. Collectives of humans and nonhumans outside of what was becoming "Europe" of course had their own practices of getting rid of their excreta, effluvia, and ejecta. But it would be a mistake to bundle them all under a (supposedly) universal "thing" called "waste". Why? For one, it is clear that the notion of "waste" was—and continues to be—used to do a lot of work in establishing relations of domination and subordination between European colonizer and colonized Other(s) (e.g., Hird 2021). "Wasteland" became a usefully flexible term to justify and enact colonial administrative rule even as the collectives of humans and nonhumans constantly "act back" and foil or confound colonizers idealized pursuit of universal applicability (de Hoop and Arora 2017; Kuletz 1998).

In an analysis of links between "infrastructural discontent" and mass movements for political change in contemporary Egypt, Arefin (2021) demonstrates the roots of that discontent to the geohistory of colonial rule and its systems of sanitation. Arefin notes Fanon's description of the Algerian experience of French sanitation schemes as always approached with suspicion: "the statistics on sanitary improvements are not interpreted by the native as progress in the fight against illness, in general, but as fresh proof of the extension of the occupier's hold on the country" (Arefin 2021, citing Fanon).

Elsewhere, while Mughal Delhi had a hydraulic system for bringing fresh, potable water into the city and for removing people's bodily excretions, the latter were not treated as "waste" in the more contemporary sense of that term. Indeed, as Prashad (2001) demonstrates installing a colonial re-ordering of social relations in Delhi under British rule was constitutively tied to the deliberate neglect of the existing hydrologic system and the installation of sanitation infrastructure devised by the British. But, of course, toileting, hygiene, morality, colonial control, and (re)ordering of Indigenous life ways were all linked in a variety of situations including German colonies of Africa and Pacific Islands. These systems also imposed regimes of regulations and surveillance and their attendant hierarchies of race between rulers and the ruled in an ostensibly civilizing mission (Walther 2017).

Cultural studies is an interdisciplinary field that emerged in the 1990s and while anthropology contributed much to this scholarship, its key representatives hailed from the humanities or targeted the humanities as their main audience. Far from the assumptions the name of this field might suggest, cultural studies' subject matter is not culture, at least not in the ordinary sense. Rather, it uses tools of cultural analysis to shed light on hidden aspects of power and inequality, mostly focusing on the symbolic and cultural register. With regards to waste, their interventions included the discourse analysis of environmental or anti-litter campaigns, visual analysis of waste-related policies, and studies of material culture and subjectivity formation. At its best, cultural studies of waste goes beyond simply dressing waste up in a positive generative robe, that is romanticizing it, and instead transcends the binary of waste/value by probing into the mentalities, practices, habitus, symbolic propaganda that guides our everyday sorting and wasting activities.

This intention to transcend the dominant dichotomies of waste/value; useless/useful and the like, means that traditional political economy approaches to waste are also critiqued. First, there is an often-implicit critique of reducing waste problems to Capitalism (with a capital C) either by pointing to a more diverse set of effects of private property on wasting, or by demonstrating how what has been presented as purely economic are indeed interwoven with things we tend to think of as purely cultural. Wonderful examples of this kind of work can be found in two anthologies: *Culture and Waste: The Creation and Destruction of Value,* edited by Hawkins and Muecke (2002), and *Economies of Recycling: The Global Transformation of Materials, Values and Social Relations,* edited by Alexander and Reno (2012). Third, there is an explicit intent to avoid the moralizing tendencies of the literature in this political economic bend, namely the condemnations of consumer waste, already present in Veblen. It is not so much that political economy or a condemnation of capitalism is avoided but rather that our understanding of capitalism is complicated with the help of key contemporary theoretical influences, such as poststructuralism, new materialism, performative perspectives on the economy and the subject. This however doesn't mean that ethics is missing. Gay Hawkins (2006) has engaged the most with the ethical implications and imperatives of disposal and recycling and argues that contemporary environmentalist projects, household recycling, reducing consumption, or anti-litter campaigns impose a rigid ethical responsibility on individuals, and as such waste economies are never just economic, but ethical projects, as well. She also suggests that such campaigns often fail because they operate through shaming humans; instead, she advocates for allowing different relationships between humans and their objects, some of which may recognize the vitality of matter and works through a more positive set of affective registers, including joy.

A key drawback of the cultural studies of waste has been its almost exclusive focus on consumer waste. When industrial wastes make an appearance, cultural studies provide poignant analyses of one's relationship to one's body and landscape narratives (Newman 2012; Pezzullo 2012) or cultural tropes embedded in nuclear waste management (Garcier 2012).

The three sections of the handbook

As we flagged at the outset, this handbook is organized into three sections as follows:

1. What questions do scholars in the field ask?
2. What methods do scholars in the field use?
3. What cases do scholars in the field investigate?

Organized as such, readers and users of this handbook are encouraged to think about the questions, methods, and cases out of which a field—waste studies—nucleates. These sections are not walls patrolled by disciplinary police, but nuclei out of which condenses a field.

Questions waste studies scholars ask

When it comes to waste studies research, we all ask questions, but the trick lies in moving from a research topic (e.g., marine plastics) to questions about that topic (e.g., How do plastics arrive in marine environments?). Questions are not neutral. One could ask questions about what threshold(s) exist beyond which harm from plastics arises or one could ask what alternatives to plastics could be instituted? The former heads into the territory of risk analyses about business as usual, the latter into reorganizing it (O'Brien 1993).

Users of this handbook are introduced to questions animating the field of waste studies. These are, of course, not an exhaustive list of such questions. They are here offered as analogia with which to think—and think beyond. In the first chapter, Liboiron treats readers to a re-examination of one of waste studies key thinkers: Mary Douglas and the concept of matter out of place. What is at stake in defining waste in waste studies? More specifically: what new interpretation of Mary Douglas is helpful to waste scholars? However, one might answer those questions. Liboiron shows us why it is crucial to disentangle waste and dirt when invoking Douglas' theories to frame our own questions about waste. All too often waste and dirt are used interchangeably, yet in Douglas' work the two are distinct and waste is *not* dirt, it is not even out of place. Waste has a proper place—we know where to put it—even if it is sometimes and temporarily displaced. Dirt, on the other hand, is that which well and truly threatens a given ordering system i.e., it is a threat to power.

Halvorson and Reno encourage readers to ask how wasting and racialization mutually constitute each other. Answering this question leads students of waste studies into both familiar and less familiar realms. On the one hand, answers lead into questions about who does the "dirty work", where, when, and under what conditions. On the other, into areas perhaps less familiar to scholars of waste studies: "… the technically sophisticated world of scientific practice and new genetics" (Halvorson and Reno, this volume) where whose bodies and whose genes get to count as worthy of investigation—and of extraction.

Next, Patrick O'Hare offers a synopsis of questions informing the social science of landfills, including what motivates people who glean from them to do so (it's not always or necessarily about desperation). O'Hare argues that the social science of landfills shows, "often in surprising ways [connections between] issues of race, nationality, class, and religion" (O'Hare this volume). It is important to ask how landfill life, as O'Hare terms it, will change as newer approaches to waste management such as zero waste and waste-to-energy (incineration) become ever more common.

Working from a case study of public signage about waste in Tunisia, Furniss shows why it matters to ask how waste is framed and with what effects in a given situation. How waste is framed is generative of identities and landscapes in Tunis. It prompts citizens to cultivate self-images—from environmental responsibility to clean, respectable, and thus deserving citizen—and landscapes associated with high or socio-economic status read from the signage present on the streets.

Weber draws our attention to questions not just of change over time (history), but to questions about the very quality of time—temporality—that can be experienced in the face of contemporary forms of wasting. For Weber a key shift—patchy and uneven in the making—occurs with the emergence of forms of waste common today. Some of today's materials, whether ostensibly remarkable (e.g., nuclear) or unremarkable (e.g., plastics) surpass any possibility of human temporal experience and usher in geologic timescales that will exceed human life (whether or not "we" also manage to deal well with any number of threats from nuclear war to climate breakdown).

Rounding out the section on questions waste studies asks is Schmidt's chapter on food waste. She asks what different, even incommensurable, practices associated with surplus food in what is today Belize can show about the cosmologies those practices index. Mayan lifeways around food continue, despite (or in spite of) colonization and invoke forms of respect unrecognizable in cash-cropping practices in the same region.

Methods

Waste studies scholars employ a wide variety of methods and criss-cross both qualitative and quantitative approaches. As Elliot (1999) advocates, the choice of methods is best guided by what the research question(s) demand. All three waste studies scholars interviewed in Chapter 2 emphasized the importance of fieldwork. It was moments of surprise on the ground in fieldwork settings, often doing research on topics these scholars did not (yet) recognize as having anything to with waste, that opened a door to their fascination with waste. Fieldwork, of course, takes many forms. Often, but not always, it is qualitative in character. It may entail long-term presence and interaction with a situation of interest, such as in ethnographic approaches, but it can take other forms as well—everything from face-to-face surveys, quantitative mapping, interviews, and waste audits among other possibilities. But there is something about "being there" that enables a recognition of the important role(s) of wasting, whether the "there" is your own university hallway where disposable, shared, and reusable options for drinking water exist side-by-side (see Liboiron 2021b,c) or someplace more distant and unfamiliar from home.

Pacheco-Vega offers readers of this handbook both a synopsis of comparative methods premised on fieldwork and a framework for designing studies to answer questions that demand such methods. Pacheco-Vega helps readers think through their own study designs by drawing out actual examples from his own comparative studies of the politics of waste governance.

Methods in waste studies are not all about research, they are also about how one might teach waste studies. Readers looking for guidance on pedagogy will find Parizeau's chapter particularly helpful. More than suggestions for syllabi, Parizeau's chapter guides readers through disciplinary and multi-disciplinary approaches to critical waste studies, how to choose classroom resources (e.g., readings, viewings, listenings, as well as options for tactility), and offers a thorough discussion of key themes that might be taught.

The next two chapters, the first from Steuer and the second from MacBride, will help readers with considerations for quantitative research in waste studies. Steuer discusses how his work on informal recyclers in China combined official state statistics with face-to-face survey questionnaires that collected data on factors such as demographics, quantities and types of materials collected daily, prices achieved, and other factors such as source locations and destinations for materials.

MacBride offers a deep-dive, insider account of how quantitative waste metrics are actually collected in waste management systems. MacBride takes readers through six categories of measurement—generation, composition disposal, diversion, capture, and contamination—and shows us what actually goes into constituting numbers such as a waste diversion rate. MacBride's point is both to help waste scholars approach quantitative data critically without simplistically dismissing that data and to advocate for two additional quantitative concepts: defilement (a measure of localized harm) and depossession (a measure of displacement of harm and opportunity).

Rounding out the Methods section is Soma et al.'s examination of the use of gamification approaches to consumer behaviors of wasting. These authors review five case studies using games as well as their own online gaming platform built to help players reduce food waste. The authors note the challenge of sustaining desired outcomes after users end their gameplay, but with the proliferation of digital tracking, there is much potential in gamifying waste, at least in some scenarios.

Cases

The third and final section of this handbook turns to a variety of cases that waste scholars have investigated. We hope to have achieved two goals in the selection of cases presented in this section. First, to illustrate breadth of analyses in the field. Second, to offer readers actual examples from which to think analogically about their own research even if the specific topic of a given case is not identical to a reader's own.

The first two chapters engage in different ways with nuclear waste. Garcier opens the section with an important intervention about the ordinariness of nuclear waste disposal in France in contrast to such waste's common reputation as exceptional. Garcier encourages readers to think with, but also beyond nuclear materials themselves to how their infrastructures and legal frameworks make them both ordinary for some and extraordinary for others. Subsequently, Potter brings readers into a nexus of nuclearity, waste, and settler colonialism in Australia. Potter helps readers understand that settler colonial worlds are premised on laying waste. While there is no possibility of return to purity from which to "clean up" those relations, there are possibilities for better futures no longer premised on the false hope of "cleaning up and moving on".

Krupar examines a very different case of waste—the redevelopment of contaminated "brownfield" real estate in the United States—but, like Potter, demonstrates its centrality to the formation of racial capitalism and settler-colonial ways of being. Out of these unjust relations between land and life chances, Krupar offers a route to antiracist brownfield research and policy. Such research would entail, among other things, a critical physical geography of soil that would treat it as part of systems of sedimented power relations, rather than mere physical and passive medium.

The international trade and traffic of waste has been of long-standing interest in waste studies. O'Neill brings readers back to an historical contextualization of the international waste trade as a matter of concern in 1980s and 1990s. It was during these decades that key global conventions, such as the Basel Convention, were negotiated and brought into force. O'Neill helps readers understand how antecedents situate and continue to inflect more contemporary attention to the global governance of waste trading (e.g., electronics, plastics).

Boetzkes asks readers to consider Kuwaiti artist Monira Al Qadiri's video *Behind the Sun* as a case through which to examine how the global economy of oil has developed an instrumental need to waste oil. Boetzkes brings ideas from Bataille on energy expenditure and Mbembe on necropolitics to interpret the burning of Kuwaiti oil fields during the 1991 Gulf War as a deliberate aesthetically charged display of energy expenditure at least as much as it was a military tactic. In Boetzkes' analysis of the deadly aesthetics of spectacular wasting of oil readers are prompted to reconsider the much slower, much less spectacular yet ongoing and relentless planetary violence of the oil economy.

Lawhon and co-authors examine waste, labor, and livelihoods in South African cites. Rather than the perhaps more familiar focus on "informal economies" in the "Global South", Lawhon et al. demonstrate how what they call the "extremes of South Africa" crystalize within a single country patterns of contestation over who will benefit from the reworking of waste-as-resource. As such, these authors suggest for readers how learning from South Africa can inform research elsewhere and on broader themes in waste studies, including into the ostensibly "green economy".

Gutberlet provides a synopsis of organizing activity by people working in the waste picking economy in Brazil. The chapter draws on 15 years of research with workers organizing a variety of collective responses to advocate for their social and economic rights, healthier working conditions, and greater recognition of the essential services they provide in the cities in which they work. As Gutberlet shows these various forms of organizing entail highly skilled knowledge of urban waste economies and political savvy held and practiced by people all too often dismissed or ignored by policy makers. Yet Gutberlet demonstrates that waste picker unions, cooperatives, associations, networks, and broader social movements can and do achieve material improvements to their lives and working conditions of people through their organizing efforts.

Rounding out the cases is an examination of prepping—the stockpiling of "stuff" by private citizens (often white, often middle class)—in anticipation of civilizational collapse. Here, Hird and Riha introduce readers to the economies of prepping and demonstrate its reliance on infrastructures of consumer capitalism—long supply chains, mass produced and standardized goods, mega-retail delivery systems—the very systems rolling the dice of catastrophe. Prepping (and preppers) offer charismatic illustrations of the limits of individualist responses to waste crises. Indeed, as Hird and Riha argue, "the best way to thrive in the world is to collectively prevent disaster".

Directions for future research

While the questions, methods, and cases the chapters in this handbook focus on and demonstrate the merits of are not meant to be exhaustive, collectively they offer future research directions in waste studies. We see these potential paths not so much in thematic terms, but rather as a series of tensions waste scholars need to recognize and navigate effectively.

An overarching tension may exist for those invested in a strong distinction between waste studies—the subject of this handbook—and discard studies. Is there a difference between the two fields? We would answer that, yes, there is but also that we are not invested in policing boundaries between them. Discard studies is somewhat different from waste studies in that the former makes a claim about systems, that is, that all systems must rid themselves of that which challenges their ordering so as to remain the systems that they are (Liboiron and Lepawsky, Forthcoming). In most instances, waste does not challenge such systemic ordering. Indeed, as Liborion reminds us in this handbook Douglas' notion of dirt—that which constitutes an existential threat to a given order—is not synonymous with waste. This ridding (or discard) can take many forms including separation, exclusion, annihilation, and elimination. Waste and wasting can, but need not necessarily be, part of any of those tactics. Consequently, there can be a relationship of non-equivalence between waste studies and discard studies. By way of quick example of how the two fields are somewhat distinct (even if they sometimes also overlap in places) is that of commercial content moderation (CCM). If you use social media of any kind, you are familiar with CCM even if you are unaware of its workings. Essentially, CCM involves people paid to make "ignore/delete" decisions about what content post by users does or does not violate a given social media platform's terms of use policies. Deleting is discarding, but it is less obviously a form of wasting (see Lepawsky 2019 for further discussion). However, insisting on a hard boundary between waste studies and discard studies is not an effort we wish to make.

As noted in several instances, the literature's focus on municipal or consumer waste is overwhelming not only in relation to other sectors' wastes, especially construction, manufacturing, and mining, but also in proportion to the amount of waste the latter generate. There are two tensions around this thematic focus we must recognize. One is that it is municipal waste that provides mostly informal waste pickers in the Global South with a relatively secure source of livelihood, and as such their struggles deserve scholarly and policy attention. Many scholars now combine their study of informal waste pickers with their participation in designing and implementing environmental policies—see Jutta Gutberlet in this volume. The second is that constructing "the waste problem" as primarily one of individual consumption means that solutions offered will necessarily target consumer and individual behavior, see an interesting example of this effort in the chapter by Tammara Soma, Belinda Li, and Virginia MacLaren. Here the tension is that reducing consumption is clearly needed, while scholars also have to recognize that this cannot be achieved only by behavior modification. Therefore, it is not so much reducing attention to household wastes that seems important to us but placing this waste stream more explicitly in a broader political economy context that recognizes the role of precarious labor, the effect of corporate strategies that severely limit consumers' discard practices, and the financialization of municipal infrastructure.

A third tension, somewhat related to the first one, arises from lack of empirical data on certain waste streams. The most accessible data, in all countries we have read about, is municipal waste. This partially explains the fields' outsized focus on households and consumer waste. Our contributors however even questioned the reliability of these data, see MacBride and Steuer in this volume. On the one hand, the unreliability and the absence of data must be recognized and in itself treated as a social phenomenon that needs analysis, on the other, we must make some sense of whatever information and statistics exist. The latter strategy is what Nicky Gregson recommended in our interview. Benjamin Steuer and Raul Pacheco-Vega, for their part, decided to generate data by skillfully measuring and tracing waste collectors in their respective sites.

A fourth tension may be said to reflect a broader friction in STS and the social sciences between a focus on materiality or nonhuman agency and social inequalities and power. In this volume, Joshua Reno and Britt Halvorson deal with this tension by demonstrating how race is constitutive of the scientific practices and material infrastructures around waste, ultimately bringing the attention back to humans, their bodies or even "junk DNAs" as abject and disposable without the epistemological recentering of the human. Shiloh Krupar in a creative intervention proposes the transdisciplinary method of antiracist soil exegesis to demonstrate through a careful investigation of materiality the sedimented power relations and color lines in brownfield development. Concepts in the biopolitics kinship have been increasingly mobilized by waste scholars to creatively and critically examine the materiality of power, and Amanda Boetzkes also reaches to a kin concept, Achille Mbembe's necropower, to demonstrate the suturing of willful destruction (burning oilfields) with the willful wasting of certain human lives as manifest in Al Qadiri's documentary.

Whatever questions, cases, and methods future waste studies scholars will pursue, tensions such as these will have to be foregrounded to maintain the field's relevance for social struggles against injustice and planetary destruction.

References

Alexander, Catherine and Joshua. O. Reno. 2014. "From Biopower to Energopolitics in England's Modern Waste Technology." *Anthropological Quarterly* 87, no. 2: 335–358.

Alexander, Catherine and Joshua. O. Reno. 2012. *Economies of Recycling: The Global Transformation of Materials, Values and Social Relations*. Zed Books.

Alexander, Catherine, Nicky Gregson, and Zsuzsa Gille. 2013. "Food, Leftovers, and Waste." In Anne Murcott, Warren Belasco, Peter Jackson (Eds.) *The Handbook of Food Research*. Berg Publishers. 471–484.

Arefin, Mohammed Rafi2019. *"Infrastructural Discontent in the Sanitary City: Waste, Revolt, and Repression in Cairo."* Antipode, 51:1057–1078 10.1111/anti.12562.

Barnard, Alex V. 2016. *Freegans: Diving into the Wealth of Food Waste in America*. University of Minnesota Press.

Bauman, Zygmunt. 2004. *Wasted Lives: Modernity and Its Outcasts*. Wiley.

Bloom, Jonathan. 2011. *American Wasteland: How America Throws Away Nearly Half of Its Food (and What We Can Do About It)*. De Capo Press.

Brook, Daniel. 1998. "Environmental Genocide: Native Americans and Toxic Waste." *The American Journal of Economics and Sociology* 57, no. 1: 105–113.

Bulkeley, Harriet, Watson, Matt, and Hudson, Ray 2007. "Modes of Governing Municipal Waste." *Environment and Planning A: Economy and Space*, 39: 2733–2753. 10.1068/a38269.

Bulkeley, Harriet, and Gregson, Nicky 2010. "Crossing the Threshold: Municipal Waste Policy and Household Waste Generation." *Environment and Planning A: Economy and Space*, 41:929–945. 10.1068/a40261.

Bullard, Robert. 1990. *Dumping in Dixie: Race, Class, And Environmental Quality.* Westview Press.

Clapp, Jennifer 1994. "The Toxic Waste Trade with Less-Industrialised Countries: Economic Linkages and Political Alliances." *Third World Quarterly*, 15: 505–518. 10.1080/01436599408420393.

Clapp, Jennifer. 2001. *Toxic Exports: The Transfer of Hazardous Wastes from Rich to Poor Countries.* Cornell University Press.

Crang, Mike, Hughes, Alex, Gregson, Nicky, Norris, Lucy, and Ahamed, Farid. 2013. "Rethinking Governance and Value in Commodity Chains Through Global Recycling Networks." *Transactions of the Institute of British Geographers* 38, 12–24. 10.1111/j.1475-5661.2012.00515.x.

Davies, Anna. 2008. *The Geographies of Garbage Governance: Interventions, Interactions and Outcomes.* Routledge.

Dinler, D. Ş. 2016. "New Forms of Wage Labour and Struggle in the Informal Sector: The Case of Waste Pickers in Turkey." *Third World Quarterly* 37, no. 10: 1834–1854.

de Hoop, Evelien, and Arora, Saurabh 2017. "Material Meanings: 'Waste' as a Performative Category of Land in Colonial India." *Journal of Historical Geography*, 55: 82–92. 10.1016/j.jhg.2016.10.001.

Elliot, Susan J. 1999. "And the Questions Shall Determine the Method." *Professional Geographer* 51, no. 2: 240–243.

Evans, Peter. 2012. "Beyond the Throwaway Society: Ordinary Domestic Practice and a Sociological Approach to Household Food Waste." *Sociology* 46, no. 1: 41–56.

Evans, Peter. 2014. *Food Waste: Home Consumption, Material Culture and Everyday Life.* Bloomsbury Academic.

Fagan, G. Honor. 2003. "Sociological Reflections on Governing Waste." *Irish Journal of Sociology* 12, no. 1: 67–84.

Foucault, Michel. 2004. *Security, Territory, Population | Lectures at the Collège de France 1977–1978.* Michel Senellart, François Ewald, Allesandro Fontana, and Arnold I Davidson (Eds.). Palgrave Macmillan.

Fredericks, Rosalind. 2009. "Wearing the Pants: The Gendered Politics of Trashwork in Senegal's Capital City." *HAGAR: Studies in Culture, Polity & Identities* 9, no. 1: 119–146.

Fredericks, Rosalind. 2018. *Garbage Citizenship: Vital Infrastructures of Labor in Dakar, Senegal.* Duke University Press.

Furniss, Jamie. 2017. "What Type of Problem Is Waste in Egypt?" *Social Anthropology/Anthropologie Sociale* 25, no. 3: 301–317.

Garcier, Romain. 2012. "One Cycle to Bind Them All? Geographies of Nuclearity in the Uranium Fuel Cycle." In Alexander Catherine and Joshua. O. Reno (Eds.) *Economies of Recycling: The Global Transformation of Materials, Values and Social Relations.* Zed Books. 76–97.

Gidwani V. (2013). "Six theses on Waste, Value, and Commons." *Social & Cultural Geography* 14, no. 7: 773–783.

Giles, David. 2014. "The Anatomy of a Dumpster: Abject Capital and the Looking Glass of Value." *Social Text* 32, no. 1(118): 93–113.

Gourlay, K. A. 1992. *World of waste: Dilemmas of Industrial Development.* Zed Books Ltd.

Goldstein, Jesse. 2013. "Terra Economica: Waste and the Production of Enclosed Nature." *Antipode*, 45, no. 2. 10.1111/j.1467-8330.2012.01003.x.

Graham Stephen and Nigel Thrift. 2007. "Out of Order: Understanding Repair and Maintenance." *Theory, Culture & Society* 24, no. 3: 1—25.

Gregson, Nicky, and Peter J. Foreman. 2021. "England's Municipal Waste Regime: Challenges and Prospects." *The Geographical Journal*, 1–13.

Gregson, Nicky, Mike, Crang, Anna Krzywoszynska, Julie Botticello, and Melania Calestani. 2016. "Doing the 'Dirty Work' of the Green Economy: Resource Recovery and Migrant Labour in the EU." *European Urban and Regional Studies* 23, no. 4: 541–555.

Graziano, Valeria and Trogal Kim . 2017. "The Politics of Collective Repair: Examining Object-Relations in a Postwork Society." *Cultural Studies* 31, no. 5: 634-658.

Gregson, Nicky. (2007). *Living with Things: Ridding, Accommodation, Dwelling.* Sean Kingston Publishing.

Gregson, Nicky and Louise Crewe. (2003). *Second-Hand Cultures.* Berg.

Gutberlet, Jutta. 2009. "Solidarity Economy and Recycling Co-Ops in São Paulo: Micro-Credit to Alleviate Poverty." *Development in Practice* 19, no. 6: 737–751

Hawkins, Gay and Stephen Muecke. 2002. *Culture and Waste: The Creation and Destruction of Value.* Rowman and Littlefield Publishers.

Hawkins, Gay. 2006. *The Ethics of Waste: How We Relate to Rubbish.* Rowman & Littlefield Publishers.

Hecht, Gabrielle. 2014 *Being Nuclear: Africans and the Global Uranium Trade.* MIT Press.

Hecht, Gabrielle. 2018. "Interscalar Vehicles for an African Anthropocene: On Waste, Temporality, and Violence." *Cultural Anthropology* 33, no. 1: 109–141. doi:10.14506/ca33.1.05.

Hird, Myra. 2012. "Knowing Waste: Towards an Inhuman Epistemology." *Social Epistemology* 26, no. 3–4: 453–459.

Hird, Myra. 2017. "Waste, Environmental Politics and Dis/Engaged Publics." *Theory, Culture & Society* 34, no. 2–3: 187–209.

Hird, Myra. 2016. "The Phenomenon of Waste-World-Making." *Rhizomes: Cultural Studies in Emerging Knowledge*. 30. doi:10.20415/rhiz/030.e15

Hird, Myra. 2013. "Waste, Landfills, and an Environment al Ethic of Vulnerability." *Ethics & The Environment* 18, no. 1: 1085–6633.

Hird, Myra. 2021. *Canada's Waste Flows*. McGill-Queen's Press.

Knowles, Caroline. 2017. "Untangling Translocal Urban Textures of Trash: Plastics and Plasticity in Addis Ababa." *Social Anthropology* 25, no. 3: 288–300.

Kuletz, Valerie L. 1998. *The Tainted Desert: Environmental and Social Ruin in the American West*. Routledge.

Lepawsky, Josh. 2019. "No insides on the Outsides." *Discard Studies*, https://discardstudies.com/2019/09/23/no-insides-on-the-outsides/

Leonard, Liam, Fagan, Honor, and Doran, Peter 2009. "A Burning Issue? Governance and Anti-Incinerator Campaigns in Ireland, North and South. *Environmental Politics*, 18: 896–916. 10.1080/09644010903345678.

Liboiron, Max. 2017. "Compromised Agency: The Case of BabyLegs." *Engaging Science, Technology, and Society* 3: 499–527.

Liboiron, Max. 2021a. *Pollution Is Colonialism*. Duke University Press.

Liboiron, Max. 2021b. Media Indigena Episode 258. "Pollution is Colonialism: Part 1." Host and Producer: Rick Harp. https://podcasts.apple.com/ca/podcast/media-indigena-indigenous-current-affairs/id1092220986?i=1000523230762

Liboiron, Max. 2021c. Media Indigena Episode 259. "Pollution is Colonialism: Part 2." Host and Producer: Rick Harp. https://podcasts.apple.com/ca/podcast/media-indigena-indigenous-current-affairs/id1092220986?i=1000523464487

Liboiron, Max, Josh Lepawsky. Forthcoming. *Discard Studies: Wasting, Systems, and Power*. MIT Press.

Liboiron, Max, Alex Zahara, Kaitlyn Hawkins, Christina Crespo, Bárbara de Moura Neves, Vonda Wareham-Hayes, Evan Edinger, et al. 2021. "Abundance and Types of Plastic Pollution in Surface Waters in the Eastern Arctic (Inuit Nunangat) and the Case for Reconciliation Science." *Science of the Total Environment* 1. doi:10.1016/j.scitotenv.2021.146809.

Melosi, Martin V . 2008. *The Sanitary City: Environmental Services in Urban America from Colonial Times to the Present*. University of Pittsburgh .

Millar, Kathleen. 2018. *Reclaiming the Discarded: Life and Labor on Rio's Garbage Dump*. Duke University Press.

Millar, Kathleen. 2008. "Making Trash into Treasure: Struggles for Autonomy on a Brazilian Garbage Dump." *Anthropology of Work Review (Wiley-Blackwell)* 29, no. 2: 25–34.

Miraftab, Faranak. 2004. "Neoliberalism and Casualization of Public Sector Services: The Case of Waste Collection Services in Cape Town, South Africa." *International Journal of Urban and Regional Research* 28, no. 4: 874–892.

Moore, Sarah A. 2012. "Garbage Matters: Concepts in New Geographies of Waste." *Progress in Human Geography* 36, no. 6: 780–799.

Muecke, Stepheen. 2009. "Cultural Science? The Ecological Critique of Modernity and the Conceptual Habitat of the Humanities." *Cultural Studies* 23, no. 3: 404–416.

Newman, Richard. 2012. "Darker Shades of Green: Love Canal, Toxic Autobiography, and American Environmental Writing." In Stephanie Foote and Elizabeth Mazzolini (Eds.) *Histories of the Dustheap: Waste, Material Cultures, Social Justice*. MIT Press. 21–48.

Oloko, Patrick. 2018. "Human Waste/Wasting Humans: Dirt, Disposable Bodies and Power Relations in Nigerian Newspaper Reports." *Social Dynamics* 44, no. 1: 55–68.

O'Brien, Mary H. 1993. "Being a Scientist Means Taking Sides." *Bio Science* 43, no. 10: 706–708. doi:10.2307/1312342.

O'Brien, Martin. 2007. *A Crisis of Waste? Understanding the Rubbish Society*. Routledge.

O'Hare, Patrick. 2020. "'We Looked After People Better When We Were Informal': The 'Quasi-Formalisation' of Montevideo's Waste-Pickers." *Bulletin of Latin American Research* 39, no. 1: 53–68.

O'Hare, Patrick. 2019. "'The Landfill Has Always Borne Fruit': Precarity, Formalisation and Dispossession Among Uruguay's Waste Pickers." *Dialectical Anthropology* 43, no. 1: 31–44.

O'Neill, K. 2000. *Waste Trading among Rich Nations: Building a New Theory of Environmental Regulation*. MIT Press.

Packard, Vance. 1960. *The Waste Makers*. David McKay Co.

Pellow, David N. and Robert J. Brulle (Eds.). 2005. *Power, Justice, and the Environment: A Critical Appraisal of the Environmental Justice Movement*. The MIT Press.

Pellow, D. N., A. Schnaiberg, and A. S. Weinberg. 2007. Putting the Ecological Modernisation Thesis to the Test: The Promises and Performances of Urban Recycling. *Environmental Politics* 9, no. 1: 109–137.

Persaud, Donny, Josh Lepawsky, and Max Liboiron. 2019. "Viscous Objects. The Uneven Resistances of Repair." *Techniques & Culture* 72: 126–129. http://journals.openedition.org/tc/12372

Pezzullo, Phaedra C. 2012. "What Gets Buried in a Small Town: Toxic E-Waste and Democratic Frictions in the Crossroad of the United States." In Stephanie Foote and Elizabeth Mazzolini (Eds.) *Histories of the Dustheap: Waste, Material Cultures, Social Justice*. MIT Press. 119–146.

Prashad, Vijay. 2001. "The Technology of Sanitation in Colonial Delhi." *Modern Asian Studies*, 35: 113–155. 10.1017/s0026749x01003626.

Reno, Joshua. 2018. "What is Waste?" *Worldwide Waste: Journal of Interdisciplinary Studies* 1, no. 1: 1–10.

Reno, Joshua Ozias. 2014."Toward a New Theory of Waste: From 'Matter out of Place' to Signs of Life." *Theory, Culture & Society*. https://doi.org/10.1177/0263276413500999.

Stephanie, Foote and Mazzolini, Elizabeth (Eds.) 2012. *Histories of the Dustheap: Waste, Material Cultures, Social Justice*. MIT Press.

Strasser, Susan. 1999. Waste and Want: A Social History of Trash. Metropolitan Books/Henry Holt and Company.

Stuart, Tristram. 2009. *Waste: Uncovering the Global Food Scandal*. Penguin.

Szasz Andrew and M. Meuser. 1997. "Environmental Inequalities: Literature Review and Proposals for New Directions in Research and Theory." *Current Sociology* 45, no. 3: 99–120.

Thompson, Michael. 1979. *Rubbish Theory: The Creation and Destruction of Value*. Oxford University Press.

Van Horen, Basil. (2004). "Fragmented Coherence: Solid Waste Management in Colombo." *International Journal of Urban and Regional Research 28*, no. 4: 757–773.

Varul, M. Z. 2006. "Waste, Industry and Romantic Leisure: Veblen's Theory of Recognition." *European Journal of Social Theory* 9, no. 1: 103–117.

Veblen, Thornstein. 2007. (1899) *The Theory of the Leisure Class*. Oxford University Press.

Volk, T. 2004. "Gaia Is Life in a Wasteland of By-Products." In. S. Schneider, J. Miller, E. Crist, and P. Boston *In Scientists Debate Gaia: The Next Century*. MIT Press.

Walther, Daniel J (2017). "Race, Space and Toilets: 'Civilization' and 'Dirt' in the German Colonial Order, 1890s–1914. *German History*, 35:551–567. 10.1093/gerhis/ghx102.

Weinberg, Adam, David N. Pellow, and Allan Schnaiberg. 2000. *Urban Recycling and the Search for Sustainable Community Development*. Princeton University Press.

Wynne, Brian. 1987. *Risk Management and Hazardous Waste: Implementation and the Dialectics of Credibility*. Springer.

Wynne, Brian. 1989. "The Toxic Waste Trade: International Regulatory Issues and Options." *Third World Quarterly* 11, no. 3: 120–146.

2

AT HOME WITH THE WASTE SCHOLAR

Zsuzsa Gille, Josh Lepawsky, Catherine Alexander, and Nicky Gregson

Waste studies is a new field. While scholars have made high-impact contributions for three decades, we are still missing a systematic overview of this literature. Our intention with this handbook has been to show the exciting possibilities waste studies offers in a structured way without imposing a canon. That said, there is much to gain from the authors of foundational texts, and we thought the best way to take stock was to interview three such scholars. We were interested in the following questions: How does one become a waste studies scholar? What are the tensions waste scholars have to work with? What theories and methodologies have been helpful or less so? What do they see as new directions in the field? We discussed these topics with an anthropologist, Catherine Alexander; a geographer, Nicky Gregson; and a historian, Gabrielle Hecht, and compiled their words of wisdom into a chapter that we hope will demonstrate the richness of waste studies and will guide newcomers to the field. Let us introduce them alphabetically.

Catherine Alexander is a British anthropologist and professor at Durham University. She conducted research in Kazakhstan, Turkey, and the United Kingdom, where she focused on changing property regimes and bureaucracy, the third sector, waste, and pollution, as well as recent environmental policies and recycling technologies, such as recycling and energy generation from waste technologies. Her many articles, a special issue concerned with the ways in which wastes are "disappeared" or "unknown", and two co-edited volumes, *Indeterminacy: Waste, Value and The Imagination* (with Andrew Sanchez, 2018, Berghahn Books) and *Economies of Recycling: The Global Transformation of Materials, Values and Social Relations* (With Joshua Reno, 2012, Zed Books) have served as foundational references for waste studies scholars.

Nicky Gregson is a British geographer and emerita professor of Durham University, whose waste-related works include the monographs *Living with Things: Ridding, Accommodation, Dwelling.* (2007, Sean Kingston Publishing) and *Second-Hand Cultures.* (co-authored with Louise Crewe, 2003, Berg.). She was the principal investigator of the ESRC-funded multi-year project "The Waste of the World" that investigated global flows of end-of-life goods in a variety of materials, including the disassembly of ships, steel, nuclear waste, used clothing, and food waste. Through her countless articles and edited special issues on waste economies, she has shaped the research repertoire of social science inquiry into waste perhaps more than anyone else.

DOI: 10.4324/9781003019077-2

Gabrielle Hecht is a multidisciplinary scholar of science and technology and currently a professor of history (and, by courtesy, anthropology) at Stanford University. Most of her earlier publications have focused on politics, culture, and technology in nuclear worlds, including *The Radiance of France: Nuclear Power and National Identity after World War II* (MIT Press 1998, 2009) and *Being Nuclear: Africans and the Global Uranium Trade* (MIT Press and Wits University Press, 2012), both of which received awards in several disciplines. Her current research centers on toxic and radioactive waste, exploring ways of approaching waste studies using lenses offered by critical theories of race, colonialism, postcolonialism, and materiality.

We transcribed and edited our conversations organizing them along a few key themes that they all emphasized. These largely center around the themes of how one enters the field of waste studies; how waste emerges; how to engage with political economy; and how to face the methodological challenges of studying something that is supposed to be out of sight and out of mind.

Becoming a waste scholar

It was fascinating to see that none of these researchers set out to become waste scholars. The phrase Catherine Alexander used, "coming to this field obliquely," is likely to resonate not just with the other two of our interlocutors, but with many of us in waste studies. They recounted either projects of their own or projects initiated by engineers or scientists that they were invited to, only to realize that there was a proverbial elephant in the room, either in the sense of wastes ignored or in the sense of a certain way of posing questions about them. Nicky Gregson had been sensitized by her research on domestic labor, and "what we would call dirty work, and" more directly in her study of second-hand consumption in the United Kingdom.

> So, the antecedents, the academic antecedents are in that consumption project...and then, I realized that when we were writing *Second-Hand Cultures*, as it got called, that we hadn't actually accounted for how all of this stuff ended up in these sites, that you actually needed to understand this process of household wasting. And that's where I really started to get into thinking about consumption as disposal, dispossession, and using things up. So, my route into waste came through consumption studies.

Alexander's moment of sensitization came from her fieldwork on a Turkish sugar beet factory. In part it was the "stench of piles of rotting beet pulp." In part it was her interlocutors' complaint about the pollution of the local waterways that destroyed crops in the nearby fields. But it was also the reuse of "unwanted beet tops that ended up as animal fodder and then the molasses byproduct from refining the sugar that went straight to another state enterprise that made alcohol." While her book on this study, *Personal States: Making Connections between People and Bureaucracy in Turkey* (2002) didn't focus on waste, the experience made her curious enough to later agree to participate in a project with engineers who were "largely working on end-of-life technologies, anaerobic digestion, and incineration".

Gabrielle Hecht's moment of revelation came in the form of plastic ducks she spotted on her research trip to a uranium mine in Gabon, where the company had filled a former open pit with water to create a "lake". They had initially stocked this with live ducks, but these regularly got eaten by pythons native to the area, and the establishment got tired of restocking them. In order to appeal to post-mining tourism, the corporation placed plastic ducks in the "lake" instead.

So, we have this hole that is being repurposed for some kind of white European tourist destination where you go and sit on a waterfront and look at ducks, which are not native to Gabon. Talk about massive cognitive dissonance! Then I realized that uranium mines elsewhere were subject to similar dynamics. Johannesburg is built on mining waste: things that look like mountains they aren't mountains, they are piles of tailings. One such mountain became a drive-in movie theatre, a popular entertainment destination for white Joburgers in the 1970s, named after the mine: the Top Star Drive-In!

However, the visceral perception of waste was only the beginning. What one does with this sensitization as a scholar is another formative step in becoming a waste scholar. Nicky Gregson was taken aback when her findings on second-hand shopping came to be labeled as "waste prevention behaviors".

> These folks had said vague things like, it's a bit like recycling isn't it, as a justification for what they were doing. But they certainly weren't green behaviors. If anything, it was 'look at how much I can buy at the charity show with all these bags!' It was a way to do more consumption, and more clever consumption cheaply. It was all about thrift and style, and obviously about poverty and getting by for some people who had a very limited budget.

Later she realized that there is even more to the structural underpinnings of second-hand cultures. At a charity shop conference, she realized that what she had been studying was only the tip of the iceberg. "80% of secondhand clothes end up in the global market. Which also tells you the problem of coming at something from the consumption end and not the economic end". It was this second revelation that led her to design the Waste of the World project.

Catherine Alexander also had to defy a certain framing of what a social scientist should contribute to a waste project. She was asked to take care of the "social things" part of a big engineering research project on waste management, but what they meant by that was to work out how to engineer "attitudes and behavior change" in consumers, especially low-income residents. Not only did she resist this perspective, but it made her realize the importance of emphasizing both uneven infrastructural provision to poorer neighborhoods and understanding the myriad ways in which people care for materials, but which are rarely visible. Research with third sector recycling organizations also showed how, in Britain at least, this sector has been increasingly enrolled in carrying out state services as being "close to the people and grassroots" but is simultaneously expected to perform like the private sector, what she calls quasi-privatization or third-wave privatization.

Governance also became a key concern for Gabrielle Hecht, leading her to think not just about waste but about residues more broadly. This led her to her current research questions:

> What counts as residual? Are residues necessarily waste? Or can they also fall into a kind of non-category, the externality that we don't even pay attention to, whether it's waste or not? What does it mean to govern residues and how is the governance of residues also governance that comes after the fact, so that the governance itself is also a kind of externality? And how does that governance treat people and places as residual, left over, and wasted?

How does waste emerge? Where does waste come from?

Through their decades of research, all three scholars came to new revelations about the kind of processes, situations or practices that generate waste. As we'll discuss below, they express reservations about the overwhelming focus on municipal or household waste, but at a higher level of abstraction, they have come to realize that there are certain waste-engendering social patterns that we haven't sufficiently interrogated yet. For them they all center around the axis of movement and stasis. Catherine Alexander identifies moments of system breakdown. "System here can signal an actual set of institutions or social order and a system of classification". In her research on postsocialist Kazakhstan, she kept hearing people talking about "choking" on car traffic, pollution, and garbage.

> It wasn't so much that these were new [after the collapse of state socialism] but that the former "ordering system screened them out."…"That physical sense of being over-whelmed came when there's no one there to regulate traffic pollution or to keep everything in the urban landscape neat and tidy. There were hulks of unfinished buildings and abandoned buildings, cannibalized buildings, the influx of plastics ev-erywhere. Along the roads people were selling all their previous means of getting by, because there was no longer the time to repair and reuse and save things. I mean there were giant glass bottles, I mean hundreds of these things that people would use, the tools, the bits, the bobs. A whole way of life disappeared.

Her other example of waste-inducing system breakdown came from post-Brexit United Kingdom (curtailing easy movement of goods between the EU and the United Kingdom) and the COVID pandemic that together created "a perfect storm". There are thousands of things, especially living or perishable things that are stranded, unable to move, including pigs. Thousands of pigs en route to slaughter were unable to be transported across a border during a particularly acute period of uncertainty around Brexit's effects on transborder commodity flows:

> The problem is that they [the pigs] are getting fatter and fatter, so a third of them now are not meeting industry standards. All those delicate just-in-time supply chains all over the world have snapped and there's all that matter that has been backing up. And it's not like things can just be released again, I don't know whether you can slim down a fat pig but once it's too old that's it!

Of course, the pigs becoming waste as a result of them being stuck joins a whole set of other examples in which food waste is the result of certain standards dictated by global commodity chains and these rubbing up against hard concrete materialities.

Nicky Gregson's concern with movement may be seen not so much as an opposite of Catherine Alexander's but a complement to it. She worries that with too much focus on households, we fail to see how the kind of mobility some of us have grown accustomed to, especially in the Global North, generates new waste streams.

> We generate waste on the move, in our office, walking along the street. It's that kind of mobility of eating and drinking and all these kinds of things, as consumption is less spatially constricted. A question that we haven't attended to is "what does that then do to thinking about waste as a product of mobility?" Furthermore, could we in fact argue that global mobility depends on disposability?

Gabrielle Hecht's concern with waste-generating movement is primarily about the movement of matter. In her new research project, she calls this problematique "inside-out Earth".

> We are turning the Earth inside out at an exponential rate and releasing all these molecules that used to be trapped in the Earth, in rocks. Now they are wreaking havoc in places that they were not originally. Are they out of place or not? I don't know, that's still a debate. For sure, we don't want carbon dioxide in the atmosphere. For sure, we don't want sulphur dioxide in our water. Et cetera, et cetera. Whether you are talking about atmospheric circulation or the tailings piles that constitute the topography of Johannesburg, you are talking about things that were once inside the Earth that are now on or above the surface. Their extraction, and the practices and residues and wastes that they leave behind, continue to *act* in ways that are shaped by power relations. You have all this toxic stuff blowing into people's homes, and who are the people in question? The people who have suffered under racial capitalism, colonialism, apartheid, and post-apartheid…Some of the uranium ended up exploded over Bikini Islands and then landed on ships that were then taken back to San Francisco, where I live, for "decontamination" at a Navy yard situated next to an historically Black neighborhood. "Decontamination" involved hosing the ships down and spreading radiological sand all over the place. This is now the subject of a huge conflict in San Francisco, as it should be. That's part of what I mean by turning the world inside out.

What about political economy?

Gregson, as some readers will know, led a major research initiative called the Waste of the World project between 2006–2011. As she describes it, one of the main motivations of that project was a frustration with a dominant political-economy tradition in economic geography that "with the labor theory of value, their point of interest *stopped* (original emphasis) at the point of sale". This presumption of an equivalence between retail and consumption ignored most of what was interesting Gregson and flew in the face of her own "searing" experiences of growing up wearing secondhand clothes. She argued that the economic geography, and we may add, other social sciences, "erased" experiences like that and

> failed to recognize that there are these circuits of value through which this secondhand retail value operates…The theoretical frames that were prevailing in these fields suffered from an inability to say anything about this, other than in the broadest sweeps of the destruction of value. The fact that loads and loads of people around the world have their livelihoods dependent on waste recovery was missing entirely. It really annoyed me that they were seeing economies as just those advanced developed world economies. Everywhere else wasn't an economy.

> So, the Waste of the World was fundamentally about trying to get economic geography to try and see the global trade in waste. And to take it seriously. And I think they have actually started to listen.

As Gregson's and her colleagues' work became more accepted as speaking to core concepts in the field of economic geography, her attention to the industry of waste management has deepened. Economic geography has a tendency to work from sectoral cases — automobile manufacturing, agri-food production, business services, textile manufacturing — to develop its understandings of "the economic". Gregson's work has, in a sense, asked what happens to our understanding of the economic if we start from flows of what at least some actors call waste? Among other things this means turning to the political economy of waste management as an industrial and a service sector. Here, of late, Gregson notes the burgeoning trend of the financialization of waste. While this is "a *huge* (original emphasis) billion-dollar industry", its sheer size is less important than what it can tell us about the implications of an economy increasingly premised on financialization, rather than "traditional" production. A financialized waste regime turns waste into assets from which new rounds of value can be extracted, either for waste-to-energy schemes or materials recovery. Ironically, though, the process is premised on generating contracted volumes of residuals (e.g., from households and municipalities) which in turn incentivizes waste generation over reduction.

Catherine Alexander shares an interest in broadscale political-economic change, and those interests brought her to the political economies of waste, albeit obliquely at first. Her research into the privatization of state infrastructures in Britain, Kazakhstan, and Turkey and how those changes profoundly altered property relations. Alexander notes her "long-standing interest in the properties of objects and I think the really rarely discussed *obligations* (original emphasis) that go with property rights gets downplayed". For Alexander, waste studies' engagements with gleaning practices that once used be common before being legislated out (see Peter Linbaugh's book *London Hanged* (Verso 2006) and Agnes Varda's classic documentary *Gleaners* (2000)) is one of the areas in which fundamental questions about property are most actively in play. Alexander notes that

> whether [something is] waste or not is very often a distinction between what is legal and illegal and what is legitimate and illegitimate. So, these things often play quite nicely off of one another. For example, when do waste pickers have a moral right to something?

Such questions bring to the fore issues of power and the political vis-à-vis the economic. Rather than waste being merely the ephemera of, or subsidiary to, the "real stuff" of economic action the line(s) between waste and value, legal and illegal, legitimate and illegitimate are struggles pertaining to power over who will benefit (or not) from the discards of economic action, where, when, and under what conditions. Alexander's work helps highlight a core idea of waste studies, that is, the unceasing tension between the political economy of waste and value.

Gabrielle Hecht is adamant that, "you can't write about waste and not write about power". She addresses the political-economic nexus by asking how seemingly disparate people, places, and things are linked together in ways that are not only shaped by power relations, but also perform and transform those relations. Relaying examples from her current research, Hecht narrates a political economy of waste connecting distinct places and people, treating waste and its toxicities not as "externalities" or "unintended consequences" but as integral parts of the systems she studies:

> How mine waste functions in Johannesburg is entirely a product of colonialism, racial capitalism, and apartheid. You have this string of waste dumps that cuts across the city. Under apartheid, black and other non-white people were forcibly moved south of these waste piles—which was also downwind and downstream of them.

Wealthier white folks lived upwind...These spatial rearrangements are very much about power. They are not "unintended consequences". They are *designed*. What's more, this isn't just about power on the urban or national scale. Power in this case operates on many different scales: the power of gold in global capitalism, and the power of nuclear things on planetary scales (because uranium and gold were in the same ore matrix).

The excerpts from our interviews with Alexander, Gregson, and Hecht all highlight, albeit in different ways, the fundamentally co-constitutive relations between waste, value, and power.

Methodology, science, data

Waste studies is an eminently empirical field. As we discussed in the Introduction, however, data are often non-existent, inaccessible, insufficient, and inaccurate. We wondered out loud how the accessibility of certain empirics has guided waste studies. Do we study household waste so much because at least we can access it, measure it, and observe its generation and collection? How do we engage with science and economic data in a critical way? How do our three interlocutors see this and related problems with empirical research?

Nicky Gregson admits that the absence of data is limiting, whether because in certain cases, we stumble upon criminal behavior, industrial secrets, or because of practices that push waste out of sight, and thus out of mind. At the same time, she argues that especially in an age of audit culture, there is a wealth of data that we have to learn to use. Data such as those she collected and analyzed from decades of UK municipal waste contracts, can be used to point to the sheer scale of a waste problem and its underestimation — just think back what the fact that 80% of UK secondhand clothing is exported had meant for her research trajectory! It is this new study that made her realize the size of the waste market and how through the financialization of waste this industry is "a massively important economic engine". More consequentially, another methodological use of even such faulty data is restoring our capacity to "speak synthetically". She worries that with defaulting to the case study, as the most common method in waste studies, as in critical social science, more broadly, which, in turn is "always reduced to an exemplification, a particular exemplification of a general pattern", we lose our ability to generalize, "but not in a very grand, and grandstanding, especially theoretical way". As an example, she brings up the early literature on municipal waste which framed the problem at hand as "'here's another case study of an anti-incineration protest.' But these case studies did not interrogate the policy behind the rollout of incinerators". In addition to seeing the scale of waste issues, another merit of synthetic quantitative data analysis is questioning the taken-for-granted grand sweeping narratives. "It is surely significant that, when we turn to examine data like this, it challenges the standard narratives and accounts that are to be found in the field, particularly those of environmental and environmental justice campaigners".

Gabrielle Hecht, whose research topics are perhaps the least accessible to researchers, gleans a lot of her empirical data from archives. And while older documents might be subject to less restriction than current ones, she warns about our reliance on natural scientists to understand the basic facts about whatever wastes we study.

The challenge is how to hold in your mind the social construction of scientific knowledge (to use an old-fashioned term) while simultaneously using scientific findings about how waste behaves. Holding those two things together is a challenge, but it's a good challenge. That's why waste research is fun.

A related problem is that when one educates oneself about the relevant science, one becomes progressively more aware of one's own ignorance. How can waste studies scholars stay at a useful level of analysis?

> Among other things, I began to think about "interscalar vehicles" as a way of not going too deeply into rabbit holes. My job is to make the connections, not to redo the science.

Whether moving downward (rabbit holes) or outward (making global connections), Hecht calls attention to the challenge of drawing boundaries around the problems we study.

> Once you start to see the world as one of ever-increasing waste, which is really how I have come to see the Anthropocene, then almost anything can be addressed with this lens. The challenge is to define the topic and figure out how you're going to draw your empirical or analytical boundaries, Yet, at the same time, one must point to the fact that the boundaries are artificial, that one could keep on going.

She argues that the methodological promiscuity and interdisciplinary nature of waste studies can be helpful in this effort.

For Catherine Alexander, whose recent work also engages with what is knowable and "brings the literatures on strategic ignorance into conversation with Waste Studies", and as having the most experience of our three interlocutors with collaborating with natural scientists and engineers, grapples with the epistemological challenges of interdisciplinary research. These engineers, in studies of mass balance and energy often suggested that "*carbon* and *energy* are the ultimate measure, because it's '*life*' (original emphasis)". However, as she noted, "then you are completely missing other economic and political scales that shape how and why materials move and end up in particular places". The methodological take-away is the value of seeing things from different perspectives but without the expectation or assumption that these will add up to "the complete picture". Nevertheless, the inter-disciplinarity of waste studies and its capacity to see things from multiple angles "reminds us that whichever way we look at something there is always another way of seeing, and those other ways may interact and affect each other".

Conclusion

While our three interlocutors nominally represent three different disciplines, they all emphasize and hail the interdisciplinary nature of waste studies. This feature of the field allows, if not demands, flexibility in research strategy and diversity in theoretical orientation, however to us and our interlocutors this doesn't mean that anything goes. The anchor of waste studies is a critical theoretical perspective and empirically grounded research. We may not always be able to produce all of our own data, however, how we use what is available and accessible should be guided by the rich conceptual directions opened up by Catherine Alexander, Nicky Gregson, and Gabrielle Hecht.

PART II

Questions waste scholars ask

3

MATTER OUT OF PLACE

Max Liboiron

Introduction

Describing and analyzing waste as "matter out of place" is a reoccurring trope in waste studies. The phrase was popularized by British anthropologist Mary Douglas' book, *Purity and Danger*, though it was used to describe waste long before the 1966 publication. Douglas wrote, "[Dirt] implies two conditions: a set of ordered relations and a contravention of that order. Dirt then, is never a unique, isolated event. Where there is dirt there is system. Dirt is the by-product of a systematic ordering and classification of matter, in so far as ordering involves rejecting inappropriate elements" (1966, 36). It is "out of place". Waste scholars often exchange "waste" for "dirt" in Douglas' theory to talk about ordering, classifying, and rejecting waste, trash, refuse, filth, and similar synonyms.

But there's a catch: Douglas specifically states that rubbish is *not* matter out of place. She writes, "rubbish is not dangerous…since It clearly belongs in a defined place, a rubbish heap of one kind or another" (1966, 160). Even when trash is not in a rubbish heap or bin, it still belongs in a heap or a bin and is merely displaced in space, not "out of place" in the way Douglas theorizes. Douglas is not making a spatial or geographical argument, but an argument about power and threats to power.

This entry focuses on questions about the way wastes is and is not related to Douglas' theory of dirt and out of place-ness. It takes the critical stance that for the most part, waste is not out of place, but it can be under specific circumstances. The categorization of trash, waste, pollution, and rubbish as "out of place" has become a truism, and the goal of this entry is to recover its theoretical power and specificity.

Fidelity to Douglas is not what is at stake in these questions, though I will certainly draw on Douglas as a key source. As other scholars have noted, "It is almost too easy to repeat [the phrase 'matter out of place'] without further critical reflection or contextualisation: it becomes reified" (Fardon 1999, 4). I believe the reification of the phrase has implications for waste studies in terms of how it can be a key source for recognizing and analyzing power relations around waste and wasting. Waste and discard studies tend to be normative fields of study, meaning that many scholars are dedicated to environmentalism and justice in our work. As such, missing or reifying concepts related to power, such as "matter out of place", may actually have effects that miss opportunities to move research towards environmental goods or social justice and may even overlook places where analyses of power are warranted.

DOI: 10.4324/9781003019077-3

I begin with a short history of the phrase "matter out of place" as "dirt" before moving on to its place in theorizing power and power relations in regard to waste and wasting.

History

There are two points in history when "matter out of place" was a term of art in scholarly and public publications. Its first popularization began in the 1850s, reaching a peak just before 1920, and the second begins after 1966 with the publication of *Purity and Danger,* with a peak in 2000.

The term was first referenced during a series of "well-lubricated" public toasts in 1852 by Lord Palmerston (1784–1865) speaking as foreign secretary (Palmerson later became prime minister of the United Kingdom, twice). Historian Richard Fardon writes that Lord Palmerston said, "I have heard a definition of dirt. I have heard it said that dirt is nothing but a thing in a wrong place. Now, the dirt of our towns precisely corresponds with that definition" (Anon 1852 in Fardon 2013, 15). Fardon further recounts that,

> The quotation, reported anonymously in *The Times*, is explicit that Palmerston claimed to have 'heard' the definition, but does not tell from whom...Palmerston's specific suggestion was that imported guano (mined bird and bat droppings) might be substituted as fertilizer by the noxious waste matter produced in Britain's cities, thereby saving farmers much expenditure, 'the dirt of our towns ought to be put upon our fields, and if there could be such a reciprocal community of interest between the towns and country—that the country should purify the towns, and the towns should fertilize the country (laughter)—I am disposed to think the British farmer would care less than he does, though he still might care something, about Peruvian guano. (Hear, hear, and cheers) (Anon 1852; Fardon 2013, 25)

After Palmerston's speeches were reported in print, the phrase begins to circulate in various texts, from science to religion, often with attribution to Palmerston. For instance, Sigmund Freud's 1908 essay on "Character and anal erotism" uses the phrase "Dirt is matter in the wrong place," as does a 1878 text entitled "The Chemistry of Dirt" that declares, "The popular mind has been long familiar with the aphorism 'that dirt is matter out of place'" before arguing that, "Very strong flavoured dirt fills and saturates the atmosphere of many loathsome manufacturers without disgusting the work people who are used to it" (Bartlett 1878, 152). The range of disciplines and meanings of the term, from psychoanalysis to chemistry, mirrors its disciplinary promiscuity today. In scholarship repositories such as Scopus, half of uses of the phrase "matter out of place" originate in the social sciences and humanities, but some also come from publications in medicine, economies, and psychology (Liboiron 2019).

As Douglas' biographer, Richard Fardon writes that Douglas first used the phrase in her field notes from her second trip to study the Lele people in the Belgian Congo in 1953. In those notes, she writes about how the Lele asked her about dirt:

> Asked to define dirt in England—Not earth, just simply [dirt]. Contrast: idea of dirt, with "good clean mud" etc. Chesterfield "Dirt is any matter displaced" e.g. hair, crowning glory etc. and hair in the soup. But child putting spoon it has licked back in the veg. tureen and told off for being "dirty". "Dirty" is much wider range than just 'dirt'. Any bodily excreta, saliva, vomit, faeces, and anything that has contact with them is dirty. Food is wholesome when served, but as soon as someone has eaten a little, and left it, it is "orts", remains, dirty. (Fardon 2010, 154)

Yet given that the phrase was already in hardy circulation, it is unlikely it has a single origin point for Douglas. One of Douglas' students, Alan Macfarlane,

> can be confident Mary Douglas did not find the full quotation from one convenient source. When he was studying for his MPhil at the [London School of Economics] in 1968, Mary Douglas used to hold small discussion groups at her Highgate home near Hampstead Heath. Feeling very honoured to be invited to one, he arrived by way of the Heath—where he was surprised to find that the sides of rubbish bins had been labelled, "Dirt is Matter Out of Place". Mentioning this in all innocence to Mary, she misconstrued his suggestion as an accusation that she had stolen the quotation from the side of a rubbish bin. Somewhat mortified, the memory stayed with him. (Fardon 2010, 25)

Whether from reading Freud, looking at rubbish bins, or having conversationd with Lele research subjects, the idea that dirt is "matter out of place" became the theoretical anchor of Mary Douglas' *Purity and Danger: An Analysis of Concepts of Pollution and Taboo*. As an anthropological text in a classic tradition, *Purity and Danger* makes universal claims about how taboo topics are understood and handled for all of humankind, drawing on a dizzying array of cultures and community customs to argue "that the difference between pollution behavior in one part of the world and another is only a matter of detail" (Douglas 1966, 35). For all people, she says, "Where there is dirt there is system. Dirt is the by-product of a systematic ordering and classification of matter, in so far as ordering involves rejecting inappropriate elements", where rejection includes a rage of actions from ignoring something to its annihilation (Douglas 1966, 36).

Power

Rejecting "inappropriate elements" isn't about sorting recycling. *Purity and Danger* is about power, first and foremost. Douglas says dirt is about "a set of ordered relations and a contravention of that order", and spends most of her time discussing the techniques that maintain the order by circumventing, punishing, or eliminating those threats to order. When she writes about how "ordering involves rejecting inappropriate elements" her examples are drowning deformed babies, strangling roosters that crow before dawn, and killing and eating Pangolins when they threaten the ordering systems that maintain power and normalcy:

> when a monstrous birth occurs, the defining lines between humans and animals may be threatened. If a monstrous birth can be labelled an event of a peculiar kind the categories can be restored. So the Nuer treat monstrous births as baby hippopotamuses, accidentally born to humans and, with this labelling, the appropriate action is clear. They gently lay them in the river where they belong…Or take night-crowing cocks. If their necks are promptly wrung, they do not live to contradict the definition of a cock as a bird that crows at dawn. (1966, 39)

Genuine threats to power must be eradicated, not sorted out and neatly laid in blue bins at the front door. Sorting something spatially, the literal meaning of "out of place", is quite different than upsetting power balances. Douglas writes that "pollution is a particular class of danger" to "sources of power" (1966, 98–99) and that "The danger which is risked by boundary transgression *is* power" (1966, 199). A theory of power in waste studies must distinguish between the stakes and politics of sorting (trash in the bin) versus eradication of threats (killing, aggressive assimilation).

As an example, think about the types of violence leveraged towards trans and gender non-binary people, both at the everyday level (where strangers demand to know someone's gender) and in newsworthy events (where trans people, particularly trans people of color, are murdered), and then read Douglas: "In these cases the articulate, conscious points in the social structure are armed with articulate, conscious powers to protect the system; the in-articulate, unstructured areas emanate unconscious powers which provoke others to demand that ambiguity be reduced" (1966, 200). To parts of society where gender binaries are central to power, gender fluidity and "ambiguity" are a threat — matter out of place, or dirt. Douglas writes that "Purity is the enemy of change, or ambiguity and compromise" (1966, 200). Purity might rely on categorization and sorting (of genders, of trash) as a tactic, but its goal is maintaining power.

Making trash in and out of place

This does not mean that litter or pressing delete on a social media post will never be "matter out of place" in the way Douglas means. Materials, practices, and their meanings shift within different arrangements in structures. Take litter, for example. By definition, litter is not where it belongs from a spatial perspective. It sits at the edge of the road, the schoolyard, or a shoreline when it should be contained in a bin, a landfill, or a recycling center. But that doesn't make litter dirt. Litter generally does not challenge systems of power or confused or contradicted cherished classifications that matter — except sometimes it does.

In the United States in the 1960s, for example, certain types of litter *were* out of place, both literally in spatial terms when litter was not in the trash bin where it belonged but also in terms of upsetting existing powerful structures. The rise in disposable containers and packaging meant that cans and bottles were appearing in ditches and cow's stomachs, and the public concern was enough to threaten the new order of industrial production. The response was first anti-litter campaigns and then the greening of recycling that sought to maintain the production of disposables while also addressing public concern. Keep America Beautiful, among countless other campaigns,

> "composed of leading beverage and packaging corporations and staunchly opposed to many environmental initiatives, sought to interiorize the environmentalist critique of progress, to make individual viewers feel guilty and responsible for the degraded environment. Deflecting the question of responsibility away from corporations and placing it entirely in the realm of individual action, the commercial castigated spec-tators for their environmental sins but concealed the role of industry in polluting the landscape" (Dunaway 2015, 81).

Reframing disposables as litter frames where to put disposables, where disposables are and are not allowed, and how to keep flows moving through space in the "right" direction. These efforts put litter *in place* by individualizing responsibility for waste not being in proper bins, eliminated the threat to industry as producers of disposables. Today, litter is still deeply "in place" based on this social order — most narratives about litter, including those that moralize litter as bad, *maintain* systems of power rather than disrupt them.

This isn't a permanent situation. Structures of power, that is, structures that uphold main-stream societal and economic norms, must maintain themselves. In 2015, New York City banned disposable polystyrene single-serve containers, also known under a brand name, Styrofoam. Within the year, container corporations and other companies sued the city and the ruling was eventually overturned, just to be reinstated again over a year later (Liboiron 2015). The back-and-forth hinged on whether or not polystyrene foam was truly recyclable or not. The industry-led arguments said it was and Dart Container Corp even offered to "pay every dime of the start-up costs for recycling" the material (DART Container Corporation 2015). Recyclability makes disposables like polystyrene "in place" and by greening them, and arguments that they were not green make them a threat to the established industrial order. Power structures and their norms and infrastructure are neither monolithic nor stable. The way waste flows in some ways and not others, to the benefit of some groups more than others, in line with some ideals and values but not others are the norms and structures that are at stake of disruption when trash becomes "out of place", a rare but not impossible occurrence.

Techniques for keeping matter in place

Douglas theorizes that when a threat to a dominant system presents itself as potential matter out of place, there are certain techniques to "to create order" once again. The example of hippopotamus babies is one such example, where "by settling for one or other interpretation, ambiguity is often reduced" (Douglas 1988, 39). She proposes five such techniques that I will outline with examples from the waste studies literature. In the examples that follow, I've aimed to make it clear that while the enumerated techniques are presented here, and in *Purity and Danger*, as if they are separate and escalating strategies, they also always work in concert and any case of keeping waste "in place" can be subject to numerous, intersecting strategies.

First, Douglas discusses how "labell[ing] an event of a peculiar kind [allow threatened] categories to be restored" (1966, 39), like hippopotamus babies, where "with this labeling, the appropriate action is clear" (1966, 39). Angeliki Balayannis' work on toxic waste removal demonstrates that such labeling and categorizing can be a lot of work (2019, 2020). Balayannis studies the myriad ways that the removal of stockpiled pesticide waste is managed in a coastal Tanzania site through contracts, inventories, photographs, categories, and protocols, as well as material sacks, linoleum, boots, and personal protective equipment. She writes that through these techniques, "the bureaucratic spectacle creates a controllable world where matter can be contained" (2020, 20). Of particular interest is the categorization of "stockpiled pesticides" that are contracted to be removed versus "pesticide-contaminated soil" that the stockpiles rest on and are not part of the contract:

> Daniel (partly) jokingly explained that decision-making about what to "bag" and what to leave behind was based on the ratio of pesticides to soil…if a patch of soil appeared to be over fifty percent constituted of pesticides, then this was a part of the stockpile. Anything less, and the material was merely "pesticide contaminated soil" and had to be left at the site. (2020, 11)

This type of categorization creates a demarcation between stockpile and spill and organizes action accordingly, is a way to keep matter "in place" by showing action had occurred to address the stockpiled pesticides: "with this labeling, the appropriate action is" not only "clear", but has demonstrably been completed. Indeed, these categories paired with other techniques such as before and after photographs that show the visual absence of a pile of chemicals, are "necessary for this heterogeneous mass of matter to become legible for globalized disposal economies" (Balayannis 2020, 18) and "creates a controllable world where matter can be contained" (2020, 20) (for more on the categorization of unruly hazardous waste as control, see Wynne 1987). Of course, Balayannis' investigation of the process uncovers spills, seeps, and residues, but for global infrastructures of environmental justice and waste disposal, the important work has occurred; business as usual is able to proceed by leveraging existing, dominant modes of categorization, understanding, and circulation.

At the same time, categorization and its bureaucracies are not necessarily the given enemies of waste justice. Plastic pollution scientist Chelsea Rochman and her colleagues, for example, made headlines in 2013 when they published a paper in *Nature* entitled, "Classify plastic waste as hazardous". The main argument was framed as a simple scientific one: plastics are currently treated the same way as food scraps and grass clippings are treated, but plastics not only absorb known hazardous industrial chemicals at a rate of one million times more than the environments around them, they also break down into tiny nanoplastics that flow more like chemicals than objects. Since plastics act, move, and harm like industrial chemicals, they should be classified as such. They argue, "With a change in plastics categorization, numerous affected habitats could immediately be cleaned up under national legislation using government funds" (2013, 170). Indeed, in 2020, Canada, where Rochman works, Environment and Climate Change Canada has indicated that the federal government will designate plastics as toxic (Baum 2020). Plastic lobby groups are contesting this decision, indicating that it is, indeed, a threat to power (Fawcett-Atkinson 2020).

"Second", Douglas writes, "the existence of anomaly can be physically controlled" (1966, 39). Her example is killing roosters that crow before dawn. A lion's share of waste studies focus on the physical control of waste and how physical containment, circulation, and deposit are central to keeping waste "away" and thus in place. The removal of stockpiled pesticides in Balayannis' work and the case of industry-sponsored recycling mentioned above are two such cases. Indeed, much of the literature theorizes "away" and its shortcomings, both in terms of justice (e.g., Bullard 1990; Davies 2018) and logistics (e.g., Beckett 2020; Gray-Cosgrove, Liboiron and Lepawsky 2015), and of course, how they intersect. For instance, in "Hidden Mountain: The Social Avoidance of Waste" (2008), a group of macroeconomists analyze the social effects of what they call "smoothing mechanisms" that keeps municipal solid waste "from becoming visible and [thus] kept in its proper place" (2008, 2) such as trash cans, curbside pick-up done in the early morning or late night, rare interactions with sanitation workers, single stream recycling that reduces the need to sort, and litter campaigns. All of these physical in-frastructural mechanisms, they write, maintain an "established social order" (2008, 2) that allows unsustainable economic system of overconsumption. Other waste literature focuses on what happens when physical control of waste fails: in his exposé of the Los Alamos National Laboratory's neglect of safe handling of radioactive waste that resulted in an exploded canister of nuclear waste in the Waste Isolation Pilot Plant in the United States, Vincent Ialenti (2018) shows how intersecting social, managerial, class, and physical systems intersect inextricably to succeed — or fail — to produce an infrastructure of assumed containment. What is common among this diverse literature is that physical control is a highly orchestrated, never complete or entirely successful, and always contingent practice.

Douglas's third method for keeping matter in place is avoidance. She writes that "a rule of avoiding anomalous things affirms and strengthens the definitions to which they do not con-form. So where Leviticus abhors crawling things, we should see the abomination as the negative side of the pattern of things approved" (1988, 39). Much research on waste describes the use of Othering, social taboos, and stigmatization of people associated with waste, but even the "smoothing mechanisms" like trash bins and night time pick-ups that allow waste and sanitation workers to be invisiblized and avoided in "Hidden Mountain" can be understood as tactics of avoidance. Some new work in waste studies goes a step further to show how rules of avoidance are often resisted, and even leveraged, to change what counts as disgusting, abominable, necessarily invisible, and wrong and to make space for often devalued social positions, showing how systems of power are never complete nor monolithic (e.g. Liboiron 2012). In a nuanced example, Waqas Butt (2019) argues that focusing on abjection and caste to understand waste workers in Lahore, Pakistan, misses not only significant factors in who becomes a waste worker such as colonialism and land settlement, but also misses how many waste workers maneuver social relations to obtain work and even rights as waste workers. He notes, monopolization of waste work by certain classes allowed for political organization: "because of the avoidance of polluting materials by caste Hindus and others, lower status groups occupied positions of power and influence as functionaries within departments of medicine, public health, and sanitation" (2019, 23). While he acknowledges that this power is fundamentally limited, he also outlines how a diversity of waste workers in Lahore are able to own land and exceed imagined rigid caste and religious categories.

Douglas's fourth rule, that "anomalous events may be labelled dangerous…But it would be a mistake to treat institutions as if they evolved in the same way as a person's spontaneous reactions. Such public beliefs are more likely to be produced in the course of reducing dissonance between individual and general interpretations…Attributing danger is one way of putting a subject above dispute. It also helps to enforce conformity, as we shall show below in a chapter on morals" (1988, 39–40). While examples of this abound in the criminalization of land protectors (Estes and Dhillon 2019) and informal recyclers (Wittmer and Parizeau 2016), I will focus on a more communal, and indeed moralized example: the tragedy of the commons. The tragedy of the commons, argued its originator, Garrett Hardin, is that humans will overuse, pollute, and otherwise trash shared space and its resources without state containment and control. Yet, rather than the foundations of this theory being about "human nature" (see Liboiron 2020), several scholars have shown that Hardin was keenly motivated by fear of low class and racialized others, promoting "an idea he called 'lifeboat ethics': since global resources are finite, Hardin believed the rich should throw poor people overboard to keep their [higher class] boat above water" and "that only racially homogenous societies could survive" (Mildenberger 2019a. Also see Cox 1985; Goldstein 2013; Fortier 2017; Mildenberger 2019b; Brinkley 2020). Through much of his work, Hardin framed poor and "diverse" populations as dangerous not just on an individual level, but as a threat to planetary stability. The racialized and class premises of the tragedy of the commons have become naturalized and are rarely known to the many environmentalists and others who take up the theory. This example points to the import for researchers of waste to carefully consider how waste is kept "in place" through such techniques.

Finally, Douglas' fifth strategy to keep threats to power in place is about how "ambiguous symbols can be used in ritual for the same ends as they are used in poetry and mythology, to enrich meaning or to call attention to other levels of existence" (1988, 40), folding it into power structures to give them more ritualistic, symbolic, or other transformative powers. A short version of this technique might be "if you can't beat them, ceremonially join them". In "Dirty hands: The toxic politics of denunciation", Amelia Fiske documents this phenomenon in action (2018). To protest dangerous and invasive oil extraction in the Ecuadorian Amazon, people who lived in toxic environments would place their hands in open pools of Chevron's spilled and abandoned oil, holding up their oil-coated hand as way to make the harm visible and embodied. As the gesture gains social and political meaning, Fisk documents how politicians and celebrities that were "politically and materially insulated from toxic violence" (2018, 19) appropriate the gesture in front of cameras, including Ecuadorian President Coreta, who "opened the door to oil extraction in one of the most bio-diverse regions of the world with one hand, and launched an international campaign to denounce the history of Texaco with the other", using his oil-soaked hand as an icon (2018, 27). This technique of ingesting symbols of resistance to annul threats to power is not unique to President Coreta. In "Protecting the power to pollute", Bell, Fitzgerald and York look at the

> central strategy of these public-relations campaigns is a process we term "Identity Co-optation," which entails appropriating and reconstructing the identities of fossil fuel industries' fiercest opponents—concerned women and mothers—in the delivery of their counterclaims. We argue that the strategic mobilization of women in defense of coal, oil, and gas is a clear example of hegemonic powers attempting to appropriate, embody—and ultimately neutralize—threats to their influence and authority. (2019, 323)

In each of Douglas' techniques of keeping matter in place, it's been clear that the techniques work together. At the same time, these five techniques can be used by activists to achieve threats to power, or at least mitigate oppression, as shown in Butt's work on waste workers in Lahore. In a similar vein, Syantani Chattergee shows how residents of Shivaji Naga, the area surrounding Mumbai's massive landfill, actively leverage demonstrations of "'failure' to their advantage to stake claims of belonging to the neighborhood, and demand state assistance, albeit often with punitive consequences" (2019, 49). One informant tells him, "You think failure is achieved by doing nothing? A lot of work goes into it…If they ask us to fail, we will fail more forcefully, so the politicians and the government can give us something in return. So, that's our way!" (2019, 51). Hijacking representations of failure to become "a medium of exchange" (2019, 51) to achieve (always limited) government services uses Douglas' fifth technique in reverse, holding powerful actors hostage to their statements of good statehood as well as their need for unofficial waste workers in the landfill. Of course, much like in Butt's work, the agency of such actors is heavily constrained, and his work also highlights how the state casts waste workers as "criminal elements and miscreants" to keep the residents of Shivaji Naga "in place".

I believe that using Douglas' theory of "matter out of place" and its opposite — the more common myriad and overlapping strategies of keeping matter "in place" — offers nuance to analysis of power in waste studies. Not only are these strategies analytical tools for looking at how power works, but also how struggles against power use these same strategies. Power, after all, is a dynamic system and not something that one group "has" over another, as the literature covered here makes clear. It seems that waste being out of place is a rarity, though it certainly exists in constant and uneven ways, and the strength of Douglas' work is to frame how contests over power are constantly being maneuvered by actors in different positions within worlds of waste.

References

1852 Anon. 1852. "The Royal Agricultural Society." *The Times* 21169, 16 July: 8.

1852 Anon. 1852. *The British Farmer's Magazine* XXII: 137.

Balayannis, A. 2019. "Routine Exposures: Reimaging the Visual Politics of Hazardous Sites." *GeoHumanities* 5, no. 2: 572–590.

Balayannis, A. 2020. "Toxic Sights: The Spectacle of Hazardous Waste Removal." *Environment and Planning D: Society and Space.* doi:10.1177/0263775819900197.

Bartlett, H. C. 1878. "The Chemistry of Dirt." *British Architect* 10, no. 16: 152.

Baum, K. 2020. "Ottawa Set to Declare Plastics as Toxic Substance." *The Globe and Mail,* March 11. https://www.theglobeandmail.com/canada/article-ottawa-set-to-declare-plastics-as-toxic-substance/

Beckett, C. 2020. "Beyond Remediation: Containing, Confronting and Caring for the Giant Mine Monster." *Environment and Planning E: Nature and Space.* doi:10.1177/2514848620954361.

Brinkley, C. 2020. "Hardin's Imagined Tragedy Is Pig Shit: A Call for Planning to Recenter the Commons." *Planning Theory* 19, no 1: 127–144.

Bullard, Robert D. 1990. *Dumping in Dixie: Race, Class, and Environmental Quality.* Routledge.

Butt, W.H. 2019. "Beyond the Abject: Caste and the Organization of Work in Pakistan's Waste Economy." *International Labor and Working-Class History* 95: 18–33.

Chatterjee, S. 2019. "The Labors of Failure: Labor, Toxicity, and Belonging in Mumbai." *International Labor and Working-Class History* 95: 49–75.

De Coverly, E., P. McDonagh, L. O'Malley, and M. Patterson. 2008. "Hidden Mountain: the Social Avoidance of Waste." *Journal of Macromarketing* 28, no. 3: 289–303.

Douglas, M. 1966. *Purity and Danger: An Analysis of Concepts of Purity and Taboo.* Routledge.

Cox, S. J. B. 1985. "No Tragedy of the Commons." *Environmental Ethics* 7, no.1: 49–61.

DART Container Corporation. 2015. "New York State Supreme Court OVertuns City's Ban on Foam." *News Stories,* September 22. https://www.dartcontainer.com/news/news-archives/news-stories/2015/09/new-york-state-supreme-court-overturns-citys-ban-on-foam/

Davies, T. 2018. "Toxic Space and Time: Slow Violence, Necropolitics, and Petrochemical Pollution." *Annals of the American Association of Geographers* 108, no. 6: 1537–1553.

Douglas, M. 1988 [1966]. *Purity and Danger: An Analysis of Concepts of Pollution and Taboo.* Ark Paperbacks.

Dunaway, F. 2015. *Seeing Green: The Use and Abuse of American Environmental Images.* University of Chicago Press.

Estes, N. and J. Dhillon (Eds.). 2019. *Standing with Standing Rock: Voices from the# NoDAPL Movement.* University of Minnesota Press.

Fardon, R. 1999. "Purity and Danger Revisited."*In Mary Douglas: An Intellectual History,* 1st ed. Routledge.69-88.

Fardon, R. 2010. 'Margaret Mary Douglas 1921–2007.' *Proceedings from the British Academy* 166: 154.

Fardon, R. 2013. "Citations Out of Place: Or, Lord Palmerston Goes Viral in the Nineteenth Century But Gets Lost in the Twentieth." *Anthropology Today* 29, no. 1: 25–27.

Fawcett-Atkinson, M. 2020. "U.S. Companies Threaten to Use CUSMA to Fight Canada's Plastics Ban." *The National Observer,* November 6. https://www.nationalobserver.com/2020/11/06/news/us-companies-cusma-canada-plastics-ban

Fortier, C. 2017. *Unsettling the Commons: Social Movements Against, Within, and Beyond Settler Colonialism.* Arp Books.

Fiske, A. 2018. "Dirty Hands: The Toxic Politics of Denunciation." *Social Studies of Science* 48, no. 3: 389–413.

Freud, S. 1997 [1908]. "Character and Anal Erotism." In A. Richard (Ed.) *On Sexuality* (Vol. 7). Penguin. 213

Goldstein, J. (2013). "Terra Economica: Waste and the Production of Enclosed Nature." *Antipode* 45, no. 2: 357–375.

Gray-Cosgrove, C., M. Liboiron, and J. Lepawsky. 2015. "The Challenges of Temporality to Depollution & Remediation." *SAPI EN. S. Surveys and Perspectives Integrating Environment and Society* 8, no. 1.

Ialenti, Vincent. 2018."Waste Makes Haste: How a Campaign to Speed Up Nuclear Waste Shipments Shut Down the WIPP Long-Term Repository." *Bulletin of the Atomic Scientists* 74, no. 4: 262–275.

Liboiron, M. 2012. "Tactics of Waste, Dirt and Discard in the Occupy Movement." *Social Movement Studies* 11, no. 3–4: 393–401.

Liboiron, M. 2015. "The Power Behind Disposability: Why New York City's Ban on Polystyrene Was Vilified, Sued, and Reversed." *Discard Studies,*September 29. https://discardstudies.com/2015/09/29/the-power-behind-disposability-why-new-york-citys-ban-on-polystyrene-was-vilified-sued-and-reversed/

Liboiron, M. 2019. "Waste Is Not 'Matter Out of Place.'" *Discard Studies,* September 9. https://discardstudies.com/2019/09/09/waste-is-not-matter-out-of-place/#easy-footnote-bottom-8-13097

Liboiron, M. 2020. "There's No Such Thing as 'We'." *Discard Studies,* October 12. https://discardstudies.com/2020/10/12/theres-no-such-thing-as-we/

Mildenberger, M. (2019a). "The Tragedy of the Tragedy of the Commons." *Scientific American*, 23 April.

Mildenberger, Matto. 2019b. "The Tragedy of the Tragedy of the Commons." *Discard Studies*, 15 July.

Rochman, C, M. Browne, B. Halpern, B. Hentschel, E. Hoh, H. Karapanagioti, L. Rios-Mendoza, H. Takada, S. Teh, and R.C. Thompson. 2013. "Classify Plastic Waste as Hazardous." *Nature* 494, no. 7436: 169–171.

Wynne, B. 1987. *Risk Management and Hazardous Waste: Implementation and the Dialectics of Credibility*. Springer.

Wittmer, J. and K. Parizeau. 2016. Informal Recyclers' Geographies of Surviving Neoliberal Urbanism in Vancouver, BC. *Applied Geography* 66: 92–99.

4

WASTE AND WHITENESS

Joshua O. Reno and Britt Halvorson

Introduction

One question that waste studies has been asking since its inception is how waste relates to race and vice versa. This is, for instance, important to environmental justice critique and reform in North America (Bullard 1990, 2005; Bryant and Mohai 1992; Pellow 2007; Carmin and Agyeman 2011) as well as broader critiques of the uneven distribution of pollution globally, and the slow violence resulting from it (Westra and Lawson 2001; Nixon 2011; Liboiron, Tironi and Calvillo 2018). Cultural notions of waste have also informed the related imagination and mistreatment of Black and Indigenous people of color (BIPoC), refugees, and migrants as if they were disposable (Bauman 2003) or, alternatively, hapless victims to be benevolently disciplined through aid (Malkki 1995; Turner 2012). Despite this attention to race and the negative impacts of waste management, one question about waste and race that waste studies for the most part hasn't asked, with few exceptions (Zimring 2015; Isenberg 2016; Solomon 2018), is about the *constitutive role of race* in the productivity of waste and whiteness. This is a key question because it concerns waste scholarship as much as waste itself. By this we do not only mean that the vast majority of waste scholars are white, including the authors of this paper, but that the polluted sites, landfills, ghettos, spaces of scavenging, and refugee camps that waste scholarship has brought to the fore are frequently tied to the constitutive production of middle- and upper-class whiteness as valuable, clean, desirable, detached from worldly concerns, and morally righteous. Seen from this perspective, race is not merely an additional form of difference, to be included in an intersectional way. All scholarship of waste is perforce constituted by concerns of race, even and especially if racial otherness appears unrelated to the subject at hand.

Race sometimes appears to be a separate concern from waste, for example, in waste approaches animated by new materialism and post-humanism, which are often (though not always) informed by discourses originating from theories in environmental studies, as well as physical and natural science (Bennett 2009; Hird 2013; Reno 2014, 2016). Yet the productivity of white supremacy, as a form of power and privilege, has arguably been central to these and other scholarly projects, not least because of the importance of whiteness in the scientific appropriation of knowledge and resources from BIPoC as well as conceptions of the environment at work in spaces and practices of environmental science and politics.

DOI: 10.4324/9781003019077-4

As we hope to make clear in this essay, the question of waste and whiteness is not merely ontological in a rational or mathematical sense, that is, it is not only a claim about a connection one finds between distinct phenomena. That is so because race has had a profound and constitutive impact on the world as "we" know and study it, including who tends to be the subject (the "we") and who the object of scholarly activity, what questions the former think to ask, which connections are perceived and which ignored, and for what ends. Put differently, race points us toward more systemic issues associated with what and who is trashed, which would put it more in line with discard studies as described by Max Liboiron (2018), rather than an analysis of waste as a specific, ontological *thing* (see Introduction to this volume). Scholars of whiteness have pointed out that one of the powerful effects of this flexible racial trope is its ability to vanish, to appear irrelevant, as not worthy of comment or attention. In that respect, at least, whiteness and waste have something epistemologically in common, since waste scholars have made similar points about the forgettability or deniability of waste practice in modern waste management systems (Nagle 2013; Reno 2016). Yet whiteness is not only something ignored because it appears insignificant; its erasure belies the extent to which racial hierarchy is thoroughly interwoven in social institutions and everyday practices, including those concerning waste and waste scholarship.

Thinking about race in terms of whiteness therefore presents a fundamental epistemological challenge to all would-be researchers (and not only scholars of discards). We want to argue, however, that conjoining waste and race in this way takes on special significance as a way to dismantle white supremacist tendencies that would otherwise deny the mutual entanglement of humanity, nature, and ideologies of human nature. In the first section, we outline the imbrication of whiteness in the history of industrial capitalism, challenging its self-erasure. This includes the manifestation of whiteness as an unevenly distributed form of oppressive and harmful pollution. At the same time, we aim to add to this literature a characterization of whiteness as a form of propertied entitlement and magnet for capital investment and political development.

Taking this insight seriously means, among other things, seeing both whiteness *and* waste where we would not expect to, including biological research and humanitarian aid initiatives. To begin with, we consider how research on wasted *things* becomes associated with the disproportionate representation of white scientists in scholarly and scientific practice wherein material excess is appropriated and/or disconnected from social relatedness and reproduction as "rubbish". This raises concerns about the role of unequal race relations in the appropriation of knowledge about racialized groups as objects of knowledge in scholarly practice. The relationship between white supremacy and discard studies introduces the need to formulate an explicitly anti-racist and anti-colonial scientific practice. In the context of genomic research, specifically, we discuss the relationship of "junk DNA" to the detachability of whiteness and consider how such biogenetic material has operated as a key ingredient for the generation of newly racialized models of 21-century identity.

In the biosciences, waste and whiteness thus become co-constitutive of the objects they observe as well as the observers' self-objectifications. The last section continues this engagement with biopower by focusing on the dirty work of care labor and the life politics of biowaste, including medical supplies and pharmaceuticals, and traces how they unevenly distribute the embodiment of life's risks and harms. We explore how, despite operating at different geopolitical scales and circuits of care, both meet in a political terrain of whiteness and waste that upholds valued, predominantly white forms of personhood and life maintenance by removing and displacing risky and unwanted substances.

Racial capitalism and/as waste

We begin with the historical fact that the growth of industrial capitalism happened in conjunction with, and was critically dependent upon, both transatlantic slavery and the colonization of much of the world by European empires. These are twin facets of what is widely considered *modernity*, but the association between these two events, between capitalism and race, has been repeatedly either dissimulated or outright dismissed. This was one of the primary lessons of the tradition of radical black scholarship that extends from the writings of W.E.B. DuBois through those of C.L.R. James to Cedric Robinson, first, that for centuries and still today, capitalism and race have been conjoined and, following from this, that a great deal of ideological and practical energy has been and is still devoted to making it seem as if it were not so (see Rabaka 2009 for a review).

Studies of waste can reconnect capitalism and race or, put differently, demonstrate that they were never separate to begin with. As a notable by-product of industrial growth, waste in its various forms offers an index of lost or irredeemable value. "The benefits and damage of industrial capitalism", in Carl Zimring's words, "produced new ways of describing race, and new inequalities based upon racial identity" (2015, 54). The deepening of capitalist control over human activity, in the form of wage labor discipline, relies on identifying unproductive or superfluous kinds of people, who nevertheless also provide a necessary pool of exploitable reserve labor (Gidwani and Reddy 2011; Yates 2011; Gidwani 2013). More than anyone else, Cedric Robinson 2000 [1983] directly connects the latter process to racialism as a product of empire-building and settler colonial operations and a driving force in its replication all over the world, which frequently entangles wasted labor with conceptions of racialized labor within capitalist dynamics.

The tripartite intertwining of denigrated labor, material excess, and denigrated humanity becomes clearest, arguably, among people exposed to waste as formal or informal laborers or as residents, living beside waste collection and disposal sites, breathing them in and becoming identified with them. Asking after the fate of waste therefore reveals a materialist foundation for the production of social difference and inequality. For example, in her discussion of Brazilian *catadores,* Kathleen Millar (2020) goes beyond both liberal environmental justice critiques that take for granted identity as something that pre-exists unequal exposure to waste, as well as purely metaphorical extensions of waste that refer poetically to wasted lives and places. Instead, Millar follows Weheliye (2014) who "understands racialization not as a category but as a set of sociopolitical relations that striate subjects according to the degree to which they conform to Man and are thus granted (or not) full human status" (2020, 6). She argues instead that garbage provides a seemingly "objective" grounds with which to anchor blackness: "the very act of sorting through waste on a garbage dump makes it such that *catadores* are perceived as not quite fully human, as evolutionarily dysselected, and therefore as 'black'" (2020, 7). Blackness thus becomes metaphorically extended in relation to waste, not the other way around.

Millar's analysis suggests how proximity to and mixture with waste can cut through distinctions that have long troubled anti-racist theory and practice. This is the case with the analogies drawn between race in the United States and caste in South Asia that were used by defenders and critics of established hierarchies in both places, but at its most productive was meant as a means of establishing lateral solidarity between differently oppressed groups against white supremacy (Slate 2011). As Waqas Butt (2019) demonstrates, for instance, low and non-caste waste workers in Lahore are similarly denigrated as a consequence of their labor yet are not beholden to the "abjectness" of religiously designated pollution or impurity (as social critics have argued for more than a century to distinguish caste from blackness, see Slate 2011).

Rather, he suggests they are better understood like the *catadores* Millar describes: as produced by and responding to historical and material conditions not of their own making.

The link between denigrated labor, devalued objects, and denied humanity is also evident among Romani in Sofia, Bulgaria. Elana Resnick describes how these connections shape the temporalities of waste work: "The time between finding a discarded object and turning it into cash is not only a product of work-time but also dependent on historical, large-scale changes in value, political regimes, social categorisation and racial, ethnic hierarchisation" (2015, 129). These hierarchies recur where waste work is concerned, but in distinct formations. Thus, Zimring discusses the representation of new immigrants to the United States in the early 20th century as part of the white/black binary: "If black people were pollutants, dusky people were as well. The new immigrants from Southern and Eastern Europe may not have been black, with the stigma blackness had in a society that equated it first with bondage and then with filth, but neither were they white" (2015, 86). These new immigrants were not only racialized in this way as abject, but as good for waste work:

> Slavs seemed to be "immune" to the dirt that "would kill a white man", so Slavic men and women did the dirty jobs Americans were unable or unwilling to do. Dirty people, in this construction, were simultaneously degenerate and physically robust, allowing whites to justify passing unpleasant yet necessary work off to new immigrants who were "biologically" suited for it. (Zimring 2015, 87)

The point of these comparisons is not to sloppily analogize migrant Slavs, Romani, low-caste Pakistani, and Brazilian *catadores* as "black", to be clear, but to suggest that the production and management of hierarchical divides between valuable and valueless things (waste) tends to be associated with those between persons (race). This makes waste work a productive way to examine race and labor cross-culturally, but as these scholars make clear it does not obviate the need for more detailed understanding of local histories and relations. What is important is not merely their particular "abjectness" as dirty workers (as Butt 2019 argues most clearly), but the common role they fill in racial capitalist orderings of value/anti-value and the disciplining of wage labor more generally.

If work with literal waste by-products help hierarchies to materialize, and provide grounds for any possible resistance, a similar set of processes linking waste and race are evident in many other social contexts beyond waste work proper. The objectification and naturalization of race through the association with waste that Millar identifies can also be compared with Rosalind Fredericks' analysis of the colonial and post-colonial governance of cities like Dakar, Senegal. As Fredericks explains, "Urban space…was produced and regulated along racial lines through ideas of dirt and disease, crystallized through pivotal moments of socio-spatial reorganization like disease outbreaks" (2018, 19). Where infrastructure is built poorly, is not maintained, or is not allowed to be built at all as a consequence of colonial occupation (Solomon 2018; Stamatopoulou-Robbins 2019), it can similarly materialize difference and distinction. Decaying and uneven infrastructure can become coded in terms of "modernity"/"development" or "backwardness" in ways that conceal the ideological and material anti-blackness they inscribe into landscapes and everyday lives. The symbolic denigration of spaces was critical for colonial projects of racial dispossession insofar as the allegation of poorly used resources and lands, "underwrites the legitimation of the expansion of white spaces into nonwhite ones" (Mills 2001, 77). While historically part of settler colonial appropriation, this has implications for how forms of discipline reinforce linkages of race with waste.

The built environment has long served to channel and reproduce white supremacist projects. Even where resistance to this material and ideological legacy is possible, Nicholas Caverley concludes that "corporate owners who hold title to increasing swaths of Detroit's built environment can rely on noxious smells, broken foundations, and virulent fires, among other impositions, to relieve [non-white people] of properties they no longer care to maintain" (2019, 13). Gentrification in the United States, Marisa Solomon argues, "makes visible how racial capitalism, through its attributions of racialized value, produces waste—objects and people—through force" (2019, 91). The force is racial as much as it is temporal, associated with being "forcibly relegated to a backward past, treated as impediments to a 'better,' whiter future" (ibid.). Not unlike Dakar as described by Fredericks, in American urban renewal discourse and policy, "the imperative and *aesthetics* of betterment…are shaped by the twin poles of respectability and authenticity that simultaneously shape and constrain the terrain of what blackness and black femininity in particular *ought to* be, and the impossible demand that it ever be 'natural'" (Solomon 2018, 113, italics in original). In between Solmon's twin fieldsites of Brooklyn and Norfolk is Baltimore, where Chloe Ahmann finds largely white residents raising very similar points about waste politics and the hoped-for whiteness of the future: "both technocratic dreams and local narratives of waste, race, and decline betray a deep ambivalence about the sorts of futures that seem plausible within a geography of 'undesirables'"(Ahmann 2019, 330).

The temporal horizons of racial geography are both a further expression of racial capitalism and also an engine of it. This is so because the imperative of whiteness, as Solomon terms it, does not merely reflect ideology but drives the accrual of value potential and the acquisition of literal wealth over time. Discussions of purity and impurity in the abstract do not do enough to disclose this material history of losses and gains. After all, the primary political and economic rationale for garbage collection and disposal in contemporary cities, whether the American ones mentioned, or Rio de Janeiro, Lahore, Dakar and Sofia, is to create desirable conditions for some, not to worsen conditions for others *per se*. At the very least this adds a level of deniability or apolitical distance between infrastructure and social inequalities such that racialization appears like an incidental after-effect of this process. Using Millar's terms, people untethered from waste become more fully human as a result of such material freedom, evolutionarily selected rather than dysselected. But their cumulative material accumulation or relative value potential remains as a lasting legacy of these positions and the racial capital formations of which they are a part. Asking after waste and race means considering how they work in concert to performatively and systemically valorize non-black spaces and people in the same move that they pollute and abject others.

To the extent that the circulation of waste leads to slow violence (Nixon 2011; Liboiron, Tironi and Calvillo 2018) and slow death (Berlant 2007) that disproportionately impacts BIPoC, therefore, this is not only about racializing them as black, for instance, but strategically creates suffering, sickness, and death in order to give life and value to people who are white (Puar 2017). "White" here means relatively non-black because not exposed (or not as exposed) to waste. This fits with Joao H. Costa Vargas' argument that "degrees of humanity accrue" as a factor of relative "distance from blackness" (2018, 42). This process can provide an obstacle to greater critical awareness:

Whether white workers…will *perceive* that their own "benefits" from racism are only relative to the oppressed conditions of Black labor, and that the social and psychological image of the Blacks-as-inferior beings actually promotes their own exploitation as well as that of Blacks, cannot be predetermined. (Marable 2015 [1983], 41, italics in original)

Whiteness, including white privilege, is therefore burdensome to more than BIPoC insofar as it denies the realization of greater critical (in this case, class) awareness and resistance to exploitation. Historian Nancy Isenberg's book *White Trash* (2016) demonstrates this point by showing how, in U.S. history, poor, landless whites were repeatedly denigrated through associations with waste and filth, but their whiteness served as cover for white elites deflecting attention from class inequality while promoting stories of rags-to-riches, individual (white) social mobility. Through infrastructures, the benefits of racial advantage are virtually attached to whiteness in a way that allows for deniability, a double erasure (people who are not close to waste/people who are not black). Such value creation is not only evident in literal waste flows, but in the decay of buildings, communities, and bodies. Moreover, as we will show in the next two sections, waste and race also raise the stakes of proprietary knowledge of, and care for, life itself.

Scientific practice and rubbish data

Many critiques of waste in capitalism are enunciated from an environmentalist position. As such, they may actually disavow hierarchical divides between particular kinds of people by embracing what appears as the opposite: the seemingly universal web of interconnected nature. This overly simplified organicist vision is belied by the complex ways non-human creatures may be enrolled in specific contexts as a source of pollution (Doherty 2019; Doron and Broom 2019) or even as mystified labor (Zhang 2020). At the same time, environmentalist discourse provides a space for some forms of scholarship and activism that name and investigate pollution or toxicity through a purely organic lens, placing all people on the side of creaturely life to be protected through enviromentally authorized biopower.

What is less commonly recognized is that this universalizing stance is supported through a close historical identification of serene nature and its enjoyment with whiteness (Finney 2014). This is clear in early conservationist efforts and representations, for example, which removed black people from images of the outdoors, as part of their general exclusion from "participating in decision-making processes and having access to resources" in general (Finney 2014, 95–96; see also Mills 2001; Zimring 2015). The actively produced yet often covert whiteness of outdoor spaces has reinforced the notion of an objective, racialized white gaze standing apart from the natural world, whether in aesthetic appreciation, spiritual contemplation or scientific study, while underwriting the white manipulation of organic life forms and land as resources.

Conceptions of a universal nature, to protect and exploit, thus sit uneasily with histories of racial oppression in an analogous way as does waste. One common analogy would classify race as a way of devaluing people and waste as a way of devaluing matter. At the same time, as our section on racial capitalism demonstrated, both forms of devaluation co-occur in efforts to discipline wage labor and extract surplus value. In a similar way, in this section we argue that the racialized positionality of whiteness can function to ideologically amplify the apparent objectivity and neutral postionlessness of "Science". Over a century ago, Du Bois "not[ed] that to be white is to be raceless" and powerful as a result (Rabaka 2007, 38), an insight with implications for both how waste studies scholarship is done and how waste is approached as an object of study. As many have noted since, the ideological racelessness of whiteness has offered a platform from which to assume a universal standpoint to grasp nature (including human nature) in a supposedly clear and impartial way. As Rabaka (ibid.) has argued,

> In the logic of the white world, race is something that soils the status of sub-humans, that is non-whites; it politically pollutes their thinking, thus rendering them powerless, irrational, and in need of clear conceptions concerning themselves and the world.

To this end, when disciplines emerged in the late 19th century specifically to understand BIPoC communities and their place in the family of humankind, they were pursued almost entirely by white men, ideologically raceless and "unpolluted" by identity.

The enduring relationship between scientific objectivity and white supremacy have inspired alternative models for examining waste in feminist and anti-racist ways. Noteworthy in this regard is Jason De León's (2015) archaeological team and their efforts to recover and revalue evidence of migrant struggle and death in the U.S./Mexican borderlands. A second example is Max Liboiron's feminist and anti-colonial "Civic Laboratory for Environmental Action Research" (CLEAR) with its dedication to a community-based, citizen science monitoring of plastic pollution (see Liboiron 2017). In different ways, these efforts demonstrate the productivity of asking questions about the world that foreground entanglements of waste and race. More precisely, De León's and Liboiron's research teams show the value of developing an ethic of accountability that is motivated by distinct histories of racial coloniality. This involves making clear the political stakes associated with which kinds of questions are asked in discard studies, by whom, in collaboration with whom, with what result, and for whose benefit, opening up clear lines of shared analysis of structural inequality. This kind of scientific practice can be contrasted with the acquisition of human and material remains as rubbish or waste to collect and study from a privileged and ideologically "raceless" distance.

Yet whiteness continues to be associated with scientific practice as a condition of possibility for extractive knowledge and exploitative propertization (Harris 1993), that is, as a means of acquiring resources (and not only as a way of rendering non-white people too partial, too particular to assume a universalizing stance). Specifically, waste has been and continues to be deployed as a way to transfer ownership of resources from BIPoC communities to ideologically and demographically white institutions. This is key, for example, in making bio-physical materials worthy of appropriation in the extractive biological sciences, which typically involves turning living cells into immortalized rubbish for the purposes of laboratory manipulation and legal privatization. We use rubbish in Michael Thompson's (2018 [1979]) sense, as a material that is neither valued nor devalued, but valueless and set aside where its value may accrue exponentially. This sense of rubbish, as processes of extreme revaluation (useless skin cells converted to profitable genetic patents) is not ordinarily extended in this way, but fits quite well. Such rubbish is more than metaphorically evident in the form of junk DNA. This term typically refers to non-coding regions of the genome, that is, parts that seem to have no direct adaptive purpose but are superfluous. Precisely for this reason, junk DNA has been thought useful for determining ancestral relationships between individuals who have similar markers, thus bearing witness to shared paternal descent (Nelson 2016, 60).

The importance of junk DNA and its revaluation is not limited to technical applications, however, but is entangled with white privilege and power at the intersection of economics, law and science. As Kim TallBear points out:

> In addition to the law, the biological technosciences are becoming increasingly important in the exercise of property claims that sustain our racial formations. Natural resources, including both black bodies and labor and Native American land and resources, constituted important forms of property in earlier centuries that drove the nation-state's civilizing project. In our twenty-first century "knowledge society", the biotechnosciences mine human bodies for raw materials and produce knowledge of them. (TallBear 2013, 136)

The bodies thought worthwhile for "mining" are not just anyone's, but more often have been the bodily materials of BIPoC, presumed to be more readily separable from their embodied humanity. Such biogenetic materials are rendered in a contradictory way as simultaneously more rare and valuable (for instance, unusual and in danger of "disappearing" through genetic admixture) and/or more easily and cheaply acquired as technoscientific "objects" for circulation and research. Swabs of cheek cells may be characterized by scientific researchers as things given away without a second thought, disposed of readily, only to be later immortalized and revalued as rubbish cells. But even this kind of seemingly trivial material transfer is made possible through a massive and transnational legal architecture that guarantees rights in intellectual property, which have long been linked to whiteness (Harris 1993). In these exchanges any would-be scientist (white or BIPoC) must "first acquire rights to translate cheek cells...into scaleable and salable products" (Reno 2018, 68). The conditions that allow for these transfers are premised on the "logic of the white world" (Rabaka 2007, 38)—and, specifically, historically organized systems of racial capitalism—where parties to exchange are already positioned along a spectrum, not only of relative humanity, but relative transcendence of particularity in their engagement with nature and human nature.

Life, risk, and dirty work in care and aid

The interrelationship of whiteness and waste can also be traced through acts of care and aid, an area of research that demonstrates distinct forms of biopower from those already considered. Doing the "dirty work" of care labor (Glenn 1992), whether bathing bodies, scrubbing floors, cleaning bodily fluids, or handling human wastes, has disproportionately been the burden of "poor, immigrant and racialized women undertaking low-waged care work for privileged others" (Murphy 2015, 723; see also Herod and Aguiar 2006). With the exception of some alternative and radical care networks (Piepzna-Samarasinha 2018), whiteness often works through such care arrangements as a structural privilege of life maintenance and life extension, of active work to shore up regenerative capacity, de-contamination, vitality, fortitude, and leisure (Davis 1981; Povinelli 2006; Puar 2017). Yet this feminized and racialized labor is not as commonly connected to waste studies, though it shares a considerable theoretical and political interest. Even less often are these scaled up hierarchies of care followed transnationally through historical forms of colonialism and empire and contemporary practices of humanitarian aid, to uncover their shared indebtedness to the political-economic conditions of white supremacy. In this section, we unsettle the work of whiteness in acts of bodily care, rescue, and compassion and offer a sense of additional directions for a critical waste studies of care and aid.

Focusing on intimate dirty work reveals how hierarchies of bodily care get reproduced, but also points to the ways care workers make their labor dignified and ennobling. In her research on elderly home care labor in Chicago, Elana Buch (2013, 2014, 2018) has shown how culturally valued independent adulthood is paradoxically sustained through a system of care workers who facilitate their clients' bodily practices of autonomy. This can include smelling and disposing of spoiled milk when an elderly client is no longer able to distinguish its putrification or digging through and cleaning up piles of garbage and mold in a hoarder's home, in order to improve their mental health. In Buch's research, home health aides, mainly women of color, actively fashion themselves as caring selves and "prioritize the bodily dispositions of their clients above their own" (2013, 639). This problem of deferred or displaced care in the bodily labor necessary to sustain life and valued personhood is, we would argue, structurally productive of whiteness. Global circuits of care play out these dynamics on a much more vast, transnational scale. Rhacel Parreñas (2015 [2001]) has shown how, for Filipina workers migrating across national borders for domestic work in wealthy homes in Rome, Los Angeles, Dubai and other places, they often must suspend their own caregiving work for their children and families at home. This results in some parlaying their pay in dollars, euros or other valuable currency to hire poorer Filipina women to perform that care work in their own homes, leading to what Parreñas (2015 [2001]) calls an "international division of reproductive labor" (see also Colen 1995 and Ginsburg and Rapp 1995 on "stratified reproduction").

Low-waged care work often entails the subtraction of undesirable, polluting substances in order to help produce dominantly valued, raced, and classed forms of embodied personhood. This insight links care labor intimately to waste work, in which the circulation and movement of unwanted or repellant materials creates and values white bodies, spaces, neighborhoods. However, when scaled up, the active, ethical work of caring, of rescuing, saving, tending to, and redeeming, is paradoxically most associated with whiteness, erasing in the process the labor and resources it takes to produce white supremacy. Claiming care in this way as an ethical and moral stance—and enlisting a range of affective motivations to sustain it—can be traced historically to the genealogy of colonial governance as care, or the moral justification that colonial occupation introduced caring institutions of, for example, health and medicine (Bornstein and Redfield 2010;Barnett 2011). Thus, whiteness and waste meet in a political terrain where the ethical choice to be recognized as wholly, benevolently caring is meted out unevenly; it is also historically motivated to conceal those very subtractions and extractions of valued bodies, resources, and materials that would enable white supremacy.

Global aid networks of various kinds scale up these racialized hierarchies of care, redistributing wastes to places devastated by neocolonial structural adjustment and often associating whiteness with acts of virtuous saving and rehabilitating (Hansen 2000). Adia Benton (2016) notes that professional aid work in many African states is overwhelmingly done by white expatriates, with a two-tier system of lesser pay and social capital for Black African expatriate aid workers. Contemporary humanitarianism, with roots in colonial institutions and the abolitionist movement, reinforces its covert whiteness through many different mechanisms, from its "moral untouchability" (Fassin 2010) to the white gaze of humanitarian photography, film and narrative testimonials (Prince 2016; Fadlalla 2007; Torchin 2006) to the affective spectatorship of the white savior complex (Cole 2012). In the case of international aid, this affective politics focuses largely on saving or redeeming BIPoC from conditions of violence, poverty and natural disaster. The tendency for waste scholars to study scavenger collectives and informal recyclers could be seen as a reaction to this global redemption narrative, given that they tend to foreground their agency and creativity rather than their victimization alone (Giles 2014; Millar 2018; Nguyen 2019; O'hare 2019). However, the moral terrain of humanitarian aid tends to conceal the central economic work performed in aid circulations that scholars of waste have sought to address (Halvorson 2012). In a reverse direction from the colonial extractions described earlier, aid often involves the global recirculation of undesirable or secondhand materials from wealthier countries to sites of humanitarian intervention. Though aid organizations would characterize these donated items as care or relief, we could flip this framing around and instead see these circulations as also aiding whiteness by removing and distancing those risky or unwanted things from predominantly white spaces and persons.

These global circuits of inequality are highly visible in forms of biowaste, in ways that bring us back to our earlier discussion of hierarchies of bodily care, longevity, and health. One example can be found in medical supply and equipment donations sent by religiously based U.S. NGOs to Madagascar (Halvorson 2018). Institutionalized U.S. medicine is governed by a complex insurance regime that makes medical items close to reaching their expiration dates both a heightened financial risk for malpractice premiums and epidemiologically unsuitable for patient care. Expelling those objects from the U.S. hospital organizes profit not only when the institution is separated from their risk but also in the form of tax credits for their "fair market price". As a result, regardless of their actual safety for patient care (Rosenblatt and Silverman 1992), thousands of medical materials, ranging from respiratory tubing to hospital beds to surgical scissors, are shipped by U.S. medical schools such as Duke and Yale and scores of non-profit agencies to hospitals in parts of Africa, South Asia, and Latin America (see e.g., Mangan 2007). Malagasy aid professionals and doctors receive and revalue U.S. medical discards in Antananarivo, often rejecting the American notion that each donation is spiritually and medically significant and instead accepting the collective lot in order to maintain a valuable foreign aid partnership (2018, 145–150; see also Halvorson 2015). But the global system of inequality ultimately motivating the transfer of U.S. medical discards to Madagascar is not at all missed by Malagasy doctors. Each donation reinforces a system in which U.S. medicine, especially the care performed in wealthier, more resource-rich white spaces, is organized around the active reduction of risk, while patient risk is normalized and even assumed as part of the routine work of Malagasy medicine.

The politics of life extension and biowaste are also starkly evident in the global trade in pharmaceuticals (see also Sunder Rajan 2012). In work on Nigeria's structurally adjusted pharmaceutical markets, Kristin Peterson (2014) examines how, after the global oil crisis of the 1970s, multinational pharmaceutical companies gradually abandoned drug markets in Nigeria, as the purchasing power of Nigerian consumers plummeted. Instead of introducing life-saving, yet expensive, patented medicines, manufacturers viewed the Nigerian market as a space in which to dump older-generation drugs, no longer considered effective and viable for wealthier consumers. Through the 1980s and beyond, they directed the resources they saved through such market abandonment strategies toward the design of blockbuster drugs for the health problems of wealthier people, such as those addressing high cholesterol, depression, anxiety and male infertility. Drug companies put little time and energy toward the most acute health issues facing most working Nigerians, because African markets were not lucrative and solving health problems was tantamount to ending revenue streams for chronic drug therapy. Peterson (2014, 109) observes that, in these circumstances, many Nigerians express "a tenacity…[…]…that speculates in the hope of attaining well-being and stability, even when economic instabilities chronically underlie life's circumstances".

In sum, when we look freshly at the role of waste in care and aid, new insights emerge about the unequal distribution of goods and harms in extending, maintaining, and nurturing life. Embodiment lies at the center of these processes and becomes a place in which we see most clearly the racialized operation and creative reworking of acts of shoring up, of creating distance from risk, of maintaining or disallowing dignity and health. White supremacy and waste work thus work in tandem to reproduce a complex system of global inequality of life and labor.

Conclusion

Pursuing questions about waste and race together can provide a critical vantage point on the uncomfortable unity that Mignolo (2011) dubs modernity/coloniality. One of the most established ways of doing so is to consider how environmental harms are unevenly meted out at various scales, but this is not sufficient to account for the many ways that power is established and maintained. As Max Liboiron (2019) argues, "environmental pollution and other forms of uneven material distribution are not an accidental by-product of capitalism, colonialism, and other power structures, but central to maintaining them as the systems that they are". In our terms, this can be seen anew by critically examining the concomitant productivity of race and waste.

Thus, for instance, environmental injustice not only harms bodies through slow violence and slow death, but helps to ensure a devalued labor pool, buildings and spaces (Marable 2015[1983]; Gidwani and Reddy 2011; Yates 2011). The disciplining of labor happens not only through literal pollution, moreover, but through rendering infrastructure and bodies in terms of backwardness and decay. In a paper that combines the literatures we have considered on racial capitalism and the biopower of care, Tamara Kohn (in press) analyzes American prisons and relates the "expectation of decay associated with lack of care and the dismissal of human value" with the "physically moulding and crumbling" built environment. Critically, Kohn argues, toxic substances like black mold can be enrolled in acts of resistance and calls for prison reform, even as they threaten to contaminate the bodies and identities of prison inmates. In these examples, neither mass-incarceration and excessive and deadly policing, nor toxic exposure, is external to issues of capitalism or waste.

Manning Marable shows, further, how efforts to produce a docile and carceral population is key to creating its opposite. Where being Black means being far more likely to be segregated by state institutions in what Kohn (in press) describes as ghost-like decay, being non-Black means magnetically attracting value-accruing and wealth-maximizing possibilities, in seemingly effortless ways as if they were "naturally" meant to be. This operates, as was noted by W.E.B. Du Bois (2013 [1935]) and later scholars like David Roediger (1991) and Cedric Robinson (2000 [1983]), as a wage of whiteness that prevents intra-class, trans-racial solidarities: "For many working class whites", Marable writes, "the Afro-American is less a person and more a *symbolic index* between themselves and the abyss of absolute poverty" (2015 [1983], 40; italics in original). By extension, relative non-blackness serves as a symbolic index of the possibility of endless accrual, a mountain of stuff rather than an empty abyss.

Discard studies and anti-racist and anti-colonial literatures are part of the same fabric of history, yet they have often developed in distinct tracks, both academically in scholarly discourse and in social and political debate. Since discard studies, as opposed to analyses of waste in particular or in general, focuses on systemic and entrenched political and moral problems (Liboiron 2018), scholars in this field arguably have an important role to play in identifying how these systems entangle. The point of questioning how race and waste relate is meant to correct this imbalance, therefore, which means both identifying those connections that exist between these domains and accounting for the misrecognition that would render them separate. We argue that pursuing these questions can lead discard studies into potentially new domains, including the high prestige and technically sophisticated world of scientific practice and new genetics. It can also bring scholars of waste further into the humble and quotidian acts of care that are a form of essential "dirty work" yet fail to gain the same attention as often (but not always) more masculine and more visible sites like landfills, incinerators, and dumps. Both population genomics and global circuits in care and aid can be thought of as distinct arenas of biopower, but our aim has been to consider how waste and race are constitutive of these social and economic practices, along with many others. The point is that asking after waste and race leads in many more directions, beyond waste sites and waste workers proper, from environmental and genomic science to care and aid networks. Nonetheless, what they share is the deep and enduring influence of racial capitalism on the world in which we live and on the many troubling forms taken by the valuation of life itself.

References

Ahmann, C. 2019. "Garbage Prospects and Subjunctive Politics in Late-Industrial Baltimore." *American Ethnologist* 46, no. 3: 1–15.

Barnett, M. 2011. *Empire of Humanity: A History of Humanitarianism*. Cornell University Press.

Bauman, Z. 2003. *Wasted Lives: Modernity and Its Outcasts*. Polity Press.

Bennett, J. 2009. *Vibrant Matter: Toward a Political Ecology of Things*. Duke University Press.

Benton, A. 2016. "African Expatriates and Race in the Anthropology of Humanitarianism." *Critical African Studies* 8, no. 3: 266–277.

Berlant, L. 2007. "Slow Death (Sovereignty, Obesity, Lateral Agency)." *Critical Inquiry* 33, no. 4: 754–780.

Bornstein, E., and P. Redfield. 2010. "An Introduction to the Anthropology of Humanitarianism." In E. Bornstein and P. Redfield (Eds.) *Forces of Compassion: Humanitarianism Between Ethics and Politics*. School for Advanced Research Press. 3–30.

Bryant, B., and P. Mohai (Eds.). 1992. *Race and the Incidence of Environmental Hazards: A Time for Discourse*. Westview Press.

Buch, E. 2013. "Senses of Care: Embodying Inequality and Sustaining Personhood in the Home Care of Older Adults in Chicago." *American Ethnologist* 40, no. 4: 637–650.

Buch, E. 2014. "Troubling Gifts of Care: Vulnerable Persons and Threatening Exchanges in Chicago's Home Care Industry." *Medical Anthropology Quarterly* 28, no. 4: 599–615.

Buch, E. 2018. *Inequalities of Aging: Paradoxes of Independence in American Home Care*. New York University Press.

Bullard, R. 1990. *Dumping in Dixie: Race, Class, and Environmental Quality*.Westview Press.

Bullard, R. 2005. *The Quest for Environmental Justice: Human Rights and the Politics of Pollution*. Sierra Club Books.

Butt, W. 2019. "Beyond the Abject: Caste and the Organization of Work in Pakistan's Waste Economy." *International Labor and Working-Class History* 95: 18–33.

Carmin, J., and J. Agyeman (Eds.). 2011. *Environmental Inequalities Beyond Borders: Local Perspectives on Global Injustices*. MIT Press.

Caverly, N. 2019. "Sensing Others: Empty Buildings and Sensory Worlds of Detroit." *Environment and Planning C: Politics and Space*. doi:10.1177/2399654419858368.

Cole, T. 2012. "The White-Savior Industrial Complex." *The Atlantic*, March 21, 2012.

Colen, S. 1995. "'Like a Mother to Them': Stratified Reproduction and West Indian Childcare Workers and Employers in New York." In F. D. Ginsburg and R. Rapp (Eds.) *Conceiving the New World Order: The Global Politics of Reproduction*. University of California Press. 78–102.

Davis, A. 1981. *Women, Class and Race*. Vintage.

De León, J. 2015. *Land of Open Graves: Living and Dying on the Migrant Trail*. University of California Press.

Doherty, J. 2019. "Filthy Flourishing: Para-Sites, Animal Infrastructure, and the Waste Frontier in Kampala." *Current Anthropology* 60, no. S20: S321–S332.

Doron, A., and A. Broom. 2019. "The Spectre of Superbugs: Waste, Structural Violence and Antimicrobial Resistance in India." *Worldwide Waste* 2, no.1: 7, 1–10.

Du Bois, W.E.B. 2013 [1935]. *Black Reconstruction in America*. Transaction Publishers.

Fadlalla, A. 2007. "The Neoliberalization of Compassion: Darfur and the Mediation of American Faith, Fear, and Terror." In J. Collins (Ed.) *New landscapes of Global Inequalities: Neoliberalism and the Erosion of Democracy in America*. School of American Research Press. 209–228.

Fassin, D. 2010. "Noli Me Tangere: The Moral Untouchability of Humanitarianism." In E. Bornstein and P. Redfield (Eds.) *Forces of Compassion: Humanitarianism Between Ethics and Politics*. School for Advanced Research Press. 35–52.

Finney, C. 2014. *Black Faces, White Spaces: Reimagining the Relationship of African Americans to the Great Outdoors*. University of North Carolina Press.

Fredericks, R. 2018. *Garbage Citizenship: Vital Infrastructures of Labor in Dakar, Senegal*. Duke University Press.

Gidwani, V. 2013. "Six Theses on Waste, Value, and Commons." *Social and Cultural Geography* 14, no. 7: 773–783.

Gidwani, V. and R. N. Reddy. 2011. "The Afterlives of 'Waste': Notes from India for a Minor History of Capitalist Surplus." *Antipode* 43, no. 5: 1625–1658.

Giles, D.B. 2014. "The Anatomy of a Dumpster: Abject Capital and the Looking Glass of Value." *Social Text* 32.1, no. 118: 93–113.

Ginsburg, F. D. and R. Rapp. 1995. "Introduction." In F. D. Ginsburg and R. Rapp (Eds.) *Conceiving the New World Order*. University of California Press. 1–17.

Glenn, E. N. 1992. "From Servitude to Service Work: Historical Continuities in the Racial Division of Paid Reproductive Labor." *Signs: Journal of Women in Culture and Society* 18, no. 1: 1–43.

Halvorson, B. 2012. "'No Junk for Jesus': Redemptive Economies and Value Conversions in Lutheran Medical Aid." In C. Alexander and J. Reno (Eds.) *Economies of Recycling: The Global Transformations of Materials, Values and Social Relations*. Zed Books. 207–233.

Halvorson, B. 2015. "The Value of Time and the Temporality of Value in Socialities of Waste." *Discard Studies*, 21 September. https://discardstudies.com/2015/09/21/the-value-of-time-and-the-temporality-of-value-in-socialities-of-waste/.

Halvorson, B. 2018. *Conversionary Sites: Transforming medical aid and global Christianity from Madagascar to Minnesota*. University of Chicago Press.

Hansen, K. T. 2000. *Salaula: The World of Secondhand Clothing and Zambia*. University of Chicago Press.

Harris, C. I. 1993. "Whiteness as Property." *Harvard Law Review* 106, no. 8: 1707–1791.

Herod, A., and L. Aguiar. 2006. "Introduction: Cleaners and the Dirty Work of Neoliberalism." *Antipode* 38, no. 3: 425–434.

Hird, M. 2013. "Waste, Landfills, and an Environmental Ethic of Vulnerability." *Ethics and the Environment* 18, no. 1: 105–124.

Isenberg, N. 2016. *White Trash: The 400 Year Untold History of Class in America*. Penguin.

Kohn, T. in press. "Decay, Rot, Mould and Resistance in the US Prison System." In G. Hage (Ed.) *Decay*. Duke University Press.

Liboiron, M. 2017. "Feminist+ Anti-Colonial Science. Civic Laboratory for Environmental Action Research (CLEAR)." Accessed July13, 2020, https://civiclaboratory.nl/2017/12/29/feminist-anti-colonial-science.

Liboiron, M. 2018. "The What and the Why of Discard Studies." *Discard Studies*. Accessed October9, 2020, https://discardstudies.com/2018/09/01/the-what-and-the-why-of-discard-studies/

Liboiron, M. 2019. "Waste Is Not 'Matter Out of Place.'" *Discard Studies*. Accessed July 13, 2020, https://discardstudies.com/2019/09/09/waste-is-not-matter-out-of-place/

Liboiron, M., Tironi, M. and N. Calvillo. 2018. "Toxic Politics: Acting in a Permanently Polluted World." *Social Studies of Science* 48, no. 3: 331–349.

Malkki, L. 1995. *Purity and Exile: Violence, Memory, and National Cosmology Among Hutu Refugees in Tanzania*. University of Chicago Press.

Mangan, K. 2007. "Duke's Medical Surplus Finds New Life in Uganda." *Chronicle of Higher Education*, 21 September, A30.

Marable, M. 2015 [1983]. *How Capitalism Underdeveloped Black America*. Haymarket.

Mignolo, W. 2011. *The Darker Side of Western Modernity: Global Futures, Decolonial Options*. Duke University.

Millar, K. 2018. *Reclaiming the Discarded: Life and Labor on Rio's Garbage Dump*. Duke University.

Millar, K. 2020. "Garbage as Racialization." *Anthropology and Humanism* 45, no. 1. doi:10.1111/anhu.12267.

Mills, C. 2001. "Black trash." In L. Westra and B. E. Lawson (Eds.) *Faces of Environmental Racism: Confronting Issues of Global Justice*. Rowman and Littlefield. 73–94.

Murphy, M. 2015. "Unsettling Care: Troubling Transnational Itineraries of Care in Feminist Health Practice." *Social Studies of Science* 45, no. 5: 717–737.

Nagle, R. 2013. *Picking Up*. Farrar, Straus and Giroux.

Nelson, A. 2016. *The Social Life of DNA: Race, Reparations, and Reconciliation After the Genome*. Beacon.

Nguyen, M. T. N. 2019. *Waste and Wealth: An ethnography of Labor, Value and Morality in a Vietnamese Recycling Economy*. Oxford University Press.

Nixon, R. 2011. *Slow Violence and the Environmentalism of the Poor*. Harvard University Press.

O'hare, P. 2019. "'The Landfill Has Always Borne Fruit': Precarity, Formalisation and Dispossession Among Uruguay's Waste Pickers." *Dialectical Anthropology* 43.1: 31–44.

Parreñas, R. 2015 [2001]. *Servants of Globalization: Women, Migration, and Domestic Work*. 2nd Edition. Stanford University Press.

Pellow, D. 2007. *Resisting Global Toxics*. MIT Press.

Peterson, K. 2014. *Speculative Markets: Drug Circuits and Derivative Life in Nigeria.* Duke University Press.

Piepzna-Samarasinha, L. L. 2018. *Care Work: Dreaming Disability Justice.* Arsenal Pulp.

Povinelli, E. 2006. *The Empire of Love: Toward a Theory of Intimacy, Genealogy and Carnality.* Duke University Press.

Prince, R. J. 2016. "The Diseased Body and the Global Subject: The Circulation and Consumption of an Iconic Photo in East Africa." *Visual Anthropology* 29: 159–186.

Puar, J. 2017. *The Right to Maim: Debility, Capacity, Disability.* Duke University Press.

Rabaka, R. 2007. *W.E.B. Du Bois and the Problems of the Twenty-First Century.* Lexington.

Rabaka, R 2009. *African Critical Theory.* Lexington.

Reardon, J. and K. TallBear. 2012. "'Your DNA Is Our History': Genomics, Anthropology, and the Construction of Whiteness as Property." *Current Anthropology* 53, no. S5: S233–S245.

Reno, J. O. 2014. "Toward a New Theory of Waste: From 'Matter Out of Place' to Signs of Life." *Theory, Culture and Society* 31, no. 6: 3–27.

Reno, J. 2016. *Waste Away: Working and Living With a North American Landfill.* University of California Press.

Reno, J. 2018. "Scaling up the Self, Scaling Down the World: Self-Objectification and the Politics of Carbon Offsets and Personalised Genomics." *Science as Culture* 27, no. 1: 44–73.

Resnick, E. 2015. "Discarded Europe." *Anthropological Journal of European Cultures* 24, no. 1: 123–131.

Robinson, C. 2000 [1983]. *Black Marxism: The Making of the Black Radical Tradition.* University of North Carolina.

Roediger, D. 1991. *The Wages of Whiteness.* Verso.

Rosenblatt, W. H. and D. G. Silverman. 1992. "Recovery, Resterilization, and Donation of Unused Surgical Supplies." *Journal of the American Medical Association* 268, no. 11: 1441–1443.

Skloot, R. 2010. *The Immortal Life of Henrietta Lacks.* Crown.

Slate, N. 2011. "Translating Race and Caste." *Journal of Historical Sociology* 24, no. 1: 62–79.

Solomon, M. 2018. *Letting Trash Talk: Garbage in the Racial Ordering of People.* Doctoral Dissertation, The New School.

Solomon, M. 2019 "'The Ghetto Is a Gold Mine': The Racialized Temporality of Betterment." *International Labor and Working-Class History* 95, 76–94.

Stamatopoulou-Robbins, S. 2019. *Waste Siege: The Life of Infrastructure in Palestine.* Stanford University Press.

Sunder Rajan, K. 2012. "Pharmaceutical Crises and Questions of Value: Terrains and Logics of Global Therapeutic Politics." *South Atlantic Quarterly* 111, no. 2: 321–346.

TallBear, K. 2013. *Native American DNA: Tribal Belonging and the False Promise of Genetic Science.* University of Minnesota Press.

Thompson, M. 2018 [1979]. *Rubbish Theory: The Creation and Destruction of Value.* Pluto Press.

Torchin, L. 2006. "Ravished Armenia: Visual Media, Humanitarian Advocacy, and the Formation of Witnessing Publics." *American Anthropologist* 108, no. 1: 214–220.

Turner, S. 2012. *Politics of Innocence: Hutu Identity, Conflict and Camp Life.* Berghahn.

Vargas, J. H. C. 2018. *The Denial of Anti-Blackness.* University of Minnesota Press.

Weheliye, A. 2014. *Habeas Viscus: Racializing Assemblages, Biopolitics, and Black Feminist Theories of the Human.* Duke University Press.

Westra, L. and B. E. Lawson (Eds.). 2001. *Faces of Environmental Racism: Confronting Issues of Global Justice.* Rowman and Littlefield.

Yates, M. 2011. "The Human-As-Waste, the Labor Theory of Value and Disposability in Contemporary Capitalism." *Antipode* 43, no. 5: 1679–1695.

Zhang, A. 2020. "Circularity and Enclosures: Metabolizing Waste With the Black Soldier Fly." *Cultural Anthropology* 35, no. 1: 74–103.

Zimring, C. 2015. *Clean and White: A History of Environmental Racism in the United States.* New York University Press.

5

LANDFILL LIFE AND THE MANY LIVES OF LANDFILLS

Patrick O'Hare

Introduction

Most people in the Uruguayan capital of Montevideo, as in many cities in the world, simply discard their waste without thinking about where it goes next. At best, some citizens might have heard of Felipe Cardoso, the landfill to which Montevidean rubbish has been taken for decades. Extremely few would ever have reason to venture there, but if they did, they would no doubt be surprised to find scores of waste-pickers, known as classifiers or *clasificadores*, recovering materials to feed back into Uruguay's productive industries or to feed their families. They would be surprised to hear that *clasificadores* refer to the landfill as "the quarry" (*la cantera*) or even as "the mother", given her capacity to provide food, clothes, and shelter to those who find their way there. They would no doubt be shocked to hear about the running battles that they have fought with police for decades in order to gain access to the treasures of the dump. Most of all, in a country nicknamed "the Switzerland of the Americas", they would perhaps struggle to believe that people would need or even choose to work in the landfill, rather than engage in other more conventional forms of labor.

Familiar theories of rubbish emphasize a definition correlated to its status as matter out of place, a notion popularized by the anthropologist Mary Douglas (2002 [1966]), despite the fact that she in fact focused on classifications of ritual dirt and pollution and not waste per se (c.f. Liboiron 2019). Indeed, Douglas is unlikely to have disagreed with the commonplace view that landfills are the right place for rubbish but for little else besides. Landfills are often designed and imagined as life-less spaces where the entry of persons, animals, and things are tightly controlled and restricted. Yet we know that far from being barren, they are often alive with human and non-human forms of life. The socio-material life of landfills includes municipal workers, waste-pickers and police, species such as birds, worms, plants, and bacteria, and forms of matter such as rubble, plastics, organic material, chemicals, toxins, and gases. This chapter explores the inter-species tensions and entanglements that social scientists have unearthed through ethnographic and theoretical engagement with spaces often assumed to be the sole preserve of sanitary engineers. Taking as a key example the author's long-term fieldwork conducted in and around Uruguay's largest landfill, it takes the reader on a tour of waste management sites as far removed as Michigan from Managua, and Rio de Janeiro from Ramallah. In doing so, it showcases the kinds of critical issues that social scientists have raised at

DOI: 10.4324/9781003019077-5

landfills and dumps, including the relationship between value and waste, commons and enclosure, safety and risk, autonomy and work, and permanence and change.

A narrow view of landfill science assumes a teleological quest, an evolutionary journey from primitive dumps to controlled spaces where socio-environmental risk is monitored and minimized. Yet this chapter shows that landfills are political spaces that serve multiple, contradictory functions. These often operate simultaneously: sacrifice zones, polluted peripheries, and contaminated borderlands but also spaces of archive, refuge, and accumulation, where certain forms of life flourish and absolute control is more of an ideal than a reality. This chapter begins by providing a brief history of how ancient dumping practices morphed into sanitary landfilling as a hegemonic waste disposal technique in the 20th century. It then discusses the themes that social scientists of landfills have explored, dividing these into spatial and temporal issues cross-cut by normative labor and hygiene regimes. As campaigns such as "zero waste" urge the diversion of all materials from landfill and incinerator, the chapter concludes by asking whether landfills will be relegated to the past or transformed into beautified techno-futures. In such circumstances, what will become of lives and livelihoods that presently depend on landfills in their current forms?

History

Landfilling implies the filling in of human-made pits and natural depressions with discards. As such it can be differentiated from simply dumping on level ground or piling up of materials in middens. At the same time the distinction is not necessarily clear-cut, as many landfills may have started as spatial "levelling" or "filling" exercises only to end in the accrual of waste mountains, sometimes known as "landraising", that tower over their surrounding landscapes. Archaeological evidence of landfills dates back to around 3000 BC Crete, where ancient waste was covered with soil, a practice that continues to be integral to landfilling to this day (Tammemagi 1999, 19–20). Before industrialization and urbanization, rural disposal practices were generally uncentralized and waste was dealt with at a household or village level. It was in the late 19th and early 20th centuries, with the growth of dense urban populations in Western Europe and the United States, and the association of waste with disease, often through a miasmatic framework centered around foul smells and vapors, that pressure grew for the creation of dumping grounds outside of urban areas. Now known as "historical landfills", these sites where there were minimal controls over what entered, weak environmental monitoring, and no remediation plans after closure, have left a dubious legacy across the world, with an estimated 100,000 in the United States and 20,000 in the United Kingdom (Brand et al 2018, 2).

Britain at the turn of the 20th century was the world's leading power, and this was an exciting time of invention and experimentation in British waste management, one that the historian Bill Luckin (2000) has referred to as embodying a "refuse revolution". The Public Health Act of 1875 had given local municipal authorities responsibility for providing waste collection and disposal services, and many raced to adopt new technologies that would better safeguard residents from industrial pollution. Initially, incineration held sway as the modern invention that could best meet the needs of the new science of public sanitation. The first operational incinerator, known as a "destructor" had been patented in 1876 and engineered in Nottingham—by 1912 there were over 300 waste incinerators distributed across the United Kingdom, with over 80 of them capable of generating electricity (Clark 2007, 260).

Yet several factors pushed the tide in favor of landfill. First, incineration was expensive, and in the post-WWI economic climate, previously generous central U.K. government loans to local authorities were curtailed. Second "controlled tipping" emerged as a modern and responsible method that began to improve the bad reputation that dumping had previously accrued. Known as the "Bradford method" after the town where it was first developed, this involved many features that are still integral to sanitary landfilling today, including daily covering with dirt and the division of the landfill into cells. Importantly, as Tim Cooper (2010) argues, proponents of controlled tipping argued that the method was not wasteful, but rather contributed to the recovery of "waste-lands" by improving soil quality. Dumps could then be turned into fertile farmlands or recreational parks. In the context of the "germ revolution" (the discovery of germs and their role in spreading disease), it was shown that natural processes of biodegradation and the work of enzymes could effectively break down waste, working with nature rather than against it, as was the case of incineration (Ibid., 1041). While these claims were contested, they were undoubtedly effective, with up to 90% of U.K. waste disposed of by controlled tipping by the 1960s (Luckin 2000).

The prestige of British sanitary science contributed to the global spread of the method. Josh Reno picks up the story in the United States, telling us that the term "sanitary landfill" was coined in the 1930s, "at a time when many American municipalities were seeking waste disposal options that could lessen the rising costs of public sanitation in a recently urbanized society" (2008, 85). The first sanitary landfill in the United States is widely regarded to have been that of Fresno, which was established in 1938 and involved the combination of four key techniques: trenching, tipping, compaction, and soil cover (Ibid.). Reno suggests that the Fresno model effectively involved a trompe d'oeil, taking waste and pollution out of sight and underground (leading to the contamination of groundwater). As in England, the popularity of landfilling in the United States also relied on what Reno calls the "transubstantiation of waste material into usable property" (Ibid., 87). Hog feeding and open dumping were still the most common US disposal methods up until the 1960s, however, when landfilling was adopted by the army as a "technologically sophisticated and cheap alternative" (Ibid., 88). With the military setting the example, sanitary landfilling soon began to be rolled out by civil administrations; as Reno writes, the landfill was "but one form of technology sanitized and legitimized by the military-industrial complex at the beginning of the Cold War era" (Ibid.).

The science behind sanitary landfill adapted to changing environmental concerns in the second half of the 20th century to include two key innovations: the capture of methane gas and leachate run-off through a network of pipes, designed to lessen greenhouse gas emissions and the contamination of groundwater respectively. Yet their increasingly technical nature has also increased operational costs, meaning that sanitary landfills, as opposed to landfills tout court, are relatively scarce at a global level. Thus, rather than a simple success story of the development and spread of sanitary landfill science, the global panorama is somewhat more nuanced. At the beginning of the 21st century, most global waste (37%) was being disposed of in landfills, of which only 8% were sanitary, while another 31% was thrown into open dumps (World Bank 2018). Of the remainder, 19% of waste was recovered through recycling and composting and 11% was incinerated (Ibid.). While important as indicators, such figures cannot always be completely relied upon. Felipe Cardoso, for instance, has the official title of a sanitary landfill and does have methane and leachate recovery and treatment facilities. Yet Montevideo's waste is not covered with soil on a daily basis and waste-pickers work there intermittently. While most sanitary engineers would see this as environmentally and socially undesirable, many waste-pickers and their organizations have a different view on the matter, as we shall see.

Class, race, and the filling of space

Landfills are fundamentally spatial undertakings: they take as their primary task the filling in of space with trash. The management of landfills involves an engineering challenge of how to fit as much material into as small a space as possible, notwithstanding environmental constraints. Spatial dimensions have also been central to the social science of landfills, often in surprising ways that connect to issues of race, nationality, class, and religion. In the siting of landfills, planners and waste managers invariably choose rural sites that have low population densities. This highlights a first tension in the spatial distribution of landfills: rural vs urban. From the days of the first landfills, rural communities have been saddled with the waste of their urban counterparts. Several strategies have been employed to defend landfills in these circumstances. First, as we have seen, sanitary landfill was initially presented as a way of recovering "wasteland" as a potential public amenity. Whilst the "improvement" element of landfilling has become less convincing over time, the practice of rhetorically devaluing the land where dumps are to sited has been more consistent, with the latter described as waste-lands, untamed spaces, or uninhabited tabula rasas. The carrot used in such circumstances is the idea that landfills bring jobs, local government income, and what Zsuzsa Gille (2015) has called "waste-dependent development".

As economic entities, there is often a cost/benefit analysis involved in deciding the geographical location of landfills. Too close to centers of population and there might be health risks and complaints from neighbors but too far and the costs of transport become prohibitive. Landfills will thus always be relatively close to *somebody* and the question then becomes who. As Robert Bullard writes in *Dumping in Dixie* (2000), "minority and low-income residential areas (and their inhabitants) are often adversely affected by unregulated growth, ineffective regulation of industrial toxins, and public policy decisions authorizing locally unwanted land uses that favor those with political and economic clout" (8). Bullard explores a series of case studies to show that, in the United States, dumps in particular are often disproportionately sited in African American areas. "Black Houston", for example, "has become the dumping ground for the city's household garbage", with only one of nine municipal landfills or incinerators sited in a white majority neighborhood (47). Well-organized and connected middle-class residents are often able to defend more successfully against the perceived threat of dumps than working class and poor communities. Four Corners, the Michigan landfill studied by Joshua Reno, for example, was initially meant to be built in the more affluent and populated Calvin Township but due to organized local opposition was soon moved to the poorer Harrison (2016a, 157). Many similar stories are told by authors such as David Pellow (2002), working within the vein of critical environmental justice studies, and are replicated at a global scale, through the redirection of waste to countries in the Global South. Nevertheless, social scientists have nuanced headlines about North-South "dumping", drawing attention to the lives and livelihoods that are sustained by the global waste trade (Lepawsky 2018), and noting the complexities even of spaces that have become emblematic of environmental e-waste irresponsibility, such as Ghana's Agbogbloshie dump (Rams 2020).

The provenance of waste and waste-pickers in landfills is another terrain of contestation. Not many folks want a landfill in their backyard but when they have one, they usually don't want it to receive waste that is deemed to come from people from across national or religious boundaries. One of the most controversial issues at Four Corners arose when the landfill began to receive waste from across the border in Canada. As Reno (2016a) writes, "for many Michiganders, the importation and burial of Canadian waste was considered a violation of American airspace" (170). At the Palestinian landfill that formed the focus of Stamatopoulou-

Robbin's (2019) book "waste siege" on Palestinian-Israel waste infrastructure and politics, conditions were even more fraught. Like elsewhere, local communities in Palestine often campaign against the siting of landfills in their vastly reduced territories: not only do they contaminate the land, they also remove it as a heritable good that can be passed down through generations as a form of wealth (35). Moreover, when the al-Minya dump was built, Israelis from nearby illegal settlements, escorted by the army, turned up to the dump their waste there (60), undermining any semblance of Palestinian control and highlighting the confluence of colonial and environmental dispossession. The national question also emerges in the work of Melanie Samson (2019) on waste-pickers (known as reclaimers) in South Africa. At the Marie-Louise landfill in Johannesburg, South African waste-pickers effectively linked their struggle to gain entry to the dump with that of national liberation. Zimbabwean immigrant waste-pickers, late on the scene, had participated in neither and were thus discursively and often physically excluded from the population deemed entitled to exploit this "national mine".

With the South Africa case we move from the question of what waste can legitimately enter a landfill space to that of what people can enter. Such a debate is tied to questions of ownership and control and is particularly salient in places that have a strong tradition of waste-picking. Sanitary landfills are often either publicly managed or managed in a public-private partnership. Yet as the case of Marie-Louise shows, landfills might ostensibly be in private or public hands but can be claimed by waste-pickers as spaces of customary access and extraction. In that particular case, reclaimers, represented by a famous anti-apartheid lawyer, were even able to win a court battle to legalize their right to access. Certification as a sanitary landfill and the presence of waste-pickers making their way between trucks, gases, and rubbish are often mutually exclusive. As such, sanitary landfills can function not only as environmental improvements on open dumping but also as technologies that facilitate the exclusion of people who may have relied on access to the bounty of the dump for generations. This was certainly the case in Montevideo where, prior to the construction of Felipe Cardoso, the local government operated a series of what *clasificadores* called "free dumps" (*canteras libres*), where vulnerable populations such as recently arrived immigrants, widows, ex-convicts, and single mothers could extract materials for sale and use. At Felipe Cardoso, however, waste-pickers have now lived through a gamut of different repressive forces that have sought to curtail their access, including regular police, mounted police, private security, and police dog units. *Clasificadores* have suffered injuries, bullets, whippings, and bites, alongside extortion and sexual harassment, as they attempted to enter the *cantera* over the last few decades. Such repression, hidden from the public eye, has been justified in terms of maintaining a cordon sanitaire around unsanitary stuff (Figure 5.1).

These dynamics led me to conceptualize certain landfills as commons and exclusionary efforts as processes of enclosure (O'Hare 2017), a point that has also been made by Zapata and Zapata Campos (2015) in their research on Managua's Chureca landfill. Most waste-pickers sell materials on recyclable commodity markets through supply chains that are often global in scale. While this excludes them from understandings of commons such as that proposed by David Harvey, who defines them as "off limits to the logic of market exchange and market valuations" (2012, 73), I choose a less restrictive definition, drawing on the characteristics of old English common lands and resources that I found replicated at Felipe Cardoso. These include customary claims made by vulnerable populations, the capacity of landfills to provide food, clothes, and shelter; the blurring of work and play; and reactive processes of enclosure. Framed as such, landfills can be seen as common spaces, under threat not only from a "hygienic enclosure" stimulated by sanitary requirements but also increasingly by new technologies, such as anaerobic digestion and capital-intensive waste-to-energy plants, that have begun to displace landfilling in the Global North.

Figure 5.1 Description: damaged waste containers lie beyond the fenced-in perimeter of Montevideo's Felipe Cardoso landfill. Image: Patrick O'Hare.

Landfill rhythms

Why would waste-pickers want to work at landfills in the first place? Surely this must only be through desperation and economic necessity? Social scientists, anthropologists in particular, have explored this question in a series of recent ethnographies. The rhythms of landfill life are one potentially surprising attraction that take us into this section, which focuses on the temporal questions that anthropologists have asked of landfills. We begin with the human, before moving on to the post-human or multi-species temporalities that together constitute landfill life.

Kathleen Millar (2015, 2018) has conducted long-term research with waste-pickers, known as *catadores*, at Rio de Janeiro's Jardim Gramacho dump. This was once the largest landfill in Latin America and reached a wide audience in Lucy Walker's et al (2010) film *Waste Land*, about the artistic collaboration between *catadores* and Brazilian artist Vik Muniz. At first, Millar tells us, she thought that the story of Jardim Gramacho was a familiar one: the horticultural garden *(jardim)* trashed by the uneven growth of a sprawling urban metropolis. Yet her opinion is changed by the way that waste-pickers talk about the space and the fact that they sometimes turn down more conventional formal sector jobs to work there. This, for Millar, can be understood by attending to the temporal affordances of landfill labor. At Gramacho, *catadores* could determine their own working day, take breaks or unannounced leave when they wanted, and enjoy a sense of autonomy and of being their own boss, while other informal sector jobs were being eliminated by municipal policies. This certainly echoed with the rhythms of landfill *clasificadores* in Uruguay: one neighbor would often take time off to look after his family, another only worked early mornings, a third would absent himself for weeks as he recovered from a particularly ferocious bender (Figure 5.2).

Figure 5.2 Description: Waste-pickers recovering metals from a truck at Montevideo's Felipe Cardoso landfill as birds fly overhead. Image: Patrick O'Hare.

Such rhythms of labor of course largely only apply to "informal" waste-pickers: organized recyclers and municipal workers have much more conventional wage labor norms and fixed working days. The work of both those who recover waste and those tasked with its burial is nevertheless partly determined by the daily flows of waste at landfills. Invariably, such flows involve a degree of regularity, even if the generation of waste might be affected by seasonal variations (more PET bottles are consumed in summer, for instance, while more packaging arrives after Christmas). At the same time, some have argued that waste management in the Global South and cash-strapped countries can also be characterized by what Dalakoglou and Kallianos (2014) refer to as "arrhythmia". Examples of arrythmia, including blockages and the unscheduled interruption of infrastructural flows, are not always accidental but can be designed to exploit the dependency of cities on just-in-time collection and disposal services. Blocking landfills is a serious business, equivalent to preventing a city from relieving itself. Such a method has been enacted by diverse stakeholders such as waste-pickers (as occurred in Montevideo), striking municipal garbage workers, and citizens groups protesting either landfills or wider political issues. One of the most famous and visually striking blockages in recent years occurred around Lebanon's capital Beirut, where in the summer of 2015 residents of Naimeh blocked the road leading to the landfill, and a multi-sect coalition emerged to protest corruption, inequality, and the mismanagement of Lebanon's core public services (Nucho 2019). Such coalitions demonstrate that landfills not only divide but also bring people together, although usually against them.

The management of landfills depends on multiple temporalities, both human and non-human. Reno (2016b) has counted at least seven different temporal cycles that must be taken into account when we think about landfills and their operation, from the daily dumping, filling, and spreading, to the barely perceived geologic shifts that create the kind of impermeable clayish soil and landscapes that make particular territories more attractive as dumping grounds. Many of the temporalities involved are neither human nor geological but involve non-human forms of animal life. Remember that one of the ways that "controlled tipping" could be sold as a disposal technique that extracted value from the rubbish was the discovery of the enzymes that broke down urban refuse into soil in a natural way. The crucial labor of what one of Reno's colleagues at Four Corners calls "spacebugs" has often been taken for granted since. As Hoag, Bertoni and Bubandt (2018) highlight, "only recently have molecular studies and advances in techniques of rapid genetic identification begun to reveal some insight into the cryptic assemblage of methano-genic bacteria and archaea that populate our local landfills" (96). Thus an engineer at AFLD Fasterholt, the Danish fieldsite where Hoag et al. conducted research, is not alone in having to be pressed into admitting that it was not exactly "waste" or "animals in the waste" but more precisely "bacteria" that produced the methane whose capture he was charged with.

For Hoag et al., the work of bacteria and archaea is nevertheless an example of what they call "undomestication": "the process whereby particular elements of human domestication are appropriated or undone by non-human species in such a way as to create novel and relatively autonomous relations of human/non-human interdependency" (88). It helps their argument that the engineer in question ingeniously adapts a paradigmatic technology of domestication—a milking pump—in order to extract methane. Microbes and archaea are encouraged to work under human supervision and in tandem with human goals. Yet they are fundamentally undomesticated and even relatively unknown. We underestimate their complexity at our peril. At Four Corners, Reno's gas technician worries that landfill managers "took for granted the activity of archaea", imagining that they could "exploit the gas field indefinitely without taking into consideration the timescales at which spacebugs operate" (2016b).

Rhythms, non-human life, and the limits of human control at landfills are themes that have been taken up in the work of Myra J. Hird (2013). As she reminds us, the engineering time involved in the planning and thinking about the lives and afterlives of landfills amounts to not much more than a hundred years. Yet at the thousands of historic landfills that dot our landscapes, often hidden beneath our feet, myriad materials, chemicals, metals, and objects continue interacting in ways that escape our knowledge and control. The unpredictability of such interactions has increased as more and more synthetic materials have been produced and released into waste streams over the course of the 20th century and into the 21st. As Hird writes, there are over 7 million known chemicals; 80,000 chemicals in commercial circulation; and over 1,000 new chemicals developed and circulated every year (Ibid., 113). While some of these will be captured by separate disposal mechanisms and receive adequate treatment, many more will end up in mixed dumps; even household waste often contains hazardous chemicals (c.f. Wynne 1987). In Uruguay, clasificadores encounter the unpredictability of landfill interactions in myriad ways. The children of waste-pickers who lived atop a closed cell of Felipe Cardoso developed lead poisoning, leading them to be evicted and rehoused. At the same time, a stroll over the old landfill also threw up plants that had surely emerged from what had been disposed of: tomatoes and potatoes that could be brought home and replanted, if one didn't mind the plastics that were entangled in their roots.

We are then in interesting times with regard to the relation between humans and their accumulated wastes. On the one hand, vast tracts of the planet have been contaminated with our rubbish and the encrusting of our geological strata with synthetic materials such as plastics has been held up as evidence of our living in an Anthropocene (Zalasiewicz and Waters 2015). On the other, scholars have pointed to the limits to our knowledge and the myriad non-human species—archaea but also birds, rats, worms, and feral species such as dogs (c.f. Tsing et al. 2020)—that intervene in often unpredictable ways in and around landfills. In response to this predicament, Hird has advocated for a feminist "inhuman epistemology" (2012) and an "ethics of responsibility" (2013) that recognizes the limits to our knowledge. Yet human intervention on a massive scale and the spiraling unpredictability that this has put into motion are in no way mutually exclusive and can even be seen to characterize climate change and the Anthropocene.

The end of landfill?

As Catherine Alexander tells us, many technologies presented as waste disposal methods in turn produce their own leftovers that must themselves be disposed of, such as the ash produced from incineration that then has to be landfilled (2016, 33). Landfills, on the other hand, have been presented as the final destination, the end point for our used and once-loved commodities, a final resting place. Yet if landfills are the end of the line for things as we know them, old criticisms of landfilling have resurfaced with a vengeance, posing the question of whether landfilling as a waste disposal method should itself be thrown onto the trash-heap of history.

Modern capitalist economies are built on mass production and consumption. As such, they have called forth massive landfills to deal with what Sarah Moore (2009) has called the "excess of modernity". New York's Fresh Kills landfill, for example, was one of the largest human engineered structures in the world, compared in Don DeLillo's Underworld to "the Great Pyramid at Giza- only...twenty-five times bigger" (2007, 184). In his monumental history of Fresh Kills, Martin Melosi (2020) notes that landfilling has been considered the disposal method that is cheapest, safest, and most suited to mass consumption and our economic model. In other words, people might not like landfills, but our contemporary mode of production has come to depend on them. New criticism of landfills has thus been tied to renewed critiques of such "linear" economic models and attempts to refashion a "circular economy" that would design out waste, build greater longevity into products and materials and challenge our "throw-away society" in the interests of a more sustainable future.

The old argument that landfills might improve waste-land is clearly no longer valid—if ever it was—given the mixed materials, including large volumes of plastic, that everyday waste streams are composed of these days. Landfilling has been portrayed as almost criminally was-teful, and has been penalized through fiscal measures such as the EU landfill tax. Alternatives to landfill include the reduction of waste generated, re-use, recycling, anaerobic digestion (AD), and waste-to-energy technologies. While many aspire to reduce waste at an individual level, moves in the direction of a more circular economy have been slow. Despite vociferous anti-plastic campaigns, for instance, global plastic waste generation is still expected to grow by at least 50% over the next 20 years (Lebreton and Andrady 2019). Recycling is dependent on there being a local market for materials, which have to compete with virgin resources and whose recovery is often costly and requires subsidy (MacBride 2011). Local re-use movements might be laudable, but they are yet to make much of an impact on reducing global waste flows. AD is increasingly popular but can only deal with organic waste. This leaves Waste-to-Energy plants as the only competing technique that can accept most waste, as landfills do, but with the benefit of being able to produce electricity. The 20th-century struggle between landfill and incineration is set to be replayed in the 21st, with each method boasting of improvements in technique, capacity, contamination control, and cost.

As the World Bank (2018) figures on global waste disposal show, however, the ability to choose between sanitary landfill and waste-to-energy is a luxury that not many countries can afford. It is principally in countries of the Global North, and particularly those with a shortage of land, that new capital-intensive waste-to-energy facilities have been constructed. In the Global South, these have been much more controversial due to capital costs, the diversion of materials away from recycling and waste-pickers, long-term path-dependency, and the high price of electricity generated by multinational operators. It thus seems that landfills will be with us for some time yet. Even when closed, landfills can give us not only trouble, but also pause for thought, opening up debates about their possible afterlives. Ancient dumps and middens have been unearthed and used by archaeologists to provide clues to past forms of production, consumption, and disposal. Archaeologists of the contemporary such as Rathje and Murphy (1992) and their celebrated Tucson Garbage Project have scoured more recent landfills to obtain data on consumption patterns that corrects information collected in more conventional social science methods such as surveys and interviews. Our current landfills will no doubt likewise provide clues for their future colleagues. Yet many landfills will not be left undisturbed until then, as the attraction of "urban mining" old dumps for valuable materials augments in tandem with the increasing scarcity of natural resources.

Some of the landfills that have not been mined have been targeted for parkland schemes like New York's Freshkills Park, a project that aims to transform this enormous wastescape into a public green space four times larger than Central Park. More than 70 years after the construction of a landfill that was expected to have an active life of at most 20 years, the residents of Staten Island may yet get the recompense that they were promised, after decades of shouldering the burden, sights, and smells of New York's rubbish. As Melosi (2020) reminds us, however, the transformation of former landfills is no simple business. Should the landfill and its materials be completely concealed and covered up as an eyesore that can finally be put out of sight and out of mind? Or should the dump's fundamental role in the history of New York during the "American Century" be alluded to, commemorated, or even celebrated?

One New York landfill to have undergone a temporary transformation that went beyond local remediation concerns to engage with issues of art and aesthetics, value and inequality, is that of Battery Park. On the closure of the Battery Park landfill in the early 1980s, the artist Agnes Denes was contracted to create a public art installation on the site, and she decided to use a large area to plant wheat, which was then harvested and distributed throughout the world. *Wheatfield* re-created the old dream that landfills might transform wasteland into useful agricultural land. Yet it did so in the heart of New York, on a property that, due to its proximity to Wall Street, was valued at the colossal sum of US$4.5 billion. As Amanda Boetzkes (2019) thus tells us, Denes' work "pitted the land's agricultural yield against its value as real estate", constituting it as a "commentary on the discrepancy between the purely economic evaluation of land and the ecological value of land in the context of world hunger" (81). The landfill's stint as a visually striking urban agricultural experiment was brief and soon gave way to the multi-million-dollar Battery Park City development.

Mel Chin is another artist who was tasked with the creation of a public art installation at a much more contaminated site: the Pig's Eye Landfill in St Paul, Minnesota. In *Revival Field*, Chin worked with a U.S. government scientist who was investigating the capacity of plants to absorb toxic metals from land, creating an art piece that also doubled up as a science experiment, with six so-called "hyper-accumulators" planted in carefully demarcated sections of the work, so that their effects could be measured and analyzed. At its most ambitious, the method explored the possibility that such plants might accumulate metals so effectively that they could be harvested for them, constituting an economic return for local authorities (Boetzkes 86).

Chin's work thus plays with the relation between waste and value, but whereas landfills were previously imagined as projects that could recover overgrown wastelands, here plants are enlisted to extract value from lands that industrial rubbish has wasted.

Denes and Chin are not the only artists who have engaged in remediation projects in contaminated spaces: as Boetzkes writes, "the first ecologically oriented artworks emerging in the 1960s were restoration projects…that aimed to remediate contaminated sites from a toxic state to one with renewed value" (79). Rather than focusing on a simple waste-value nexus, however, such works and those from more recent generations of artists also constitute critiques of consumption, misplaced *values*, and ecological vandalism. The critical nature of these installations, as well as their short time frames, experimentalism, and global reach, can be contrasted with both highly localized parkland projects and the new architectural aesthetics of Waste-to-Energy plants, which often espouse a straightforward endorsement of the capacity of the latter both metabolize waste and improve urban aesthetics (see Alexander 2016). Transformation into art, or through artistic installations, is thus one of ways that the lives of landfills continue to stimulate and challenge even after they have been decommissioned.

Conclusion

This chapter has explored some of the key issues that social scientists have raised in and around landfills. For many citizens in the Global North, landfills and their waste exist "out of sight and out of mind". For many others, however, including the poor, ethnic minority communities that are often chosen for waste sites, and many residents of the Global South, relationships with landfills and dumps have been much more intimate. Whereas sanitary landfills have been managed in such a way as to control and minimize the life forms that might flourish there, we have seen how human and nonhuman life often find a way of getting in and out, from microscopic toxins, to the waste-pickers whose livelihoods depend on it.

The question of how landfills will evolve over the course of this century is an open one. One possible scenario has been raised here: a re-enacted battle between dumping and incineration, now framed as sanitary landfill and waste-to-energy facilities, in the context of attempted shifts towards more circular economies. At the heart of this debate are questions not only of health, safety, and environment, but also of profit margins and municipal budgets, where landfills are spaces of economic, as well as inter-species, accumulation (Stamatopoulou-Robbins 2019). Even if we see the phasing out of mass landfilling, which is unlikely, we have also highlighted how landfills continue to provoke controversy and debate, acting as archives for past ways of life, current environmental headaches, and canvases onto which architects and artists project critique and imagine how we might live otherwise.

References

Alexander, Catherine. 2016. "When Waste Disappears, or More Waste Please!" *RCC Perspectives* 1: 31–40
Boetzkes, Amanda. 2019. *Plastic Capitalism: Contemporary Art and the Drive to Waste*. MIT Press
Brand, James H., Spencer, Kate L., O'shea, Francis T., and Lindsay, John E. 2017. "Potential Pollution Risks of Historic Landfills on Low-Lying Coasts and Estuaries." *WIREs Water*, 5: 1–12.10.1002/wat2.1264.
Bullard, Robert D. 2000. *Dumping in Dixie: Race, Class, and Environmental Quality*. Westview Press
Clark, J. F. M. 2007. "'The Incineration of Refuse Is Beautiful': Torquay and the Introduction of Municipal Refuse Destructors." *Urban History* 34: 255–277

Cooper, Timothy. 2010. "Burying the 'Refuse Revolution': The Rise of Controlled Tipping in Britain, 1920–1960." *Environment and Planning A* 42: 1033–1048

Dalakoglou, Dimitris and Yannis Kallianos. 2014. "Infrastructural Flows, Interruptions and Stasis in Athens of the Crisis." *City* 18: 562–532

DeLillo, Don. 2007. *Underworld: A Novel.* Scribner

Douglas, Mary. 2002. [1966]. *Purity and Danger.* Routledge

Gille, Zsuzsa. 2007. *From the Cult of Waste to the Trash Heap of History: The Politics of Waste in Socialist and Postsocialist Hungary.* Indiana University Press

Gille, Zsuzsa. 2015. "Ecological Modernization of Waste-Dependent Development? Hungary's 2010 Red Mud Disaster." In Ruth Oldenziel and Trischler Helmuth (Eds.) *Cycling and Recycling: Histories of Sustainable Practice.* Berghahn.

Harvey, David. 2012. *Rebel Cities: From the Right to the City to the Urban Revolution.* Verso Books

Hird, Myra J. 2012. "Knowing Waste: Towards an Inhuman Epistemology." *Social Epistemology* 26, no. 3–4: 453–469

Hird, Myra J. 2013. "Waste, Landfills, and an Environmental Ethic of Responsibility." *Ethics and the Environment* 18, no. 1: 105–124.

Hoag, Colin, Filippo Bertoni, and Nils Bubandt. 2018. "Wasteland Ecologies: Undomestication and Multispecies Gains on an Anthropocene Dumping Ground." *Journal of Ethnobiology* 38, no. 1: 88–104

Kaza, Silpa, Lisa Yao, Perinaz Bhada-Tata, and Frank Woerden. 2018. *What a Waste 2.0: A Global Snapshot of Solid Waste Management to 2050.* Urban Development Series. World Bank

Lebreton, Laurent and Anthony Andrady. 2019. "Future Scenarios of Global Plastic Waste Generation and Disposal." *Palgrave Communications* 5: 1–11.

Lepawsky, Josh. 2018. *Reassembling Rubbish: Worlding Electronic Waste.* MIT Press

Liboiron, Max. 2019. "Waste Is Not 'Matter Out of Place'." https://discardstudies.com/2019/09/09/waste-is-not-matter-out-of-place/.

Luckin, Brian. 2000. "Pollution in the City." In Martin Daunton (Ed.) *The Cambridge Urban History of Britain* (Vol. 3). Cambridge University Press. 1840–1950.

MacBride, Samantha. 2011. *Recycling Reconsidered: The Present Failure and Future Promise of Environmental Action in the United States.* MIT Press

Melosi, Martin V. 2020. *Fresh Kills: A History of Consuming and Discarding in New York City.* Columbia University Press

Millar, Kathleen M. 2015. "The Tempo of Wageless Work: E.P. Thompson's Time-Sense at the Edges of Rio de Janeiro." *Focaal- Journal of Global and Historical Anthropology* 73: 28–40

Millar, Kathleen M. 2018. *Reclaiming the Discarded: Life and Labor on Rio's Garbage Dump.* Duke University Press

Moore, Sarah A. 2009. "The Excess of Modernity: Garbage Politics in Oaxaca, Mexico." *The Professional Geographer* 61: 426–437

Nucho, Joanne Randa. 2019. "Garbage Infrastructure, Sanitation, and New Meanings of Citizenship in Lebanon." *Postmodern Culture* 30: 1.

O'Hare, Patrick. 2017. *'Rubbish Belongs to the Poor': Hygienic Enclosure and Montevideo's Waste Commons.* Doctoral Thesis, University of Cambridge

Pellow, David Naguib. 2002. *Garbage Wars: The Struggle for Environmental Justice in Chicago.* MIT Press

Rams, Dagna. 2020. *Agbogbloshie: Dumping No More.* https://discardstudies.com/2020/03/09/agbogbloshie-dumping-no-more/

Rathje, William and Cullen Murphy. 1992. *Rubbish!: The Archaeology of Garbage.* University of Arizona Press

Reno, Joshua O. 2008. *Out of Place: Possibility and Pollution at a Transnational Landfill.* PhD Thesis, University of Michigan

Reno, Joshua O. 2016a. "The Life and Times of Landfills." *Journal of Ecological Anthropology* 18, no. 1: 5.

Reno, Joshua O. 2016b. *Waste Away: Working and Living with a North American Landfill.* University of California Press

Samson, Melanie. 2019. "Trashing Solidarity: The Production of Power and the Challenges to Organizing Informal Reclaimers." *International Labor and Working-Class History* 95: 34–48

Stamatopoulou-Robbins, Sophia. 2019. *Waste Siege: the Life of Infrastructure in Palestine.* Stanford University Press

Tammemagi, Hans. 1999. *The Waste Crisis: Landfills, Incinerators, and the Search for a Sustainable Future.* Oxford University Press

Tsing et al (Eds.). 2020. *Feral Atlas: The More-than-Human Anthropocene.* Stanford University Press.

Walker, Lucy, Jardim, Joao, and Harley Karen. 2010. Waste Land. London and Sao Paulo: Almega Projects and and 02 Filmes, DVD.

Wynne, Brian. 1987. *Risk Management and Hazardous Waste: Implementation and Dialectics of Credibility.* Springer

Zalasiewicz, Jan and Colin Waters. 2015. *The Anthropocene. Environmental Science.* Oxford Research Encyclopedia. 15 pp.

Zapata, Patrik and María José Zapata Campos. 2015. "Producing, Appropriating and Recreating the Myth of the Urban Commons." In Christian Borch and Kornberger Martin (Eds.) *Urban Commons: Rethinking the City.* Routledge

6

READING THE SIGNS: SOME WAYS WASTE IS FRAMED IN TUNISIA

Jamie Furniss

Introduction

"With Monoprix, I commit to protecting the environment"
"The city is a mirror that reflects society, so protect its cleanliness"
"Whomsoever puts garbage here shall receive neither blessing nor reward"

You might not think so, given their radically different linguistics registers—and, as we shall see, different sources of authorship and social geographies—but these phrases are in fact three ways of saying "don't litter". They are slogans employed in Tunisia to try, often in vain, to channel people's comportment with respect to waste. Each was coined by a different group of people and reflects a different conception of the type of problem waste is and/or the most effective lever to pull in order to influence people's behavior with respect to it. If you litter, you will have failed to protect the environment and to be a responsible consumer; you will damage the honour and good name of the society to which you belong; you will deprive yourself of divine reward. We might call these, for short, the environmental, reputational, and religious framings. Although what the third phrase tells us is less that waste is a specially religious problem, and more that religious language is a common idiom for reviling others, but I will get to that later.

The goal of this chapter is, using public signage observed in Tunisia as the main source material,[1] to demonstrate some different construals, or what I call framings, of waste. My hope is that this method, as well as the way of thinking and posing questions that I apply to it, can be of use to waste studies scholars from multiple disciplines, in whatever context they study. Across social class, geography, culture, and whatever else, waste is a shared concern, but one that takes on many different meanings and leads in varied directions. Contextually specific explorations of such meanings and directions are a central way in which anthropology, and social science-based waste studies more broadly, can contribute to a better understanding of waste and how to deal with it. The simple, even simplistic, method employed in this chapter to get at those meanings is to examine public signage. This approach is particularly fruitful in Tunisia where such signage often takes the form of a principle or maxim seeking to justify why littering is bad, or an argument about whose responsibility it is, but I believe it can be useful elsewhere also.

DOI: 10.4324/9781003019077-6

After some elaboration on the concepts of framing and mediation the chapter proposes to employ, I seek to unpack the signs quoted above through an analysis of the phrases they contain and the context of their production. This is meant as an illustration of some different ways of conceiving of, problematizing, and enrolling waste in aspirational projects, and how these can be usefully interpreted in relation to sociological and institutional features of their producers. I then explore framings of the waste problem in Tunisian signage more broadly, showing how cleanliness, religion, responsibility, and environment are invoked in different ways and contexts. Connecting these framings to sociological, political, and historical contexts of production, it is possible to present some hypotheses about why and for whom certain framings are popular, and how ideas about waste spread. I also attempt to show that neither of the two framings one might conjecture would be predominant in Tunisia—environment or religion—is hegemonic or even central.

A few points of reference may help to position this chapter relative to existing waste studies literature. First, it is a critique of Hawkins' previous work, but an admiring one in the sense that I attempt to adapt and expand her approach rather than discard it, which would be too ironic a fate for research on waste. In *The Ethics of Waste,* Gay Hawkins argued on the basis of anti-litter campaigns that over the course of the 20th century the concern with waste has shifted from maintaining the purity of the self to a moral concern aimed at "protecting the purity and otherness of the environment" (2006, 31). Anti-litter campaigns are indeed a productive and easily accessible entrée for studying shifting problematizations of waste. And Hawkins' argument seems to me to be true as far as it goes.[2] However as we move outside the frame of 1970–2000s era Australo-American anti-litter campaigns, it begins to unravel, or perhaps to get tangled up. My critique pertains equally to concepts of waste, and of nature and the environment, which are closely associated in Hawkins' analysis. This point is made at length below on the basis of my fieldwork, but a quick illustration from a different context provides a sense of what I'm getting at. In her ethnography *The Mushroom at the End of the World,* Tsing discovered that while human disturbance of forests is what American foresters think they should combat, Japanese foresters seek to encourage it (2015, 222). Their concepts of "nature" and humans' relation to it are totally inverted. Tsing noted a similar inversion with respect to waste: the same left-behind trash that the white hikers and the Forest Service rely on to argue that South Asian mushroom pickers should be excluded from U.S. forests is used by the pickers to align their movements with others who have passed before them (2015, 247–48). My point here is that when you look cross-culturally (as well as within "cultures", which are of course never monolithic or homogeneous) humanity is far, far away from having a consolidated or unified view of why or how waste can be adequately dealt with, or how its accumulation is relevant to other notions such as nature and the environment.

In arguing that it is not true, cross-culturally, to claim that "environmentalism in all its varieties dominates representations of wasted things" (Hawkins 2006, 7), and in suggesting that waste in fact constitutes a different "type of problem" to different audiences (see Furniss 2017), this chapter echoes Gille when she suggests that there exist different "waste regimes", that is, historically specific ways of producing, conceptualizing and politicizing waste (Gille 2007, 9). The notion of waste regimes places greater emphasis on economic, political, and material dynamics, compared to this chapter's greater focus on factors such as language, practice, culture, and group ties. However both views emphasize how waste is simultaneously contingent and indeterminate, but not wholly subjective either. Relative to a classical canon in waste studies, this perhaps represents a middle path between Thompson's *Rubbish Theory* (2017 [1979]) and Mary Douglas's *Purity and Danger* (2002 [1966]). In making that statement, I make Thompson's argument, that value forms are transient and constantly shifting (yesterday's kitsch or slums become today's masterpieces or architectural heritage), stand for a subjective account of what constitutes waste, and Douglas' argument stand for an "objectivist" view. Such a characterization of Douglas may seem misplaced or at least ironic applied to an anthropologist (the discipline of relativism par excellence) who devotes a significant part of her work to recounting "curious" practices from around the world and arguing that Western societies' hygienic or scientific framings of their pollution taboos are "sheer fantasy" (2002, 98). However, for Douglas, who sought "theories and 'analytical schemes' that apply to all societies at all times" (Thompson 2017, 43), the exploration of the *apparent* variety of ways of seeing and understanding waste ultimately serves a structuralist attempt at unifying all forms of waste and ways of thinking about it into a single logical operation that applies equally to everyone: the tabooing of category errors, summed up in the famous phrase "waste is matter out of place" that has become synonymous with Douglas' work. While both authors are indispensable in developing our thinking about waste, neither approach seems to me sufficient to address the waste problem in a manner that is intellectually or politically tenable today.

Framing and mediation, or how we understand waste and what we can do with it

If people believe waste makes them sick, they are framing it as an epidemiological or biomedical problem. If they see it as a sign of state failure, perhaps adding that the private sector would be more efficient, they are framing it in political or politico-economic terms. If they believe that waste accumulations reveal bad behavior and a lack "civility" or, more self-critically, the excesses of consumer capitalism, those might be termed ethico-moral framings. And so forth. The term *framing* is used here to refer to the action of devising, shaping and articulating a problem: the process by which waste comes to be "framed" as epidemiological, aesthetic, or environmental is an active, historic one through which waste is rendered problematic in terms of a particular conceptual scheme. As anthropologist François Laplantine reminds us, comparing the framing process that occurs in the cinema to that in the social sciences, "there is never a single pre-established frame, but a large number of possible frames, never a definitively closed (or open) system, but only the progressively closed and the progressively opened" (2015, 42). Thus "there is no frame without an effort to frame, de-frame, reframe" (Laplantine 2015, 42), or to borrow from Susan Sontag's analysis of photography, an art based largely on framing, "the number of photographs that could be taken of anything is unlimited", (Sontag 1977, 17), as one walks around an object looking at it through the viewfinder from different angles: below, high to low, and so forth.

The term "to frame" has two additional meanings that while not my reason for choosing it, are worth mentioning. First, those who espouse a particular framing may seek to "enclose" the waste problem within it, that is, completely envelop it within a single scheme to the exclusion of others. When it is not a form of obfuscation or outright error, this kind of framing (as enclosure and exclusion) is at least a rhetorical or aesthetic choice. We must bear in mind the artificiality of the boundary between what is shown and the off-screen ("*hors champ*" as Laplantine refers to it) that results from framing: what is presented to the viewer is but a part of the vaster space that extends beyond the frame. In these terms, my argument is that there are multiple coeval framings that may either compete or overlap, and that framings tend to overlay diachronically like sediments rather than replace one another. This is an argument against the scholarship and popular opinion to the effect that the contemporary Western political discourse's "environmental" framing is a master narrative. The second additional meaning of framing is to falsely pin a crime on someone. In North Africa, "poor behaviour", a "lack of education and civility", and so forth are very routinely "framed" for the waste problem in this way (see e.g. Bonatti 2015 for a similar phenomenon in southern Italy). While the point of this chapter is precisely to examine different framings of the waste problem within signage, on a higher level of abstraction, all of the material examined shares this premise insofar as it seeks to intervene on individual behaviours. It thus arguably has a depoliticizing effect that kneecaps more radical (and arguably more persuasive) explanations that emphasize the role of consumer capitalism, industrial design, infrastructural failure or inadequate public budgets in generating excess and disposability, and poor waste management (see e.g., MacBride 2012).[3] My desire to critique such an approach comes, certainly, from my belief that changing consumer behaviour alone is not sufficient. But it also, and above all, comes from a desire to disrupt and context the conclusions people draw when they connect visible waste accumulations solely or primarily to flaws in individual behavior. If you are introduced in a work or social setting around Tunisia as researcher working on waste, expect some version of the comment "you really have your work cut out for you" (i.e., the waste problem here is somehow much worse than in Europe or North America). The existence of this judgment nicely reveals the predominance of the occularcentric approach to waste, that is, of a criterion emphasizing the visual presence of waste in public spaces rather than volumes produced (much lower on a per capita basis than in the countries speakers consider as having mastered the waste problem) or recycled (much higher, thanks to the work of the "informal sector", than people assume on the basis of the absence of source separation). Where things get ugly is the way such comments imply judgments and hierarchies of a moral and "developmental" character. Indeed if the conversation continues, an unrelenting, rather predictable series of complaints and armchair theories about how individual behavior and worse, Tunisian or "Arab" "culture", are the primary explanation for why the cities, countryside and beaches are littered with plastic water bottles, disposable bags, and other rubbish, is sure to follow. Thus, at the same time they deplore the accumulation of waste, many people (including Tunisians) see in it a confirmation and reiteration of Western superiority—not just economic or technical superiority, but also moral superiority.

Waste, I also want to claim, has the power to "mediate". By way of illustration, consider how since, say, the 1970s, the "environment" has become a "value sphere" or realm of ethical activity in which many people strive for excellence. Which aspects of one's daily practice allow for making visible and recognizable, to oneself and others, efforts to behave ethically in this abstract and potentially distant domain (see Lambek 2010, 29)? How can I make my connection to the hole in the ozone layer, global warming, the great pacific garbage patch, deforestation of the Amazon or the extinction of species tangible and actionable? Waste may play a role here, much the way materials like bodily substances (blood, semen, breast milk…), money, gifts, ghosts, etc. flow between people and practically or symbolically significant domains, creating and shaping relations, such as those we have to kin, the cosmos, or the afterlife (see Carsten 2011). The visible material presence of waste and the regularity of its recurrence in our lives make it an easily graspable—in both senses of to comprehend and to use—way of expressing and rendering visible ethical striving with respect to "the environment". This point has been nicely made by Gay Hawkins (2006), who is absolutely correct to argue that waste management is a cultural performance enrolled in self-cultivation within a contemporary virtue system (compare Keane 2010, 65 discussing Protestantism according to Weber). I consider that Hawkins' invitation "to open up another way of making sense of waste beyond the trope of environmentalism" (2006, 3) remains open: what *are* some of the non-environmental ways of making sense of waste that can be observed through fieldwork? The research I have conducted in settings such as in Tunisia and Egypt reveals is that waste also mediates people's relationships to other "virtue systems" ranging from medical to moral, personal, political, national, and religious. One way of getting at this is to ask: what do people think they are "saving" by picking up litter or sorting out the recycling? Beyond the planet or the environment, the possibilities include national pride, personal dignity, health, even one's soul.

Three billboards

One simple, easily accessible, and surprisingly rich source for exploring different framings of the waste issue—at least in the Arabic-speaking southern Mediterranean countries—are signs. This isn't about a semiotic (re)turn, but about actual public signage: billboards, graffiti, metal and wooden panels, stickers, and so forth. In Tunisia you have to keep your guard up not to bump your head on one: they are everywhere. Their authors are numerous and varied: municipalities, neighborhood or national associations, branches of the bureaucracy, private companies, property owners, and so forth. While some are rusted and peeling, and hardly legible, others have paint that looks like it is still wet. While it is often difficult to know exactly when they were put up, this temporal depth is discernible not just in the physical marks of time, but also in the language and ideas they use, which correspond to different moments in national and global environmental history and different genealogies of ideas and vocabulary. Frequently, in the way people throw waste in the teeth of the very signs that purport to prohibit it, or in little dialogues that appear as scribbles added to the signs, a societal and political dialogue pertaining to appropriate behavior, authority, and power unfolds through and around them. For example on a sign saying "It is prohibited to put waste in front of the school" in front of a school, someone, presumably a teenage student, tagged in fat black letters "this is where I throw my garbage". Such moments remind us that we as academics, no more than the social actors who create the signs, can confer on them direct legislative authority. Rather, they participate in and contribute to a micropolitics of space and behaviour, both reveal and seek to shape a series of "structures of attitude and reference" through which people understand waste (see Said 1994, 75 and *passim*). Indeed, apart from their sheer frequency, what makes these signs particularly interesting is that almost none of them are mere prohibitions. Whereas a sign like the speed limit is generally "presented as being in need of no other support but itself" (Fish 2016, 14),[4] signage pertaining to waste in Tunisia functions differently. It almost always gives a reason, which often takes the form of a principle or maxim one should meditate and seek to interpret, rather than a measurable, external norm to which one must conform. Thus, pure injunctions, such as "Don't litter" or "No dumping", which are common in North America, are relatively rare in Tunisia. This is not a general feature of signage or rules in Tunisia, since outright prohibitions (often relating to police or military: no stopping, photography prohibited, etc.) certainly exist. This section returns to the three examples presented at the beginning of this chapter before trying to sketch the key concepts that I believe are deployed for framing waste in contemporary Tunisia.

"With Monoprix, I commit to protecting the environment", is a phrase printed on reusable cloth bags sold by a supermarket chain in place of disposable plastic bags. Its producers are some combination of the marketing and public relations departments of a French multinational corporation, whose framing of the waste problem as "environmental" is a familiar one in the Euro-American context. The phrase *je m'engage* has a "Frenchness" to it that becomes apparent when it comes time to translate it into Arabic. In France, *je m'engage* is a conventional, formulaic way of invoking notions of solidarity, politicization, and revolt, used here to try to flatter clients by construing their purchase as a form of politics and to assist them to construct a narrative about themselves. However to say "I engage/enlist/commit" would sound linguistically odd in

Arabic, much the way it does in English. The notion of *engagement* disappears in the Arabic version, replaced by the verb "to protect" conjugated in the first person: *aḥfaẓ* [I protect – أحافظ]. The audience of the message are the store's clients as well as anyone who might see the bags, which become a form of advertising for the corporation's progressive values. Its bilingualism, which is not a feature of the two other slogans quoted above, conveys both a sociological fact, and a yearning, pertaining to branding and clientele. Monoprix undoubtedly uses French for the practical purpose of makings its messages accessible to non-Arabophone clients, of whom there are a certain number in Tunisia, particularly among elites. But it also uses the French language to cultivate an image of sophistication and to convey its prestige as an international brand. Despite its colonial origins, and perhaps *because* it is less spoken today than in previous generations, French is a formidable source of class distinction in Tunisia. This makes the message "With Monoprix, I protect the environment" a certain expression of sophistication and cosmopolitanism which the brand, the French language, one's political awareness and commitment, and finally the environment itself, all play a role in constructing.

"The city is a mirror that reflects society, so protect its cleanliness", which appears in Arabic only, is a phrase coined by the Tunisian "National Programme for Cleanliness and Support for the Environment". It is printed on decals applied to large waste bins placed in public space. The example I found was located in the vicinity of Bourguiba Avenue (the capital's most exalted boulevard), near *Place de la monnaie* ("Currency" or "Mint" Square). This geography is not insignificant. The slogan appeared on a waste bin located in the rectilinear *ville nouvelle* that was juxtaposed with the curvilinear Medina during the colonial era (cf. Messick 1992, 247), and which remains home to much of the administration, 70s' era hotels with modernist concrete architecture, the headquarters of many banks, *etc.* It would be funny—in both senses: humorous but also peculiar—to see a slogan of this kind in one of the poor, "low-class" peripheral neighborhoods of the city like Tadammun, Douar Hicher, or Wadi Ellil. Not only does no one of significance risk passing by such an area, they are not parts of the city with the capacity to represent the larger whole of Tunisian society. Not so much to represent Tunisian society as it is, since downtown is no more—and arguably a good deal less—representative than any other part of the city. But to represent it as a certain group of people wish it to be, in the eyes of a certain set of observers. Staged micro-spaces of spectacle and order, whose role it is to represent the larger whole and exemplify positive traits like cleanliness, is a point to which I will return to later. As for this specific slogan, produced by the employees of a public sector initiative and formulated without detour through the language of foreigners, the phrase seems almost to create a moment of "cultural intimacy" (Herzfeld 2004), as if to say: we may not be able to solve the waste problem, but at least let it not become a source of embarrassment. However the phrase remains the voice of the state, with its hint of nationalistic vanity and a certain pedantry that reveals the persistence of the public sector's sense of its "mission" to edify the masses, or perhaps simply a certain haughtiness toward them.

"Whomsoever puts garbage here shall receive neither blessing nor reward", is spray-painted on the cement wall of a side street where people dump their trash on the ground in a "popular" (i.e., low/working class) neighborhood. Its author is a person who lives nearby—perhaps the owner of the wall or the closest resident—and its audience is likely the neighbors. The phrase does not mention any societal or overarching problem that is caused by or associated with waste, and it never states what is wrong with dumping garbage. It is assumed that the reasons are clear. As Darwish (2020) points out analyzing a similar phrase "God will not have mercy on the parents of those that burn and dump garbage here", – these formulations are essentially "curses". Curses in the sense of language serving to assail or scornfully abuse someone, not formal conjurations of the wrath of spirits. One of the interesting features of swearing at people as a general matter, which seems to apply here, is that it is typically directed at those who have wronged us once it is too late to change their behavior. This makes swearing more like a way of letting them know we noticed, and of conveying condemnation and moral censure, rather than a genuine effort to dissuade. And indeed, the slogans are generally painted *a posteriori* in places where waste has started to be habitually dumped. This, combined with the fact that they are often ineffective, gives rise to a major irony, as well as the dialogic aspect mentioned previously: the signs almost invariably appear next to giant heaps of waste, as though they had the effect of *attracting*, rather than repelling garbage. They therefore bear a striking similarity to the humorous image described by Sudipta Kaviraj, of a Calcutta sign reading "Commit no nuisance", beneath which a row of people were photographed urinating. While Kaviraj's argument, that "those who promulgated the notice had one conception of what public space meant, and those who defiled it had another" (1997, 84), is rich with possibilities in the Tunisian context, I would prefer in this particular case to see the issue as concerning who has authority over the space, and the limits of private property ownership. The issue is as much over whether the space is public or not, as it is about appropriate behaviors in public space. One must also never neglect, in interpreting waste accumulations in a place like Tunisia, the brute material force of the lack of alternatives. Many people would prefer not to dump waste in the places and the ways they do, but believe they have no choice. There is a wonderful moment in a chapter by John Collins that makes this point, underscoring the political and intellectual hazards of isolating the act of dumping from the constraints that give rise to it, and ascribing it to moral, cultural, or behavioral features of the person doing the dumping. A Brazilian NGO is seeking to characterize defecating in public as immoral, unhygienic, incorrect, etc in the course of a program to educate and "civilize" residents of Salvador (Bahia). One of the participants interrupts the nurse to ask "What if I should need to defecate in *your* neighborhood, Madame?" "If I am going to shit all over myself and the street" because there is no public toilet, "and instead I walk up to the door and knock, and you answer the door [pause] No, sorry, not you, but your maid answers the door. [...] Is she going to let me in to use to use your bathroom?" (Collins 2013, 179).

Some waste framings

Some themes and clusters have emerged as I have compiled an archive of public and private waste-related signage in a range of settings in Tunisia from 2016 to 2020. Such signage employs a variety of tropes ranging from duties and responsibilities of the citizen, to cleanliness, aesthetic beauty, religious obligation/threat of punishment, and environment. This makes it difficult to generalize or even to disentangle these threads. For instance in describing the proper disposal of waste as necessary to maintain "cleanliness", campaigns often intermingle religion, civility, or aesthetic values. Cleanliness can be used alone, as in "Keep the Beach clean" (Municipality of Le Kram, Tunisia) or "Let's collaborate so that we can work in a space that is clean" (Tunis Transport Authority). But it is often associated with notions of "civility" such as in the phrase "cleanliness is a civilized habit", (especially common in Egypt, but also to be found in schools in Tunisia, cf. Argyrou 1997 for Cyprus). Or religion, most famously in the phrase "cleanliness is part of faith", which is common in all Arabic-speaking countries. Cleanliness is also often invoked in relation to the "ranking" of nations and individuals on the scale of their degree of civilizedness, and as a standard by which a society is judged, as in the phrase analyzed earlier, "The city is a mirror that reflects society, so protect its cleanliness". And it can be associated with beauty and aesthetic values, for instance in the phrases "A clean station is a beautiful station" and in the title of Tunisia's government-sponsored *National programme for the cleanliness of the environs and the beauty of the environment*, or the name of Cairo's waste collection agency, *The Cairo Cleanliness and Beautification Authority* (Figure 6.1).

Figure 6.1 Tunisian Transit Authority: Let's collaborate together, so that we can work in a space that is clean. Photo by the author.

Cleanliness

However it is clear that cleanliness is transversal in discourses on waste in Tunisia: a line that intersects several other lines. Take for instance the "Recycle Me" project, funded from approximately 2015 onward by German and Swiss International development agencies as well as the Danone corporation, with the aim of "organizing" and supporting plastic and aluminium collection and recycling by the *berbécha,* "scavengers" who collect selectively from public waste accumulations in Tunisia (the name comes from the verb "to rummage" or "to search"). One of the project's initiatives was to place deposit points for recyclables in public space on which the slogan "Let's clean our country" was printed. However the slogan was printed overtop an image of planet earth in different shades of green, with trees and plants growing out of it, creating a syncretic blend between environmental and cleanliness messaging, that nicely summarizes the intersection between domestic and international discourses on waste. Messaging around "environment" is often used when the creators of a campaign have transnational ties, making it a marker of flows of ideas through NGOs or foundations with international funding, ideological, or institutional links. Cleanliness, meanwhile, is more prominent in the discourse of institutions that are in less direct contact with "the outside" (as "abroad" or "international" is said colloquially): for example schools, local councils, or national supermarket chains. Illustratively, Magasin Général (MG), a large national competitor of Monoprix, also produces reusable bags like those analyzed earlier. However MG's slogan is "Official sponsor of our neighbourhood's cleanliness", (in Arabic and French). Recall that Monoprix's slogan was "With Monoprix, I commit to protect the environment". The slogan used on the reusable bags of the other French multinational supermarket chain, Carrefour, is: "For the sake of our children, I protect the environment [*bi'a* - بيئة]" (in Arabic only, interestingly). If we compare Carrefour and Monoprix on the one hand, with MG on the other, not only does MG employ the more "local" idiom of cleanliness rather than the more transnational idiom of environment, but it also frames the problem on a more local scale: the supermarket sponsors the cleanliness of *the neighborhood*, not the more abstract and large-scale sphere of "the environment". The visual imagery on its bag is of a Tunisian flag, national monuments, and a camel. In comparison to imagery of, say, a blue planet, this choice is not insignificant in terms of both the scale on which it defines the issue, or its enrollment of stylized nationalistic tropes and the sentiment of pride to which they speak.

Another feature of the discourse on waste that reveals the centrality of cleanliness in framing the problem is the choice of word for refering to waste. Like English, Arabic is a language with a motley vocabulary for things we dispose of and find useless. From among these, the most commonly used in Tunisian signage are *fadlat* [فضلات], *ziblé* [زبلة] and *awsakh* [اوساخ]. I will not speculate here as to why the terms *nifayat* [نفاية] and *qimama* [قمامة] are rarely used, nor get too deep into the finer distinctions between these terms. What is remarkable though is the frequency with which the term *awsakh* is used by residents in their signs. Meaning "dirty things", close to what in French are referred to as *ordures*, the term has no real equivalent in English, except perhaps filth. It does not mean garbage (while all *awsakh* is waste, not all waste is *awsakh*) and I would not translate it as "dirt" in this context since soil and mud each have their own terms and dirt lacks the connotations of disgust that *awsakh* inspires. The term is twinged with moral overtones and can involve a certain amount of visceral revulsion. It is like dirty dishwater or the contents of a dustpan on one end of the spectrum, and what you find down the outhouse hole or in the bin of a butcher's shop after a day or two in the sun, at the other. The focus on this narrower category in the language of spray-painted signs gives us a sense of the importance of dirtiness/cleanliness in the problematization of dumping. The term *awsakh* is not used in official slogans, where it would be inappropriately colorful, a bit like if the Park Service put up a sign saying "Thanks for your visit and don't forget to pack your scum back out with you".

Citizenship and responsibility

One genus of signage in Tunisia frames waste in terms of the duties and responsibilities of citizenship. The types of slogans belonging to this category include "The cleanliness of the city is matter of awareness and responsibility" (Municipality of Soukra, in the Greater Tunis area), "Cleaning and caring for the environment is the duty of the state and the role of the citizen" (Municipality in the Sahel region), "Cleanliness is a collective concern and a shared duty" (Tunis Transit Authority). These tend to emanate from state or parastatal bodies, such as the public transit authority and local councils. While they occasionally acknowledge a "shared" dimension or even explicitly mention a "duty" on the part of the state, what distinguishes them is the emphasis they place on responsibility and "awareness" on the part of individuals.

One striking feature of these signs is that they do not make statements about what people should or should not do. They are totally different from ordinance signs often issued by the equivalent institutions in North America, which state such things as "no dumping" "tenants' trash only", "leave no trash on the ground", "garbage only, no recyclables". They instead act like statements of position with respect to the public controversy over who is responsible for "maintaining cleanliness" or dealing with waste, and who bears responsibility for upkeep of non-privately-owned spaces. This is a much-debated issue in Tunisia, as in other countries in the region. For instance in a focus group I conducted with representatives of civil society organizations and local government in Qasserine in November 2019, there were lengthy exchanges over the causes of the waste problem, which broke down largely into an opposition between "lack of awareness and education" on the part of residents, or insufficient efforts by the municipality (often due to insufficient budgets). One speaker summed this up by saying it was "as if we are asking which came first, the chicken or the egg. The situation will not change. The municipality will blame the citizen, and the citizen will blame the municipality, to infinity". He then went on to note how local councils and administrations have seized the new vocabulary of "participatory democracy" that has taken root since the 2011 revolution, using it as a device to shift responsibility for issues like waste management onto citizens, who must now "participate" in cleaning the city. Members of the public push back against this in various ways, not least by continuing to throw their waste in public, particularly when they find no bins in which to deposit it. In the "picture of the day" rubric on October 31, 2019, the Tunisian newspaper *Al-Shuruq* published a photo of a pile of trash, captioning it as follows: "one of the accumulations of waste a citizen sees on a daily basis. The need for awareness on the part of citizens is no substitute for the responsibility of the municipal officials". It almost seemed like a direct responses to the claims made in the municipal signs (Figure 6.2).

Figure 6.2 La Soukra (Tunis): The cleanliness of the city is a matter of awareness and responsibility. Photograph by the author.

Thus, public authorities are simultaneously involved in trying to pick up waste, and in developing a meta-level commentary about whether they should be doing this and whose fault it is when they fail. This dialogue arguably conveys an awareness of the way the "performative role in cleaning the city through managing urban waste" (Fredericks 2014, 534, see also Fredericks 2013) establishes or undermines the legitimacy of public/state actors. The Arabic-speaking southern and eastern Mediterranean is full of examples of how waste accumulations gather politicized publics and can telescope into much broader movements, both before and after the "Arab spring". A good pre-2011 example is the "Forget the government, help yourself on your own" campaign in Egypt in 2009. By using the apparently "depoliticized" issue of poor waste management as a kind of Trojan horse to sneak past censorship, this campaign was able to express a political critique by publishing photos of piles of garbage in the newspaper accompanied by the name of the official or institution that was supposed to pick it up. A good, recent example of this is the "You Stink" movement in Lebanon, mentioned previously. In this case, the closure of Beirut's main dumpsite prior to the siting of a new dump led to massive accumulations of waste. The resulting crisis became a focal point of dissatisfaction with the status quo in which the accumulation of waste was seen as a material manifestation of broader failures in governance (Abu-Rish 2015, Geha 2019). Tunisian public authorities seek to manage this "moral hazard" through signage that specifically seeks to reframe the garbage problem away from one of governance and toward one of citizenship, awareness, responsibility, etc. on the part of residents.

Religion

Discussions of the relevance of religion to the waste question in Muslim societies must endeavour to be very specific about who uses religious discourse, in which contexts, to say what.[5] Take, for example, the book *Garbage Citizenship*, one of the few examples of recent scholarship to take this bull by the horns. Compared to general statements, such as waste collection is "profoundly meaningful" in "a culture where cleanliness of body and soul is of deep spiritual import", (Fredericks 2018, 3), or "notions of purity in Islam ascribe particular importance to the act of cleaning" (Ibid., 143), the specific observations about the way male waste collectors employed by the municipality of Dakar found their work embarrassing and tried to hide what they did from their family and girlfriends while simultaneously seeking to "frame their labor as an act of Muslim piety rooted in the spiritual value of cleanliness" (Fredericks 2018, 13) in order to earn a measure of respect, are more informative, and more defensible. Female workers, for their part, rarely bothered trying to claim they were doing something of religious value when they collected trash (Fredericks 2018, 145–146). The profession was no less stigmatizing for them than for men, only their gender made it impossible for them to escape the association with dirty work.

In Tunisia, religious language is deployed with respect to waste in two main ways. One mode, which is recognizable from other settings such as Egypt or Senegal, consists of affirming that cleanliness is part of faith. This phrase is taught in religious education classes and is used on signage in schools. Outside of pedagogical contexts it is uncommon and one gets the sense an adult couldn't address the phrase to another adult without coming across as patronizing. I have never seen a sign employ the phrase outside of a school. If trying to square the notion that "cleanliness is part of faith" with the massive public accumulations of household waste in North Africa is a valid research question, then the most useful extant hypothesis is Michèle Jolé's. Her argument is that the apparent "dirtiness" of public space in Rabat and Tunis does not arise in spite of the emphasis on cleanliness and purity, but precisely *because* of it: the strict standards of bodily and household cleanliness demand regular elimination of wastes beyond the threshold of the home (1989, 1991). However the maxim "cleanliness is part of faith" and the Muslim faith of residents may be superfluous factors in explaining the phenomenon: the same exact thing happens in Marseille or Naples when municipal collection stops for a few days. In other words, rather than translating a particularly Muslim notion of cleanliness that separates the public and private spheres, accumulations of waste in public space may simply reflect infrastructural inadequacies and breakdowns. Another view might be that it is an error to connect religious purity to waste accumulations in public space. Emic efforts to make such a connection between Islam and household garbage do exist, for instance in Egyptian textbooks (Starrett 1998). However could they not be contrived efforts on the part of elites who believe that because of their purported piety the most effective way to reason with "the uneducated masses" is through religious principles? The principal evidence for this interpretation is the simple fact of how unsuccessful slogans such as "cleanliness is part of the faith" are at overcoming the waste problem that they purport to address.

The second way in which religious language is used with respect to waste, which I have not encountered in fieldwork in Egypt, is that mentioned earlier: basically, to say "God damn anyone who throws trash here". As noted, such phrases are spray-painted on walls by residents; such profanities are far too crass and inappropriate to be employed by an institution. They take different forms, such as "God has no mercy on the parents of those who put garbage [here]", "God has no mercy for those who throw dirty things [*awsakh*] here" (sometimes adding "It's the person who throws their dirt here who is the dirty one") or "Whoever puts garbage here shall receive neither blessing nor reward". Some more genteel individuals try to turn the phrase inside out and write phrases like "May God bless the parents of those who do not throw garbage here". In Sfax, a city reputed for its conservativism and industry, I have observed walls decorated with calligraphied Qur'anic verses, and phrases such as "thank you for not throwing waste here" then written in small letters below. The text is supposed to cause would-be litterers to hesitate out of fear of defiling the sacred word of the Lord. A friend once told me that he had seen a sight where the author had painted an arrow indicating where people should not put litter, causing there to be litter pile just a few centimeters to the side of it. The author had therefore added another arrow, causing the litter pile to be moved a few more inches to the right. And so forth until a dozen arrows appeared all along the wall in a spectacle that became a kind of running gag in the neighborhood. Variants on signs of this kind are most common, even characteristic of "popular" neighborhoods throughout the country. For instance one appears in the music video of the Tunisian rap song *Houmani*, which became famous for the way it captured the look and feel of the shit-life of low-class neighborhoods. (*Houmani* meaning roughly a "chav" in British English, or maybe a hooligan: what in French are called *banlieusards*.) However, they are not the exclusive domain of the poor, and one can find them in a neighborhoods with various of socio-economic profiles (Figure 6.3).

Analyzing them within a broader social matrix of religious beliefs and practices pertaining to impurity and danger, and some fascinating and quite original observations about odors, Darwish has argued that the problem of "garbage and malodor" in Tunisia is that they remove the protection of guardian angels and make people vulnerable to Jinn (spirits), the evil eye, and so forth (Darwish 2020). Taken as a whole, this approach is a good illustration of what I have in mind when arguing for attentiveness to local waste framings. However I have two objections to the specific way Darwish analyzes such signs. First, while Darwish notes briefly that "there were some graffiti that just read, 'no dumping'", he does not analyze those signs, or any others, choosing instead to focus exclusively on the ones that "referred to God and solicited his punishment on those that discarded their waste in public space" (Darwish 2000, 7). Since overemphasizing

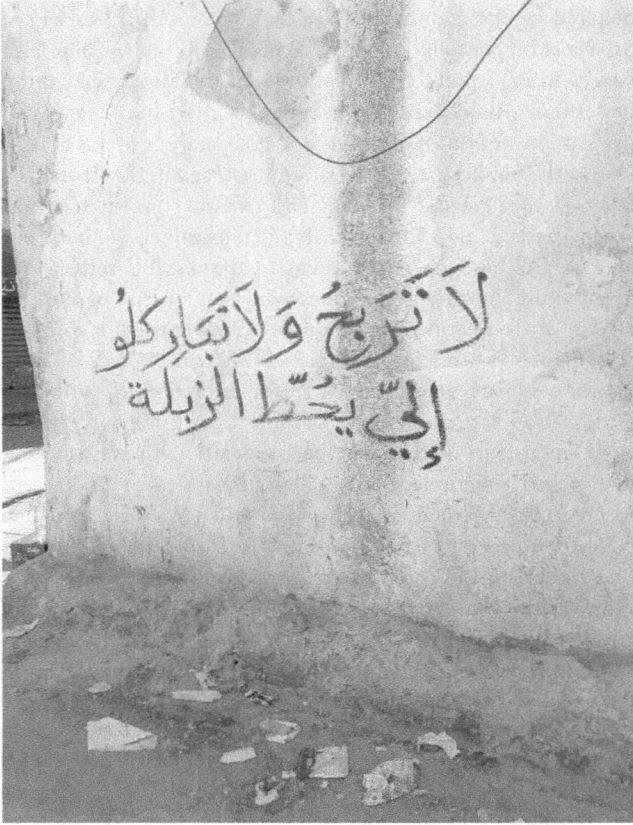

Figure 6.3 Bab al-Khadra (Tunis): Those who put garbage (here) will not receive (God's) reward or blessing. Photograph by the author.

Islam as the explanation and determinant of every facet of Muslim majority societies bears the risk of essentialism, I believe it is critical to underscore that these are but one of several types of signage, which when taken as a whole convey a diversity of framings and conceptions, as this chapter has shown. Second, asking God to punish people for doing something (or simply saying "God damn bastards …") does necessarily confer on the offending behavior a peculiarly religious status. The same language, can be used to swear at someone who harassed a woman, drove dangerously, cheated in a game of dominoes, or ripped you off in business. Such phrases are evidence of the reliance on religious idioms to tell people off, not of the specifically religious nature of waste for Tunisians.

Environment

By now it should be apparent that environmental discourse it is not at all the most common register for framing waste in Tunisia. The institutional sources and geographic distribution of messages relying on the environment demonstrate that nowadays it has mainly penetrated the upper classes and is being driven largely by organizations with international ties. Thus, it is in upscale neighbourhoods of Tunis like La Marsa where one finds signs such as this one (Figure 6.4).

Figure 6.4 La Marsa: "Our environment is our life. So let's all cooperate together in protecting it". Photograph by the author.

Only at the MG branch in the picturesque village of Sidi Bou Said, one of Tunis's most famous attractions, does one find containers inviting shopper to deposit their empty glass bottles, saying this is an "Environmental initiative". Such containers are rare-to-inexistent at branches of the store in more down-market neighborhoods.

The environment and environmental protection are political and historic categories with a fascinating and complex role in Tunisia over the past 30 years that can only be sketched in the vaguest terms here (see Guillaumet 2019, 145–300). As an administrative and political category, "environment" is still very much in the process of being consolidated. Tunisia's Ministry of the Environment was created in 1991 (and Morocco's only in 1996). In 2017, an entirely new branch of the police called the Environmental Police was created in Tunisia. The translation of the term from/into Arabic is not settled and continues to fluctuate between [*bi'a* – بيئة] and [*muhit* – محيط]. The latter term, which is also the term used for "ocean", refers to environment in the sense of what immediately surrounds a person, more like the English "environs". Thus, the Tunisian institutions known in English/French as the *Ministry of Local Affairs and the Environment*, and *National Agency for the Protection of the Environment* use different terms for environment in Arabic. This isn't just a matter of synonyms. A number of people who employ the shared signifier "environment" are referring to different underlying objects and concerns. While term *bi'a* does not have a pejorative connotation in Tunisia the way it does in Egypt, where it is used as an insult to describe people from low class social backgrounds (see Furniss 2017), the word *muhit*, because of its association with "what immediately surrounds a person", does often encompass poor people, drugs and alcohol, hooliganism, sex, and so forth within the scope of "environmental" problems (see Hopkins et al 2001, Loukil-Tlili 2013). The moral tinge of the term environment can be seen in the associations made between waste, environment, and certain types of behaviors. Messages that mix civility and manners with environment and waste are common in upper-class Tunis neighborhoods. On the walls in the stairwell of a building housing several doctors' offices, one sees side-by-side signage stating "No Smoking" (in Arabic and English) as well as "Please maintain silence and cleanliness—Thank you" (in Arabic and French). A sign near the seaside walkway in La Marsa, produced by the municipality, says "Please respect the environment and the neighbors. I protect the environment. I protect quietness". The environmental protection message depicts someone throwing a paper into a wastebasket, while the protection of quietness message shows a musical note with red bar across it. And so forth. The "environs" (surroundings) meaning of the term *environment* is detectable in the mandate of the environmental police, which primarily pertains to "local" issues, including hygiene. In 2020, for example, the environmental police were deployed to raid cafés that violated closure regulations that were adopted to halt the spread of COVID-19. Many people refer to them simply as the municipal police. Similarly, the Tunisian Ministry of Environment is combined institutionally with the Ministry of Local Affairs, which supervises municipalities.

The class and political cleavages that are created by and reflected in the deployment of the category environment are perhaps nowhere more visible than in the term's pre-2011 usages. Virtually every neighborhood in Tunis and every city in Tunisia has a "Boulevard of the Environment". Had he been aware of them when writing his book *Seeing Like a State*, James Scott might readily have used these as one of the examples of the "microenvironments of apparent order [such] as model villages, demonstration projects, new capitals, and so on" (Scott 1998, 225), designed to transform society. Meant to play an exemplary role and spur a transformation of conduct and mentalities, the boulevards functioned, as Timothy Mitchell once said of urban techniques of power in Egypt, to provide an "appearance of order" (Mitchell 1991, 63–93) that is "at the same time an order of appearance" (1991, 60). Set up as though they were the picture of something, they could also be the first thing shown during a visit by a foreigner or a representative of the central bureaucracy, serving to whisk visitors quickly past or around neglected neighborhoods a few streets over. While the Boulevards of the Environment are artifacts of the pre-2011 political era, the concept of "model spaces" continues to have a certain appeal in the country.[6]

At one end of a Boulevard of the Environment, one typically finds a statue of a desert fox, known as Labib (Figure 6.5).

Under the rhyming slogan *Labib, lil bi'a habib* (Labib, friend of the environment), this mascot was used throughout the 1990s and 2000s to promote "environmental awareness", particularly among children and young people. Imagine a government-sponsored version of Yogi Bear, who instead of being known for his comic antics was supposed to be "simultaneously an image of a sincere friend and a rigorous guide for children and the public", to quote a document produced in 1996 by the Ministry of Environment. In addition to the statues in public spaces, he appeared in short, animated sequences on television and comic book–style vignettes in textbooks.

In 2011 many statues of Labib were attacked, their arms and ears broken off. In the eastern countryside near Al Kaf, one person told me she saw a statue of Labib that had been desecrated in the genitals. The reason for these seemingly surprising acts was that Labib, and because of its political uses, the environment in general, became a symbol of the regime of former Tunisian dictator, Ben Ali. Enrolled in a project of legitimation, the environment played a role similar to that of the "women's rights issue". Both were instrumentalized to portray the authoritarian regime as a progressive and politically open, particularly in the eyes of foreigners for whom

Figure 6.5 Enfidha, on the highway between Tunis and Sousse: Labib, the Mascot of the Environment. Photograph by the author.

these were priority issues. This also explains a line from an article on public parks in Tunisia that long puzzled me. In it, the author notes that from the start of the revolution until December 2012, "public parks were subject to numerous attacks and desecrations" (Loukil-Tlili 2013, 120). The author regards these as yet another manifestation of Tunisian's "incivility", noting, apparently to prove that under some circumstances Tunisians are capable of being civil, that protesters did sweep up garbage from Kasbah Square after their political sit-ins. In fact, the parks and green spaces drew fire because they were part of the same system of authority as Labib and the environment more broadly.

Pure injunctions

While I noted that pure injunctions are an uncommon form for signs to take in Tunisia, one does see them from time to time. Some examples of this type of sign in poor neighborhoods are "It is prohibited to put waste [*fadlat*] in front of the school. –Municipal order" (Tadamun, Greater Tunis) and "It is prohibited to throw dirty things [*al-awsakh*] and waste [*fadlat*]" (Tozeur). And some examples of it in rich neighborhoods are "Depositing waste [*fadlat*] prohibited / Interdit de déposer les ordures" and "Do not throw waste [*fadlat*] and dirty things

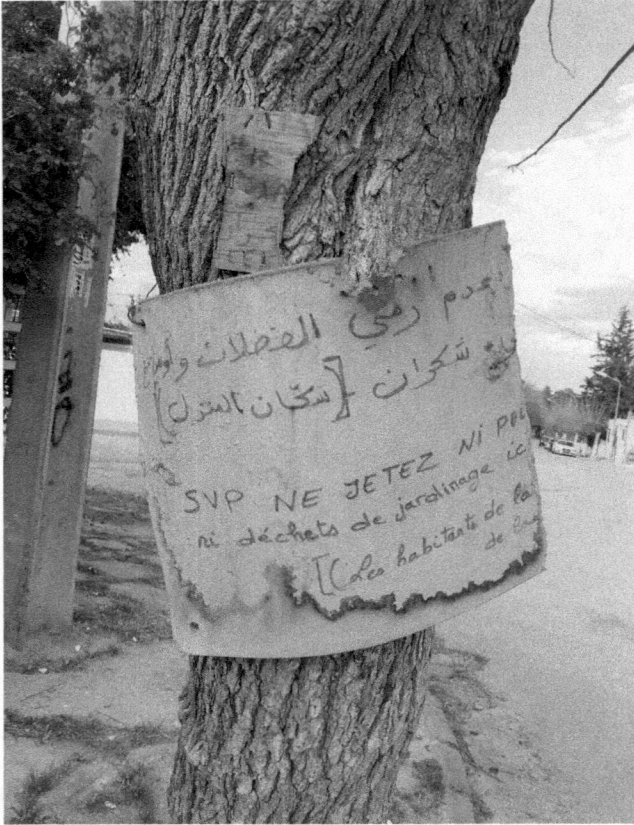

Figure 6.6   Mutuelleville (Tunis): Do not throw waste and dirty things here. Thank you. –the residents of the house. Photograph by the author.

[*al-awsakh*] here. Thank you.—the residents of the house / SVP ne jetez ni poubelles ni déchets de jardinage ici.—les habitants de la casa". The latter two appeared in Mutuelleville, a faded upper-class neighborhood of the capital. One of them contained an error in Arabic, spelling *shukran* with the letter *nun* at the end rather than the accusative case ending. The bilingualism and the error in Arabic are significant reminders to us that the producers were at least as comfortable expressing themselves in French as they are in Arabic. While there is an extraordinarily high level of bilingualism in the country, this nevertheless clearly marks them as members of a very small elite minority. Although, and indeed *because*, they provide no justification, the key thing these signs can reveal is which groups of people feel empowered to edict signs that proceed directly from authority. The identification of the speaker (municipal order, residents of the house) is an acknowledgment of the need to establish by what legitimacy the order is issued (Figure 6.6).

Conclusion

There is no "consensus" or standard way of making a public sign about waste in Tunisia. A diversity of registers coexist and compete with one another. Through attention to the locations and sources of these different approaches, it becomes possible to discern patterns in the way the issue is framed. For instance, high socio-economic status groups with transnational ties tend to view the environmental framing positively, while lower-class groups associate it with pre-2011 state repression. Public authorities tend to portray the waste issue as being a problem of citizens not living up to their "duties" and "responsibilities". People in poorer neighborhoods more frequently use profanity to insult and "swear" at litterers. The registers vary as one moves between rich and poor, central and peripheral, historic and new neighborhoods in a way that makes the city legible through the different ways people frame waste: shown a sign, you can venture a guess as to what type of neighborhood you are in, who its likely author is, and about what period it dates from. As accumulated sediments of different political tides that have swept over the country, the palimpsest of writing overlaid by later writing reveals a series of evolving conceptions and uses of waste and environment, as well as histories of how ideas have traveled and the roles of different actors in their spread. The kinds of arguments and justifications the signs provide, or the way they invite people to contemplate their behavior with respect to waste, provide temporally, sociologically, and geographically specific views of waste, i.e., "framings".

When I conduct interviews or give talks in Tunisia, people usually want to turn the tables, and ask me, "What solutions do *you* propose to the waste problem?" They seem to find my efforts to avoid this question unconvincing, and at times so do I: few waste studies scholars choose the topic out of a love of littered landscapes and landfills, even if we develop a certain morbid fascination with, or professional impassiveness toward, their sight and smell. With that in mind, in order to make clear two things this chapter is not trying to do, I must concede a shortcoming. If we think of the signs as media, this chapter has little to say about the question of "reception", despite it being what "anthropologists are probably best prepared to study", according to Abu-Lughod (1997, 112). It does not take a position on the effectiveness of different framings relative to one another. I believe that is an important question, and one that is answerable with the right methods, though I have not conducted the research necessary to do so. Second, while I am not against trying to figure out which framing is most effective (in a specific context) to motivate an audience to alter their behavior with respect to waste, I am also not arguing that raising awareness and changing people's attitudes is the best way to deal with waste, in Tunisia or elsewhere. As a municipal councillor put it during a workshop on waste I facilitated with local government and NGOs in a provincial Tunisian city, "sure, we can write "cleanliness is part of faith" and "God damn you if you put trash here" on the walls, but do you think that will solve the problem? We need trash pickup operations organized by the municipality, not slogans".

Beyond the specifics of how waste is conceived and problematized in contemporary Tunisia, I believe this article has three potentially useful propositions for waste studies scholars working in other settings. First, that people interpret waste accumulations and comprehend the threat they pose through different conceptual schemes or "framings". Second, that waste has a capacity for "mediating", that is, acting as an intermediary or vector linking people to spheres outside themselves, making waste a substance through which we can engage in self-cultivation or demonstrations of excellence within various virtue systems. Third, that there exist numerous non-environmental ways of making sense of waste. I have tried to show some ways of getting at these methodologically, and to argue that we need to stay alert and open to the way people know and interact with waste through situated and variable framings. The framings of waste in one place or time are not necessarily identical to those in another. Trying to uncover different framings and spheres to which waste connects us is an anthropological project insofar as it involves developing the categories with which to name and understand the waste problem through fieldwork, rather than in advance or by reference to supposed intrinsic, general properties, or supposed immutable essences. This approach has a practical relevance that has been significantly underestimated in both local waste-related campaigns and global political discourse on the environment, which rarely contemplate what language and concepts are already being "locally" employed to grasp the issues they wish to tackle.

Notes

1 I have been attentive to this issue in the course of fieldwork on waste and waste collectors in Egypt (ongoing since 2007, for a total of about 2 years) and Tunisia (ongoing since 2016, for just under 2 years). Tunisia and Egypt share some commonalities but are by no means identical with respect to this—or any other—issue.
2 It is most persuasive on the level of the individual citizen. There is a tension in Hawkins' work between her claim that waste disposal began as a "technical" issue (2006, 29) and the claim that from the 1960s onward it became a moral and environmental issue. Arguably, it was over that time that a vocabulary of expertise and techno-politics emerged and increasingly sought to ring-fence the waste issue in a technical and managerial sphere.
3 This point, which relates to the way framings (de)politicize the waste issue in one direction or another, has of course not escaped the attention of the actors themselves. In Tunisia, civil society movements addressing landfill siting in Djerba and Tunis in the 2013–2015 period sought to use the category of "environment" to "define themselves as apolitical" and emphasize the "the non-political nature of their initiatives" (Loschi 2019, 102, 107). In Lebanon the exact opposite occurred during the 2015 garbage crisis when activists built the "You Stink" movement "on the framing of the garbage crisis as a failure of governance," one of them telling the researcher that "we had to make it clear that this was *not* an environmental issue" (Geha 2019, 89, 84 emphasis added).
4 Occasionally an argument is given to justify the speed limit, such as "Kids a play" or "School zone".
5 Edward Said's classic observations that Islam is often used in scholarship about Muslim majority societies to "signify all at once a society, a religion, a prototype, and an actuality" (Said 2003 [1978], 299) and that "unlike any other religion Islam is or means everything" (Ibid., 279), remains sadly a propos. The relevance of religion to understanding people's behavior and ways of seeing the world in Muslim majority societies seems at times so overstated that it can be hard to believe that once upon a time, prior to the 1970s, "representation of Islam was almost always brief" in anthropological studies of the region (Varisco 2005: 16). This risk—of overstating the pervasiveness of practice and belief, as well as its relevance to people's behaviour or ways of thinking about in other realms (politics, work, the household, etc.)—is heightened when touching on the issue of cleanliness, due to the temptation to posit, "logically" rather than ethnographically, that since "Islam is a religion of purity,' and so forth, waste must have a special significance or receive special treatment. Rather than draw too quick and too straight a line from waste to Islam, the precise ways in which Islam is—or is not—mobilized in debates about waste or considered relevant to apprehending the problem must be examined *in situ*.
6 At the entrance to Rue de Marseille, a pedestrianized side-street running north from Bourguiba Avenue, the city of Tunis has erected in recent years a sign saying "Model street, so let's take care of it together". The French translation of *namuzaji* (model) as *témoin* (witness) nicely captures the dual purpose of the street, both allowing visitors to witness the spectacle of cleanliness and order, as well as well as witnessing in the sense of furnishing evidence or proof of the orderliness, attractiveness, etc. that can be achieved.

References

Abu-Lughod L. (1997). "The Interpretation of Culture(s) After Television." *Representations, Special Issue The Fate of "Culture": Geertz and Beyond* 59, 109–134.
Abu-Rish, Z. (2015). "Garbage Politics." *Middle East Report* 277.
Argyrou, V. (1997). "'Keep Cyprus Clean': Littering, Pollution, and Otherness." *Cultural Antrhopology* 12, no. 2: 159–178.
Bonatti, Valeria. 2015. "Mobilizing around Motherhood: Successes and Challenges for Women Protesting against Toxic Waste in Campania, Italy." *Capitalism Nature Socialism* 26, no. 4: 158–175.

Carsten J. (2011). "Substance and Relationality: Blood in Contexts." *Annual Review of Anthropology* 19, 19–35.

Collins, J. (2013). "Ruins, Redemption, and Brazil's Imperial Exception." In Ann Laura Stoler (Ed.) *Imperail Debris: On Ruins and Ruination.* Duke University Press. 162–193.

Darwish S. (2020). "Flowers in Uncertain Times: Waste, Islam, and the Scent of Revolution in Tunisia."*Ethnos* 8, no. 2:1–22. doi: 10.1080/00141844.2019.1696859.

Douglas, M. (2002 [1966]). *Purity and Danger: An Analysis of Concepts of Pollution and Taboo.* Routledge.

Fish S. (2016). *Winning Arguments: What Works and Doesn't Work in Politics, the Bedroom, the Courtroom, and the Classroom.* Harper.

Fredericks R. (2013). "Disorderly Dakar: The Cultural Politics of Household Waste in Senegal's Capital City." *The Journal of Modern African Studies* 51, no. 3: 435–458.

Fredericks R. (2014). "Vital Infrastructures of Trash in Dakar." *Comparative Studies of South Asia, Africa and the Middle East* 34, no. 3: 532–548.

Fredericks R. (2018). *Garbage Citizenship: Vital Infrastructures of Labor.* Duke University Press.

Furniss J. (2017). "What Type of Problem Is Waste in Egypt?" *Social Anthropology* 25, no. 3: 301–317.

Geha C. (2019). "Politics of a Garbage Crisis: Social Networks, Narratives, and Frames of Lebanon's 2015 Protests and Their Aftermath." *Social Movement Studies* 18, no. 1: 78–92.

Gille Z. (2007). *From the Cult of Waste to the Trash Heap of History: The Politics of Waste in Socialist and Postsocialist Hungary.* Indiana University Press Bloomington.

Guillaumet A. (2019). *La place de la nature dans la société tunisienne post-révolution entre politiques de protection et exploitation touristique: représentations, approches institutionnelles et pratiques sociales.* Doctoral dissertation, Department of Geography, Avignon University.

Hawkins G. (2006). *The Ethics of Waste: How We Relate to Rubbish.* Rowman & Littlefield Publishers.

Hopkins, N. S., Mehanna, S. R., and el-Haggar, S. 2001. *People and Pollution: Cultural Constructions and Social Action in Egypt.*The American University in Cairo Press.

Herzfeld M. (2004). *Cultural Intimacy: Social Poetics in the Nation-state.* Routledge.

Jolé M. (1989). "Le déchet ou « l'autre côté de la limite »." *Maghreb-Machrek* 123, 207–215.

Jolé M. (1991). "Gérer ses résidus en public. R'Bati, Slaoui, et habitants de Temara aux prises avec leurs déchets." *Les Annales de la Recherche Urbaine* 53: 32–39.

Kaviraj, S. (1997). "Filth and Public Sphere: Concepts and Practices about Space in Calcutta." *Public Culture* 10, no. 1: 83–113.

Keane W. (2010). "Minds, Surfaces, and Reasons in the Anthropology of Ethics." In M. Lambek (Ed.) *Ordinary Ethics: Anthropology, Language, and Action.*Fordham University Press. 64–83.

Lambek M. (2010). "Introduction." In M. Lambek (Ed.) *Ordinary Ethics: Anthropology, Language, and Action.* Fordham University Press. 1–36.

Laplantine, F. (2015). *The Social Life of the Senses: Introduction to a Modal Anthropology.* Bloomsbury.

Loukil-Tlili B. (2013). "Parcs et jardins de Tunis: gestion et usages des espaces paysagers." M. Bourgou and A. Hatzenberger (Eds.) *Des Paysages.* Centre de Publication Universitaire/École normale supérieure de Tunis. 115–124.

MacBride S. (2012). *Recycling Reconsidered. The Present Failure and Future Promise of Environmental Action in the United States.* MIT Press Cambridge.

Messick B. (1992). *The Calligraphic State: Textual Domination and History in a Muslim Society.* University of California Press Berkeley.

Mitchell T. (1991). *Colonising Egypt.* University of California Press Berkeley.

Said, E. (1978). *Orientalism.* Pantheon Books.

Said, E. (1994). *Culture and Imperialism.* Vintage Books.

Scott J. C. (1998). *Seeing Like a State: How Certain Schemes to Improve the Human Condition Have Failed.* Yale University Press New Haven.

Sontag, S. (1977). *On Photography.* Penguin Modern Classics.

Starrett G. (1998). *Putting Islam to WorkEducation, Politics, and Religious Transformation in Egypt.* University of California Press Berkeley.

Thompson, M. (2017 [1979]). *Rubbish Theory: The Creation and Destruction of Value.* Pluto Press.

Tsing, A. L. (2015). *The Mushroom at the End of the World on the Possibility of Life in Capitalist Ruins.* Princeton University Press Princeton.

Varisco D. (2005). *Islam Obscured: The Rhetoric of Anthropological Representation.* Palgrave Macmillan New York.

7

UNMAKING THE MADE: THE TROUBLED TEMPORALITIES OF WASTE

Heike Weber

Introduction

In May 2020, two cooling towers of a German nuclear power station in Philippsburg were blown up in the process of the station's dismantling and demolition. The removal of Germany's nuclear power plants currently challenges nuclear, construction, and waste engineers, but except for such stunning explosions, most of the dismantling operations take place behind the scenes. According to press reports, the collapse of the 152-meter-high towers was "unspectacular" and without critical incidence, and due to the coronavirus pandemic, spectators were not allowed on ground.[1] Nevertheless, the abrupt moment of destruction was highly visible. But in addition to the visible portion, the dismantling has and will continue to involve lengthy, invisible operations of preparations, cleanup, and removal. Over months, construction workers had systematically weakened the roughly four-decades-old cooling towers and their ferroconcrete structure. Tons of construction debris, including contaminated material and the previously removed nuclear core, await their fate in what environmental historian Joel Tarr has coined as the "ultimate sink" (1996): a final disposal site. But what is more, in respect to the so-called nuclear *Endlagerung* — i.e., the "final storage" of nuclear waste — the German nation has not even decided yet where and how to install such an "eternal" waste storage site.

This brief glance at a single episode of nuclear power plant dismantling demonstrates the complexity of any unmaking in modern societies. In our technical world, "unmaking" the "made" — be it infrastructures or factories, weapons or consumer goods — is a task at least as complex and nearly as frequent as "making" it. It requires labor, scientific knowledge and technical skills, technological equipment and infrastructures, and material and energy (Weber 2014a; Salehabadi 2014). But what is more, unmaking does not conclusively terminate with acts of dismantling, discarding, or simply abandoning the infrastructures, buildings, or materials defined as obsolete or out of use, as indicated dramatically by the current challenge of where and how to store nuclear waste for centuries to come.

Under the header of "unmaking" — a verb form which helps to highlight the active doing, rearranging and redefining required for any abandoning, discarding, or disposing — this chapter calls for a temporal perspective on waste. Producing, handling, and unmaking waste represent material and temporal (re)arrangements of our made world and the environment. How waste is defined, perceived, produced, or treated is interlinked with temporal processes and

DOI: 10.4324/9781003019077-7

temporalities, some of which are obvious, others of which operate rather subtly, and many of which conflict. Any "making" of our material world has the flipside of aging, decaying, and unmaking—temporal processes of material transformation. Organic materials are subject to rapid decomposition and putrefaction; with prolonged use, artefacts show signs of wear and tear and are eventually considered "obsolete". But the very moment of declaring something as "waste"—as expired, aged, obsolete, or otherwise worthless—does not conclude the story of waste. Rather, this classification inaugurates subsequent waste disposal operations such as disassembly, recycling, or landfilling and incineration. Again, these operations involve questionable temporal concepts. Waste disposal techniques emerged from the idea of an "ultimate sink" that could solve the waste crisis, whereas recycling methods pretend an "eternal" circulation of materials in closed material loops.

The field of waste studies has underlined the spatial, material, and sociocultural dimensions of waste. It has critically analyzed the power hierarchies and the forms of social or environmental injustice at stake when waste is traded, sorted, or treated. Historical studies on urban sanitation have focused on rising waste amounts in emerging consumer societies and on the spaces and technologies of waste management, reaching from the domestic dustbin to the sanitary landfill (Melosi 2005). Consumer historians such as Susan Strasser (1999) have described how consumer culture gradually shifted from a traditional "stewardship of objects" to throwaway consumption patterns once mass consumption and municipal disposal infrastructures became the norm. From an ethnographic perspective, Nicky Gregson et al. have emphasized that wasting as a form to get rid of personal belongings involves social practices and meanings (2007). More recently, Josh Lepawsky has widened the predominant perspective on post-consumer waste by describing the manifold "discardscapes" which emerge along producing, using, and discarding e-waste (Lepawsky 2018). Waste and waste disposal infrastructures are increasingly discussed as drivers of injustice, and the global waste trade has been defined as a novel, "toxic" form of colonialism (Müller 2022).

Adding to such perspectives, the following foregrounds the temporal dimensions of waste. Focusing on time and temporality calls us to reflect on both long-term developments as well as the temporal processes, sequences, and concepts on the micro level of waste making and waste treatment. The first section is devoted to waste's historicity and the fundamental material changes of discards in human societies over historical time; it brings these historical changes into play with the "deep history" of geological times. The second section approaches waste from the angle of "unmaking". It urges humanities to follow waste from its point of origin through time and space along the diverse waste work processes such as sorting, dismantling, scrapping, reprocessing, and disposal and discusses the problematic vision of recycling as closing material loops. The third section questions common concepts that try to grasp the "age" of things and their appropriate time of use, including their "end-of-life". The fourth section concludes the temporal perspective on waste by considering long-term effects: the so-called afterlife of things and stuff and waste legacies.

Questions of time and temporalities have been recurrent themes in history—per se a discipline concerned with time and its passing—and they recently entered social sciences and the humanities at large (Gange 2019; van der Straeten and Weber 2021). The dimension of how societies experienced, perceived of, or modified time has frequently been studied, while global history scholars have nuanced the modernist acceleration narrative by carving out the uneven and heterogenous appropriation of modern time around the globe (e.g., Ogle 2015). Adding to this, scholars from environmental humanities, history of technology, anthropology, and science and technology studies have turned to the question of non-human actors and their "temporal agency". This latter strand of temporal thinking brings the temporalities of materials, of the "made", of the environment, and of the entire Earth into play. The ephemeral plastic bag, for instance, links past and future to current throwaway practices—a deep past of hundreds of thousands of years through which fossil resources have accumulated connect to an unknown future that will be confronted with plastic's global environmental consequences (Westermann 2020). Likewise, seen from a distant future, the "ultimate sinks" of the past and present turn into material repositories and eventually, fossil records—these sinks are merely temporary cessations of material flows and transformations, and many of them might cause long-term environmental hazards. Sanitary landfills represent sites in need of long-term aftercare, and nuclear waste storage facilities enclose the residues for nuclear depletion to take place, which must be measured on geological time scales rather than on human ones. Drawing on the temporal reflections of Barbara Adam, Gabrielle Hecht, and others (Adam 1998; Hecht 2018; van der Straeten and Weber 2021), the plastic bags or nuclear sites can be said to be situated in different, even conflicting or colliding, time scales when they are discarded or dismantled.

This chapter suggests that many challenges of waste in modern societies have their origin in such troublesome temporalities. In their fetish for innovation, novelty, and mass consumption, modern societies have neglected the material and temporal issues at stake when the "made" is considered to have become waste and is ready to be "unmade" since (waste) matter does not simply disappear.

"Technofossils" and the historicity of waste

According to the Anthropocene thesis, "humans" have become a geological factor in what is being called the epoch of the Anthropocene. The Anthropocene premise prompts us to reason along geological times and in respect to waste, the question arises of what distant future generations will have received from current times, even if we experience our material world as transient. Under the heading of "technofossils", geologists are currently investigating man-made techno-material legacies. In addition to volatile emissions, such as carbon dioxide or methane—the key drivers of climate change—such material discards have risen rapidly since industrialization. The amount of technofossils is currently estimated at around 30 trillion tons of material, representing about 170,000 different types of "synthetic mineral substances" (Ellis 2018, 148). Jan Zalasiewicz and others refer to the "technofossil record of humans" (Zalasiewicz 2014) which ultimately consists of the sedimented stuff of the past. Plastics and radiation signatures from nuclear testing, but also broiler chicken bones preserved in landfills as remainders of today's chicken-based diet, are currently discussed as the single most identifiable markers of this "Age of Humans" (Bennett et al. 2018).

Industrialization, urbanization, and 20th-century mass consumption have resulted in increasing amounts of urban waste; remnants which could not easily be channeled back to the spheres of nature, production, or consumption accumulated in urban agglomerations and became a threat to urban hygiene. Zooming in on the last hundred years or so, studies on the socio-economic metabolism of societies have underlined that the globe's resource extraction has been increasing, and while a growing share of material consumption has flown into the formation of stocks, waste amounts have been vastly increasing (Krausmann et al. 2018). From the 1950s onwards, environmental historians claim that a "great acceleration" (Engelke and McNeill 2016) occurred in resource extraction, consumption, and waste amounts in societies characterized by mass consumption and mass production. But even since then, not every region or society has produced similar amounts or sorts of "technofossils", and in today's globalized world, the geographies of waste disposal diverge largely from the geographies of waste production.

As a geological indicator, the claimed "technofossil record of humans" helps to identify the Anthropocene, but this perspective tends to ignore the historicity of waste and the fact that societies of different regions and historical periods have differed in how much and what kind of residues they generated and how they handled them. In a more nuanced historical approach, the varying amounts and kinds of past discards can be used as indicators of the changing and diverging material basis of past societies and of differences in how they handled resources and residues.

To uncover this nuanced dimension, we need to turn to archaeology as for large parts of human history, only archaeological finds can inform us about past cultures and societies. Frequently, such finds have been passed down to us as remnants from manure pits, dumps, or rubble heaps. Famous material testimonies of Roman antiquity include major ruined monuments such as aqueducts, but also the Monte Testaccio in Rome—a 45-meter-high waste heap created through the dumping of debris;, in particular, broken amphorae which were used to carry oil and other foodstuffs into the city. Among the largest material testimonies of modern waste are the Fresno sanitary landfill which, as the initial American landfill of the 1930s, was designated a National Historic Landmark in 2001 (Melosi 2002), and Staten Island's Fresh Kills landfill, which was reopened to absorb the ground zero debris of 9/11. Europe's turn from landfilling to incineration has created the idiosyncratic effect that potential future material relics are being passed down predominantly as indecipherable ashes and slags.

Wilhelm Rathje emulated the archaeological method for modern waste sites when he launched his "garbology" project, which is facilitated by the fact that highly compacted modern landfills developed anaerobic environments that have unwittingly preserved much of the dumped stuff. Garbologists have secured and evaluated remnants from American sanitary landfills as records of postwar consumption, while other researchers have excavated yard pits which served as dumping grounds before the late-19th-century sanitary reform (Rathje and Murphy 1992; Crane 2000). The photograph in Figure 7.1 represents relics of waste which were found on former dumping grounds on the periphery of Berlin.

The reclaimed debris (Figure 7.1) stems from early-20th-century Berlin households. Back then, a certain amount of Berlin's waste was dumped in that region for meliorating the land to grow vegetables. Reusable materials such as scrap, leather, or textiles would have been picked out by waste-pickers or rag-and-bone men at earlier stages in the waste's transfer from households to dumping ground or would have decomposed over time. Glass or porcelain shards or parts from objects such as pottery, bottles, buttons, or electric plug sockets, switches, or fuses were left behind. The lettered shard of an ash-tray is a relic of the famous Berlin cigarette-maker Garbáty. As in many other cities of the early 20th century, Berlin households still sold rags, old leather, metal objects, or bones to the rag-and-bone trade; paper was burned in domestic fireplaces. In smaller towns, ashes or organic matter were used in gardens, and kitchen scraps were fed to cats or rabbits. The technofossil record provides a rough outline for how the amount of waste, and the diversity and complexity of its composition, increased in industrialized regions and mass consumer societies: Organic materials gave way to non-organic ones, chemicals and compounds diversified, and by the last third of the 20th century, plastics overtook steel as the most widely used material.

Not only did discards—their amounts and material composition—and their treatment or reuse differ over time, but the notions and terms for waste also changed. Umbrella terms such as *waste*, the French *ordures ménagères*, or the German *Abfall* and *Müll* came into use in the late 19th century and at first referred to the waste generated by cities and their households; back then, this mainly meant market leftovers, street sweepings, waste from small businesses, and household rubbish (Kuchenbuch 1988; Barles 2005). Previously, substance-specific terms—excrement, rubble, scrap,

Figure 7.1 Debris which has recently been recovered from a site on Berlin's periphery where waste from Berlin households had once been used to meliorate the grounds.

Copyright: Oswin Nikolaus, 2021.

rags, garbage (for kitchen waste), or sweepings, to name a few—were common, and much of these residues were reused, whether by farmers, rag-and-bone men, or industry. By contrast, it is the possessor's overall intention for discarding, rather than any material categorization, that underlies the meaning of the novel terms for waste. In Western urban agglomerations, the decades around 1900 represent a caesura also in respect to waste handling. Urban waste was less and less channeled to specific destinations of reuse and was perceived as a threat to urban health and hygiene. When cities installed municipal waste services, these were defined as public health infrastructure charged with the systematic collection and fast removal of urban waste.

What started as an urban sanitation initiative turned into a global, environmental challenge over the course of the 20th century. The way current societies deal with and discuss waste has been shaped by the era of urban sanitation, the rise of mass consumption, and an ecological awareness emerging since the late 20th century. In postwar North America or Western Europe, municipal waste services began serving rural households causing traditional reuse or disposal methods to be hampered or lose their economic sense. Since then, mass consumer goods found global distribution, and the waste amounts generated by their production have now surpassed those of postconsumer waste. From the late 20th century onwards, societies have thus been confronted with a shortage of sinks to get rid of the stuff we have made so far and keep on making—from waste to carbon dioxide emissions. The "reduce, reuse, recycling" formula has been a reaction to this ongoing waste crisis. Recycling became an element of waste management over the last decades but it has not yet disrupted the dual trends of increasing waste amounts and ecological costs created by mass consumption.

Colliding temporalities: the reverse logistics of unmaking and the closed-loop recycling vision

Every social formation generates residues when consuming and producing, archaic as well as modern ones. Declaring something to be a discard is an important cultural means to re-classify things in the cycle of creation and decay. As Zsuzsa Gille has argued with the term "waste regime", any economy, including the socialist Hungarian case she studied, is characterized by specific ways of identifying, defining, organizing, and treating waste (Gille 2007). While largely neglected in economic history, waste has always been a source of economic activity. Even if the waste business was often dominated by informal markets and undesirable waste work was left to the poor or otherwise socially marginalized people, the scrap trade and the rag-and-bone trade were thriving economic fields before enterprises such as Veolia or Remondis grew into global players in the late-20th-century recycling markets (Barles 2005; Denton and Weber 2021). Unmaking was and is a common process, though largely happening behind the scenes, in informal markets or even combined with illegal activities.

As foundations of mass production and consumption, the logistics of resource extraction, production, and distribution are well known and described. Waste engineering and waste management provide the opposite logistic services: they collect, accumulate, and remove widely dispersed stuff once it has been declared as waste. The operations of collecting, dismantling, and reprocessing waste for recycling follow a "reverse logistics", to use current business management terminology. Municipal waste services, for instance, collect wastes from their multiple places of origin, transport and sort them, and then channel them back as resources either to industry or to their seemingly "ultimate sink".

Inside humanities, such reverse unmaking processes have been described for selected cases such as urban demolition work, municipal waste disposal, ship dismantling, or e-waste reclamation in the Global South. In dense inner cities, it is only demolition that makes room for novel buildings (Hommels 2008). Ships, which still figure as the infrastructural backbone of global trade, are nowadays most likely to end their life in the shipbreaking yards of Bangladesh and India, where they are scrapped through arduous human labor and turned into various reusable and non-reusable materials, toxic waste among them (Crang et al. 2012; Dhawan 2021). As potential source of hazardous effects, but also as potential resource for salvageable metals, electronic discards have problematic afterlives along their rampant scrapping chains from rich to poor regions. Besides, as recent research has underlined, residues begin accumulating from the very beginning of their production, starting with the ore mining and manufacturing process of electronics (Lepawsky 2018; Gabrys 2011).

The reverse logistics of the waste business make it both labor and transportation intensive. Waste collecting relies on high numbers of collecting agents, while sorting or dismantling for recycling requires a highly-trained workforce with an appropriate, tacit knowledge of materials—a knowledge that waste pickers and the sorting workforce acquire only over time. Even today, recycling often relies on exploiting manual sorting labor rather than applying high-tech separation methods such as optical sorting. A means to economize was and is to save on labor costs. As socially stigmatized work, waste collecting or sorting was often left to the urban poor and done within informal markets and labor conditions. Moreover, the history of sanitary waste disposal demonstrates that its infrastructures were consistently lagging behind technical progress made in production and distribution. Plastic, for instance, was widely appropriated in postwar consumer goods as a malleable and hygienic material, but with time, it caused serious problems inside waste incineration and composting plants. Specific channels for battery disposal were implemented only once the heavy-metal pollution of municipal waste had been detected as an environmental hazard. Even today, novel consumer products are not designed with disassembly in mind, even if they are composed of complex composite materials.

In the case of waste disposal, reverse logistics discharge the waste material into dumps, landfills, or incineration sites. Recycling, by contrast, relies on further reverse operations, such as sorting and dismantling, to produce "secondary materials" for production. Any successful recycling depends on the demand for such recovered materials in production. For most of human history, recycling was the norm. Rags served as the raw material for the paper industry before the use of groundwood pulp in the late 19th century; dyestuffs and semi-finished chemicals such as glycerin, stearin, or bone ash were obtained from coal tar residues and animal bones, respectively, before petrochemicals took hold. New branches of waste utilization emerged along with new industries—e.g., when steel slag (so-called Thomas slag) was made into fertilizer in the late 19th century or when PET bottles became outdoor wear a century later.

Around 1900, the traditional rag-and-bone man was still a prominent figure of the urban economy, making a living selling discards from households and small businesses to industry (Barles 2005; Strasser 1999; Denton and Weber 2021). They worked at their own risk and expense and at the end of the day, delivered their pickings to the local scrap dealer, who then did a rough sorting and traded the different waste components to the next agent in the hierarchy, until a wholesaler did the final sorting and channeled the waste back to the site of production. In this chain of sorting and reprocessing, paper was sorted into more than 70 categories, celluloid into 10–15 categories, and bones into 6–8. Rags and textiles were sorted into several hundred categories, most of which were bought up by cardboard, roofing, and textile industries. The most important material in the waste salvage business was scrap metal; since this was still the "age of steel". The second largest one was rags, and the third, bones. Cans were exploited for tin, and up to the 1920s, even light bulbs were dismantled into brass sockets and tungsten filaments. During World War I and II, warring states restructured these waste economies and enforced waste salvage for the purpose of resource mobilization. In Nazi Germany, forced labor and victims of concentration and extermination camps were systematically exploited for such dismantling or waste sorting work (Weber 2021a).

There is no clear-cut historical caesura at which reuse, repair, or recycling gave way to disposal. Reuse practices were situated in specific regional or temporal settings of a respective waste regime and depended on economic relations or production and recovering technologies as much they did on moral economies. When virgin material prices in postwar North America and Western Europe sank in parallel to rising labor costs, disposal eventually outpaced waste salvage, particularly for household waste, which was widely dispersed and highly heterogenous.

Historical records of past unmaking activities are scarce, in part, because much of the work had been informal and thus invisible. However, the dismantling and scrapping business has been studied for metal scrap, ship dismantling, and car scrapping. (Zimring 2005, 2011). Cars—the main consumer good of the 20th century—have been disposed of in scrapyards where parts of them were discarded while others were re-traded, reused, or recycled as scrap. But what happened to electrical appliances once they turned into mass consumer items? On the one hand, discarding supposedly obsolete electric devices is not a novelty of the digital

Figure 7.2 Dismantling mainframes in Switzerland, 1968 by courtesy of SRF.

age—it accompanied mass consumption in affluent societies from the outset. On the other hand, such equipment was regularly repaired, reused, or resold (Weber 2021b). As a film report from 1968 on the first Swiss "computer graveyard" suggests (Figures 7.2 and 7.3),[2] manual dismantling of mainframes was common practice to recover metals such as iron, aluminum, zinc, copper, and other reusables—a practice we should not romanticize as potentially contaminating materialities were present in computer manufacturing from the outset. As the report stressed, the mainframes of the first and second generation were only five to ten years old but had lost their economic value as computers, reduced to the material value of their extractable scraps of metal at the scrap dealer's junkyard. Obviously, the laborious and unhygienic or hazardous waste work was gradually outsourced from rich to poor countries. Manual dismantling and sorting still characterize today's global waste business, but they have been shifted to the formal and informal markets of the Global South (Corteel and Le Lay 2011; Lepawsky 2018).

As the "unmaking" coinage underlines, waste incurrs further actions and processes which involve tedious and often hazardous sequences of waste collecting, waste processing, or dismantling and sorting. Unmaking takes place in human time and space and it requires human labor, time, transport, and energy as well as complex infrastructures. Not only does it result in further remainders, but also, often, in hazardous longterm effects. The time arrow of human agency, entropy, and dissipation cannot be reversed, nor can a previous state be fully restored (Georgescu-Roegen 1971). Any making and unmaking has irreversible effects and involves remainders.

Modern waste engineering created the idea of an "ultimate sink" which would absorb the residues of production and consumption and forever unmake the made. Over the course of the

Figure 7.3 Dismantling mainframes in Switzerland, 1968 by courtesy of SRF.

20th century, the shortage of sinks, as well as arising problems of waste legacies, have rebutted this idea, from the hazardous waste scandal of the Love Canal disaster, to the present-day challenges of microplastics in the sea and our bodies and the growing levels of carbon dioxide in the Earth's atmosphere.

The temporalities of unmaking oppugn the ultimate sink ideal, but they are likewise inconsistent with the closed-loop vision of recycling as the latter is bare of any temporal dimension. This metaphor has a long history reaching back to the archaic Ouroboros symbol. Since the 1970s, recycling was politically promoted as an ecological strategy to reduce waste; in European countries, the recycling movement was also driven by landfilling capacities becoming scarce. When the American Environmental Protection Agency (EPA) published an initial study on the waste recycling market in 1972 (Figure 7.4), the study's cover depicted the Ouroboros of Abraham Eleazars' *Uraltes chymisches Werk* from 1760. The study explored the most important salvage markets of the day—namely paper, metals, glass, textiles, rubber, plastics, slaughterhouse waste, and construction waste—all of which, except for rubber and plastics, were fields in which reuse had been the norm for centuries. The ancient symbol was seen to have a modern counterpart in ecological recycling, which is now frequently represented by a circular symbol. As a "timeless" figure, the circle visually promises a complete circularity of material flows. But as the more critical "downcycling" term underlines, the complex reverse logistics of recycling—multi-tiered labor and transformation processes anchored in time and space—involve material and energy losses and decreasing material qualities. Moreover, nature's "cycles" differ from those that we can imitate through recycling technologies as much as nature's times differ from human ones (Weber 2020).

Figure 7.4   The Ouroboros icon on the cover of an EPA study on recycling markets.

Source: Arsen Darnay, William A. Franklin: *Salvage Markets for Materials in Solid Wastes*, U.S. Environmental Protection Agency, Kansas City 1972.

Temporalities of the made: product obsolescence in mass consumer societies

Innovations and novelty characterize modern society, but at the same time, the made—infrastructure, buildings, capital goods, cars, or computers—requires constant care, maintenance, or repair (Graham and Thrift 2007; Anand, Gupta and Appel 2018). Steve Jackson has thus argued for a "broken world thinking" which takes "erosion, breakdown, and decay, rather than novelty, growth, and progress" as a starting point (Jackson 2014, 221). This broken world thinking grasps only a certain set of temporalities of the made, namely those linked to natural or mechanical wear and tear. Further ones include the heterochrony of our material world: An inhabitant of New York, for instance, might use a brand new cell phone, but century-old infrastructures such as the metro, streets, or canalization are also part of his or her daily routines. Moreover, societies ascribe temporalities to their material culture, be it in the form of innovation cycles, of values such as "novelty", or of durations considered to be an appropriate time of use.

Next to material wear and tear, cultural concepts of obsolescence and society's expectations of technical progress, newness, and durability define whether, when, and why artefacts are to be considered "aged", or even "antiquated", "outdated", or "obsolete" and how long, on average, they should last in use. The example of houses might illustrate the point: Modern buildings often last only for some decades and next to construction methods and materials, regional and cultural settings have the biggest influence on such durabilities of use; an average house in Japan, for instance, stands for 30 years before it is demolished; in the United States, the average life span of a house is 55 years; in Britain, it's 77 years (Cairns and Jacobs 2014, 127).

Naturally, these diverse life spans come with differing practices of use and maintenance, of repair and breakdown, of renewal and demolition.

Modern mass production and consumption brought about novel temporal concepts of the made. New institutional, organizational, and sociocultural temporal constructs emerged to dictate the appropriate time for substituting the old with the new and for deeming a technology or artifact "aged", "mature", or "obsolescent" (Weber 2018). "Old" began to refer not to a thing which is worn out, but to one which has been overtaken by better-performing technologies. Economists of the second half of the 20th century developed innovation cycle theories; engineers began to calculate the durability of the made in quantifiable use frequencies and "lifespans". Based on average use frequency and driving habits, most cars, for instance, were designed to last for a range of 7–12 years; at the same time, we know from the history of socialist East Germany that its automobile models were used three times as long as their originally projected life span of 8–10 years (Möser 2012, 218). Current mobile phones are constructed to last around ten years, but are regularly considered outdated after less than two years due to ever-accelerating digital innovation cycles. In the 1970s, consumer advocates accused the car industry of building predetermined breaking points—built-in "corroding spots"—into automobile design; today's printers or cell phones are said to have a similar built-in obsolescence.

Indeed, the history of mass production and mass consumption has been accompanied by a regular debate and constant negotiation of product durabilities, particularly during critical turning points such as the American Depression of the 1930s, the ecological movement of the 1970s, and today's multiple environmental crises (Weber et al. 2018). While serviceable hours, the rhythms of upkeep and maintenance, and durabilities are known facts for capital goods such as airplanes or medical and military technology, they have stayed obscure for most consumer goods: Producers hardly give an account of engineers' and designers' calculations and construction with respect to prospective durability, and many consumers hardly reflect upon how long they expect or want to use so-called "consumer durables".

By the early 21st century, we routinely applied anthropomorphic metaphors such as "age", "life span", "technological generations", or "death" to refer to such temporalities of the made, as if the "biography of things"—to use Arjun Appadurai's metaphor (1986)—resembled that of a living being. Human life has been commonly thought of in a model of different stages of age, and in the early modern period, the image of stairs that first ascend and then descend from cradle to grave became popular. Figure 7.5 shows one of the widespread broadsheets of this "life stairs" *sujet* from the 19th century. But in many ways, it seems inappropriate to create a parallel between the temporalities of the material word and that of human life and biological time.

It might be time to reflect on more appropriate ways to analyze material culture's temporalities, especially as such bio-metaphors have become increasingly common in engineering with the concepts of lifecycle assessments and the cradle-to-cradle design approach. As shown previously, the idea of eternal sinks and eternal circular material flows are misleading. In the end of the unmaking process, recycling will have destroyed an artefact to recover its material resources for future production. Repair, reuse, and refurbishment, by contrast, represent strategies which prolong the human-object interaction. And "ultimate sinks" come down to being stockpiling sites, in which wastes are stored to last for future times and potentially endure further transformation and decay.

Temporal chasms: "afterlife", "aftercare", and waste legacies

For a long time in history, waste was either reused or just left to itself to decay and ruin. Demarcation lines between reuse and ruin were often blurry. For instance, when so-called

Figure 7.5 Nineteenth-century visualization of the "life stair", illustrating the stages of human life from cradle to grave in ten-year stages.

Source: Museum Europäischer Kulturen der Staatlichen Museen zu Berlin - Preußischer Kulturbesitz, D (33 R 497) 488/1977.

"wastelands", such as marshes or badlands, were reclaimed by waste, urban waste was left behind as fertilizer and filling material and would decompose into the soil. Gradually metabolizing, it turned into earthly matter so that nature seemed to incorporate the waste in its eternal cycles of becoming and passing. When urban waste amounts turned into a hygienic and engineering challenge for cities around 1900, sanitary engineering and municipal waste services aimed at quickly removing waste and the hygienic risk of uncontrolled putrefaction and contagion out of the urban agglomeration. "Out of sight, out of mind", to use Melosi's term, ruled most of 20th-century waste management (Melosi 2005). In the name of hygiene, waste engineers dumped waste into pits or rivers, into wetlands, lakes or the sea, relying on nature's so-called "self-cleaning power". Most cities refrained from strong financial and infrastructural investments in waste disposal and either experimented with waste's potential to recultivate land or simply dumped the waste on the urban periphery until changing waste composition rendered such traditional practices dangerous.

Defining waste as a danger to nature is historically relatively new. Even the sanitary landfill of the 1970s was conceptualized as artificial sedimentation: an "ultimate sink" in which waste would degrade while being isolated from its surroundings. Landfilling was considered a technological fix to the ensuing waste crisis. Instead, it turned into an environmental problem. As Boudia and others have underlined, residues "disobey boundaries, appear where they shouldn't appear, alter environments, and enter communities and bodies without permission" (Boudia et al. 2018, 167). What is more, they have their own material-specific temporalities which have often been ignored.

By the late 20th century, remediation and aftercare engineering emerged as legitimate fields of scientitic and professional expertise to deal with the aftereffects of earlier production and consumption—from abandoned coal mining sites to contaminated sites of previous industrial activity or waste dumping. Sanitary engineers now redefined landfills as "eternal reactors", rather than natural or artificial sediments, which need around 200 or more years of aftercare (Weber 2014b). Today, the waste hills of 20th-century consumer society are prominent urban and suburban landmarks that testify to the non-circularity of the metabolisms of past consumer societies (Weber 2019).

The persistence of waste and its sinks of the past have resulted in temporal chasms. The long time horizon of environmental hazards defies the short-term logics of politics and policies (Adam 1998) as well as the modern fetishism of novelties. The environmental historian Verena Winiwarter and others have recently classified abandoned technological sites according to the longevity of their respective legacies and the prospective burdens of aftercare that future generations will have to take on (Winiwarter and Schmid 2020; Winiwarter et al. 2016); "wicked legacies" is their coinage for those among them with more or less eternal need for care. But unmaking (and making) operations also causes diverse problematic short-term effects—the "slow violence" of creeping social and ecological damage, e.g., when hazardous waste is outsourced to poor or disempowered regions (Nixon 2011). The Anthropocene concept is often construed as calling for new temporalities. It can also be read as a narrative about the neglected temporality of wastes left behind by past and current generations.

Notes

1 FAZ, report (without title), 15 May 2020, p. 7.
2 Film stills from a Swiss report on a "computer graveyard" (*Computer-Friedhof in der Schweiz*. N/A, CHE 1968, TC: 00:01:20; 00:01:38), https://www.srf.ch/play/tv/archivperlen/video/1--computer-friedhof-in-der-schweiz-1968?urn=urn:srf:video:07fb8dbe-bdc6-4024-a905-2ad39dcf645c). I want to thank Regine Buschauer for sharing this report with me.

References

Adam, B. 1998. *Timescapes of Modernity: The Environment and Invisible Hazards*. Routledge.
Appadurai, A (Ed.). 1986. *The Social Life of Things: Commodities in Cultural Perspective*. Cambridge University Press.
Anand, N. , A. Gupta, and H. Appel (Eds.). 2018. *The Promise of Infrastructure*. Duke University Press.
Barles, S. 2005. *L'invention des déchets urbains, France: 1790–1970*. Champ Vallon.
Bennett, Carys E. et al. 2018. "The Broiler Chicken as a Signal of a Human Reconfigured Biosphere." *Royal Society Open Science* 5, no. 12. doi:10.1098/rsos.180325.
Boudia, S., A. N. H. Creager, and S. Frickel et al. 2018. "Residues: Rethinking chemical environments." *Engaging Science, Technology, and Society* 4: 165–178.
Cairns, S. and J. M. Jacobs (Eds.). 2014. *Buildings Must Die: A Perverse View of Architecture*. MIT Press.
Corteel, D. and S. Le Lay. 2011. *Les travailleurs du déchets*. Eres.
Crang, M., N. Gregson, F. Ahamed, R. Ferdous and N. Akhter. 2012. "Death, the Phoenix and Pandora: Transforming Things and Values in Bangladesh". In A. Catherine and J. Reno (Eds.) *Economies of Recycling: The Global Transformation of Materials, Values and Social Relations*. Zed Books. 59–97.
Crane, B. C. 2000. "Filth, Garbage, and Rubbish: Refuse Disposal, Sanitary Reform and Nineteenth-Century yard Deposits in Washington, DC." *Historical Archaeology* 34: 20–38.
Denton, C. and H. Weber (Eds.) 2021."Rethinking Waste Within Business History: A Transnational Perspective on Waste Recycling in World War II." Special Issue, *Business History*. https://urldefense.com/v3/__https://doi.org/10.1080/00076791.2021.1919092

Dhawan, A. 2021. "The Persistence of SS France: Her Unmaking at the Alang Shipbreaking Yard in India." In S. Krebs and H. Weber (Eds.) *Histories of Technology's Persistence: Repair, Reuse and Disposal*. Transcript. 265–287.

Ellis, E. C. 2018. *Anthropocene: A Very Short Introduction*. Oxford University Press.

Engelke, P. and J. R. McNeill. 2016. *The Great Acceleration: An Environmental History of the Anthropocene Since 1945*. Cambridge University Press.

Gabrys, J. 2011. *Digital Rubbish: A Natural History of Electronics*. University of Michigan Press.

Gange, D. 2019. "Time, Space and Islands: Why Geographers Drive the Temporal Agenda." *Past & Present* 243, no. 1: 299–312. doi:10.1093/pastj/gtz013.

Georgescu-Roegen, N. 1971. *The Entropy Law and the Economic Process*. Harvard University Press.

Gille, Z. 2007. *From the Cult of Waste to the Trash Heap of History: The Politics of Waste in Socialist and Postsocialist Hungary*. Indiana University Press.

Graham, S. and N. Thrift. 2007. "Understanding Repair and Maintenance." *Theory, Culture & Society* 24, no. 3: 1–25.

Gregson, N., A. Metcalfe, and L. Crewe. 2007. "Moving Things Along: The Conduits and Practices of Household Divestment." *Transactions of the Institute of the British Geographers* 32, no. 2: 187–200.

Hommels, A. 2008. *Unbuilding Cities: Obduracy in Urban Socio-Technical Change*. MIT Press.

Hecht, G. 2018. "Interscalar Vehicles for an African Anthropocene: On Waste, Temporality, and Violence." *Cultural Anthropology* 33, no. 1: 109–141.

Jackson, S. J. 2014. "Rethinking Repair." In T. Gillespie, P. J. Boczkowski, and K. A. Foot (Eds.) *Media Technologies: Essays on Communication, Materiality and Society*. MIT Press. 221–239.

Krausmann, F., C. Lauk, W. Haas, and D. Wiedenhofer. 2018. "From Resource Extraction to Outflows of Wastes and Emissions: The Socioeconomic Metabolism of the Global Economy, 1900-2015." *Global Environmental Change* 52: 131–140.

Kuchenbuch, L. 1988. "Abfall: Eine Stichwortgeschichte". In H. G. Soeffner (Ed.) *Kultur und Alltag*. Schwartz. 155–170.

Lepawsky, J. 2018. *Reassembling Rubbish: Worlding Electronic Waste*. MIT Press.

Melosi, M. V. 2002. "The Fresno Sanitary Landfill in an American Cultural Context." *The Public Historian* 24, no. 3: 17–35.

Melosi, M. V. 2005. *Garbage in the Cities: Refuse, Reform, and the Environment*. University of Pittsburgh Press.

Möser, K. 2012. "Thesen zum Pflegen und Reparieren in den Automobilkulturen am Beispiel der DDR". *Technikgeschichte* 79, no. 3: 207–226.

Müller, S. M.2022.*Hazardous travels: A ship's tale of US waste and the global environment*. University of Washington Press.

Nixon, R. 2011. *Slow Violence and the Environmentalism of the Poor*. Harvard University Press.

Ogle, V. 2015. *The Global Transformation of Time: 1870–1950*. Harvard University Press.

Rathje, W. and C. Murphy. 1992. *Rubbish!: The Archaeology of Garbage*. Harper Collins.

Salehabadi, D. B. 2014. Making and Unmaking e-Waste: Tracing the Global Afterlife of Digital Technologies in Berlin. PhD diss., Cornell University.

Strasser, S. 1999. *Waste and Want: A Social History of Trash*. Metropolitan Books.

Tarr, J. A. 1996. *The Search for the Ultimate Sink: Urban Pollution in Historical Perspective*. University of Akron Press.

van der Straeten, J. and H. Weber. 2021 "Technology and Its Temporalities: A Global Perspective". In G. Carnino, L. Hilaire-Pérez, J. Lamy, and L. Zakharova (Eds.) *Global History of Technology (19th-21st centuries)*.

Weber, H. 2014a. "Entschaffen": Reste und das Ausrangieren, Zerlegen und Beseitigen des Gemachten. (Einleitung)". *Technikgeschichte* 81, no. 1: 1–32.

Weber, H. 2014b. "Von wild zu geordnet?: Konzeptionen, Wissensbestände und Techniken des Deponierens im 20. Jahrhundert". *Technikgeschichte* 81, no. 2: 119–146.

Weber, H. 2018. "Made to Break?: Lebensdauer, Reparierbarkeit und Obsoleszenz in der Geschichte des Massenkonsums von Technik". In S. Krebs, G. Schabacher, and H. Weber (Eds.) *Kulturen des Reparierens: Dinge—Wissen—Praktiken*. Transcript. 49–83. https://www.transcript-verlag.de/media/pdf/de/5e/c0/oa9783839438602GnXIQBOBhdVzB.pdf.

Weber, H. 2019. "20th Century Wastescapes: Cities, Consumers, and Their Dumping Grounds". In T. Soens, D. Schott, M. Toyka-Seid, and B. De Munck (Eds.) *Urbanizing Nature: Actors and Agency (Dis)Connecting Cities and Nature Since 1500*. Routledge. 261–289.

Weber, H. 2020. "Zeit- und verlustlos?: Der Recycling-Kreislauf als ewiges Heilsversprechen". *Zeitschrift für Medienwissenschaft* 23 Zirkulation, no. 2: 19–31. doi:10.25969/mediarep/14821.

Weber, H. 2021a. "Nazi German Waste Recovery and the Vision of a Circular Economy: The Case of Waste Paper and Rags." *Business History* doi:10.1080/00076791.2021.1918105.

Weber, H. 2021b. "Mending or Ending?: Consumer Durables, Obsolescence and Practices of Reuse, Repair and Disposal in West Germany (1960s–1980s)". In S. Krebs and H. Weber (Ed.) *Histories of Technology's Persistence: Repair, Reuse and Disposal*. Transcript. 236–263.

Westermann, A. 2020. "A Technofossil of the Anthropocene: Sliding Up and Down Temporal Scales With Plastic." In D. Edelstein, S. Geroulanos, and N. Wheatley (Eds.) *Power and Time: Temporalities in Conflict and the Making of History*.University of Chicago Press. 122–144.

Winiwarter, V., M. Schmid, H. Haberl, and S. J. Singh. 2016. "Why Legacies Matter: Merits of a Long-Term Perspective." In H. Haberl, M. Fischer-Kowalski, F. Krausmann, and V. Winiwarter (Eds.) *Social Ecology: Society-Nature Relations Across Time and Space*. Springer. 149–168.

Winiwarter, V. and M. Schmid. 2020. "Socio-Natural Sites". In S. Haumann, M. Knoll, and D. Mares (Eds.) *Concepts of Urban-Environmental History*. Transcript. 33–50.

Zalasiewicz, J., M. Williams, and C. N. Waters. 2014. "The Technofossil Record of Humans."*Anthropocene Review* 1:34–43.

Zimring, C. 2005. *Cash for Your Trash: Scrap Recycling in America*. Rutgers University Press.

Zimring, C. 2011. "The Complex Environmental Legacy of the Automobile Shredder." *Technology and Culture* 52, no. 3: 523–547.

8

COMMODIFICATION AND RESPECT: INDIGENOUS CONTRIBUTIONS TO THE SOCIOLOGY OF WASTE

Michelle Schmidt

Introduction

This chapter explores Indigenous animistic perspectives on food and disposal to argue that food waste is a social construct of capitalist cosmologies and commodity food systems. I began my study of food waste in San Jose, a Mopan Maya community in Southern Belize, with the goal of understanding waste as a reflection of cultural values through observations of food waste practices and paradigms in a contemporary Indigenous community. I was immediately struck by two things: (1) I was studying the absence of a phenomena and (2) my inquiry was limited by my preconceptions of what I understood to be "waste". I learned that the English word for "waste" itself poses limitations for understanding relevant distinctions in resource use categorization and practice in the Maya case (see Brown 2020). Treating "waste" as a universal concept with inter-cultural variability precludes a deeper understanding of waste as a cultural construct of capitalism, built in to capitalist livelihoods and lifestyles.

The idea of "waste", as I set out to study it, presupposes human-environment relationships characterized by culture/nature, subject/object dualisms that are specific to the capitalist worldview. In capitalist cosmology, "waste" is used as a signifier for inanimate excess, refuse, rubbish, or goods otherwise not used to their full potential. Food waste is an externality to be calculated into the costs of production, distribution, and consumption of food. While exploitation and unusable excess are endemic to objectified views of the environment and material world, they are not cultural universals.

In traditional Maya cosmology, food is part of a relational system that understands food as an agentive subject in social relations with humans and the spiritual-natural world across the production—consumption cycle. Mopan people daily enact an understanding of themselves as dependent upon and subordinate to the supernatural through the careful, respectful treatment of food. The Mopan concept of *tzik* describes actions that display deferential respect towards superordinate persons and entities through greeting, prayer, ritual, and offering. Food "waste" is the antithesis of *tzik* and spiritual retribution is an expected consequence.

While food use is highly conservative and what I would identify as "food waste" is minimal, the analogous Mopan term for food excess or refuse (*k'as*) holds very different cultural

DOI: 10.4324/9781003019077-8

connotations from the English word "waste". The Mopan word for trash (*k'as*) does not apply to food excesses except circumstances wherein it also denotes a spiritual, moral transgression. While *pul k'as* is the term for the village dump wherein all manner of non-putrescible waste would be collected (tires, plastics, metals etc.), compostable food debris or excesses are not regarded as trash or waste. Instead, they are fertilizer for future crops and part of a relationship with the landscape. If proper respect is not shown, for example food is heedlessly discarded, allowed to rot, or left unused, it is considered disrespectful (*ma tzik*) rather than a waste.

Food waste is part of commodified food systems in which food is an object in cost-benefit analyses of monetary worth. Whereas farm foods that have been allowed to spoil are *k'as* for the disrespect that has been shown to the cosmological cycle of growth, improperly used commodity foods are considered "a waste". Many staples of the modern Maya household are bought commodities with an unclear relationship to the traditional principles of respect governing food disposal. These global commodity foods are disarticulated from the local landscape and traditional practices of disposal. Commodity foods represent new forms of materiality and essence that are governed by an alternate, capitalist environmental cosmology. When people rely on purchased rather than grown food, as is the case in the global food system, food is a fungible household resource to be used efficiently, or conversely, wasted.

As a concept, waste reflects a specific, albeit highly globalized, set of social arrangements that supplant local cosmologies to masquerade as objective material facts. "Waste" is an ideational and material construct that turns people into consumers, nature into objects, and surplus into trash. Food waste provides a window into the processes of environmental alienation, demystification, and commodification that have accompanied greater participation in contemporary forms of globalization. While foods in traditional Maya farming are spiritually vested, capable of retribution, and deserving of respect, the divestment of meaning from the formerly sacred enables alteration of material practices of food disposal. As practices of *tzik* are demystified and melded with capitalistic livelihood and valuation, the cultural underpinnings of historical food disposal practices are changing. Maya people are reconstituting waste through new processes of moral categorization, constructed through engagement with global capitalist markets and ideologies.

My research focuses on food disposal as part of differing environmental cosmologies that beget alternative paradigms and practices of resource use. Engaging Maya perspectives on respect-based food and waste practices offers new possibilities for theoretical exploration and action. The Mopan respect-based paradigm routinizes practices that minimize food excesses and maintain an integrated human-environmental approach reminiscent of the "cradle-to-cradle" aims of some eco-modernist policy proposals (Mol, Sonnenfeld and Spaargaren 2020). The lived experiences of people in San Jose show that alternative modalities of waste management are not only part of an eco-modernist future but long pre-exist it as sovereign Indigenous ways of life in cultural models that emphasize integrated sustainability over limitless growth, and respect over maximization. Through a focus on food waste practices as expressions of cosmology and value, this chapter contributes to understanding the cultural specificity of "waste", and the practice-materiality-value complexes by which it is created.

Waste studies

This research contributes to waste studies scholarship on the social construction of waste by exploring how changes in livelihood, material networks, and broader cosmology alter waste practices. In San Jose, as in most societies, disposal is a central aspect of everyday human and material existence (Gregson, Metcalfe and Crewe 2007a; Hetherington 2004; O'Brien 2008). This is especially true in the case of food use (Evans 2012), which is a routinized enactment of human-environmental relationships and cultural values. This research explores how animistic cosmologies that view nature and the natural as actants work to minimize waste.

Maya traditional environmental knowledges (TEK) integrate humans into the supernatural world that sustains them, with food as an important locus of interaction (Berkes 1999). Forming relationships of respect with the landscape through food is shared by many Maya peoples across language groups and national borders (Anderson, Tzuc and Chale 2005; Danziger 2001; Faust 1988; Wilk 1997), as well as other Indigenous animistic worldviews in Central and South America (see Costa and Fausto 2010; Viveiros de Castro 1996; Descola 2013; Halbmayer 2012; Hornborg 2006). Animism is a relational ontology that decenters nature/culture distinctions (Descola 2013, Halbmayer 2012, Viveiros de Castro 1996). In animistic perspectives, humans communicate and establish understandings with other-than-human persons and the material world is a manifestation of the supernatural. The natural world has agency and is divergent from one context to the next (multi-natural). Mononaturalism, the assumption of a single, unifying and universal nature, is an opposing worldview predominant in capitalist cultures (Haraway 2013; Latour 2004).

Actor-network-theory (ANT) (Callon 1999; Latour 2005; Law 1992) and new materialisms (Bennett 2004; Coole and Frost 2010) broach agentive materialities—the ways in which the non-human world acts upon the social—from secular, sociological perspectives. These theories work to subvert predominant capitalist interpretations of the material world as inert matter to be acted upon by humans, but have yet to be incorporated into everyday routines, economic, or political configurations.

Ethnographies of waste practices provide evidence that acts of disposal are morally fraught even in industrial and consumer societies wherein waste is endemic to cycles of production and consumption (e.g., Evans 2014; Hetherington 2004). The stress and guilt felt by many consumers at disposing of food waste suggests cross-cultural relevance for respect-based paradigms of food use. However, this anxiety over waste is the secular sentiment of a consumer culture, rather than a spiritual concern that centers the agency of the natural. The respectful management of food excesses is an enactment of Maya identity, virtue, and a deservingness to continue to receive (see Gregson, Metcalfe and Crewe 2007b for ridding and identity).

While the relationships of status and meaning associated with waste are subject to political and economic processes (O'Brien 2008), waste is not simply a matter of production, consumption, inputs, and outputs. Waste regimes emerge from interdependent processes of waste and value circulation (Gille 2012). Processes of valuation are active, dynamic, and negotiated expressions of worth (Thompson 1979). Disposal is a source of social integration, identity making, and membership that constitutes social and ethical activity (Hetherington 2004). "Waste" is an expression of value and morality that is intimately situated in everyday and institutional contexts (O'Brien 2008). Consequently, as institutional constraints and value systems change, so do practices of wasting.

In Maya society, the rhythms of food provision and everyday domestic life work to configure food as non-waste, but practices of disposal change with social and material reorganization (Evans 2012). Peoples cross-culturally who engage directly with their localized food systems have respect for food, but these systems of food management are difficult to maintain when engaged in capitalist food and labor regimes (Friedmannn 2005; McMichael 2006). The physical waste that results from incorporation into global commodity chains (Landecker 2019) is part of larger transformations in environmental cosmologies and systems of value.

This case study responds to a need for waste research outside of the Global North that gets beyond conventional understandings of waste as "object" to instead conceptualize "living waste" (Bell 2019). Scholarship on food waste has primarily focused on the Global North (Cappellini and Parsons 2012; Evans, Campbell and Murcott 2013; Evans 2012, 2014; Fraser and Parizeau 2018). However, research on food waste in the Global South has shown the cultural specificity of food provisioning practices and categorizations of waste (Soma 2017), resonating with the findings of waste studies more generally that trash is created through socially defined processes of sorting (Strasser 1999).

Understanding the cultural logic of alternative food use paradigms is of great importance in a time when a third of the world's food is wasted across the supply chain with an increasing share at the consumer level (Alexander, Gregson and Gille 2013; FAO 2011). Food losses are a social and environmental concern, with consequences for food security, conservation of water, land, and soil, and control of methane pollution (Kummu et al. 2012; Stuart 2009). Addressing the global problem of waste requires alternatives to systems that abstract the value of the material world to finance, governed by the logic of externalities. Maya respect-based food disposal offers an alternative paradigm for resource management and a new perspective on waste as a social construct. Studying the interface of these paradigms with new materials, lifestyle constraints, and processes of valuation, provides an understanding of how global food systems change practices of wasting.

Methods

This chapter draws on three terms of ethnographic research in the southernmost Toledo district of Belize: 9/2010–1/2011, 1/2015–5/2015, and 12/2015–5/2016, as well as recurring (c ten-day) visits in the intermittent years, primarily to San Jose Village.[1] It is a Mopan Maya community (pop c 1200) and is one of 21 villages in the Maya Mountains Reservation (MMR) system, which spans the eastern quarter of the Toledo district along the Guatemala border.

Mopan is the primary language spoken in San Jose and the Maya language used throughout this text. I undertook language training in Mopan from local tutors and used it conversationally while living as part of the community. While most of my research was conducted with Mopan speakers, eight ethnographic interviews I conducted were with Q'eqchi' speaking Maya people, and all but three of the government and development representatives I interviewed were non-Maya.

Participant observation—Residing in the village allowed for the ethnographic practice of "deep hanging out" and experience of "being there", which provided me with first-hand experiences and observations of the practices and material routines of daily village life (Geertz 1988). I conducted participant observation research with the 13 largest family groups in San Jose (approximately 48% of the community) with attention to gathering multiple perspectives on food use practices and values as they intersect cultural change. I documented the spatial categorizations of food disposal and discussed participants' views on what constituted waste in varying social and spiritual contexts. I participated in all aspects of peasant farming and food consumption including planting, field maintenance, harvest, hand-processing grain, food preparation, and disposal. I joined another household in the village for at least one meal per day. I most frequently enjoyed meals with my host family and two other families with whom I had weekly meals, but also joined meals at 27 other households in the village from different family groups. During these meals, I participated in different stages of food preparation including, cleaning chicken/meat, grinding grains, harvesting, and cleaning ground foods, gathering seasonings, baking tortillas, and disposing of food excess. Participating in daily food preparation gave me a sense of routine consumption and disposal patterns as both a guest and regular member of the household. To better understand formalized and high-volume food use practices, I attended village and district cultural activities, school fundraisers, holiday festivals, and family parties (birthday, baptism, wedding). I attended village public meetings, each with a segment about maintaining community health through careful waste management.

Interviews—I conducted a series of semi-structured interviews with Maya community members and professionals with positions relevant to food waste management. Respondents included 53 residents of San Jose Village, the Government of Belize (GOB) Minister of Agriculture, District Health Counsel Administrator, San Jose Community Health Worker, representatives from two International Non-Government Organizations (INGO) operating in Toledo District (Sustainable Harvest and Plenty Belize), a member of the Maya Leaders Alliance who conducted Master of Arts research on waste in Belize. Village interview respondents ranged in age from 18 to 72, with 28/53 female and 25/53 male participants. I gathered emic interpretations of the relationship between food practices and changing social and material relationships in Maya communities. In each interview I asked respondents for four types of information: (1) categorization and Maya vocabulary for different forms of waste, with emphasis on food; (2) respect-based paradigms of food use and disposal; (3) major changes observed in food use practices; and (4) the reasons for changing cultural practices and values. Interviews were conducted primarily in English, with intermittent Mopan as my language skills progressed. English is the national language of Belize, and the mutually intelligible mode of communication for the country's nine language groups (including Mopan and Q'eqchi' Maya communication with one another). Maya women above the age of 40 and men above the age of 60, however, would often bring in a younger family member with whom they were comfortable to provide translation. Respondents' names were used by request, otherwise names are excluded. Interviews were selectively transcribed, coded, and analyzed using AtlasTi software.

Ethnographic background

Maya peoples have been in the lowland area of the Yucatan Peninsula, present-day Belize, Guatemala, Western Honduras, and Southeastern Mexico, since around 2500 BCE. There are approximately 5 million Maya speakers, from 25 language groups in these areas.[1] There are about 25,000 Mopan, Q'eqchi', and Yucatec Maya people in Belize. Mopan people are Indigenous to Belize, with about 5,000 Mopan speakers currently residing in the southernmost, Toledo District (Statistical Institute of Belize 2013).

Belize was formally British Honduras and achieved independence from the United Kingdom in 1981. Because of its mercantile beginnings and later de-colonization, Belize has always been a neoliberal state, with an economy defined by transnational interests (Sutherland 1998). The economic, political, material, and social realities of Belize are highly globalized. A diverse array of INGOs filled the post-independence gap in governmentally provided social services, giving Belize one of the highest per capita allotments of developmental aid in the world (Merill 1993). Toledo Maya communities like San Jose have been a favored target of both development and religious intervention programs designed to incorporate Maya peoples into global material and ideological networks (Wainwright 2008). The result has been notable cultural changes over the last 40 years.

Formal independence from Britain offered a similar inroad to Protestant missionaries who began to proselytize in Maya communities beginning in the late 1970s, gaining community acceptance by providing goods and services (Steinberg 2002). Since establishment of the first Nazarene church in the village in 1978, San Jose residents have gone from 1 syncretic

Maya-Catholic faith to 7–8 Protestant denominations and many non-practicing residents. Evangelism in San Jose advocates market-oriented agriculture and individualism while obstructing collective rituals and positioning the neo-traditional Maya animist cosmology in which food is a sacred vessel as antithetical to economic development (Cook 1997).

San Jose village was founded in 1955 as an offshoot of neighboring San Antonio (founded in 1890s) and is one of the more remote Mopan villages in Belize. Located at the end of a 20-mile gravel spur off the Southern Highway (completed to Punta Gorda (PG) in 1990, paved in 2010), San Jose is a two-hour bus ride from the district capital, PG, where many residents commute to attend school, work, sell farm goods, and purchase commodities. The village was not connected to PG by gravel road until 1980, which in turn was not connected to the rest of the country until 1990, so many residents have recollections of carrying goods into town or across the Guatemala border by foot. While 67% of households have at least one member working outside of the village for part of the week, most households (89%) reported "farming" as a primary occupation on the 2015 census. Peasant farming provides staple crops (corn, beans, rice), fruits and vegetables for subsistence, and surplus to sell in PG markets. Since the 2010 completion of the International Highway that runs through neighboring Santa Cruz, San Jose villagers can now walk a two-mile mountainous path to reach the paved road and more regular bus service.

Many changes to cultural practice and materiality have come with better road connectivity to PG and the rest of Belize. People can more readily commute for labor, goods, and services, which has translated to dramatic transformations in food production and consumption, along with a host of other cultural practices. With more time spent away from peasant farming, more money has been spent on commodity foods from outside the village food system. These new materialities and practices are causing residents to readjust disposal practices and the underlying value systems.

Ethnographic findings

Food disposal as relational respect

Waste, rubbish, debris, and trash of all kinds are kept to a minimum in San Jose, as much out of necessity as ethical obligation. Municipal waste disposal services are only available twice a year when a dumpster truck lumbers down the gravel road from Punta Gorda to collect whatever debris villagers could not otherwise dispose.[2] When I toured the village loop to examine these trashed items, they were almost exclusively medium to large metal objects and some durable plastics deemed of no use value to people in the community (e.g., busted washing containers, old wiring, dilapidated plastic chairs). In between dump truck visits, smaller pieces of trash (e.g., old cans, bottles, busted buckets) are dropped at an informal dump site (approx. 200′ × 20′ × 4′) between San Jose and the next village either by bus or personal vehicle. Papers, thin plastics, Styrofoam, and cardboard are kept in household trash piles to be burned weekly. Each of these items disposed of by fire, dumpsite, or truck, are objects that, once used, are considered "trash". Food refuse, on the other hand, is never considered trash, because it is part of a sacred and practical relationship to the landscape.

Neo-traditional Maya paradigms understand all humans as embedded in an agentive landscape that is responsive to human action (Baines 2014; Danziger 2001; MLA Maya Leaders Alliance 2010; Wilk 1991). *Tzik* is a paramount virtue that communicates an attitude of religious deference to the gods, (or God) through relationships of reciprocal obligation (Anderson, Tzuc and Chale 2005; Danziger 2001). Disrespect *(ma tzik)* results in spiritual

Plentiful food, clean water,
fertile soil, good weather,
healthy people

Food scarcity, dirty water, soil
depletion, bad weather,
sickness

SACRED
non-commodity-
subject

PROFANE
commodity-object

PEOPLE   RESPECT
(*TZIK*)   NATURE   DISRESPECT
(*MA TZIK*)   PEOPLE

CLEAN/PRETTY/GOOD
(*KI'ICHPAN*)

POLLUTION/WASTE/
SIN
(*K'AS*)

Proper disposal, ritual
offering, prayer, clean
landscape, collective,
relational

Improper disposal, control,
exploit, pollute, waste
individual, external

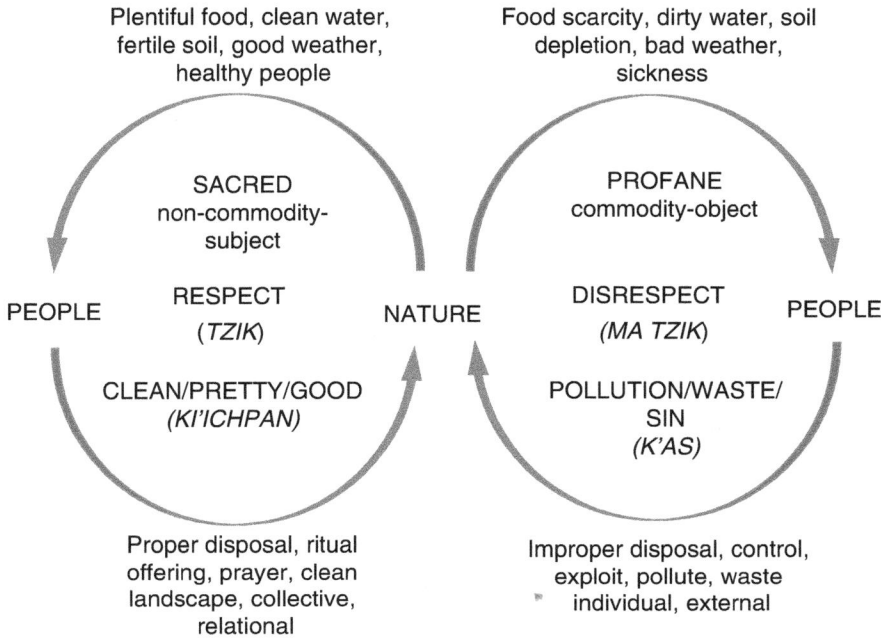

Figure 8.1   Author's diagram depicting respect-based relationships with nature.

pollution *(k'as),* and causes all manner of supernatural consequences, physical ailments and mishaps including bad harvest, headache, cancers, possession by evil spirits, death, fire, and storms (Anderson, Tzuc and Chale 2005; Danziger 2001; Faust 1988). This paradigm reflects the sometimes-harsh realities of food and medical insecurities that characterized the not-too-distant past.

The Mopan word for trash or rubbish (*k'as*) carries strong negative moral connotations that imply spiritual pollution as well as referring to physical materials. In the Maya peasant agricultural practices used by most San Jose residents, food is grown for home or village use following modern renditions of traditional cosmology. As an integral part of human-nature relationships that is essential for survival, food is grown, harvested, and consumed with respect (*tzik*). (Figure 8.1)

The related concepts of *k'as* and waste are applied to items within different consumptive realms. *K'as* was used primarily as a spiritual, moral descriptor prior to the introduction of persistent waste materials to Maya society. The English word "waste" is most often applied to goods purchased from outside of the home or village food system. Until recently, this has explicitly excluded food excess, surplus, refuse, etc. Whereas insensitive disposal of corn is a failure to exercise proper *tzik* and can be considered *k'as*, money, sugar, or pump water used in excess is considered a *waste*. Similarly, "waste" is usually only applied as a verb to commodity objects, not food. A person can waste a resource in which money has been invested, but improper use of food is instead considered disrespect (*ma tzik*) and a source of spiritual contamination (*k'as*). While "to waste" carries a negative moral connotation somewhat analogous to *k'as*, "wasting" is a sin of the secular, commodified material world, rather than the spiritual.

Food is an especially important focus of respect because it is embedded within natural-spiritual relationships that have material consequences. Respondents universally expressed that respecting food is important because "That food is your life. That corn and tortilla is every-thing". Respect is daily enacted to one's garden, field, plants, yard, home, and creek. Corn, cacao, local coffee, orange trees, and other home-produced goods must be shown *tzik* through careful treatment over the life span of the food. Thomas, one of the key participants in my research, explained to me:

> Well, the only way to respect left over food is by giving it to the people, animals, or creek. Maybe if you have a neighbor you could give it to them, and they could use it or maybe they have animals you don't have so you give it to them. Because you show respect. Otherwise, you give to the creek. The creek is the only thing you usually give food to because of the fish and when you go to the creek you still show respect. But if you just throw it out that means you are not respecting that food.

A great deal of energy is invested in the preservation and respectful use of food. Leftover food is uncommon from individual plates, guarded against by the expectation that you are to eat what you are served. *Caldo* (traditional soup), meat, and beans are carefully reheated so as not to spoil and neighbors are gifted with excesses of the feast. Allowing food to spoil is itself a form of disrespect, but the negative implications of this negligence can be mitigated through accepted means of disposal. Rolanda explained to me that "you can't throw food away; you have to respect it…but never throw it away in the garbage". Spoiled *masa* (ground corn and lime used to bake tortillas) is served to small chickens on banana leaves, meat that has gone bad is given to hungry dogs, and vegetables that have turned rotten can be used to nourish other plants in the garden. Traditionally sacred crops such as cacao and corn can only be disposed of properly at the creek (also see Steinberg 2002; TMCC Toledo Maya Cultural Council 1997; Tedlock 1996).

Ritual observations of respect are particularly formalized during planting and harvest. During these events, the cultural importance of food as an agentive component of the human life cycle is explicitly recognized and enacted through prayer, offering, and careful treatment of food.

The first time I participated in a corn planting was at Pablo and Elvira's when they invited me to "learn the traditional way of the Mayas". Corn planting begins with prayer and offering of food to both the Gods and animals, asking for permission to plant and protection for the fields. Rolanda's parents place the corn on the table and kneel, praying to God for the safety of their field. Thomas' grandfather is one of the few *uchben pol*, or "old heads", in the village who continues to practice ritual respects when planting. He has a saint to whom he prays for a fertile field. When asking for protection of his crops, his grandfather prepares a meal which is cooked separately and prayed over to honor the saint and the spirit. "They go there and feed the saint and pray…they feed the saints and spirits so they offer that for the saints. They say 'this is your food and tortilla…we are thanking you for what you are doing". Food is a gift to be given and received in material and symbolic exchange.

As we shelled the corn for planting and the day's tortillas, my unskilled hands scattered corn to either side of my basket. Elvira's daughter carefully picked up each of the corn pieces I had mislaid, sorting out the larger kernels for planting and the smaller for the day's tortillas. It is especially important to treat all kernels of corn that originate from the same ear as those you are planting with respect "because its brothers will not grow" if kernels are heedlessly tossed on the ground. Kernels not selected for planting are consumed by the family on the day of planting to complete the cycle of growth. Evarista described to me how powerful life force (*k'inam*) would wither plants in the garden or fields when not shown proper respect, leading to the general wisdom that crops would not grow if food was disrespected. Rolanda's father shells all of the corn for planting himself in a ritually secluded space and disposes of any unused materials associated with the seed corn in the creek.

Failure to show respect results in poor crops, familial misfortune, and sometimes even haunting from the voices of the food itself. Groves must be tended, pods gently pulled, and corn properly consumed with respect and care, or the perpetrator risks spiritual retribution. Several respondents described events following a careless treatment of food (three of corn, one with cacao) wherein they could hear the spirit of the disrespected food cry. Marcelo explained, "It's not good for the things because if I just throw my leftover there, maybe something [food/spirits] will talk to me so that means it's not good for earth". In the Old Days, Pablo reflected, dropping corn and beans on the ground during shelling would cause restlessness and noise amongst the seeds. Though the seeds would not directly cause harm, they might refuse to grow and haunt the transgressor with their voices. Food dropped on the ground during planting may become animate and attack the farmer's crops. Andrea recounted an experience from childhood wherein her father upset the spirits of the land with his failure to provide them the annual offering of food. For weeks afterwards the house shook, bats attacked the children at night, farm animals became sick, the garden withered, and the spirits whispered their discontent.

Cultural change and the creation of waste

In response to my inquiries about food waste, a couple of research participants told me the story of *The Hill God and the Hunter,* as an explanation of waste as cultural change. Variations of this story have been circulating in Maya communities of Southern Belize since the early 1990s. It reflects themes of respect for nature recorded in the sacred text of the *Popol Vuh*, Maya ethnographies (e.g., Baines 2014; Fink 1987; Thompson 1930; Vogt 1993; Wilk 1997), missionary texts, and contemporary articulations of Maya culture (TMCC Toledo Maya Cultural Council 1997; MLA Maya Leaders Alliance 2010). The village origins of the protagonist hunter vary between narrators, as do the details of his abduction and communicative vessels he encounters in the cave (some say corn smoke, some say a rabbit, others say it was the hunter's neglected quarry or the bird that carried him to the cave), but the overarching message remains the same: Maya culture is changing. People are forgetting the central principle of respect for the earth (*tzik*), which will come with material and spiritual consequence.

The Hill God and the Hunter—Told to me by Thomas Coh and Lucia Bol

There once was a hunter from down South in the Q'eqchi' villages. One day he had just gone out into the forest when a bird flew into his face. When he looked at the bird he suddenly fell faint and the bird carried him off. When he awoke, he was in a cave with an old man and many animals, all hurt and maimed, some shot in the paw, some missing a foot. The old man said that before, people used to give respect to the animals and make their prayers and offerings to the animals, asking permission from the master of the animal before they hunt, but now they just shoot for sport. In front of the old man were bowls of beans and dry corn. When the visitor asked what they were for, the old man said that people these days don't show respect for their crops anymore and just throw them aside like nothing. The food in the cave represented all of the spoiled and wasted corn, the dirty water, and the abandoned animals. The old man told the hunter that because people fail to respect the old hunting traditions and the food that comes from the land, the food and animals would no longer grow as they had in the past. The old man sent the hunter back to this world to tell the people the things that he had learned.

The post-independence (1981) resurgence of *The Hill God and the Hunter* is a response to dramatic changes in cultural practice, materiality, and cosmology. The philosophy of food use and disposal in *The Hill God and the Hunter* has been eroded by economic and spiritual change over the lifetimes of research participants. Food waste, be it spoiled corn or injured game, is the antithesis of respect (*tzik*), a paramount virtue in traditional Maya cosmology (*Kustumbre*). With his foreboding message of impending food shortage, the Hill God (voiced through the old man) urges his people back to the "Old Ways"[3] of ritual and respect that distinguish them from the world's more profligate occupants.

When the village predominantly relied on a subsistence agricultural economy, *tzik* was the foundational value governing disposal of food. However, with an influx of new foods, time constraints, and cosmologies, many people are concerned that respectful food use practices are fading, despite the centrality of *tzik* to Mopan identity. Reflecting on the cultural changes over his lifetime, Francisco stated:

> It is our way. We have to respect our food because that is what we believe. Right now, our Mayan people are kind of confused. We don't know who we are anymore. A lot of people don't show respect because they throw food away or waste it. It is our belief that we have to respect everything, the food, the rain, the land, everything. We are lost.

The anxiety over loss of cultural identity expressed in the above quote is an underlying message of *The Hill God and the Hunter* and motivation for the circulation of this story in recent decades. For some, this story is a warning message from the spirits of Maya cosmology against the disrespectful treatment of food in the modern era. As Ernesta explained, people are becoming increasingly negligent of respectful food practices which is "why now the spirits have come back. God gave us the corn and if we don't respect it and just throw it away, God will be vexed and feel sad".[4]

While some advocate for a return to the Old Ways to protect the future of Maya food systems, others, especially devout Protestants, see the Old beliefs as dangerous and best left in the past. The agenda of progress requires freedom from agentive nature. This freedom allows for exploitation of natural resources, including food, which are re-framed as objects rather than vessels of the sacred. Alberto, a Nazarene minister, explained regarding departure from the old food ways, "We used to worship the things of this earth, but now we worship the one who made it directly". Several Baptists, a Jehovah Witness, and a Pentecostal minister affirmed this explanation with varying levels of condemnation for what they perceived to be idol worship. For many in San Jose, Jesus offers a more predictable, less vengeful form of spiritual appeal, free of the dangerous complexities of ritual knowledge transmission associated with the Old Ways.

Some respondents considered commodity foods to be more like objects. While Andrea notes in relation to commodity foods that, "god makes all of it", she finds complication with the application of respect to these goods. In purchasing food there is no way to ensure that the proper ritual processing of the good was performed in either planting or harvesting, so it is futile to attempt to preserve those virtues. In the case of commodity foods, respect is therefore granted to the work rather than the food itself because while labor is indicative of community and familial contribution, commodities are removed from the relational system of sustenance.

Rolanda and Andrea perceive the food one buys, though not part of the sacred relationship to the land, as still deserving of respect owed to the hard work necessary to make the money for its purchase. Laughing, Andrea told me, "Money is hard to make and fast to go!" She described her frustration with people who waste sugar and other bought commodities. "Treating food like nothing" is a failure to honor what has been given or earned and carries a similarly negative weight to disrespect.

Other respondents saw village-grown and commodity foods to occupy very different spheres of value, with *"k'as"* applying to the former, and "waste" being more accurate for the latter. Elvira explained to me that "waste" could be applied to describe misuse of bought food, but not to home grown corn. If food is part of a relational food system, it is never considered a "waste", as it is valued even when discarded. Surplus fruit and seeds are left on trees or vine for birds and other animals to consume as a routine offering for the other inhabitants of the earth, while food excesses or leftovers are used to "feed" the soil for other plants. These small-scale surpluses are not externalities to the food system (waste). Rather than detracting from capitalist values of efficiency and cost-effectiveness, they contribute to the overall sustainability of the system. According to this logic, one may waste imported, purchased flour, however, improper disposal of village-produced corn meal is disrespectful (*ma tizk*) and a source of spiritual contamination (*k'as*).

The seed selection process for each of the dozen planting days I attended reinforced the distinction between commodity and non-commodity foods. There is no guarantee that food will be shown respect once it is sold, so farmers have adapted selection practices to separate seeds for market from those for planting and home consumption. Seeds to be planted are shown ritualized respect through prayer, offering, and physical separation from their profane counterparts prior to planting. If the purchaser does not show respect, the actions of the consumer could have negative bearing on the future of the farmer's crop. Lucio, an avid vegetable farmer, explained, "So if I sell you beans, I don't know what you are going to do with it, maybe you will throw it away…if I don't separate the seeds for planting from the ones I sell then maybe I'll have bad luck the next time I plant". The substance of food crops purchased from known farmers similarly contains and therefore must be shown respect. But foods from unknown origin have an unknown essence that may not warrant such treatment.

Lucio's concern over the future of his crops once sold reflects larger anxieties about the changing cultural practices surrounding food use and disposal. While new forms of work, education and spirituality offer excitement and opportunity, they alter the relational food system in ways over which community members often feel they have little control. Though people have less time to farm, they have more capital to expend on commodity equivalents to traditional foods. Nescafe replaces local coffee, flour tortillas subsidize the stock of corn, sugar is brought down from the Northern Districts rather than ground in the neighbors' sugar mill. As commodity foods become more prevalent in the village food system, research participants express ambivalence about when and how the concepts of *tzik, k'as,* and waste should be applied to commodity foods.

Some of the most illustrative observations of how the tensions over changing values are reflexively articulated in modern culture were at the three Maya Day cultural celebrations I was able to attend. At each of these events, material culture was widely celebrated, and themes of Maya pride were embraced. However, the spiritual underpinnings of Old cosmologies were entirely excluded. Along with traditional music, clothing, dance, and food, Maya Day festivities feature a set of "culture competitions" for Maya youth to demonstrate their skills at shelling corn, making tortillas, chopping wood, and performing traditional dances. These competitions are of particular interest to me because they demonstrate a major shift in the handling of food as part of cultural expression.

As I watched children race to be the first to shell five ears of corn, kernels flew in all directions, scattered by the frenzy of hurried hands. Not only was corn haphazardly strewn across the ground, but it was an object in a culture competition that valued time efficiency over ritual sanctity. I thought back to my experiences shelling corn and participants' accounts of the deliberate, respectful procedures for both ritual and routine consumption. Visions of Elvira's daughter, meticulously picking up my carelessly dropped kernels accompanied me the following morning when I revisited the festival site to see corn debris from the day before still strewn across the ground, one of my few observations of food waste in San Jose.

Discussion and conclusion

San Jose perspectives on food and waste highlight the cultural particularism of commodification and the ways in which it is advanced through transformations in livelihood, spirituality, and material networks. From the Mopan respect-based cultural perspective, treating food as a fungible commodity and food excess as waste is an oddity of capitalist culture. Food essential for our survival and produced through the good will of nature, so it is a cultural object worthy of reverence. Disposal is an explicitly relational phenomenon, bound up with essential questions of Mopan identity and paramount value of *tzik*.

When food is part of a relational system, excess is not the same as "waste". Food excesses are of benefit to other, non-human elements of the natural cycle of which food is an integral component, rather than an externality to an otherwise efficient system. Food that is not consumed at the dinner table will feed hungry animals, or the soil for future crops. As it is used for convivial practices amongst companions, food is used in ritual practice to extend this relational value to the spirits of the natural world. Food is an important vessel for communication of *tzik*, as part of the intimate life cycle relationship between Maya people and the spiritually vested landscape that sustains them. From earth to body to earth again, the routine symbolic usage and disposal of food serves as a daily enactment of *tzik* in animistic cosmology. Food must be used in a very specific way or it is considered disrespectful and a form of spiritual pollution that may bring sickness upon the land or the people themselves.

Before the commodification of food, goods were not quantified and subject to reductionist evaluations of worth. Rather than efficiency and "waste", resources were handled with respect. The physical volume of "waste" was minimal, and the concept itself did not apply to food, which was never seen as garbage. The idea of "waste" follows demystification and an understanding of food as a commodity-object, rather than a subject in a complex system of relationships that must be maintained.

The resurgence of the *Hill God and the Hunter* in oral tradition expresses anxieties about the loss of traditional respect in food use and disposal practices. This Maya perspective on "appropriate disposal" is entwined in larger cultural narratives surrounding relationships to the landscape and things of the earth. The differences between locally produced and commodity foods are not limited to material substance but extend to the essence of the food and the spiritual and labor relationships in which it is suspended. The material introduction of ubiquitous food commodities necessitates the re-framing of Mopan relationships to the supernatural world. While commodity capitalism supplants the relational (closed system) paradigm with a disconnected idea of waste as an externality to the system, Protestant theology removes the spiritual/moral element of *"k'as"* as *"ma tzik"*.

"Waste" exists outside of extant cultural-ecological systems. "Waste" is a creation of capitalist modes of valuation that accompany incorporation into global commodity chains. It is excess or refuse of a system designed around monetary efficiencies and profit margins rather than long term sustainability and respect of the entire life system. "Waste" represents the secularization and commodification of culture in globalized capitalist modernity. Whereas the ideas of *tzik* and *k'as* are concerned with relationships to a spiritually vested landscape and the food it provides, "waste" is an external problem that comes in from outside the village.

The tensions between commodity and agentive nature, waste and *k'as* are indicative of larger tensions of cultural hybridization. Capitalist waste regimes are an extension of global food regimes and reflect a capitalist cosmology—a worldview built on environmental maximization, demystification, and alienation—in which food, water, and land are commodities to be bought, sold, conserved, and wasted. This is a culturally specific set of values, beliefs, practices, and materialities that has been globalized through colonialism. Backed by colonial political economies and territorialization, these powerful regimes work to marginalize and transform local practices and values.

Engagement in outside employ and missionization is promulgating the literal dis-integration of existent logical cosmologies in San Jose. As villagers increasingly engage in external markets, education, and theology, changes in values parallel the rise in a cash economy and creation of food as a commodity object. The fundamentally different materiality and essence of commodity foods beget new use and disposal practices. Respect-based food use paradigms are difficult to maintain when faced with the time constraints of wage labor, the secular valuation of economy (rather than spirituality), and inundated with foods that are disconnected from observable relationships with local landscapes.

Food purchased from the village store, sourced from beyond its boundaries, is not a vessel for communication with God. Packaged food does not carry the essence of the farmer's labor, self, and life as demonstrated by the care, skill, and respect needed to produce and process the food. It instead embodies a new system of value, symbolized by the monetary investment necessary to procure the good. While some people incorporate commodity goods into definitions of *tzik*, others replace respect with the more commodity-congruent value of efficiency. Though financial value can be understood as an object of respect, those more practiced in its exchange tend to see money as a component of a less sacred and disconnected value system. Within this system, *tzik* does not play as integral a communicative role, nor does the food that embodies this virtue.

Once subject to the cost-benefit analysis of capitalist consumption, efficiency becomes the dominant paradigm through which food waste practices are governed. This shift is more complex than the supplanting of one value "*tzik*" for another "efficiency" and relies in part on accompanying theological change. Though syncretic with valuation inherent in the application of *tzik*, efficiency does not accommodate, nor does it depend upon, connectedness between the earth, the people, and their food. Rather than a culturally specific, always subjectivized and relational understanding, the concept of efficiency is instead is based on the idea of food as abstract, fungible, and non-specific. Food waste can then become a present absence enabled by abundance and convenience that, excluded from the daily enactment of the sacred, can be forgotten by the whim of the well-meaning but absentminded consumer (Hetherington 2004).

Rather than a universal category that can be locally constructed all over the world (cultural variable view), it is helpful to consider waste as a contextual social process. Practices of disposal are a routine enactment of relationships with and around the object-world. These relationships can be central to identity making and re-making in the face of cultural changes brought about by colonial social and economic development. The juxtaposition of *"k'as"* as relational, spiritual/moral with "waste" as externality, secular/moral, demonstrates both the cultural particularism of waste as a construct of commodification and the limitations of the paradigm that relegates "waste" problems to resource management.

Maya food paradigms have lessons for the current problem of global food waste. Farming that recognizes the cycle of production-consumption-disposal as an elemental part of human existence can lead to more sustainable human-environmental relationships. Participation in localized food systems can foster cultures of respect for food and the land and resources through which it is produced. Respect-based models of resource use can minimize "waste" provided they are structurally supported with land, time, and cultural investment.

Notes

1 While the linguistic terminology used throughout this chapter are specifically Mopan, much of the ethnographic literature on human-environmental relationships refers to broader Maya cultural cosmologies not unique to the Mopan language group. Likewise, Mopan people often refer to themselves as "Mayas" and their practices as "the Maya way" reflecting membership in a larger pan-Maya identity cultivated through cultural exchange between Maya groups. The articulation of Maya and Indigenous identity is an ongoing project of the Maya Leaders Alliance, an organization that provides Indigenous educational materials and advocates for Maya rights in the region. Contemporary Maya identity is built from both Indigenous cultural continuities and interaction with outside cultural forces (Medina 2003; Fischer 1999). It seeks to maintain cultural specificity of sub-groups while invoking essential characteristics of what it means to be Maya (Hervik and Kahn 2006).
2 The infrequency of these visits has become more of a problem with the influx of commodities into the village. Waste disposal more generally is an ongoing topic of concern at monthly village council meetings with the community. Whereas old-style food wrappers were bio-degradable (banana peels, corn husks, banana leaves), new foods come packaged in plastics for which there is no convenient means of disposal.
3 The "Old Ways" are a nostalgic reflection on ancestral Maya traditions. These traditions were rooted in melding of pre-Columbian and Catholic cosmologies that recast many Maya deities as Saints. These "Old Ways" are defined in contrast to newer, modern practices, demarcated by the post-independence rise of Protestantism and increased participation in external markets.
4 Community members still sometimes interchangeably use God and Gods, when referring to the agentive power of nature, though typically people will use the term "spirits" to refer to supernatural powers of wind, hills, water, and food. However, the modern, culturally sanctioned religious narrative is of one, Christian God.

References

Alexander, Catherine, Nicky Gregson, and Zsuzsa Gille. 2013. "Food waste." *The Handbook of Food Research* 1: 471–483.

Anderson, E. N., Felix Medina Tzuc, and Pastor Valdez Chale. 2005. *Political Ecology in a Yucatec Maya Community*. University of Arizona Press.

Baines, Kristina. 2014. *Embodying Ecological Heritage in a Maya Community: Health, Happiness, and Identity*. Lexington Books.

Bell, Lucy. 2019. "Place, People and Processes in Waste Theory: a Global South Critique." *Cultural Studies* 33, no. 1: 98–121.

Bennett J. 2004 "The Force of Things: Steps Towards an Ecology of Matter." *Political Theory* 32: 347–372.

Berkes, Fikret. 1999. *Role and Significance of 'Tradition' in Indigenous Knowledge: Focus on Traditional Ecological Knowledge*. Indigenous Knowledge and Development Monitor.

Brown, Steve. 2020. *The Sociological Review Monographs 2020* 68, no. 2: 385–400

Callon, Michel. 1999. "Actor-Network Theory—the Market Test." *The Sociological Review* 47, 1(suppl): 181–195.

Cappellini, B., and Parsons, E. 2012. "Practising Thrift at Dinnertime: Mealtime Leftovers, Sacrifice and Family Membership." *The Sociological Review* 60, no. 2S: 121–134.

Coole, Diana and Samantha Frost. 2010. "Introducing the New Materialisms. In: Diana Coole and Samantha Frost (Eds.) *New Materialisms: Ontology, Agency, and Politics*. Duke University Press. 1–43.

Cook, G. 1997. *Crosscurrents in Indigenous Spirituality: Interface of Maya, Catholic, and Protestant Worldview*. Brill

Costa, Luiz, and Carlos Fausto. 2010. "The Return of the Animists: Recent Studies of Amazonian Ontologies." *Religion and Society* 1, no. 1: 89–109.

Danziger, Eve. 2001. *Relatively Speaking*. Oxford University Press.

Descola, Philippe. 2013 *Beyond Nature and Culture*. University of Chicago Press,.

Evans D. 2012. Beyond the Throwaway Society: Ordinary Domestic Practice and a Sociological Approach to Household Food Waste. *Sociology* 46, no. 1: 41–56.

Evans, D. 2014. *Food Waste: Home Consumption, Material Culture and Everyday Life*. Bloomsbury Publishing.

Evans, David, Hugh Campbell, and Anne Murcott. 2013. *Waste Matters: New Perspectives of Food and Society*. Wiley-Blackwell.

FAO. 2011. *Global Food Losses and Food Waste—Extent, Causes and Prevention: Study Conducted for the International Congress SAVE FOOD! at Interpack2011 Düsseldorf, Germany*. FAO

Faust, B. 1988. *Cosmology and Changing Technologies of the Campeche Maya. Anthropology*. Syracuse University.

Fink, Ann. 1987. "Shadow and Substance: A Mopan Maya View of Human Existence." *The Canadian Journal of Native Studies* 7, no. 2: 399–414.

Fischer, E. (1999). "Cultural Logic and Maya Identity: Rethinking Constructivism and Essentialism." *Current Anthropology* 40, no. 4: 473–499.

Fraser, C. and K. Parizeau. 2018. "Waste Management as Foodwork: A Feminist Food Studies Approach to Household Food Waste." *Canadian Food Studies* 5, no. 1: 39–62.

Friedmann, Harriet. 2005. "From Colonialism to Green Capitalism: Social Movements and Emergence of Food Regimes." *Research in Rural Sociology and Development* 11: 227–264.

Geertz, C. 1973. Thick Description: Toward an Interpretive Theory of Culture. In C. Geertz (Ed.) *The Interpretation of Cultures*.Basic. 3–30.

Geertz, C. 1988. Being "Here: Whose Life Is It Anyway?" In C. Geertz (Ed.) *Works and Lives: The Anthropologist as Author*. Stanford University Press. 129–149.

Gille, Z. 2012. "From Risk to Waste: Global Food Waste Regimes." *The Sociological Review* 60, 27–46.

Gregson, Nicky, Alan Metcalfe, and Louise Crewe. 2007a. "Moving Things Along: The Conduits and Practices of Divestment in Consumption." *Transactions of the Institute of British Geographers* 32, no. 2: 187–200

Gregson N, Metcalfe A, Crewe L, 2007b, "Identity, Mobility, and the Throwaway Society." *Environment and Planning D: Society and Space* 25, 682–700.

Halbmayer, E. 2012. *Amerindian Mereology: Animism, Analogy, and the Multiverse. Indiana* 29, 103–125.

Haraway, Donna. 2013 *Simians, Cyborgs, and Women: The Reinvention of Nature*. Routledge.

Hervik, P. A. and Kahn, Hilary. 2006. "Scholarly Surrealism: The Persistence of Mayaness." *Critique of Anthropology* 26: 209–232.

Hetherington, Kevin. 2004. "Secondhandedness: Consumption, Disposal, and Absent Presence." *Environment and Planning D: Society and Space* 22, no. 1: 157–173

Hornborg, Alf. 2006. Animism, Fetishism, and Objectivism as Strategies for Knowing (or Not Knowing) the World. *Ethnos* 71, no. 1: 21–32.

Kummu, Matti, Hans De Moel, Miina Porkka, Stefan Siebert, Olli Varis, and Philip J. Ward. 2012. "Lost Food, Wasted Resources: Global Food Supply Chain Losses and Their Impacts on Freshwater, Cropland, and Fertiliser Use." *Science of the Total Environment* 438: 477–489.

Landecker, Hannah. 2019. "A Metabolic History of Manufacturing Waste: Food Commodities and Their Outsides." *Food, Culture & Society* 22, no. 5: 530–547.

Latour, Bruno. 2004 *Politics of Nature*. Harvard University Press.

Latour, Bruno. 2005. *Reassembling the Social: An Introduction to Actor-Network-Theory*. Oxford University Press.

Law, J. 1992. *Notes on the Theory of Actor Network: Ordering, Strategy and Heterogeneity*. Center for Science Studies.

McMichael, Philip. 2006. "Global Development and The Corporate Food Regime." 11: 265–299.

Medina, Laurie Kroshus. 2003. "History Culture and Place-making: 'Native' Status and Maya Identity in Belize." In Matthew Gutmann, Feliz Matos Rodriguez, Lynn Stephen, and Patricia Zavella (Eds.) *Perspectives on Las Americas*. Black. 195–212.

Merill, T. (Ed.). 1993. *Guyana and Belize. Country Studies, Library of Congress*. Federal Research Division.

Mol, Arthur PJ, David A. Sonnenfeld, and Gert Spaargaren (Eds.). 2020. *The Ecological Modernisation Reader: Environmental Reform in Theory and Practice*. Routledge.

MLA (Maya Leaders Alliance). 2010. *Supreme Court Affirms Maya Customary Rights for All Maya Communities in Southern Belize*. MLA [press release] June 29, 2010.

O'Brien, Martin. 2008. *A Crisis of Waste?: Understanding the Rubbish Society*. Routledge.

Sahlins, Marshall. 2005. "The Economics of Develop-man in the Pacific." In Joel Robbins and Holly Wardlow (Eds.) *The Making of Global and Local Modernities in Melanesia: Humiliation, Transformation and the Nature of Cultural Change*. Ashgate Publishing, Ltd.

Soma, Tammara. 2017. "Gifting, Ridding and the "Everyday Mundane": The Role of Class and Privilege in Food Waste Generation in Indonesia." *Local Environment* 22, no. 12: 1444–1460.

Statistical Institute of Belize. 2013. *Census Reports*. Retrieved on August 10, 2016, www.statisticsbelize.org.bz/dms20uc/default.asp

Steinberg, Michael K. 2002. "The Second Conquest: Religious Conversion and the Erosion of the Cultural Ecological Core Among the Mopan Maya." *Journal of Cultural Geography* 20: 91–105.

Strasser, Susan. 1999. *Waste and Want: A Social History of Trash*. Metropolitan Books/Henry Holt and Company.

Stuart, T. 2009. *Waste: Uncovering the Global Food Scandal*. W. W. Norton.

Sutherland, Anne. 1998. *The Making of Belize: Globalization in the Margins*. Bergin & Garvey.

Tedlock, Dennis. 1996. *Popol Vuh: The Maya Book of the Dawn of Life (Revised Edition)*. Touchstone.

Thompson, Gregory. 1930. *Ethnology of the Mayas of Southern and Central British Honduras*. Field Museum of Natural History.

Thompson, M. 1979. *Rubbish Theory: The Creation and Destruction of Value*. Oxford University Press.

TMCC (Toledo Maya Cultural Council). 1997. *Maya Atlas: The Struggle to Preserve Maya Land in Southern Belize*. North Atlantic Books.

Viveiros de Castro, Eduardo. 1996. "Images of Nature and Society in Amazonian Ethnology." *Annual Review of Anthropology* 25, no. 1: 179–200.

Vogt, Evon. 1993. *Tortillas for the Gods: A Symbolic Analysis of Zinacanteco Rituals*. University of Oklahoma Press.

Wainwright, Joel. 2008. *Decolonizing Development: Colonial Power and the Maya*. Blackwell Publishing.

Wilk, Richard R. 1991. *Household Ecology: Economic Change and Domestic Life Among the Kek'chi Maya in Belize*: University of Arizona Press.

Wilk, Richard R. 1997. *Household Ecology: Economic Change and Domestic Life Among the Kek'chi Maya in Belize*. Northern Illinois University Press.

PART III

Methods waste scholars use

9

COMPARATIVE METHODS FOR THE STUDY OF WASTE

Raul Pacheco-Vega

Introduction

Current scholarship on the governance of waste has focused on various ways in which waste can be used, reused, and reintegrated into the production cycle. While much has been written about technologies, communities, actors, and regulatory approaches using a broad range of methodological approaches, many of these are case studies where theorization has been limited. There is less work available on comparison and the comparative method as a methodological approach (or series of approaches) to improve our understanding of waste across regions, nations, and cities. In this chapter, I survey the literature on comparative approaches to waste studies. I explore the ways in which comparison as a method can improve our understanding of waste as an object of investigation. I also offer a framework that should help scholars think about how to design and implement a comparative study across temporal and spatial scales, cities, regions, and countries.

Methodological examinations of comparative studies in the social science of waste are scarce and therefore this chapter helps fill in a gap in the literature. Comparison, particularly across countries or regions, is a powerful methodological strategy to draw valuable insights from the way in which other jurisdictions tackle issues that would pose a challenge if analyzed separately or individually. Comparative methods offer structured strategies to examine how phenomena occur across time and space. And comparing how different methods are applied to the comparative study of waste from a social science perspective is an important and undervalued endeavor. In this chapter, I undertake a systematic review of the literature on comparative methods and comparisons within waste studies and propose an innovative approach to investigate how comparison is undertaken at various temporal and spatial as well as cross-regional scales.

Throughout the chapter, I focus on two different, but interrelated elements of waste studies that have an impact on how we conceptualize comparisons across systems. The first one is *the recognition that waste studies across disciplines will all eventually require comparative methods to offer insights that could not be drawn from single-discipline or single-case analyses*. There is an extremely broad range of studies across the social sciences, engineering, and humanities where comparison is at the core. Disciplinary differences will have an impact on how comparative studies will be undertaken but ultimately, comparative research design always focuses on contrasts and

DOI: 10.4324/9781003019077-9

differentiation (Morlino 2018). Comparing waste treatment and disposal systems can be done from a technological, engineering, or natural sciences viewpoint. But at the same time, we can also investigate and contrast the social impact of specific technology choices. There are also valuable "lateral" (or ancillary) social-science-research insights that derive from socio-technical comparisons. For example, studies that compare the cost of landfill disposal versus incineration can yield important insights into whether communities will be receptive to specific site choices for these facilities, and the negative impacts that they may face, even if these were not the original intentions for these types of studies.

The second element that I emphasize throughout the chapter is the *importance of carefully making the correct research design choices that will lead us to the successful undertaking of research projects on comparative studies.* While it is possible to compare cases and phenomena across countries, cities, regions, or temporal/spatial scales in a haphazard, random way, designing comparative studies with a focus on explanation of outcomes and phenomena might be much more productive and helpful. To this end, we should also be able to distinguish between explanatory studies and descriptive studies. Are the projects we launch intended to highlight similarities and differences across units of analysis or are we looking to explain how variation across cases occurs? Deciding which approaches we want to take is useful in developing and implementing a systematic methodology for the comparative study of waste.

To make this chapter more useful from a pedagogical standpoint, I present two vignettes that focus on different aspects of the comparative politics and policies related to informal waste picking. In the first, I present a summary of my study of the dynamics of informal labor and how recyclers operating in the informal economy contribute to a circular economy. I am particularly interested in highlighting the methodological choices required to understand whether informal waste-pickers are recognized across different jurisdictions at the national and subnational levels, and especially whether this recognition (or lack thereof) has an impact on their livelihoods. In the second, I use my cumulative research and fieldwork experience across 8 countries and 13 different cities to discuss how different governance systems may impact the types of technologies, locational choice and picking strategy. These will vary accordingly as governments engage in more collaborative or more confrontational approaches in their interactions with waste pickers. While I do not present the entire specifics of the fieldwork and empirical methodologies that I used in both projects, my goal here is to use these vignettes as examples of the types of questions we can ask ourselves as researchers, and the strategies we may decide to follow to engage in comparative waste studies.

The chapter is structured as follows: after this brief introduction, I set the stage for the remainder of the chapter in Section 2. I briefly overview what we know about comparison as a method and survey the literature on comparative methods as it applies to waste studies. It is important to note that there are subtle yet important differences in how comparisons are undertaken across disciplines, and I highlight these throughout the chapter. I discuss fieldwork and case selection in Section 3 of the chapter. Section 4 offers the framework I posit for comparative studies, which can obviously then be applied to waste research. In the fifth section I present two vignettes on my research on waste studies, whereas the sixth section presents a synthesis of my framework and offers a comparative view of two studies as well as a summary of how the conceptual model I present may be engaged. Finally, in the seventh section, I conclude by setting a research agenda on comparative waste studies that can be launched based on this chapter and suggest the various ways in which research design becomes important in the way we design comparative studies.

1.   What do we understand for comparative methods?

Comparison is at the core of any scholarly enterprise, regardless of the number of cases. Even if we do not think we really are comparing, we always look for difference, even within a single unit of analysis. Intellectually, we are humans wired to contrast and detect variations across systems and establish patterns that help us explain how variation occurs. Comparing across different units of analysis is also an exercise in systematization and organization of information that can provide a foundation for structured learning about different production models, organizational systems, institutional arrangements, countries, regions, and cities. Comparative methods have increasingly gained popularity in policy analysis (Peters and Fontaine 2020), public administration (Jreisat 2005), public policy (Engeli and Rothmayr Allison 2014), but they have been a staple of the main field of political science (Collier 1993). This chapter engages mostly with comparative politics/policy/public administration because those are the main areas/disciplines that I was trained in and that my research and teaching contribute to, but the various fields and disciplines in which I publish (history, geography, methods) are also closely related. Nevertheless, I also draw examples and ideas from the fields of anthropology, sociology, and even literature.

I want to note that there is no single "comparative method". When I indicate comparative method(s) and whether I use plural or singular is a deliberate choice that signals the broad range of words that attempts to express how different types of comparisons can be undertaken across case studies, variables, and methods. One can contrast press releases and analyze text production across different organizations to discern how anti-waste-facility non-governmental organizations frame their messaging to achieve specific objectives. We can also compare different treatment processes and the decision-making processes of engineers in charge of waste management facilities. All of these strategies use comparisons. I use "the comparative method" or "comparative approaches" interchangeably to note that there are various strategies, methods, approaches, and techniques to compare across cases, but also that I use the broad umbrella term "comparison" to denote the actual act of contrasting across units.

Structured comparisons help us understand the differences and similarities between cases, actors, and more broadly, phenomena. Depending on whether the comparison is descriptive or explanatory, we will select the type of comparison to undertake and the data collection method as well as the methodological and data analytical choice. When we engage in comparison, we contrast different units of measurement and their respective values. For example, we can say "Mexican citizens produce 400 grams of household refuse every day, whereas U.S. citizens generate 550 grams". This comparison makes an implicit assumption about the homogeneity of units (the weight of garbage) across different measuring techniques and instances. Unfortunately, comparisons like this generalize without facilitating contrasts that provide insights. We could ask ourselves a question such as: "Do French residents, on average, produce more waste per household per day than individuals in the United States or other countries?" This question could deliver a much more meaningful answer because what we compare is average household production per person per day. There is no average French citizen, but *on average*, with this hypothetical example, we could say that standard production levels reach approximately 600 grams per person per day. Comparing properly requires us to discern and decide on the specific metrics and measurement and evaluation methods to obtain, calculate, or understand these metrics.

While there is no single "comparative method", there's general agreement on what needs to be done to compare. Recent volumes on comparative methods for policy analysis (Peters and Fontaine 2020) have synthesized how comparisons need to be designed and undertaken so we can elucidate the sources for variation across various cases in policy analysis. For example, we can potentially compare how hydraulic fracturing processes occur in Europe and North

Table 9.1 Five questions to consider for robust research design of comparative studies

Question	Explanation
Why do we compare?	Rationale for the study and for engaging in comparison
What can we compare?	Case studies, variables, outcomes, processes, mechanisms
How do we compare?	Methods for comparison, data collection and analysis strategies
Which time periods are we focusing on?	Temporal comparison (cross-sections or diachronic)
How do we choose which comparisons to undertake?	Sites for fieldwork, case studies, data availability

America (Weible et al. 2016). We could also study global environmental meetings by comparing their processes and outcomes (Lima and Gupta 2013). Examining waste governance across different times and scales can also be done, for example, when studying garbage production and management across four different cities over a period of 100 years (Strach, Sullivan and Pérez-Chiqués 2019). There are many approaches to comparative studies and we can learn from a variety of disciplines and methods (Licbbach and Zuckeman 1997; Lichbach and Zuckerman 2007; Rose 1991; Steinmo, Thelen and Longstreth 1992).

The paradigmatic divergences in comparative methods for comparative politics usually emphasize large-N and small-N studies. There are two general streams of thought on the number of cases for comparative study. While Landman (2008) suggests that it is equally relevant and frequent to do large-N comparative analyses, Lijphart proposes that comparative methods are more popular when doing small-N studies (Lijphart 1971), but in fact, many of the comparisons we have seen in the literature, particularly in comparative politics, emerge from large-N comparisons.

While we can use a broad range of methods for comparison, I will settle on a brief discussion of two specific ones that can be discerned from the type of data and method of analysis being used. Quantitative comparisons will often involve a substantially large number of cases, whereas qualitative comparisons involve a much smaller case count. Both are equally important, depending on the specific issue under analysis. There are issues for which a large-N, cross-national comparison may make more sense whereas in other cases we might settle for comparing across three or four countries or cities. For Lim, comparative analysis is mostly or primarily qualitative (Lim 2016), although Landman suggests that one could potentially place exactly the same weight on qualitative or quantitative methods (Landman 2008), depending on the research question and the number of cases under analysis. From a systematic survey of the literature, large-N studies are quite rare at least in the social sciences and humanities, particularly because obtaining data for waste collection, treatment and disposal across multiple locations ends up being quite difficult. While not impossible, obtaining quantitative data from many cases could prove challenging and therefore, many scholars settle for comparing a relatively low (less than ten) number of cases. Nevertheless, we must never dismiss the power of large-scale quantitative studies and that is part of the reason why I included them in this discussion.

In one of the most systematic and popular accounts for comparative methods in political science, Landman lays out a simple framework to justify why do we compare, and how comparisons work. In his book, Landman argues that one can compare multiple countries, just a few, or even simply discuss a single country in comparison to others (Landman 2008). The core issue at heart remains that we compare to understand variation and explain what causes it. Comparison does have its problems and it is not without its critics. Particularly when the number of cases increases, it is hard for the researcher to discern at which point the comparison is robust enough. How many cases are enough and when should we stop looking for cases, and more importantly, which factors should be included in the selection of specific units of analysis for comparison. Nevertheless, there are powerful reasons for comparison (Geddes 2003; Landman 2008; Lim 2016; Morlino 2018), including the ability to learn from the successes and mistakes of others, the inherent insight from contrasting across different countries, cities, and regions, and the versatility of comparative methods across time, space, and scale.

Single-country studies can certainly be comparative in nature (Pepinsky 2019). In the same vein, single case studies are also ripe for comparison (Flyvbjerg 2006, 2011), though they also have their own insights that do not derive from contrasts. Comparative case studies are powerful methodological strategies to draw conclusions about how different systems work. Comparisons can be horizontal (across cases of the same type) or vertical (across different levels of analysis), depending on the direction of comparison (Bartlett and Vavrus 2017). While methodologically one could think that it is much more challenging and perhaps unsound to compare across levels of analysis, we do this type of contrast usually when a public policy program or a policy instrument operates across different levels. For example, we could compare different types of collection fees or disposal taxes across types of waste. Hazardous waste is usually a federal-level (national) responsibility, but in theory, federal governments can impose collection fees in a remarkably similar manner as subnational governments do. We could potentially then compare across scales or levels of government.

Case studies can also be used to develop a series of hypotheses and craft a causal mechanism explanation for the specific phenomena under study. These causal case studies offer a systematic approach to understanding the causal processes through which specific factors impact the development and evolutionary trajectory of a phenomenon. Causal case studies aim at discerning the logic of inference and the different steps and mechanisms through which a factor causes specific outcomes.

Key reasons why comparison is important across time and spatial scales include the fact that there is inherent variation when we shift the boundaries of the system under analysis from one scale to another. Waste production in Pittsburgh does not look the same right now than it did in the early 1800s, for example. Literature on comparative history and comparative historical analysis offer systematic methodologies to study the role of historic events and critical junctures (Mahoney 1999; Sewell 1967; Skocpol and Somers 1980) that can be applied in the comparative study of waste production and governance (Table 9.1).

Comparison at the subnational level is most popular in studies of comparative urban governance, particularly because it facilitates contrasts between relatively different units but within the same national context. For example, Duan et al. (2020) compare different disposal systems across three Chinese "representative" cities. Their approach focuses specifically on how waste is disposed and what types of technology variation we can observe across these cities (Duan et al. 2020). As I indicated previously, these comparisons help contrast different units (cities) but within the same national context. This comparative strategy keeps national frameworks and contexts constant, all the while varying the individual unit of analysis (disposal strategy and technology at the city level).

In contrast to same-country, different-city analyses like Duan et al., Lee-Geiller and Kutting compare how governance differs across two different countries (South Korea and the USA) and cities (Seoul and NYC). This type of comparison is useful to understand processes associated with waste production and disposal even if the units of analysis differ substantially (country and city). There are potentially lessons for New York City to draw from the Seoul case and vice versa (Lee-Geiller and Kütting 2021). Subnational comparisons are valuable regardless of the level at which we do them. Some scholars will compare entire cities, while others will limit the unit of analysis to the neighbourhood level within the same city, for example. In particular, I side with Snyder in arguing that "because a focus on subnational units is an important tool for increasing the number of observations and for making controlled comparisons, it helps mitigate some of the characteristic limitations of a small-N research design" (Snyder 2001, 93). Ethnographic methods can also help with subnational comparisons (Pacheco-Vega 2020).

2.   How do we select cases for comparison and how should we conduct fieldwork?

Case selection in comparative studies is one of the most important elements of research design. How we select cases will influence the choice of methods and the types of data we seek to obtain. Which cases shall we study and how should we choose those so that the comparison will yield useful explanations? The process of selecting cases starts with the research question we intend to answer. At the core, in comparative studies we seek to explain variation, and comparative waste research is no different. Case selection can be deliberate and systematic and yet leave enough leeway and flexibility to facilitate diversions from original research design trajectories. For example, if we are looking to understand how changes in technology of waste collection systems can possibly impact treatment processes and technologies, we would want to select a broad range of technologies both for collection and treatment. However, we would need to make sure that the number of cases is manageable for this study to be operationalized in a manageable way. This approach implies that for the research design to be workable it should seek to explain variation in technology across at least two cases or over time. We could also compare technological choices across two different points in time.

Selecting cases in comparative research requires thinking about the specifics of the variables we are looking to examine to better understand the overall system. As I have mentioned, the overall goal of comparative research is to understand systems on their own, but also to explain variation across cases and/or temporal or geographical scales. In the words of Morlino, "it is essential that explicit and clearly defined criteria are used when choosing the countries or areas on which to conduct research" (2018, 9). Carefully considering which cases we select is important regardless of the type of data collected and the method of analysis. Whether it is a dataset of cities and their individual production rates across multiple countries or a set of fieldnotes from ethnographic work undertaken within a single city suburb, devising rigorous principles and methods for selecting the cases we are seeking to examine is fundamental for our improved understanding of the phenomena under study, in this case waste in its various physical and social forms.

Establishing criteria and principles for case selection that yield the correct is a first step to engage in comparative waste studies. Which cases can be examined that will enable researchers to explore variation across cases and the factors that explain said variation? I argue that this can be a core principle for case selection. Once we have systematically chosen which cases we have decided to investigate further, we need to decide on which strategies we will use to collect data. As a scholar of mixed methods who sees himself as an ethnographer myself, my preferred mode of data collection is conducting research on the field. My comparisons are fundamentally guided by an ethnographic sensibility (Pacheco-Vega 2020; Simmons and Smith 2017) and thus, I strongly believe that fieldwork-based comparison is better than traditional, non-field-based contrast.

Fieldwork-based scholarly research on waste is extraordinarily important, particularly when discussing research about vulnerable and highly marginalized communities (Pacheco-Vega and Parizeau 2018). There is some comparative scholarship on informal waste picking practices in Latin America, of which Marello and Helwege are the most representative. These researchers purport to comparatively examine informal waste-picking across three Latin American countries. Among the research insights that these scholars offered was the relative importance of collaborative (waste-picker–government) relationships. While I agree with Marello and Helwege that collaborative practices do not necessarily lead to improvement in waste pickers' lives (Marello and Helwege 2018), it is not clear from their article that they actually conducted any fieldwork in Mexico City, but instead only followed Castillo Berthier's 1978 book as a description of what happens in Mexico for waste-pickers. This is an important point to make, as much of the literature on informal waste-pickers relies on second-hand accounts and published articles and other varied non-scholarly literature. Conducting fieldwork, interviewing, and conversing with recyclers help us understand more realistically the phenomenon and rationales for waste-picking. Critiques of inclusive practices for waste-pickers must be scrutinized because their approach often fails to adequately reveal positive elements of inclusion, as those can only be deduced from speaking with waste-pickers themselves and learning from their experiences.

The challenge may reside in how inclusion is defined, and much like the formality-informality continuum, inclusion can also refer to a spectrum of approaches. For Marello and Helwege, inclusive practices are nothing more than tokenism, whereas for other authors, these strategies reside at the core of long-term welfare improvement trajectories (Gutberlet et al. 2016). In the end, it is possible to reconcile both views by recognizing that here are different views and definitions of inclusion, and that we ought to first define our terms clearly so that there are fewer confusions and confrontations when the overall pursued goal is the same. Fieldwork is important for comparative studies because the choice of research site informs the line of inquiry and which questions we want to answer and the kinds of insights we may be able to pursue in our studies.

3. How do we undertake systematic comparisons in the field of waste studies? A proposed framework

Comparing systems, processes, individual units of analysis, approaches, and paradigms is an activity that we undertake on a regular basis, yet we seem to rarely reflect on what it entails. In this section, I intend to shed light on the process by building a systematic methodology for undertaking comparative studies in the waste field. Whereas earlier sections of this chapter were more focused on explaining and describing how other authors have undertaken comparisons, in this section my intention is to present a systematic, organized approach to comparing where I focus on seven factors.

My framework suggests that whenever we design comparative studies, we need to design the proper research question using an explanatory approach where our main goal is to explain variation across cases, units of analysis, time periods, and/or scales. We also need to set up a structure for comparison that considers an examination of which factors influence the outcome and a potential explanation for how these factors impact process and product. While not every comparative study needs to be causal, nor should it be, it is important for researchers to try and tease out the inner workings of each phenomenon as we compare across instances and cases. Finally, I assert that comparing needs to be an exercise in case selection where the proper cases can be examined to yield insights into the processes we are studying. Thus, I suggest that even before we settle on which cases we will formally compare, we need to do some preliminary comparison in order to properly select which ones will merit in-depth comparing. This exercise in case selection methods facilitates the decision-making process involved in deciding the definite set of cases to study.

Comparative history uses systematic analyses that may contrast events across two or more time periods, or it can explore the different and potentially converging or diverging trajectories of events that lead to multiple outcomes. Just as an example, in an examination of the cases of the Mexican states of Nuevo León and Veracruz, Bess shed light on the ways in which the state was built in Mexico along a very similar pathway, at least from a political standpoint (Bess 2017). However, as Bess demonstrates, there are extremely clear differences in how each state developed its railroad infrastructure. These divergences in developmental strategies also signaled variation in state formation maturity. Whereas Veracruz showed a continuous model of infrastructure development, Nuevo León's was a lot more stop-and-go. We can also compare how China has managed e-waste in contrast to the United States (O'Neill 2018, 2019) or how waste trading occurs across rich nations (O'Neill 2000). O'Neill's work demonstrates that contrary to expectations that hazardous waste is traded from rich countries to poor ones, we can find this type of trading across wealthy nations, particularly Great Britain as an importer and Germany as an exporter. What O'Neill compares is the rationale across different countries for trading hazardous waste without having economic need to do so.

Over many years of studying comparative politics, comparative public policy and comparative public administration, I've come up with a framework that approximately summarizes the kinds of questions I ask myself when designing a research project where I expect to engage in comparisons across cases or temporal/spatial scales. This framework is not intended to be static or "the last word" on comparison, but it offers pointers and considerations as to what we need to think about when comparing (Table 9.2).

Table 9.2 A proposed framework to think about how to conduct a comparative study

Theme	Types of questions we might ask ourselves when designing the study
Research question	• What is the phenomenon (or which phenomena) we are trying to understand? • Am I seeking to explain a phenomenon, or do I just want to describe and systematize similarities and differences across cases?
Case selection	• Which cases would enable me to answer the research question I am interested in? • How many cases do I need to study in order to have enough comparisons to explain the variation across cases that I am trying to understand?
Variation	• What is the variation I am seeking to explain with this comparative study? • If I am not seeking to explain variation, then what am I looking to study comparatively?
Explanation	• Is this a descriptive study or an explanatory one? • How is variation a component of this descriptive or explanatory analysis?
Alternative explanations	• Of the factors I am considering, which ones could potentially explain the phenomenon I am looking to understand? • Which explanations could be eliminated? • What kind of mechanisms, strategies, and tactics do we use to discriminate between different alternative explanations?

My framework can be used to guide comparative waste studies by thinking through the questions I pose for each one of the elements that need to be considered. For example, if we are comparing two different end-of-pipe technologies (incineration versus landfilling), we may be seeking just to describe both systems side by side, in which case we are not seeking to explain something but just engage in pure description. Other times we might seek to investigate how a particular choice of engaging in informal waste picking activity is the result of various specific societal factors.

4.   Two vignettes of comparative studies in the social sciences of waste

While an in-depth, broad overview of the literature on waste from a comparative perspective would fall beyond the scope of this chapter and would be extremely time-consuming and intensive, and perhaps not all-encompassing, I wanted to showcase the different types of comparisons that are usually present in the literature. I identify at least two streams of research. The first focuses on comparing productive systems across countries (Alabi et al. 2012; Delgado, Ojeda-Benítez and Márquez-Benavides 2007; Wilson et al. 2012). The second one centers around comparing waste governance systems (Bull, Petts and Evans 2010; Muñoz-Cadena, Arenas-Huertero and Ramón-Gallegos 2009; Pacheco-Vega 2013; Scheinberg et al. 2016; Thomashausen, Maennling and Mebratu-Tsegaye 2018; Uddin et al. 2020) and the social components of waste (what Parizeau refers to as "the social life of waste"). Comparing productive systems helps us understand variation across technologies and treatment methods. Comparing the social life of waste helps us better understand the ways in which individuals look at and associate with waste. Along the same lines of thinking, comparing waste governance systems across various jurisdictions enables us to better design policy instruments aimed at minimizing, reducing, and perhaps eliminating waste at the source.

One of the best ways to teach about comparative methods in the field of waste studies is to describe how we design and conduct our analyses. In this section, I present two vignettes of comparative work that I have undertaken in the field of waste studies. I choose vignettes instead of full case studies as my intent with these brief descriptions of my research projects is simply to showcase how I designed the study, how I chose the cases, and how I carried out the fieldwork. All the comparative waste studies I have conducted include a broad range of methods, techniques, and theoretical paradigms in the social sciences, though my research design strategies are deeply rooted in comparative politics.

In this section, I employ a narrative approach to discussing the two vignettes that showcase how comparative studies of waste can be undertaken. I chose this specific approach to explaining the importance, value, and strategies for the application of comparative study methods to understanding waste for pedagogical purposes. Given that this handbook is focused on summarizing and explaining a large and broad body of work on waste studies, I decided to use this section to explain in a more narrative way how I approached the decision-making processes that my research design and implementation entailed.

At the beginning of my scholarly career, I worked on the comparative politics of environmental policy instrument choice. My research was primarily aimed at understanding the factors that drove adoption of voluntary instruments for pollution control. I closely followed the work of Dr. Kathryn Harrison in this area (Harrison 1998, 2002) and developed an interest in examining information-based policy instruments, which culminated with a study of policy transfer of pollutant release inventory programs across North America (Harrison, Pacheco-Vega and Winfield 2003). While the earlier parts of my work focused more on toxics and solid waste, over the years I moved in the direction of studying wastewater treatment and water access.

Nevertheless, because of the highly polluting nature of the industries I studied for my doctoral dissertation (leather-manufacturing and shoe-making), I always retained an interest in the governance of solid waste and municipal refuse, including various types of discards.

I began work on the comparative politics of informal waste in earnest in early 2012, when I started studying how the government of the city of Aguascalientes connected with their informal recycler population. Informality as an area of study has always interested me, having done extensive work on formal and informal rules in the design and implementation of sanitation and wastewater policy in subnational contexts. My analysis explored whether and if so, how wastewater policies in the five states that include the Lerma-Chapala river basin differed. This study inaugurated my subsequent comparative research in other policy issues, such as water access and wastewater treatment for 15 years, and more recently, solid waste management (Pacheco-Vega 2015, 2017).

At the beginning of my career, while my research design choices were a lot more intuitive I already tried to include socially consequential issues of comparison, such as the choice of the study sites, decisions on heuristics or case study selection, and the potentially negative impact of my study on the research subjects. Beginning this new project taught me about the importance of ethics and the duty of a researcher to protect populations facing a strong degree of vulnerability. The following two vignettes draw on my scholarly research on the comparative politics of informal waste picking, but the first is more focused on the cooperation-conflict dynamic between governments and informal recycling communities.

The first vignette draws from my comparative work between two cities, Vancouver, British Columbia, Canada and León, Guanajuato, México). While it could be argued that my research design and case selection were biased towards a convenience sampling strategy that would yield comparisons across rather dissimilar systems, in fact, Vancouver and León share similar demographics (about 1.7 million inhabitants) and geographical dispersion. I did have the experience of having lived in both cities for many years, which that benefitted me in discerning specific critical junctures. Between 2012 and 2018, I conducted fieldwork in Vancouver for several weeks at a time, with key stays in November of 2015 and 2018.

While I had also lived in Vancouver for over 20 years, I did most of the field research after I had moved away. Throughout my fieldwork in Vancouver, I mostly focused on those informal recyclers who do binning and collect cans to be exchanged for cash, usually with a non-governmental organization. Vancouver binners have been studied before (K. Parizeau 2016; Tremblay, Gutberlet and Peredo 2010; Wittmer and Parizeau 2018), though my specific focus was comparative in two ways. First, I wanted to explore the ways in which binning happened in Vancouver and compare it with the types of wastes and strategies that waste-pickers used in León. And second, I wanted to examine what explained the whether the local government chose a confrontational or a collaborative relationship. with waste-pickers. In León, I undertook extensive participant observation and informal conversations (usually over weekends) across the city, particularly in middle-income neighborhoods like León Moderno, La Azteca, and San Isidro. I also did field visits to the El Caracol landfill (northeast of León) and observed waste-picking in the downtown area of the city. What I found in Vancouver was a cooperative relationship where binners thrive and have a non-confrontational relationship with the city government. In León, I found a combative waste management agency and a non-cooperative relationship where informal waste-pickers have been completely shunned from accessing municipal refuse at the landfill stage. What this has meant for binners in Vancouver and informal waste-pickers in León is that where collaboration exists, quality of life has vastly improved for this marginalized population. Even though the comparison is conducted across countries and cities, the lessons drawn from contrasting confrontational and collaborative relationships are durable and permeate across.

In a second vignette, I present a brief description of one of my medium-N research projects in hopes that it will help researchers with their research designs. For this specific project, my intent was to explore how waste-pickers in different countries and cities would choose specific technologies. When I started this project (Pacheco-Vega 2018), I was originally looking for descriptions of different informal waste-picking processes. Inductive (and abductive) learning allows us to start establishing working hypotheses and tentative research questions. What I wanted to do with this specific project was to accumulate knowledge and understanding about the practices that informal waste-pickers engaged in. My research question centred on how recyclers chose sites, technologies, locations, etc.

This project intended to explain variation in waste-picking practices in six countries (Uruguay, Argentina, Spain, Mexico, Canada, Japan). I chose three specific variables to study: *locational decision* (where pickers decided to search for and extract garbage pieces), *technology choice* (which type of method for picking do recyclers use), and *material prioritization* (which discards were most valuable for each community of informal waste-pickers). I started with fieldwork in Montevideo (Uruguay) and Buenos Aires (Argentina). This choice was informed by my interest in earlier work on *cartoneros bonaerenses* (K Parizeau 2013). Reading Parizeau, I became convinced that there were valuable insights to gain from understanding how waste-pickers decided to choose where to collect garbage and how they transported it to inter-mediaries. Comparing cardboard recycling in two different cities (Buenos Aires in Argentina and Leon in Mexico) made a lot of sense and I set out to engage in this comparison. While I began fieldwork in Montevideo, I traveled to Buenos Aires to immerse myself in the social practices of informal recyclers.

During my fieldwork, I noticed a lot of similarities in how waste-picking occurred: where cartoneros chose to pick cardboard, the time of day, and the type of method they used to transport their materials. What then became interesting was: how were these activities done across different cities and countries. I began considering other countries for comparison, and within those, specific cities. This decision took me to Madrid, Tokyo, and Paris. While the lengths of fieldwork varied in each one of these sites, my research was always fundamentally ethnographic.

Ethnography is a fieldwork-based strategy of scholarly inquiry that helps us understand cultures in specific sites (Brewer 2000). As a qualitative research method, ethnography helps us learn in detail from communities and individuals located in specific geographical locations (Ladner 2014). Given its use of field-based data collection strategies, ethnography is particularly useful to help us understand processes of discarding, wasting and producing refuse. While comparison is said to be intrinsic to the ethnographic enterprise (Vogtz 2002), I find explicitly-designed comparative ethnography especially valuable for studies of waste that require con-trasting across cases (Schnegg and Lowe 2020; Simmons and Smith 2019), particularly when analyzing waste-specific public policies (Pacheco-Vega 2020).

From undertaking comparative ethnography of informal waste-picking across two countries and two cities, I drew important lessons on the different types of relationships that governments and informal workers could have. These insights could not have been drawn without engaging in explicit comparison. Without contrasting at least two cases, I would not have been able to make inferences on the impact (positive or negative) that collaborative (or confrontational) relationships between informal waste pickers and city governments would have on the ability of these workers to survive and maintain a livelihood. These lessons are extremely relevant learnings that derive from undertaking comparative studies of waste across cities and nations.

Among the key lessons I learned from both projects was the importance of carefully choosing sites for fieldwork, of improving and refining my research question as I listened to the stories of informal waste-pickers, to triangulate across different sources (from newspaper articles to recyclers to intermediaries to government officials to experts), and more importantly, to elevate the voices of those who are systematically forgotten or whose voices are not listened to because they are not as privileged as members of other communities. I also learned a lot about what some call "ethnographic refusal", that is, what is not and should not actually be reported in research, particularly done to protect the communities under examination (Tuhiwai Smith 2008; Zahara 2016).

5. Synthesis: What can we learn from cumulative experiences studying the social elements of waste from a comparative perspective?

Comparison can be a powerful a methodological strategy to advance social science inquiry in the field of waste studies. From the body of works I have systematically reviewed for this chapter, it is clear that comparative methods are subtly taking over the field of waste studies. However, this takeover is less visible in terms of methodological improvements and innovations on methods to conduct comparisons and more perceptible on actual comparisons across case studies, spatial and temporal scales, populations and communities, jurisdictions, and processes. While there is much cumulative work on how waste circulates across countries, regions, and cities, and comparative examinations of labor processes in the production, transformation, and disposal of waste, there is less inquiry on why scholars choose comparison as a productive research method that offers explanations that others would not be able to provide.

Which decisions need to be made when and in which sequence? How do we decide which cases we examine and the depth to which we need to engage in fieldwork at each specific location? What are the technological, socio-technical, and infrastructural features of each context and how do they impact the studies we launch? In this section, I offer a comparison of the two vignettes I presented and discuss how I applied my framework to these case studies, the kinds of questions I asked myself, and the things that worked and the ways in which I could have done these research projects better.

While there are many more elements that I considered in designing these two cognate research projects, Table 9.3 serves more as a brief comparative summary of the types of decisions I had to make when undertaking my research projects. In hindsight, had I developed this framework *before* I started my fieldwork would have been much more helpful to my research. This is partly the reason why I developed this chapter for future and current scholars of waste studies, so their comparative studies can be more easily and better designed.

Studying waste requires understanding a broad range of interactions across material elements and humans. The study of informal waste-picking practices requires much more than just an understanding of the diverse interactions and relationships that exist between those who produce waste and those who process it and reintegrate it into the system. Waste-pickers are frequently extremely vulnerable individuals because they are poor, marginalized, and their work involves accessing, classifying, picking apart, sorting, and reorganizing municipal refuse. For obvious reasons, public health concerns regarding informal waste-pickers are at the top of the mind for scholars. *How can we study waste-picking practices of informal workers in a way that does not bring harm to them?* This ethical approach to ethnographic research is what Pacheco-Vega and Parizeau call "Doubly Engaged Ethnography". A doubly engaged ethnographic approach considers the

Table 9.3 Comparing two different research projects on waste and the decisions made in the research design of comparative analyses

Analytical FactorCase study/Vignette	Informal waste pickers' relationships with local governments in Vancouver (Canada) and León (Guanajuato)	Waste picking practices, technological choices, and locational decisions across six different cities (Madrid, Vancouver, Aguascalientes, Montevideo)
Site choice	Similar cities with metropolitan characteristics, with dissimilar relationships between local governments and recyclers (most different cases)	Snowball and targeted sampling strategy, choosing cities where informal waste-picking is considered quite relevant and where I could do fieldwork
Case study selection criteria	Different (and specifically, opposite approaches to governing the relationship with informal recyclers)	Convenience case selection (countries and cities where I had identified relevant cases from the literature where waste picking has been already studied
Explanation sought/research question	Which factors could explain the type of relationship between government representatives and bureaucrats with waste pickers?	Which elements explain the technological choices and picking sites across different locations?
Selected findings	In cities with collaborative relationships with their informal waste-pickers, these manifest improvements in welfare and well-being. Regions where this relationship is contentious and confrontational, there is an increasing degree of precarization of informal workers.	Technology choices were driven by the type of location where waste-pickers had to engage in their activities. Where recyclers could access landfills, technologies did not matter as much because they were able to participate in informal work because accessibility was not an issue.

responsibility that researchers have towards communities and individuals under study, and engages in practices that reduce the probability of harm by deeply considering issues of representation, insider-outsider roles, positionality, and reflexivity (Pacheco-Vega and Parizeau 2018).

How could a researcher interested in the comparative study of informal waste-picking practices harm a community? This risk is inherent to the ethnographic, fieldwork-based study of any informal practice. Given the fact that these informal practices are sometimes considered illicit or illegal (Kate Parizeau and Lepawsky 2015; Rosa and Cirelli 2018), researchers need to be careful about not imposing unnecessary burden on workers who depend on being surreptitious and hidden from traditional regulatory channels. Undertaking these comparative studies requires us (scholars) to make specific case study/site/community choices where a burden can be reduced, minimized, or at least, mitigated. However, we also need to understand that comparisons based on fieldwork (such as the ones I espouse and present in this chapter) are inherently complex and complicated precisely because site/case study choice may be limited, depending on the research question being asked and the types of insights that we are seeking to gain. I encourage fieldwork-based comparativist scholars to consider the potential risk to subjects that embedding ourselves inside their community can bring along to those we seek to better understand.

6.   Conclusion: Towards a research agenda for the application of comparative methods to the study of waste

In this chapter, I reviewed the comparative studies of waste governance. Previous work has emphasized the procedural and technical components of waste, including the various ways in which waste can be used, reused, and reintegrated into the production cycle. More recently, there has been a significant push to explore comparison and the comparative method as a valid and useful approach to improve our understanding of waste across regions, nations, and cities.

Throughout the chapter, I explored several ways in which comparison can improve waste studies. I offered a framework to engage in comparison that draws a few elements to consider across different units of analysis. While I agree that there are elements of "art" and "craft" in how we design comparative studies (Boswell, Corbett and Rhodes 2019), I am more inclined to believe that comparison can be done in a systematic, rigorous and analytical way even if it does not involve causation. Engaging in serious consideration of the various elements involved in analytical comparisons helps researchers reduce bias (Bennett 2004; Geddes 2003; Mahoney 2007) and establish a more systematic, coherent, and cohesive approach to understanding variation across cases.

The vignettes I presented draw from my experience researching the comparative politics of waste governance, with a special focus on informal waste picking. My studies shed light on three important elements for comparative studies of waste: (a) the value of undertaking comparative studies of waste governance across cities, countries, geographical, and temporal scales; (b) the multiple challenges facing researchers who study waste in undertaking comparisons across seemingly disparate and heterogeneous case studies; and (c) the opportunities that comparative methods offer for the advancement of waste studies.

The two vignettes I use to illustrate the importance of comparative methods in waste studies draw from two inter-related projects on informal waste-picking and comparative public policy. In the first one, I compare the experiences of informal recyclers in two cities of approximately the same size: Vancouver (Canada) and León (Mexico), in order to reveal the types of confrontational (exclusionary) or collaborative (inclusive) models of waste governance that are available to local governments. In this chapter, I showed how I studied the impact of developing inclusive strategies, which may have proved a better model to follow for bureaucrats and politicians interested in implementing a circular economy paradigm.

In the second vignette, I showcased the diversity and broad range of strategies that informal waste-pickers engage across six different countries. These strategies enable waste-pickers to survive and thrive even within contexts that are tremendously heterogeneous and sometimes hostile. The choices that waste-pickers make regarding technology for collection and transportation, disposal site, and picking location frequently depend not only on their relationship with the local governments but also with the local communities where informal recycling takes place. My decisions regarding which countries to visit and the cities I would embed within were in this second case, a lot more intuitive and haphazard.

The vignettes I presented demonstrate that developing inclusive strategies may prove a better model to follow for bureaucrats and politicians interested in implementing a circular economy paradigm. Moreover, I found that the great diversity and broad range of strategies that informal waste-pickers engage across six different countries enable them to develop coping tactics that over time prove extremely useful. Survival mechanisms for recyclers across countries are extremely varied and revealing them signals a powerful and valuable lesson for those who study waste governance, and waste more generally. A comparative approach to studying the formal and informal rules involved in the governance of waste helps us reveal nuances, and the challenges that informal recyclers face across wildly different and diverging contexts.

Comparative methods allow us to understand how waste-pickers survive and thrive even within contexts that are tremendously heterogeneous and sometimes hostile. But this is just one of many types of applications of the comparative method to waste studies. Waste itself is extremely heterogeneous, in both sources and composition. To explain differences across diverse types of waste we need to compare various properties of each and examine them using a systematic approach. Comparative research can also be applied to understanding how these property differences signify new and improved ways in which one can understand their social impact. Understanding how different contexts produce distinct ways of managing and governing waste is an unique insight that cannot be gained from single-case studies, and therefore requires comparison.

My broad goal with this chapter was to showcase how applicable comparative methods can be, and how they can be used to better understand the inner workings of waste management both within formal and informal contexts. As the two vignettes I offered demonstrate, understanding the context and characteristics of each individual case study as well as the broader context helps us not only design better comparisons but also establish research questions that are much more systematic and analytical.

The approach I present here should appeal to scholars of comparative politics, comparative public policy and administration, and more generally comparative methods, but it should be generally applicable to scholarship derived from other disciplines, including history, sociology, and anthropology. These last three disciplines have their own closeness to the comparative method, and I have borrowed from all three for this chapter.

The decisions we need to make about research design are equally important as the implementation ones. Without a clear pathway and template to determine which variables are important in the comparative study, how will the comparison take place, and to what extent can we discard alternative explanations for the phenomena we are studying, it will become much more difficult to undertake rigorous comparative analyses. Research design is frequently undervalued because of the potentially serendipitous nature of scholarly work. Sometimes we "walk into a new topic" or "wander into a new case study". But the oftentimes haphazard nature of academic research does not substitute for solid comparative research design. In the words of the incomparable Barbara Geddes, "we have made our own fate through our inattention to basic issues of research design" (Geddes 2003). Robust comparative research design will always yield better insights than simply doing research as it comes.

References

Alabi, Okunola et al. 2012. "Comparative Evaluation of Environmental Contamination and DNA Damage Induced by Electronic-Waste in Nigeria and China." *The Science of the Total Environment* 423: 62–72. http://www.ncbi.nlm.nih.gov/pubmed/22414496.

Bartlett, Lesley, and Frances Vavrus. 2017. *Rethinking Case Study Research: A Comparative Approach.* Routledge.

Bennett, Andrew. 2004. "Case Study Methods: Design, Use, and Comparative Advantages." In Detlef F. Sprinz and Yael Wolinsky-Nahmias (Eds.) *Models, Numbers, and Cases: Methods for Studying International Relations.* The University of Michigan Press. 19–55.

Bess, Michael K. 2017. *Routes of Compromise Routes of Compromise: Building Roads and Shaping the Nation in Mexico, 1917–1652.* University of Nebraska Press.

Boswell, John, Jack Corbett, and R. A. W. Rhodes. 2019. *The Art and Craft of Comparison*. Cambridge University Press.

Brewer, John D. 2000. *Ethnography*. Open University Press.

Bull, Richard, Judith Petts, and James Evans. 2010. "The Importance of Context for Effective Public Engagement: Learning from the Governance of Waste." *Journal of Environmental Planning and Management* 53, no. 8: 991–1009.

Collier, David. 1993. "The Comparative Method." In Ada W. Finifter (Ed.) *Political Science The State of the Discipline II*. American Political Science Association. 105–119.

Delgado, Otoniel Buenrostro, Sara Ojeda-Benítez, and Liliana Márquez-Benavides. 2007. "Comparative Analysis of Hazardous Household Waste in Two Mexican Regions." *Waste Management (New York, N.Y.)* 27, no. 6: 792–801. http://www.ncbi.nlm.nih.gov/pubmed/16820287

Duan, Ning et al. 2020. "Comparative Study of Municipal Solid Waste Disposal in Three Chinese Representative Cities." *Journal of Cleaner Production* 254: 120134. doi:10.1016/j.jclepro.2020.120134.

Engeli, Isabelle, and Christine Rothmayr Allison. 2014. "Conceptual and Methodological Challenges in Public Policy." In Isabelle Engeli and Christine Rothmayr Allison (Eds.) *Comparative Policy Studies: Conceptual and Methodological Challenges*. Palgrave Macmillan. 1–13.

Flyvbjerg, Bent. 2006. "Five Misunderstandings About Case-Study Research." *Qualitative Inquiry* 12, no. 2: 219–245.

Flyvbjerg, Bent. 2011. "Case Study." In Norman K Denzin and Yvonna S. Lincoln (Eds.) *The Sage Handbook of Qualitative Research*. SAGE. 301–316.

Geddes Barbara. 2003. "How the Cases You Choose Affect the Answers You Get." *Parasigms and Sand Castles*. University of Michigan. 89–129.

Geddes, Barbara. 2003. *Paradigms and Sand Castles: Theory Building and Research Design in Comparative Politics*. The University of Michigan Press.

Gutberlet, J. et al. 2016. "Socio-Environmental Entrepreneurship and the Provision of Critical Services in Informal Settlements." *Environment and Urbanization* 28, no. 1: 1–18. http://eau.sagepub.com/cgi/doi/10.1177/0956247815623772.

Harrison, Kathryn. 1998. "Talking with the Donkey: Cooperative Approaches to Environmental Protection." *Journal of Industrial Ecology* 2: 51–72.

Harrison, Kathryn. 2002. "Ideas and Environmental Standard-Setting: A Comparative Study of Regulation of the Pulp and Paper Industry." *Governance* 15, no. 1: 65–96.

Harrison, Kathryn, Raul Pacheco-Vega, and Mark Winfield. 2003. "The Politics of Information Dissemination: International Institutions, Transnational Coalitions and Policy Change." In *2003 Fall Conference of the American Public Policy and Management Association*. American Public Policy and Management Association.

Jreisat, Jamil S. 2005. "Comparative Public Admininstration Is Back in, Prudently." *Public Administration Review* 65, no. 2: 231–242.

Ladner, Sam. 2014. *Practical Ethnography*. Left Coast Press, Inc.

Landman, Todd. 2008. *Issues and Methods in Comparative Politics. An Introduction*. 3rd Edition. Routledge-Taylor & Francis Group.

Lee-Geiller, Seulki, and Gabriela Kütting. 2021. "From Management to Stewardship: A Comparative Case Study of Waste Governance in New York City and Seoul Metropolitan City." *Resources, Conservation and Recycling* 164: 105110. 10.1016/j.resconrec.2020.105110.

Licbbach, Mark Í., and Alan S. Zuckeman. 1997. "Research Traditions and Theory in Comparative Politics: An Introduction." In Mark Irving Lichbach and Alan S. Zuckerman (Eds.) *Comparative Politics. Rationality, Culture and Structure*. Cambridge University Press. 3–16.

Lichbach, Mark Irving, and Alan S. Zuckerman. 2007. In Mark Irving Lichbach and Alan S. Zuckerman (Eds.) *Comparative Politics. Rationality, Culture and Structure*. Cambridge University Press. http://cs5538.userapi.com/u11728334/docs/e36ddc6ff1fc/Mark_Irving_Lichbach_Comparative_Politics_Rat.pdf#page=278.

Lijphart, Arend. 1971. "Comparative Politics and the Comparative Method." *American Political Science Review* 65, no. 3: 682–693. doi:10.2307/1955513.

Lim, Timothy C. 2016. *Doing Comparative Politics. An Introduction to Approaches and Issues*. 3rd edition. Lynne Rienner Publishers, Inc.

Lima, Mairon G Bastos, and Joyeeta Gupta. 2013. "Studying Global Environmental Meetings." *Global Environmental Politics* 13: 46–64.

Mahoney, James. 1999. "Nominal, Ordinal, and Narrative Appraisal in Macrocausal Analysis." *American Journal of Sociology* 104, no. 4: 1154–1196.

Mahoney, James. 2007. "Qualitative Methodology and Comparative Politics." *Comparative Political Studies* 40, no. 2: 122–144. http://journals.sagepub.com/doi/10.1177/0010414006296345.

Marello, Marta, and Ann Helwege. 2018. "Solid Waste Management and Social Inclusion of Wastepickers: Opportunities and Challenges." *Latin American Perspectives* 45, no. 1: 108–129. http://journals.sagepub.com/doi/10.1177/0094582X17726083.

Morlino, Leonardo. 2018. *Comparison. A Methodological Introduction for the Social Sciences*. Barbara Budrich Publishers.

Muñoz-Cadena, C. E., F. J. Arenas-Huertero, and E. Ramón-Gallegos. 2009. "Comparative Analysis of the Street Generation of Inorganic Urban Solid Waste (IUSW) in Two Neighborhoods of Mexico City." *Waste Management* 29, no. 3: 1167–1175. 10.1016/j.wasman.2008.06.039.

O'Neill, Kate. 2000. *Waste Trading Among Rich Nations: Building a New Theory of Environmental Regulation*. The MIT Press.

O'Neill, Kate. 2018. "The New Global Political Economy of Waste." In Peter Dauvergne and Justin Alger (Eds.) *A Research Agenda for Global Environmental Politics*.Edward Elgar Publishing. 87–100.

O'Neill, Kate. 2019. *Waste*. Polity Press.

Pacheco-Vega, Raul. 2013. "Geographies of Wastewater: A Comparative Analysis of Urban Sanitation Governance in the Mexican Municipalities of Aguascalientes (Aguascalientes) and Leon (Guanajuato)." In *2013 Meeting of the American Association of Geographers* (Vol. 25). American Association of Geographers.

Pacheco-Vega, Raul. 2015. "Governing Garbage? An Application of the Institutional Analysis and Development (IAD) Framework to Understanding Conflict and Collaboration Dynamics between Municipal Governments and Informal Waste Pickers." In *International Conference on Public Policy*, Milan, Italy. International Public Policy Association (IPPA). 1–25.

Pacheco-Vega, Raul. 2017. "Applying Qualitative Methods to Comparative Public Policy Analysis: Insights from Multi-Site Ethnographies of Informal Garbage Governance." In *International Conference on Public Policy (ICPP)*, Singapore. International Public Policy Association (IPPA), 1–20.

Pacheco-Vega, Raul. 2018. "Policy Styles in Mexico: Still Muddling through Centralized Bureaucracy, Not yet through the Democratic Transition." In Michael Howlett and Jale Tosun (Eds.) *Policy Styles and Policy-Making: Exploring the Linkages*Routledge. 89–112

Pacheco-Vega, Raul. 2020. "Using Ethnography in Comparative Policy Analysis: Premises, Promises and Perils." In B. Guy Peters and Guillaume Fontaine (Eds.) *Handbook of Research Methods and Applications in Comparative Policy Analysis*.Edward Elgar Publishing. 308–328.

Pacheco-Vega, Raul, and Kate Parizeau. 2018. "Doubly-Engaged Ethnography: Opportunities and Challenges When Working with Vulnerable Communities." *International Journal of Qualitative Methods* 17, no. 1: 1–13.

Parizeau, K. 2016. "Witnessing Urban Change: Insights from Informal Recyclers in Vancouver, BC." *Urban Studies*. http://usj.sagepub.com/cgi/doi/10.1177/0042098016639010.

Parizeau, K. 2013. "Formalization Beckons: A Baseline of Informal Recycling Work in Buenos Aires, 2007–2011." *Environment and Urbanization* 25, no. 2: 501–521.

Parizeau, Kate, and Josh Lepawsky. 2015. "Legal Orderings of Waste in Built Spaces." *International Journal of Law in the Built Environment* 7, no. 1: 21–38. http://www.emeraldinsight.com/doi/10.1108/IJLBE-01-2014-0005.

Pepinsky, Thomas B. 2019. "The Return of the Single-Country Study." *Annual Review of Political Science* 22, no. 1: 187–203.

Peters, B. Guy, and Guillaume Fontaine (Eds.). 2020. *Handbook of Research Methods and Applications in Comparative Policy Analysis*. Edward Elgar Publishing.

Rosa, Elisabetta, and Claudia Cirelli. 2018. "Scavenging: Between Precariousness, Marginality and Access to the City. The Case of Roma People in Turin and Marseille." *Environment and Planning A: Economy and Space*. doi:10.1177/0308518X18781083.

Rose, Richard. 1991. "Comparing Forms of Comparative Analysis." *Political Studies* 39: 446–462. doi: 10.1111/j.1467-9248.1991.tb01622.x.

Scheinberg, Anne et al. 2016. "From Collision to Collaboration—Integrating Informal Recyclers and Re-Use Operators in Europe: A Review." *Waste Management and Research* 34, no. 9: 820–839. doi:10.1177/0734242X16657608.

Schnegg, Michel, and Edward D. Lowe (Eds.). 2020. *Comparing Cultures. Innovations in Comparative Ethnography.* Cambridge University Press.

Sewell, William H. Jr. 1967. "Marc Bloch and the Logic of Comparative History." *History and Theory* 6, no. 2: 208–218.

Simmons, Erica S., and Nicholas Rush Smith. 2017. "Comparison with an Ethnographic Sensibility." *PS—Political Science and Politics* 50, no. 1: 126–130.

Simmons, Erica S., and Nicholas Rush Smith. 2019. "The Case for Comparative Ethnography." *Comparative Politics* 51, no. 3: 341–359.

Skocpol, Theda, and Margaret Somers. 1980. "The Uses of Comparative History in Macrosocial Inquiry." *Comparative Studies in Society and History* 22, no. 2: 174–197.

Snyder, Richard. 2001. "Scaling Down: The Subnational Comparative Method." *Studies in Comparative International Development* 36, no. 1: 93–110.

Steinmo, Sven, Kathleen Thelen, and Frank Longstreth (Eds.). 1992. *Structuring Politics: Historical Institutionalism in Comparative Analysis.* Cambridge University Press.

Strach, Patricia, Kathleen Sullivan, and Elizabeth Pérez-Chiqués. 2019. "The Garbage Problem: Corruption, Innovation, and Capacity in Four American Cities, 1890–1940." *Studies in American Political Development* 33, no. 2: 209–233.

Thomashausen, Sophie, Nicolas Maennling, and Tehtena Mebratu-Tsegaye. 2018. "A Comparative Overview of Legal Frameworks Governing Water Use and Waste Water Discharge in the Mining Sector." *Resources Policy* 55: 143–151. doi:10.1016/j.resourpol.2017.11.012.

Tremblay, Crystal, Jutta Gutberlet, and Ana Maria Peredo. 2010. "United We Can: Resource Recovery, Place and Social Enterprise." *Resources, Conservation and Recycling* 54, no. 7: 422–428. doi:10.1016/j.resconrec.2009.09.006.

Tuhiwai Smith, Linda. 2008. *Decolonizing Methodologies: Research and Indigenous Peoples.* Zed Books/University of Otago Press.

Uddin, Sayed Mohammad Nazim, Jutta Gutberlet, Anahita Ramezani, and Sayed Mohammad Nasiruddin. 2020. "Experiencing the Everyday of Waste Pickers: A Sustainable Livelihoods and Health Assessment in Dhaka City, Bangladesh." *Journal of International Development.* doi:10.1002/jid.3479

Vogtz, Franziska. 2002. "No Ethnography without Comparison: The Methodological Significance of Comparison in Ethnographic Research." In *Debates and Developments in Ethnographic Methodology* (Vol. 6). Elsevier Science Ltd. 25–32.

Weible, Christopher M., Tanya Heikkila, Karin Ingold, and Manuel Fischer. 2016. *Policy Debates on Hydraulic Fracturing: Comparing Coalition Politics in North America and Europe.* Palgrave

Wilson, David C. et al. 2012. "Comparative Analysis of Solid Waste Management in 20 Cities." *Waste Management and Research* 30, no. 3: 237–254.

Wittmer, Josie, and Kate Parizeau. 2018. "Informal Recyclers' Health Inequities in Vancouver, BC." *NEW SOLUTIONS: A Journal of Environmental and Occupational Health Policy.* doi:10.1177/1048291118777845.

Zahara, Alex. 2016. "Refusal as Research Method in Discard Studies." *Discard Studies.* https://discardstudies.com/2016/03/21/refusal-as-research-method-in-discard-studies/ (April 16, 2021).

10

TEACHING CRITICAL WASTE STUDIES IN HIGHER EDUCATION

Kate Parizeau

Introduction

Teaching is an opportunity to spark curiosity about the social and material arrangements that we take for granted in our lives. Western societies and cultures have naturalized the "take-make-dispose" cycle of resource extraction, manufacturing, and consumption that undergirds our economy and many of our social relations. As Lang (2016) notes, "we have to know things … to think critically about them" (15). Teaching critical perspectives on waste may therefore first involve processes of excavating hidden truths about the social lives of discards in our societies.

In this chapter, I explore my experiences of teaching critical waste studies at the postsecondary level in Canada. These experiences are drawn from teaching first- to fourth-year classes (including entire courses focused on waste studies, and those where it features as one topic among many), as well as student supervision at the undergraduate and graduate levels. Following are some of the questions that animate my teaching on critical waste studies: Why do some societies waste so much? Who decides how waste is managed, and what influences those decisions? When does something become "waste"? Does "waste" mean the same thing to everyone? Why are some people affected by waste and pollution more than others? What can different actors do to prevent waste?

Answering these questions requires a collective and collaborative exploration of the ideas that underpin our systems of production and consumption. I begin this chapter with a brief discussion of approaches and strategies for teaching critical waste studies in higher education. In the following section, I provide an overview of the theoretical concepts that can support a denaturalization of wasteful systems through classroom conversations, including understandings of waste/discards, power, materiality, metabolism and flows, and governance/regimes. As an element of each concept, I discuss key topics and case studies that can support the illustration of these theoretical ideas in the classroom.

Approaches to teaching critical waste studies

Pedagogical approach

My teaching methods are inspired by critical pedagogy (in the tradition of Freire 2005). This approach centers the relationship between teachers and learners, and aims to enable students to

DOI: 10.4324/9781003019077-10

act for social change as they learn about themselves and their worlds. Giroux (2013) argues for "pedagogy as part of an individual and ongoing struggle over knowledge, desires, values, social relations, and most important, modes of political agency" (5). In this method of teaching, students are invited to learn about the production of knowledge as a social phenomenon, whether through an examination of texts or an examination of lived personal experiences. Motta (2013) argues that cultivating compassionate, generous, and empathetic relationships with students is foundational to critical pedagogical approaches, and that it is important to recognize teaching as a fundamentally embodied and affective endeavor.

Critical pedagogy as a teaching method is well suited to waste studies because this subject usually involves the unlearning of normative social "truths" about waste and its familiars, and also because power dynamics are so central to the ways that waste is categorized, made visible/invisible, and allowed to circulate through our societies.

Multi- and interdisciplinarity

The study of waste is a broad and multi-disciplinary endeavor. Scholars from STEM fields, social sciences, the humanities, and arts have produced knowledge about wastes from diverse epistemic traditions. It is important to acknowledge this disciplinary breadth in a class on critical waste studies in order to assess how different scholarly approaches can reveal and obfuscate. The interdisciplinary field of science and technology studies can be used as an entry point for learning about reflexivity on intellectual projects and the social construction of knowledge more generally (e.g., Hackett et al. 2008). In the 2020 Discard Studies Twitter conference, King (2020) advocated for a systems-thinking approach to teaching, which "considers wholes over parts and emphasizes "#relations" as the unifying and connective tissue within every structure". This pedagogical approach emphasizes dynamic interconnections within systems, including human connections. Considering the social inequities of the connections within most waste systems, I include ethical reasoning as a learning outcome when teaching critical waste studies.

Choosing classroom resources

Given the historic marginalization of many voices in academia, I advocate for providing students with a diverse reading list including the scholarship of Black, Indigenous, and other scholars of color of different gender identities. These authors have made vital contributions to critical waste studies, as evidenced by the references provided in the following sections. The representation of these perspectives can also be enhanced by including non-academic course resources authored or created by racialized scholars.

News articles, blogs, films, and art can provide students with accessible and evocative entry points into complex ideas surrounding waste and its management. Many relevant phenomena have also not been broadly researched, perhaps because of the obscurity or invisibilization of some aspects of waste (e.g., abject social experiences or proprietary commercial information). Artistic and journalistic approaches may provide new windows onto such understudied phenomena.

Because many of us have had visual, tactile, olfactory, and auditory experiences of waste in our lives, introducing these kinds of visceral experiences into the classroom can also help to consolidate learning around the material dimensions of waste. This might include sharing video footage of wasted landscapes or bringing obsolete objects into the classroom (I have brought my old cell phones into the class to have students experience how easy it used to be to remove and replace the batteries).

Experiential learning can also provide new vantage points on waste, including field trips to waste management sites, or food recovery and rescue operations. Case studies and palpable examples of the phenomena we discuss in the classroom can help bridge students to more theoretical and abstracted discussions of waste, which is often both concrete and marginal in our lived experiences.

Co-creation of the learning agenda

I work to involve students in the design of class content and experiences, including polling them on topics they would like to learn about, having students research and present on topics of their interest, and beginning classes with interactive activities (e.g., mind-mapping, visioning activities, inquiry approaches that encourage students to determine the learning process) that set the agenda for our subsequent discussions. Many aspects of waste and its management are hidden to us, and so centering classes around these kinds of sounding activities helps me to assess where students are in their intellectual and emotive journeys to learning about wastes. I have also found that engaging with liminal topics (e.g., stigmatized waste work, exportation of hazardous materials) and objects (e.g., excreta, long-lived nuclear wastes) can require imagination, and creating space for student voices and non-traditional resources can enable visionary moments in the classroom.

Central theoretical concepts in critical waste studies

Waste/discards

The materials that become waste are diverse. They include post-consumer wastes (materials thrown out after people are done using them, which can occur in homes, workplaces, schools, retail sites, etc.) as well as production wastes (including mining wastes, construction wastes, pollution from factories, over-produced goods, faulty products, etc.). O'Neill (2019) differentiates between "waste as an object" and "waste as streams," invoking the paths that can be traveled by waste objects depending on their provenance, their materiality, and their treatment or management options. While what has been designated as "waste" may have lost value to its users, it may be considered reusable, recoverable, or recyclable by others.

In order to understand what waste means, we must consider how waste becomes. The decision of when to dispose of something is socially and culturally constructed, meaning that "waste" is not universally definable, nor is it static. Many observers draw attention to the widespread systemic and cultural factors that can encourage consumption and disposability. Cloke (2013) invokes "vastogenic (waste producing, waste dominated and profiting from waste) systems of production, processing, distribution and sales" (628). However, Evans (2011) also encourages us to look "beyond the throwaway society" to consider how daily routines, social relations, and material contexts also reinforce high rates of waste generation. This UK-based research demonstrates that the overprovisioning (and therefore wasting) of food is widespread, but also anxiety-inducing for individuals, suggesting that they have not been entirely subsumed by a culture of mindless consumption.

Waste and wasting can be intimately related to identity formation. Hawkins (2006) discusses the myriad ways that waste informs our ethical sensibilities as people who waste. Gregson (2007) observes that "ridding, along with holding and keeping, is every bit as much part of identity work as acts of expenditure and acquisition" (165). Gregson describes how ridding work is also implicated in the gendered social relations of households, demonstrating the importance of gendered analyses in waste studies (see also Balayannis and Garnett 2020 and Krupar 2012 on applying feminist and queer analytical lenses to pollution and waste). Soma's (2017) research on food gifting and ridding in Indonesia highlights how interclass dynamics can also impact wasting practices.

The difficulties associated with deciding when and how to waste extend to the very nature of waste itself: is this material redeemable? Might it have value for someone else? Is it safe to keep, and is it safe to pass along to someone or somewhere else? Alexander and Sanchez (2019) suggest that the concept of indeterminacy can be useful for understanding that which resists definitive classification as either value or waste (Alexander and Sanchez 2019): "Waste matter often appears as indeterminacy, a form that can be terrifying because it suggests dissolution and indecipherability, something that is either unknowable or uncanny in its hints at previous forms" (3). As an example of the unsettled nature of wastes, Coles and Hallett (2012) trace the fluctuating values of salmon heads at a fish market, describing how they can vacillate between the categories of food and waste. Moore (2012) explores multiple conceptualizations of waste, considering how these center waste "whether because of its inherent qualities (risk, hazard, filth), or because of its indeterminacy (as out of place, disorder, abject), as that which disturbs or disrupts sociospatial norms" (781). She catalogs a variety of framings of waste, concluding that despite their differences, they are similar in their attention to how waste is kept out-of-sight and how it is related to sanitation policies and other social ordering mechanisms.

The contributions of critical waste studies extend beyond the study of materials as wastes; this field also considers how landscapes (e.g., Dillon 2014; Rosa 2016; Runyan 2018) and people (e.g., Wright 2006; Gidwani and Reddy 2011; Oloko 2018) can be considered disposable or waste-able. The designation of disposability can carry stigma, social exclusion, maltreatment, or repressed visibility of these people and places within our societies.

Case studies that may be useful in exploring the diversity and malleability of what counts as waste include examples of circular economies. The Ellen MacArthur Foundation (2017) features a series of case studies of circularity ranging from individual products to industrial parks to regional policy initiatives. Related concepts such as closed-loop systems, eco-industrial parks, industrial ecology, industrial symbiosis, extended producer responsibility, or ecological modernization may also proffer examples of the recovery of waste as a resource. These examples should be balanced with critical conversations about the limitations and underlying assumptions associated with circular economies, however (for example, see Gregson et al. 2015; Baxter, Aurisicchio and Childs 2017). The Zero Waste movement as practiced in municipal policy (e.g., Cole et al. 2014), business models (e.g., Rathoure 2020), or as a social movement (e.g., MacBride 2012; Hannon and Zaman 2018) may also be useful in problematizing what societies waste and why.

Classroom resources for exploring the definition and ambiguities of waste could include Chapter 2 of O'Neill's (2019) book *Waste*, entitled "Understanding Wastes". The Discard Studies blog is an excellent resource for exploring what "waste" means, and Liboiron and Lepawsky's (2018) post "The What and Why of Discard Studies" provides an excellent introduction to the field of critical waste studies, encouraging readers to denaturalize many commonly held misconceptions about waste. They argue for a closer examination of our discards: "it becomes crucial for the humanities and social sciences to contextualize the problems, materialities, and systems that are not readily apparent to the invested but casual observer. Otherwise, waste seems like a technical problem rather than a social, cultural, economic, and political problem". The introductory chapter to Ekström's (2015) edited collection *Waste Management and Sustainable Consumption: Reflections on Consumer Waste* also provides an accessible overview of key issues in contemporary waste studies. Liboiron's (2016) exploration of marine microplastics is an excellent teaching resource that challenges preconceived notions of what waste/pollution is and how it behaves. This paper addresses the difficulties of measuring and assessing the impacts of plastic chemicals, our cultural inability to conceptualize the extent and nature of plastic pollution in marine environments, as well as the difficulties of regulating this type of waste.

In terms of classroom activities, I have used a word association exercise in early meetings to encourage students to excavate their pre-conceptions about "waste" and related terms. I ask them to write down what comes to mind when they think of the words waste, trash, disposable, reduce, reuse, recycle, etc. We then employ a think-pair-share approach where students first consider the discussion prompt on their own, then partner with one other student to compare, and finally return to the larger group to share their insights. I then ask students to draw a conceptual map of waste generation in our society ("where does it come from, and where does it go?"), and we then go through a similar sharing exercise. Some common themes emerge in these exercises, including students' perception that waste is a technocratic issue (rather than a socio-cultural phenomenon), and that material flows are straightforward and legible, modeled around recycling and landfilling systems. Making these assumptions visible at the outset of a learning experience about waste sets the stage for future conversations that demystify and re-socialize the phenomenon of waste.

Power

A critical approach to waste studies implies that we must understand power dynamics in order to understand waste systems, behaviors, and cultural values. Scholars and activists in the environmental justice movement have made important contributions to this field, underscoring the inequitable distribution of environmental harms and environmental benefits across societies. In particular, Black, Indigenous, and other communities of color are more likely to experience the negative environmental impacts of waste. Bullard (1994) defines environmental racism as "any policy, practice, or directive that differentially affects or disadvantages (whether intended or unintended) individuals, groups, or communities based on race or color" (451). In a Discard Studies blog post on "Waste colonialism", Liboiron (2018) articulates how colonial power relations are about access to land: "The assumed entitlement to use Land as a sink [for pollution and waste], no matter where it is, is rooted in colonialism…Exporting these [extractivist] models to other places and then blaming the local people for not properly managing colonial sinks is colonialism". Liboiron's (2021) forthcoming book (*Pollution is Colonialism*) promises to be an important volume on this topic, exploring not only the unequal distribution of pollution's harm through colonial systems, but also the colonial nature of the science that undergirds predominant epistemologies of pollution and waste.

Pellow (2000) posits a theory of Environmental Inequality Formation that centers processes and relationships as the determinants of environmental injustices. In particular, scholars are encouraged to consider procedural and historical factors, the relationships between multiple stakeholders (including resistance), and a life-cycle approach to hazards in order to understand how environmental inequalities arise. When teaching about environmental justice, it is important to frame uneven environmental exposures as the result of structural and societal factors, and not just an accident of materials distribution. Potential case studies for teaching about the power dynamics of waste include siting studies, such as Dhillon's (2017) analysis of how citizen science contributed to environmental justice goals during the siting of a waste transfer station in California. Kemberling and Roberts (2009) investigate the strategic and contextual factors that impacted activists' calls for environmental justice in siting decisions and contamination cases in Louisiana during the 1990s.

While certain places may bear unequal exposures to waste and its stigma (such as the sites of landfills or waste treatment facilities), certain people may also be unequally exposed through their work activities. Case studies of waste workers can be used to investigate the power dynamics of who and what places are exposed to and associated with wastes. This can include formal waste workers in affluent places (e.g., Pellow 2002; Nagle 2013), or informal recyclers working in the Global North (e.g., Wittmer and Parizeau 2018; Chikowore and Kerr 2020) or the Global South (e.g., Gutberlet 2015; Akese and Little 2018; Samson 2019). The power dynamics explored in these case studies are complex, and can include government-community tensions, corporate exercises of power, gendered and racialized dynamics, and interpersonal struggles based in social positioning. Case studies can provide students with accessible windows into the nuance and complexity of these power dynamics.

In class, we have watched segments of the "Indian Givers" film (Alexander and Mallon 2012), a documentary created with the Sarnia/Aamjiwnaang-based Kiijig Collective of Indigenous and non-Indigenous youth. This film depicts the contamination of Indigenous communities and lands by the oil refineries and chemical plants in neighboring Chemical Valley, the heart of Canada's petrochemical industry. It centers the perspectives of Indigenous youth on these unequal distributions of harms. We've also watched videos about examples of environmental racism, such as a Global News story that explained a proposed bill to address environmental racism in Nova Scotia, Canada, giving examples of the legacy of an open dump surrounding the community of Africville, the health impacts of a dump near the community of Lincolnville, and protests against pulp mill wastes mounted by the Pictou Landing First Nation (Pace 2015).

The Story of Stuff Project has produced a series of videos that can be useful in starting classroom conversations. We've watched "The Story of Bottled Water" (2010) to explore the how environmental harms are unevenly distributed, in concert with local and regional stories about the socio-political dynamics of bottled water use. The video focuses on the idea of manufactured demand for bottled water, and the disposal and downcycling of most plastic water bottles. Pairing this video with place-based stories of the paths traveled by bottled water encourages students to understand the myriad social contexts and power dynamics associated with this disposable product. For example, we talk about the Wellington Water Watchers: an environmental organization based in Guelph-Wellington, Ontario, that opposes the extraction of local aquifer water by Nestlé bottling plants. We also discuss the reliance of predominantly Black communities in Flint, Michigan, on bottled water after government mismanagement and disinvestment resulted in lead contamination of the drinking water supply (NRDC Natural Resources Defense Council 2016), and the persistence of dozens of long-term drinking water advisories in Indigenous communities in Canada, which the federal government has perpetually failed to remedy (Stefanovich 2020). These case studies connect some communities' reliance on bottled water (and the associated waste) to the environmental injustices of unequally distributed pollution that flows toward racialized bodies, and governmental abdication of their environmental protection responsibilities in these communities.

Materiality

The materiality of waste refers to its inherent properties: these are the characteristics that influence how waste circulates, how it is disposed of and treated, and how it is understood by the people who encounter it. Some wastes can be hazardous to humans and environments (including nuclear wastes, toxic industrial wastes, biohazardous wastes, and household wastes like paint, some cleaners, and batteries). These wastes require equipment, technological systems, and infrastructures in order to remediate them, and their management can have social, cultural, and political repercussions (e.g., Kinsella 2001; Khoo and Rau 2009). Gregson et al. (2010) trace the fluctuating noxiousness of asbestos through shipbreaking activities: while this insulator is inert when stable and contained, it becomes a hazardous waste when it is disrupted during demolitions, requiring changes to the labor, techniques, and financing of shipbreaking activities. Waste matter can therefore influence systems, social relations, bodies, economic systems, and more, demonstrating a form of agency related to its materiality (Bakker and Bridge 2006; Gille 2010; Liboiron 2016).

Even when wastes are not hazardous, they may evoke reactions that are emotional or affective, such as the visceral disgust that many experience when they smell or see rotting food (e.g., Watson and Meah 2013; Waitt and Phillips 2016). Negative visceral associations with waste are often extended to the people who are read as proximate to waste as a form of stigma (e.g., Reno 2009; Carenzo and Good 2016). Gregson and Crang (2010) warn that an in-attention to the materiality of wastes threatens to abstract waste into the metrics used to describe it, leaving the technologies and practices of waste management to stand in for waste itself: "Caught within a teleological fix, that which is managed as waste is waste, and that which is waste is what is managed" (1026).

Case studies for exploring materiality as a theme in critical waste studies could include wastes that present management challenges, thus asserting their materiality in a disruptive way. This could include toxic substances like nuclear wastes whose materiality requires perpetual management (e.g., Cockerill et al. 2017), or dyed PET bottles that are considered less recyclable, making it harder for informal recyclers in Brazil to participate in the recycling market (Pereira da Silva 2019). Our research group has conducted research on the materiality of food waste that may be of interest to students. Fraser and Parizeau (2018) present a case study of the visceral and affective impulses that lead people to decide that food has become waste in their homes. We have assessed how materiality influences the management of food and food waste across the value chain in Van Bemmel and Parizeau (2020), and Millar, Parizeau, and Fraser (2020) investigates the material and logistic challenges associated with food rescue operations, arguing that this practice should not be seen as a primary solution to either food waste or food insecurity.

An activity that can enliven the materiality of waste for students is a field trip to a waste management center. We have visited the Waste Resource Innovation Centre in Guelph, Ontario, where the municipality manages organic waste, recyclables, and residual wastes. Students can observe separated drop-off sites for hazardous and non-hazardous materials, see the transition of food scraps to soil amendment in an industrial composting facility, and observe the different machinery and technologies required to separate the recyclable stream into its constituent parts. This kind of tour can also highlight the disruptive agency of waste: students see the sorting room where workers pull contaminants off the recycling lines, potentially exposing themselves to hazards. They learned about the disordering effect of plastic contaminants on composting processes, and the different end-uses for compost meeting different standards. Students heard from city staff about shifting materials markets, and how policies in distant places can impact which materials are perceived as recyclable or disposable.

Metabolism and flows

The paths that wastes travel are often obscure or obfuscated. There is a great mystique in contemporary societies about what happens after something is designated as waste. The invisibility of waste generation and treatment has led to widespread misconceptions about modern waste management systems. Many people believe that waste systems are guided by the principles of reducing, reusing, and recycling, but the "Three Rs" are largely a myth in an economic system that encourages increased consumption and that does not incent recycling. For example, the USEPA (2021) has estimated that only 8.7% of all plastics generated in the United States were recycled in 2018 (however, it is not clear whether exported plastic wastes were accounted for in this figure).

The NPR and PBS Frontline documentary "Plastic Wars" (2020) reveals how the plastic industry touted recycling as an environmental solution in the 1980s and 1990s, although industry officials knew that a large-scale recycling system would be costly, ineffective, and likely unfeasible. Creating and selling new plastics was always the primary goal of the industry. The blue box thus became a greenwash for the fossil fuel industry of plastics manufacturing. This documentary would be a compelling classroom resource for students who have been taught that recycling is an important sustainability behavior. It recenters resource extraction and usage as a corporate responsibility rather than an individual one and encourages viewers to look upstream to the systems that generate waste rather than focusing on the moment of disposal.

Metabolism is a concept that describes the interconnected flows of materials, wastes, and beings across ecosystems. Waste generally flows from more to less affluent/privileged places. For example, in the United States, landfills are often located in peri-urban areas where there are higher levels of poverty and higher proportions of people of color (Cannon 2020). Scholars have used the concept of metabolism to demonstrate the connections between political economy and the materiality of systems. For example, Demaria and Schindler (2016) investigate the alliance of middle-class residents and informal waste workers that was forged to resist an urban waste incinerator in Delhi, India. Residents were concerned about the material impact of toxins on their bodies, and informal waste workers were concerned about the prospective loss of income that incineration would represent for them. The flow of waste through incinerators would threaten both group's material well-being, and these tensions precipitated political confrontations with the municipality.

The concept of metabolism can also help us to understand interrupted flows of waste, such as China's recent decision to stop accepting most imported plastics due to the high levels of contamination in this waste stream. This political decision has had material and economic implications for many North American and European municipalities that used to send their recyclables overseas for processing. In some cases, this shift in metabolism has led to more landfilling or incineration of potentially recyclable materials. Some observers suggest that this disrupted flow has the potential to create new material circuits, and perhaps even change sociocultural conditions in locales that have grown dependent on the export of post-consumer materials (e.g., Katz 2019).

In teaching waste flows and metabolisms, MacBride's (2012) book *Recycling Reconsidered: The Present Failure and Future Promise of Environmental Action in the United States* provides an excellent overview of the flows of waste through U.S. society, as well as the policies, practices, and social movements that shape this metabolism. A concrete example that can help students to understand the afterlives of their consumer goods is provided in a CBC News story entitled, "Here's where your donated clothing really ends up" (Jay 2018). This exposé charts how high proportions of donated clothes (which many believe are being resold locally) are actually shipped overseas for resale, used as rags, ground up for reprocessing, or landfilled.

Another persistent narrative that requires excavation is the idea that e-wastes flow linearly from the Global North to "pollution havens" in the Global South. Lepawsky's research complicates the narrative of rich-to-poor waste export by investigating the global paths of e-waste. For example, the case studies in Lepawsky and Mather (2011) explore the "boundaries and edges of the geographies of e-waste" (247) in an engaging way, describing how the authors follow e-waste components through Bangladesh and Canada. This reading can be used to introduce students to the difficulties of tracing the circulation of commodities/wastes across their life spans, especially as waste materials circulate through new cycles of production and consumption. This analysis also highlights the intermingling of "waste" and "value" through the recovery of e-waste, as certain materials fluctuate in and out of usefulness, and in and out of demand.

As a classroom activity, we have analysed text and images from screenshots of Greenpeace's e-waste campaign websites (since removed; some pages are still retrievable via the Way Back Machine). For example, a page entitled, "Where does e-waste end up" depicts a map with one-way arrows traveling from the USA and Europe toward India, reinforcing the linear narratives of global e-waste flows as framed by Western organizations like Greenpeace (2009). This page also includes a photo of a person in Delhi working with metal with their bare hands, accompanied by a caption describing how "e-waste…is broken up by unprotected workers and children". This narrative positions residents of the Global South as victims in e-waste flows, rather than active agents. We analyze the messages presented through this campaign and compare these representations to the flows described in the academic readings we've read on this topic.

Planned obsolescence offers another entry point for thinking about waste metabolisms and flows. Articles like Satyro et al.'s (2018) overview of how planned obsolescence is designed into products can provide examples and rationales for this economic approach to waste-oriented production. In the classroom, we investigate planned obsolescence in our own lives by recording how many cell phones, laptop computers, and desktop computers we have each owned and used in our lives. We also note how and when they were disposed of, and we then compare our answers to Statistics Canada (2016) data that show that the highest proportion of old cell phones are not disposed of at all: they are kept by users who presumably do not know what to do with waste electronics, or do not find disposal alternatives accessible. We also discuss the charts from Lepawsky's (2018) book *Reassembling Rubbish: Worlding Electronic Waste* (Figures 6.5, 6.6, and 6.7) that depict the amount of CO_2 generated at different life cycle stages for specific models of cell phones, laptops, and desktop computers. In most cases, the majority of CO_2 emissions occur during the production of these technologies, rather than in their use or their disposal. Regardless, much of our preoccupation with e-waste pertains to its disposal. This exercise encourages students to consider the upstream impacts of consumption choices, rather than focusing on disposal as the defining environmental moment of a product's life. I've coupled discussions of planned obsolescence with examples from the Right to Repair movement to highlight consumer resistance to this trend. For example, Massachusetts has recently passed a law requiring car manufacturers to provide an open data platform, enabling third-party repairs of their products and making visible data that is being collected and transmitted by car systems (Robertson 2020).

Waste governance and waste regimes

The governance and regulation of wastes and their management are influenced by both state and non-state actors, including governments, scientists and technical experts, citizen scientists and activists, engaged community members, and private companies (such as waste collection companies, waste treatment facility operators, and alternative users of recovered materials). O'Neill (2019) observes that waste governance and regulations have required adaptation to the new global waste economy, and innovations in governance are on the rise. While international agreements have increased over time, most of them cannot address the complexity of contemporary waste flows. Waste activism has become an important influence on waste governance, with social movements and informal waste workers at the forefront of these efforts.

Gille's (2010) theorization of waste regimes represents a macro-level engagement with the "production, circulation, and transformation of waste as a concrete material" (1056). This approach to understanding waste systems and waste governance centers the analysis of power across scales, historical processes, social relations influencing waste generation, and material culture. The regimes framing also highlights the dynamism of waste systems and their governance: as conditions shift, new regimes may emerge. Regimes provide a compelling framework for investigating the place-based governance of diverse waste materials. Gille (2010) explores three waste regimes in Hungary as a case study of this concept.

Additional case studies that could support the discussion of waste governance include Corvellec and Hultman's (2012) discussion of the impact of societal narratives on waste governance in Sweden. This paper unpacks how the narratives of "less landfilling" and "wasting less" re-ordered activities, infrastructure, and business models in Sweden. Similarly, Corvellec (2016) describes the designation of food waste treatment facilities as "sustainability objects", which then are understood as agents of sustainability because they have been designated as such. This act of performative definition gives fluctuating agency to the food waste treatment products, and enables a form of mundane governance of the many actors involved in food waste generation and management. In a case study of urban waste governance, Tuçaltan (2020) highlights how waste governance and urban development are intertwined in Ankara, Turkey, suggesting that this case is an example of eviscerating urbanism. This case study demonstrates how the municipality and private-sector actors were complicit in the wasting of urban land, the subsequent dispossession of informal waste workers and poor residents, and finally the development of a prestige project to transform the area (thus facilitating the inequitable production of value).

In recent years, there has been increased attention to case studies of food waste governance. Mourad (2016) considers competing approaches to food waste management and governance in the United States and France. This analysis reveals tensions between different stakeholders as they championed competing solutions to wasted food across the value chain. Mourad demonstrates that a key distinction can be made between solutions that sought to prevent food waste (epitomizing a "strong sustainability" approach), versus those focused on "weak" efforts to promote efficiencies, recycling, and recovery of food wastes. These competing solutions offered divergent visions of governance, with the "strong sustainability" approach requiring more radical transformations to the regulation of food waste, as well as changes to the power relationships between different stakeholders in the food system. Morrow (2019) is also concerned with food recovery, using the idea of urban food commons to interrogate the governance of food sharing and the associated risks using a case study of public fridges in Berlin, Germany.

Additional readings that could be useful for classroom settings include the case studies of national waste management systems in Ireland and New Zealand presented in Davies' (2008) book *The Geographies of Garbage Governance: Interventions, Interactions, and Outcomes*. Wastewater management is another key topic in waste governance: Pacheco-Vega (2015) analyzes the governance of wastewater in Latin America, emphasizing the need to consider cultural, economic, and technical contextual factors (including social inequalities) when designing wastewater management systems.

The degrowth paradigm makes for interesting classroom conversations around waste regimes. Degrowth can be thought of as "production without economic or material growth, and it encompasses a great diversity of types of economies that might achieve this" (Liboiron 2015). Less growth implies less waste, as well as reimagined systems and social relations vis-à-vis resources and consumption. Proponents of degrowth sometimes articulate visions for how our societies and economies could function differently. For example, Trainer (2012) advocates for a "Simpler Way, centring on the enjoyment of non-affluent lifestyles within mostly small and highly self-sufficient local economies under local participatory control and not driven by market forces" (590).

A classroom activity that I have used to spur conversations about degrowth involves asking students to return to the conceptual mapping exercise that we completed at the beginning of the course, this time as a capstone exercise to consider how waste is governed in our society. The students work in small groups to revisit the waste maps they created at the beginning of the course: "Draw a conceptual map of waste generation in our society. Where and why does this waste occur?" We synthesize these mental maps to create a collaborative class map of waste, which is posted on the wall. The students tend to depict more complex flows in this revisited exercise, and they often note the social, cultural, political, and economic forces influencing these flows. They then work individually to write an answer to the following questions on (salvaged) sticky notes: "How could we redesign this system to create as much waste as possible? (These are worst-case scenarios.) How could we redesign this system to create as little waste as possible? (These are best-case scenarios.)"[1] Students bring up their sticky notes for the worst-case scenarios and apply them to the shared concept map of waste generation where they most apply. We use the extreme examples that they generate (e.g., "all publicly-served food must be served in disposable containers;" "federally enshrine bottled water as a solution to boil-water advisories") to discuss the conditions, procedures, and relationships that contribute to vastogenic systems. We then turn to the best-case scenarios, which often have been honed by first thinking about their opposites. In this exercise, students have described community-based, equity-oriented interventions that reduce our reliance on material goods and decenter consumption as the cultural anchor-weight for our economies.

Conclusion

Just as wastes and their afterlives are diverse and fluctuating, the critical study of waste is also dynamic. I've suggested some approaches, key theoretical ideas, resources, and case studies for teaching and learning about waste in higher education. There are surely additional and emerging themes, voices, and examples that can contribute to a critical approach to teaching waste studies, and I so am committed to renewing my course materials regularly as part of a critical pedagogical approach to this subject.

At the beginning of any new learning endeavor (whether a class, an independent study project, or a thesis), there is often a moment of uncertainty and trepidation—both for me and for the students. What if they are not interested in learning about waste in this way? What if the many demands that they are facing as students in contemporary academe impede them from engaging creatively in this kind of learning process? What if they do not see themselves in relation to vastogenic societies? I have experienced false starts in teaching and advising relationships, and have learned that when given an open invitation, most people can find a connection between themselves and waste, although it may not be in a direction that I would have anticipated. Creating opportunities for students to check in and provide both anonymous and identified feedback on their learning experiences has become a cornerstone of my pedagogical approach.

Centering undergraduate and graduate students in the learning process has been extremely rewarding for me as an educator. The projects that students choose to investigate, the resources that they bring to our conversations, and the approaches they take to researching waste have been constant sources of inspiration and learning for me. My teaching, research, and community service work have all been influenced by the insights and contributions of students. For me, teaching critical waste studies has been an exercise in excavating, illuminating, and demystify the social lives of waste.

Note

1   These questions were inspired by the work of Dave Kranenburg and Caitlyn Colson of the Rhizome Institute, who facilitated a multi-stakeholder workshop on reimagining food waste flows for the project "Building a Research Agenda for Reducing Food Waste in Ontario".

References

Akese, G. A. and P. C. Little. 2018. "Electronic Waste and the Environmental Justice Challenge in Agbogbloshie." *Environmental Justice* 11, no. 20: 77–83. doi:10.1089/env.2017.0039.

Alexander, C. and A. Sanchez. 2019. "Introduction. The Values of Indeterminacy." C. Alexander and A. Sanchez (Eds.) *Indeterminacy: Waste, Value. and the Imagination.* Berghahn. 1–30.

Alexander, I. and S. Mallon (directors). 2012. *Indian Givers.* Rocketship Productions Documentary.

Bakker, K., and G. Bridge. 2006. "Material Worlds? Resource Geographies and the 'Matter of Nature.'" *Progress in Human Geography* 30, no. 1: 5–27. doi:10.1191/0309132506ph588oa.

Balayannis, A. and E. Garnett. 2020. "Chemical Kinship: Interdisciplinary Experiments with Pollution." *Catalyst: Feminism, Theory, Technoscience* 6, no. 1: 1–10. http://www.catalystjournal.org.

Baxter, W., M. Aurisicchio, and P. Childs. 2017. "Contaminated Interaction: Another Barrier to Circular Material Flows." *Journal of Industrial Ecology* 21, no. 3: 507–516. doi:10.1111/jiec.12612.

Bullard, R. 1994. "The Legacy of American Apartheid and Environmental Racism." *St. John's Journal of Legal Commentary* 9, 445–474.

Cannon, C. 2020. "Examining Rural Environmental Injustice: An Analysis of Ruralness, Class, Race, and Gender on the Presence of Landfills Across the United States." *The Journal of Rural and Community Development* 15, no. 1: 89–114. https://journals.brandonu.ca/jrcd/article/view/1737/407.

Carenzo, S. and C. Good. 2016. "Materiality and the Recovery of Discarded Materials in a Buenos Aires Cartonero Cooperative." *Discourse* 38, no. 1: 85–108. https://muse.jhu.edu/article/626073.

Chikowore, N. R. and J. M. Kerr. 2020. "A Qualitative Inquiry into Collecting Recyclable Cans and Bottles as a Livelihood Activity at Football Tailgates in the United States." *Sustainability* 12, 5659. doi:10.3390/su12145659.

Cloke, J. 2013. "Empires of Waste and the Food Security Meme." *Geography Compass* 7/9, 622–636. doi:10.1111/gec3.12068.

Cockerill, K., M. Armstrong, J. Richter, and J. G. Okie. 2017. "The Unpredictable Materiality of Radioactive Waste." In K. Cockerill et al. (Eds.) *Environmental Realism.* Springer International Publishing. 67–87. doi:10.1007/978-3-319-52824-3_4.

Cole, C., M. Osmani, M. Quddus, and A. Wheatley. 2014. "Towards a Zero Waste Strategy for an English Local Authority." *Resources, Conservation and Recycling* 89, 64–75. doi:10.1016/j.resconrec.2014.05.005.

Coles, B. and L. Hallett. 2012. "Salmon Heads, Waste and the Markets That Make Them." *Sociological Review* 60, no. 2: 156–173. doi:10.1111/1467-954X,12043.

Corvellec, H. 2016. "Sustainability Objects as Performative Definitions of Sustainability: The Case of Food-Waste-Based Biogas and Biofertilizers." *Journal of Material Culture* 21, no. 3: 383–401. doi:10.1177/1359183516632281.

Corvellec, H. and J. Hultman. 2012. "From 'Less Landfilling' to 'Wasting Less.'" *Journal of Organizational Change Management* 25, no. 2: 297–314. doi:10.1108/09534811211213964.

Davies, A. 2008. *The Geographies of Garbage Governance: Interventions, Interactions, and Outcomes*. Routledge.

Demaria, F. and S. Schindler. 2016. "Contesting Urban Metabolism: Struggles Over Waste-to-Energy in Delhi, India." *Antipode* 48, no. 2: 293–313. doi:10.1111/anti.12191.

Dhillon, C. M. 2017. "Using Citizen Science in Environmental Justice: Participation and Decision-Making in a Southern California Waste Facility Siting Conflict." *Local Environment* 22, no. 12: 1479–1496. doi:10.1080/13549839.2017.1360263.

Dillon, L. 2014. "Race, Waste, and Space: Brownfield Redevelopment and Environmental Justice at the Hunters Point Shipyard." *Antipode* 46, no. 5: 1205–1221. doi:10.1111/anti.12009.

Ekström, K. M. 2015. *Waste Management and Sustainable Consumption: Reflections on consumer waste*. Routledge.

Ellen MacArthur Foundation. 2017. "Case Studies." https://www.ellenmacarthurfoundation.org/case-studies.

Evans, D. 2011. "Blaming the Consumer—Once Again: The social and material contexts of everyday food waste practices in some English households." *Critical Public Health* 21, no. 4: 429–440. doi:10.1080/09581596.2011.608797.

Fraser, C. and K. Parizeau. 2018. "Waste Management as Foodwork: A Feminist Food Studies Approach to Household Food Waste." *Canadian Food Studies* 5, no. 1: 39–62. http://canadianfoodstudies.uwaterloo.ca/index.php/cfs/article/view/186/224.

Freire, P. 2005. *Pedagogy of the Oppressed: 30th Anniversary Edition*. Bloomsbury.

Gidwani, V. and R. N. Reddy. 2011. "The Afterlives of 'Waste': Notes from India for a Minor History of Capitalist Surplus." *Antipode* 43, no. 5: 1625–1658. doi:10.1111/j.1467-8330.2011.00902.x.

Gille, Z. 2010. "Actor Networks, Modes of Production, and Waste Regimes: Reassembling the Macro-Social." *Environment and Planning A* 42, no. 5: 1049–1064. doi:10.1068/a42122.

Giroux, H. A. 2013. *On Critical Pedagogy*. Continuum.

Greenpeace 2009. "Where Does E-Waste End Up?" https://web.archive.org/web/20190905160947/https://www.greenpeace.org/eastasia/campaigns/toxics/problems/e-waste/.

Gregson, N. 2007. *Living With Things: Ridding, Accommodation, Dwelling*. Sean Kingston Publishing.

Gregson, N. and M. Crang. 2010. "Materiality and Waste: Inorganic Vitality in a Networked World." *Environment and Planning A* 42, no. 5: 1026–1032. doi:10.1068/a43176.

Gregson, N., M. Crang, F. Ahamed, N. Akhter, and R. Ferdous. 2010. "Following Things of Rubbish Value: End-of-Life Ships, Chock-Chocky Furniture and the Bangladeshi Middle Class Consumer." *Geoforum* 41, no. 6: 846–854. doi:10.1016/j.geoforum.2010.05.007.

Gregson, N., M. Crang, S. Fuller, and H. Holmes. 2015. "Interrogating the Circular Economy: The Moral Economy of Resource Recovery in the EU." *Economy and Society* 44, no. 2: 218–243. doi:10.1080/03085147.2015.1013353.

Gutberlet, J. 2015. "Cooperative Urban Mining in Brazil: Collective Practices in Selective Household Waste Collection and Recycling." *Waste Management* 45, 22–31. doi:10.1016/j.wasman.2015.06.023.

Hackett, E. J., O. Amsterdamska, M. E. Lynch, and J. Wajcman (Eds.). 2008. *The Handbook of Science and Technology Studies*. 3rd Edition. The MIT Press.

Hannon, J. and A. U. Zaman. 2018. "Exploring the Phenomenon of Zero Waste and Future Cities." *Urban Science* 2, 90. doi:10.3390/urbansci2030090.

Hawkins, G. 2006. *The Ethics of Waste: How We Relate to Rubbish*. Rowman and Littlefield.

Jay, P. 2018, May 29. "Here's Where Your Donated Clothing Really Ends Up." *CBC News*. https://www.cbc.ca/news/canada/ottawa/here-s-where-your-donated-clothing-really-ends-up-1.4662023.

Katz, C. 2019, March 7. "Piling Up: How China's Ban on Importing Waste Has Stalled Global Recycling." *Yale Environment 360*. https://e360.yale.edu/features/piling-up-how-chinas-ban-on-importing-waste-has-stalled-global-recycling.

Kemberling, M. and J. T. Roberts. 2009. "When Time Is on Their Side: Determinants of Outcomes in New Siting and Existing Contamination Cases in Louisiana." *Environmental Politics* 18, no. 6: 851–868. doi:10.1080/09644010903345637.

Khoo, S. M. and H. Rau. 2009. "Movements, Mobilities and the Politics Of Hazardous Waste." *Environmental Politics* 18, no. 6: 960–980. doi:10.1080/09644010903345710.

King, J. P. 2020, November 16. #ThinkingInSystems Is the Best Way to Help Students Understand and Critique Any Structure. A Program Focused on #SystemsThinking Considers Wholes Over Parts and Emphasizes "#relations" as the Unifying and Connective Tissue Within Every Structure (https://bit.ly/3kA0iLt). 7/10. Retrieved from https://twitter.com/jpkinggg/status/1328370843861258240.

Kinsella, W. J. 2001. "Nuclear Boundaries: Material and Discursive Containment at the Hanford Nuclear Reservation." *Science as Culture* 10, no. 2: 163–194. doi:10.1080/09505430120052284.

Krupar, S. R. 2012. "Transnatural Ethics: Revisiting the Nuclear Cleanup of Rocky Flats, CO, Through the Queer Ecology of Nuclia Waste." *Cultural Geographies* 19, no. 3: 303–327. doi:10.11 77/1474474011433756.

Lang, J. M. 2016. *Small Teaching: Everyday Lessons from the Science of Learning.* Jossey-Bass.

Lepawsky, J. and C. Mather. 2011. "From Beginnings and Endings to Boundaries and Edges: Rethinking Circulation and Exchange Through Electronic Waste." *Area* 43, no. 3: 242–249. doi:10.1111/j.14 75-4762.2011.01018.x.

Lepawsky, J. 2018. *Reassembling Rubbish: Worlding Electronic Waste.* The MIT Press.

Liboiron, M. 2015. "An ethics of Surplus and the Right to Waste?: Discards and Degrowth." https:// discardstudies.com/2015/05/25/an-ethics-of-surplus-and-the-right-to-waste-discards-and-degrowth/.

Liboiron, M. 2016. "Redefining Pollution and Action: The Matter of Plastics." *Journal of Material Culture* 21, no. 1: 87–110. doi:10.1177/1359183515622966.

Liboiron, M. 2018. "Waste Colonialism." https://discardstudies.com/2018/11/01/waste-colonialism/.

Liboiron, M. 2021. *Pollution is Colonialism.* Duke University Press.

Liboiron, M. and J. Lepawsky. 2018. "The What and Why of Discard Studies." https:// discardstudies.com/2018/09/01/the-what-and-the-why-of-discard-studies/.

MacBride, S. 2012. *Recycling Reconsidered: The Present Failure and Future Promise of Environmental Action in the United States.* The MIT Press.

Motta, S. C. 2013. "Pedagogies of Possibility: In, Against and Beyond the Imperial Patriarchal Subjectivities of Higher Education." In S. Cowden and G. Singh (Eds.) *Acts of Knowing: Critical Pedagogy In, Against and Beyond the University.* Bloomsbury Academic. 85–124.

Millar, S., K. Parizeau, and E. D. G. Fraser. 2020. "The Limitations of Using Wasted Food to 'Feed Hungry People.'" *Journal of Hunger & Environmental Nutrition* 15, no. 4: 574–584. doi:10.1080/1932 0248.2020.1730292.

Moore, S. 2012. "Garbage Matters: Concepts in New Geographies of Waste." *Progress in Human Geography* 36, no. 6: 780–799. doi:10.1177/0309132512437077.

Morrow, O. 2019. "Sharing Food and Risk in Berlin's Urban Food Commons." *Geoforum* 19, 202–212. doi:10.1016/j.geoforum.2018.09.003.

Mourad, M. 2016. "Recycling, Recovering and Preventing 'Food Waste': Competing Solutions for Food Systems Sustainability in the United States and France." *Journal of Cleaner Production* 126, no. 10: 461–477. doi:10.1016/j.jclepro.2016.03.084.

Nagle, R. 2013. *Picking Up: On the Streets and Behind the Trucks with the Sanitation Workers of New York City.* Farrar, Straus and Giroux.

NRDC (Natural Resources Defense Council). 2016. "Flint Water Crisis: Everything You Need to Know." https://www.nrdc.org/stories/flint-water-crisis-everything-you-need-know.

O'Neill, K. 2019. *Waste.* Polity Press.

Oloko, P. 2018. "Human Waste/Wasting Humans: Dirt, Disposable Bodies and Power Relations in Nigerian Newspaper Reports." *Social Dynamics* 44, no. 1: 55–68. doi:10.1080/02533952.2018. 1441111.

Pace, N. 2015, April 29. "Environmental Racism Plagues Low-Income and Minority Communities Across Nova Scotia." *Global News.* https://globalnews.ca/news/1968889/environmental-racism-plagues-low-income-and-minority-communities-across-nova-scotia/.

Pacheco-Vega, R. 2015. "Urban Wastewater Governance in Latin America." In I. Aguilar-Barajas, J. Mahlknecht, J. Kaledin, M. Kjellén, and A. Mejía-Betancourt (Eds.) *Water and Cities in Latin America: Challenges for Sustainable Development.* Routledge. 102–108.

Pellow, D. N. 2000. "Environmental Inequality Formation: Toward a Theory of Environmental Injustice." *American Behavioral Scientist* 43, no. 4: 581–601. doi:10.1177/0002764200043004004.

Pellow, D. N. 2002. *Garbage Wars: The Struggle for Environmental Justice in Chicago.* The MIT Press.

Pereira da Silva, T. M. 2019. "A Materially Contextualised Account of Waste Pickers' Marginalisation in Brazil: The Case of 'Rubbish PET'." *Worldwide Waste: Journal of Interdisciplinary Studies* 2, no. 1: 1–9. doi:10.5334/wwwj.26.

Plastic Wars. 2020, March 31. https://www.pbs.org/wgbh/frontline/film/plastic-wars/.

Rathoure, A. K. (Ed.). 2020. *Zero Waste: Management Practices for Environmental Sustainability.* CRC Press, Taylor and Francis Group.

Reno, J. 2009. "Your Trash is Someone's Treasure: The Politics of Value at a Michigan Landfill." *Journal of Material Culture* 14, no. 1: 29–46. doi:10.1177/1359183508100007.

Robertson, A. 2020, November 4. "Massachusetts Passes 'Right To Repair' Law to Open Up Car Data." https://www.theverge.com/2020/11/4/21549129/massachusetts-right-to-repair-question-1-wireless-car-data-passes.

Rosa, B. 2016. "Waste and Value in Urban Transformation: Reflections on a Post-Industrial 'Wasteland' in Manchester." In C. Lindner and M. Meissner (Eds.) *Global Garbage: Urban Imaginaries of Waste, Excess, and Abandonment*.Routledge. 181–206.

Runyan, A. S. 2018. "Disposable Waste, Lands and Bodies Under Canada's Gendered Nuclear Colonialism." *International Feminist Journal of Politics* 20, no. 1: 24–38. doi:10.1080/14616742.2017.1419824.

Samson, M. 2019. "Trashing Solidarity: The Production of Power and the Challenges to Organizing Informal Reclaimers." *International Labor and Working-Class History* 95, 34–48. doi:10.1017/S0147547919000036.

Satyro, W. C., J. B. Sacomano, J. C. Contador, & R. Telles. 2018. "Planned Obsolescence or Planned Resource Depletion? A Sustainable Approach." *Journal of Cleaner Production* 195, 744–752. doi:10.1016/j.jclepro.2018.05.222.

Soma, T. 2017. "Gifting, Ridding and the 'Everyday Mundane': The Role of Class and Privilege in Food Waste Generation in Indonesia." *Local Environment* 22, no. 12: 1444–1460. doi:10.1080/13549839.2017.1357689.

Statistics Canada. 2016. "Trash Talking: Dealing With Canadian household E-Waste." https://www150.statcan.gc.ca/n1/pub/16-002-x/2016001/article/14570-eng.htm.

Stefanovich, O. 2020, September 28. "COVID-19 May Delay Liberal Pledge to End Long-Term Boil Water Advisories on First Nations Social Sharing." *CBC News*. https://www.cbc.ca/news/politics/stefanovich-reconciliation-throne-speech-2020-1.5738098.

The Story of Stuff Project. 2010. "The Story of Bottled Water." https://www.storyofstuff.org/movies/story-of-bottled-water/.

Trainer, T. 2012. "De-Growth: Do You Realise What it Means?" *Futures* 44, no. 6: 590–599. doi:10.1016/j.futures.2012.03.020.

Tuçaltan, G. 2020. "Waste and Metropolitan Governance as Vehicles of Eviscerating Urbanism: A Case from Ankara." *Capitalism Nature Socialism* 31, no. 4: 76–90. doi:10.1080/10455752.2019.1692050.

USEPA (United States Environmental Protection Agency). 2021. "Facts and Figures about Materials, Waste and Recycling. Plastics: Material-Specific Data." https://www.epa.gov/facts-and-figures-about-materials-waste-and-recycling/plastics-material-specific-data.

Van Bemmel, A. and K. Parizeau. 2020. "Is it Food or Is it Waste? The Materiality and Relational Agency of Food Waste Across the Value Chain." *Journal of Cultural Economy* 13, no. 2: 207–220. doi:10.1080/17530350.2019.1684339.

Waitt, G. and C. Phillips. 2016. "Food Waste and Domestic Refrigeration: A Visceral and Material Approach." *Social & Cultural Geography* 17, no. 3: 359–379. doi:10.1080/14649365.2015.1075580.

Watson, M. and A. Meah. 2013. "Food, Waste and Safety: Negotiating Conflicting Social Anxieties into the Practices of Domestic Provisioning." *The Sociological Review* 60, no. 2: 102–120. doi:10.1111/1467-954X.12040.

Wellington Water Watchers. (N.d.). http://wellingtonwaterwatchers.ca.

Wittmer, J. and K. Parizeau. 2018. "Informal Recyclers' Health Inequities in Vancouver, BC." *NEW SOLUTIONS: A Journal of Environmental and Occupational Health Policy* 28, no. 2: 321–343. doi:10.1177%2F1048291118777845.

Wright, M. W. 2006. *Disposable Women and Other Myths of Global Capitalism*. Routledge.

11

HUNTING FOR HIDDEN TREASURES: A RESEARCH METHODOLOGY ON CHINA'S INFORMAL RECYCLING SECTOR

Benjamin Steuer

Introduction

In the field of environmental management in China, waste and waste management have gained enormously in attention over the past decades. After reform and opening, the focus of environmental governance and legislation primarily centered on natural resource degradation, air-, land-, and groundwater pollution (Edmonds 1999; Mu et al. 2014) and it was only in the mid-1990s that the first solid waste law was issued. After this first step, the progress notably accelerated in the early 2000s when the institutional, i.e., rule-based, waste management (WM) system was vigorously promoted by central and local governments (Steuer 2017). Another 20 years later, WM has gained in importance to an extent that the country not only remoulded its own development model and largely centred it on waste recovery, i.e., the circular economy (McDowall et al. 2017); moreover, national WM-related policies, such as the very recent waste import ban, have substantial repercussions on the waste recycling infrastructure of exporting nations.

This particular development trajectory of course bears the question regarding the drivers that propelled the entire dynamic. The common argument regarding the macro-perspective of things is that the People's Republic of China (PRC) was forced to adopt sustainable strategies due its past growth model. Waste recycling, among many other measures, is considered a means to gradually counter the effects of China's traditional resource-intensive and export-driven growth model. Given low per capita availabilities of domestic resource reserves—steel, aluminium, and copper rank around 44–49% of the global average (Zhu et al. 2015)—resource efficiency in production and secondary resource recovery are a must to sustain economic growth. In an economy like China, in which state planning still dominates over market coordination, the government perceives resource security as a key task for safeguarding its envisaged development. Testimony to this strategic outline is the PRC's impetus behind resource conserving policy approaches: Be that the circular economy as national development strategy or the 2019 law for the taxation of mineral resources to manage resource efficiency at the corporate level. Evaluated from this perspective, it comes to no surprise that China's modern WM has been assigned the role of promoting

DOI: 10.4324/9781003019077-11

material recovery (State Council of the PRC 2013). The second motivation for the government to develop its WM system, simply stems from the sheer magnitude of waste generated in urban and rural areas. Between 2003–2018, China's municipal solid waste (MSW) (as measured in treated quantities by the public sector), increased by around 8% (Ministry of Housing and Urban Rural Construction of the PRC 2003–2019). In comparison, the EU-27 only exhibited a median MSW growth rate of 0.4% (Eurostat 2020). The challenging issue here is that waste generation is positively correlated to consumer affluence: So given the PRC's continued economic growth a further rise in waste quantities has to be expected. As unsustainable treatment predominates China's public WM—landfilling (52%), incineration (45%) (MOHURD 2019)—and available space for new landfill sites is limited (Xiao et al. 2018), inducing a shift towards recycling or even non-deformative recovery approaches is evidently a must.

Against this background, it may appear somewhat stunning that China has no coherent recycling industry in the formal sense. While there clearly were efforts to develop an encompassing public WM structure during the late Mao years, initial efforts were thwarted during the reform and opening period of the 1980s. This essential turnaround emerged as a byproduct of system change and inherent governance logic. Similar to former planned economies in Eastern Europe, the PRC initially strived to maximise resource recovery to counter endemic resource scarcities. With economic reforms encompanied by capital-intensive growth and one-dimensional profit-maximisation, recovering and conserving domestic resources had apparently become a costly burden. Formal structures were gradually disbanded and rural-urban migrants soon began to emerge as new dominant players in the recovery of recyclable MSW (Li 2002; Goldstein 2020). The implication is that modern-day WM in China resembles a fragement system, in which stakeholders with different levels of "formalisation" control different stages of recovery and, depending on the recyclable fraction, cooperate or compete with each other. In e-waste management, formal dismantling yards have been emerging since the early 2000s (Schulz and Steuer 2017), however the collection infrastructure is essentially dominated by non-registered collectors (Gu et al. 2016b; Steuer, Ramusch and Salhofer 2018a). A similar picture is reflected in urban recyclable waste management: While formal private and public processing yards for recyclable waste streams exist, research shows that collection is predominated or at least strongly influenced by self-organized waste collector groups (Linzner and Salhofer 2014; Xiao et al. 2018; Steuer et al. 2017a and Steuer, Ramusch and Salhofer 2018a). This constellation, which is frequently challenged by governmental intervention at the municipal level (Tong and Tao 2016), entails substantial implications for the entire WM system: Given standard material recycling techniques, this stage of collection is proportionally much more sensitive for achieving a high recycling rate than the downstream processing steps (Goorhuis 2014). By implication, recyclable waste collection being predominately operated by the "informal recycling sector" (IRS) indicates the sector's potential to co-determine the fate of China's urban recycling industry.

The above-outlined system characteristics of China's urban MSW management, clearly underscore the need for research to go deeper into the specific agency dynamics within. In particular, there appears to be a strong link between the IRS, a term discussed in more detail later, and general waste recyclables such as paper, plastics, metals, glass, and fabrics. Due to these characteristics, the following sections will address two pivotal questions: Firstly, what is known about recyclable material quantities in China and the impact by the IRS? Secondly, given the undisputed influence of the IRS and the need for an assessment of capacities, what are viable ways to study and survey these self-organized stakeholders?

In line with this setting, the following paragraphs start by addressing aspects related to recyclable waste data reliability on the basis of data outputs from government and research. To complement this analysis from the opposite angle, the discussion will then shift to what is known about dimension and recovery performance of the IRS in collection. As this segment's specifics are only little understood, the subsequent section proposes a methodology for surveying the IRS in urban China, which is based on personal field research experience. The final section thereupon concludes that a meaningful explanation of the IRS' recovery effectiveness has to take the inherent institutional, i.e., rule-based, mechanisms into account.

Official Chinese MSW data: reliability and obstacles for verification

Given various references to figures and data below, a word of caution is necessary at the outset: A main issue with MSW numbers from Chinese official repositories and scientific research relates to the way that data is gathered, compiled, and cited. Very often, yet not in all cases, Chinese sources remain opaque on how quantitative WM data has been produced, which in turn demands particular caution from researchers. Acknowledging this crucial issue, the following analysis aims to shed light on the mechanics and difficulties in MSW data generation in China.

From the perspective of generation, urban China is massively contributing to global MSW increases. Based on official records regarding treatment by landfills, composting, and incinerators, the country accounts for approximately 30% of the global total waste generation in 2012 (Gu et al. 2016a). In the PRC, MSW is composed of anything but industrial solid waste, medical waste and construction waste (ibidem) and exhibits a rapid annual growth of around 8%. The major source of generators are private households, which accounts for about 70% of MSW (Gu et al. 2016a), and thus ranks Chinese cities at the upper end of international benchmarks (Christensen 2011). This important source of MSW however presents the first real conundrum in regard to data reliability: Surveys on the same or similar cities based on per capita income indicate an enormous variation in the composition of household waste (see Figure 11.1). Temporal and survey-specific differences may

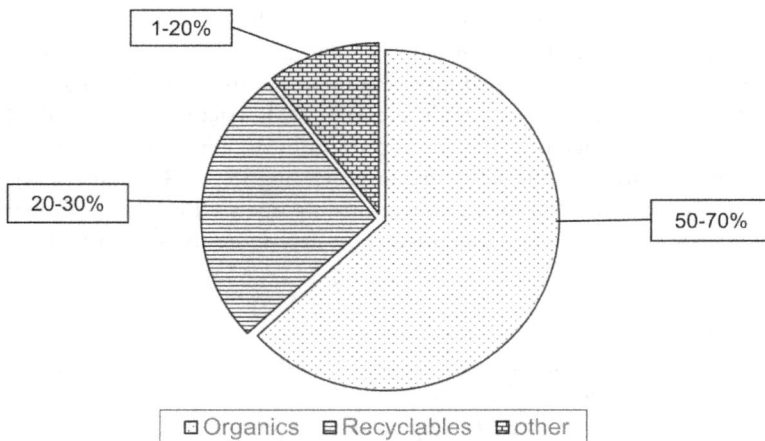

Figure 11.1 Composition of household waste in urban China (adapted from Steuer, Ramusch, and Salhofer 2017b).

explain part of these variations, yet the possibly largest factor of influence is the impact of the IRS and its relation with households. While this aspect will be subject of the next section, these apparent data inconsistencies constitute a substantial problem for WM service efficiency: Without a clear quantitative picture on household waste generation, it becomes virtually impossible to properly design and set up a WM network that is adequately synchronised with waste quantity and streams.

In light of this uncertainty, authorities face the task of improving WM official data collection efforts via administrative means. What on first sight appears to be a basic statistical procedure, is in fact one of the major problems for modernizing the PRC's WM. Essentially, governmental capacities at central and local levels suffer from several, system-specific challenges. Firstly, owed to various socio-economic transitions in the PRC since 1949, the official statistical system has had to repeatedly adjust the means of measuring waste generation. Linked to this problem is the second factor, which was entailed by the above-mentioned dismantling of the official WM system. Due to the cost intensity of WM structures and the political reforms that prioritized economic growth, the public waste re-cycling system was gradually disbanded along with its inherent data provision mechanisms (Li 2002; Fei et al. 2016). Simultaneously, with the engagement of rural labor migrants in waste recovery, recyclable materials left the formal and entered the informal, statistically undocumented realm (Steuer 2018).

Besides this parallel operation of two to three systems—formal public, formal private, and non-registered/informal—the third major impediment pertains to the multiple channels in formal data collection. At the core of the issue is the division of WM-related tasks among several industries. Duties for MSW management, for example, are divided among four central ministries and by implication also among the corresponding subordinated departments at the municipal level. This division not only entails typical infighting for budget allocations, but also impediments for cooperation and data collection on the subject matter (Steuer 2018). The direct impact of this particular arrangement on MSW governance has been discussed by Liu et al. (2017) for the case of Beijing. The process of gathering data on recyclable waste is conducted independently by three agencies and the local statistical bureau, yet without merging the separate findings into one comprehensive set. Neither these administrative units, nor the enterprises they collect the data from follow a unified standard on what actually counts as recyclable waste. Moreover, the authors indicate that only statistical data that suits the interest of recyclers and agencies becomes transmitted and documented. The vested interests in this respect pertain to subsidy provisions, budget allocation and tax arrangements. By implication, data on recycled quantities or fractions that is not relevant to these goals is not collected or reported, which implicitly entails statistical distortions. Potentially the most problematic aspect pertains to the scope of direct data gathering vis-à-vis what is extrapolated. Data acquired by the local statistical bureaus, for example only covers quantities documented by recyclers above a certain size, which are then amended by guess-estimates regarding the generation by small and medium-scaled companies. Similarly, the local Commission of Commerce only requests data reports from large companies, smaller (normally informal) entities can report data on a voluntary basis. As a result, the final instance of recyclable quantities documented by the Commission are composed of amounts reclaimed by large-sized recyclers (50%) and extrapolations from less reliable, non-registered small and medium entities (50%). To make matters worse, this statistical information does not separately record trading among informal collectors and companies or other exchanges (Liu et al. 2017).

An alternative approach, adopted by local Environmental Protection Bureaus under the Ministry of Ecology and Environment is to gather data on quantities reclaimed from residential areas. In principle, quantities delivered by waste collection trucks are weighted at transfer stations so as to assess generation by households covered by the collection route. While sound in theory, this practice falls victim to the realities in China. Direct comparisons between such official and scientific measurement approaches indicate generation differences of up to 0.7 kg/cap/day (Steuer 2018). Tracing the explanation behind it—the point of measurement being communal waste bins for the former and the household level for the latter—one inevitably arrives at the conclusion that a sizable quantity of MSW never made it from the household to the waste bin. Assuming that most of the 0.7 kg/cap/day could be recyclable waste, it is very likely that doorstep collection services offered by the IRS divert this substantial amount, before it ever reaches the formal channels. Relevant observations on this practice were provided in the past (Li 2002; Tong and Tao 2016; Steuer et al. 2017a) and for the larger context of WM data reliability it underscores the significant role of the non-registered collector segment. Such collection activities outside of the formal realm, in turn, diminish the value of most official publications on waste quantities: Statistical compilations on MSW for example appear to completely ignore recyclable waste streams as these only cover the treatment methods of landfilling, incineration and composting (MOHURD Ministry of Housing and Urban Rural Construction of the PRC, 2003–2019). Similarly, the highly laudable annual reports on the recycling industry by the Ministry's of Commerce (Ministry of Commerce of the PRC, 2012–2019) suffers from a basic flaw: Assuming that data is compiled as in the case of Beijing, i.e., 50% direct reporting from enterprises, 50% guesstimates, then there is no way to ascertain the actual dimension of recyclable waste in urban China.

The challenge of obtaining reliable data on recyclable waste in the PRC mirrors what Samantha MacBride has outlined in this book as a general, pervasive hurdle. Jurisdictions, or governance regimes at different levels, exhibit broad variations on how data is collected. Second to that, data is often aggregated up from reports and subject to arbitrary approaches in waste categorization (MacBride, in this volume), which in turn may further distort the means of how generating sources are covered. MacBride's suggestion of "digging down to the jurisdiction of origin…[to] really witness how much is being generated" perfectly aligns with the method proposed in this chapter: Clearly, the missing link to more complete WM data in the PRC is to be found from IRS, yet given the sector's operative nature, conducting research on it requires accordingly adapted approaches. One feasible method to do so is outlined in the subsequent sections.

Dimension, performance, and mechanisms of the IRS

While the previous discussion has indicated various difficulties for official stakeholders to acquire reliable WM data in the past, more recent approaches indicate efforts to improve the situation. Not only have policy experiments through WM pilots, which are testing new modes of MSW recovery, increased over the years (Xiao et al. 2018). The recent push on household waste sorting and separation initiated by the central government moreover attests to the awareness in the formal domain that households play a key role in MSW generation. Interestingly, both instances of activities have to some extent included the IRS by either forcefully or via incentives integrating this particular segment (Tong and Tao 2016; Steuer 2020a).

So what and who constitutes the segment of informal stakeholders in Chinese WM, how did this system come into being, and why does it continue to operate until the present day? Before attending these central questions, a short discussion on the terminology is imperative: The notion "informal" (*fei zhenggui*) in official Chinese discourses on waste recycling, especially discarded electronics or e-waste, has been predominately framed in a negative way. The stigmatisation of this group foremost depicts it as inefficient due to its small scale and "backward" technology as well as environmentally harmful. This in turn leads to the merging of two perspectives, rendering informal as illegal and, by implication, designating a social group as undesirable (Schulz 2019). This problematic generalization features significant shortcomings: First, it does not distinguish between material processing (extraction, dismantling, hydrolysis and pyrolysis), which indeed entails negative externalities for the environment and human health, on the one hand, and collection, transfer and pre-processing, on the other. The latter are widely performed by the IRS and do not necessary lead to secondary pollution. Statements by officials made to the author as well as colleagues (Schulz 2019) that "there exists no informal sector" (Personal communiction with Zhang M., Hubei Province, September 2014) further bear evidence of an intentional ignorance regarding informal WM operations. Second, discussions among official and scientific circles feature a clear obsession with formal operations as opposed to informal ones. The term informal, *fei zhenggui*, loosely translates as "not being of correct standard or rule". Such terminology appears to originate in the compulsive idea that the economy has to be administratively coordinated on the basis of pre-set standards. Anything outside of this regulatory realm can by implication be neither desirable nor effective. The position held in this chapter in part aligns with and in part rejects the official stance. The designation of non-registered waste collectors as parts of an informal recycling system stems from their use of non-formal means of self-regulation and self-governance. Instead of codified laws and rules, the IRS uses informal institutions, such as habits, routines, and conventions to interact and organize their operations. Analytical frameworks centering on institutional dynamics show that it is these bottom-up emerging informal rule structures that enable the IRS to achieve high recovery rates (Steuer et al. 2017 and 2018a; Steuer 2020a). Despite their non-conformity with officially sanctioned rules and technology these stakeholders continue to operate at high levels of effectiveness. This performance achieved on the basis of informal rule structures has been of use for the formal WM domain and in fact created monodirectional dependencies for the entire recycling system. Local administrators have in instances permitted collection by the IRS as their operations helped municipalities to economise on their WM budgets (Li 2002; Steuer, Ramusch and Salhofer 2018b). At the processing stage, the provision of recovered materials, e.g., WEEE, have been vital for sustaining recycling activities at formally licensed plants (Steuer, Ramusch, and Salhofer 2018a). These realities clearly prove that effective alternatives to the official canon not only exist, but furthermore are indispensible to maintaing the recycling economy.

The selection of an analytical framework for assessing the dynamics between the formal and the informal becomes more understandable, when looking at background and emergence of the IRS. Largely composed of rural-urban labor migrants, these stakeholders have left the formal sphere in the wake of agricultural decollectivization in search of opportunities in the liberalized urban environment during the early years of reform. For some, recyclable waste turned out to be the hidden treasures left behind by an increasingly consumerist society. What further facilitated their entry into WM was the dismantling of formal recovery structures to which they offered a broadly welcomed replacement option (Li 2002). Their lasting presence has up until now been explicitly buttressed by formal regulations at the national level: In the mid-1980s and early 1990s, the Ministry of Commerce as well as the State Council, China's chief administrative body, issued regulations that enable the IRS to operate after a simple registration (Ministry of Commerce of the PRC 1985; State Council of the PRC 1991). With few requirements in place and no respective enforcement agency IRS networks began to spread all over the entire country. While municipal governments have oscillated in their regulatory vigour between prohibiting and enabling the IRS, recent policy documents at the central level have again stated efforts for the "integration of the sector". While the State Council has been in favor of a standard formalization approach (SC State Council of the PRC 2011, art. 9), the Ministry of Commerce proposed a more lenient integration that should "make full use of existing waste collector structures" (*chongfen liyong shihuang renyuan*) (MOC Ministry of Commerce of the PRC 2016, art. 6).

This recent stance of the central government becomes more understandable, when looking at key data on performance and dimension of the sector. It is part of China's informal economy that Philip Huang estimated at around 260 million people in 2006 (Huang 2009). As for exact numbers on the IRS, various studies offer guesstimates either for the entire PRC or specific regions. Such assumed dimensions range between 18–20 million people, with annual fluctuations leaving or entering the sector of up to 2 million (Wang et al. 2013; MOC 2015). More sophisticated approaches as offered by Linzner and Salhofer (2014) compare international analyses with research on the sector in China and conclude that that around "0.2–2.5% of the urban Chinese population...[are] involved in informal waste collection and recycling". Similar difficulties arise for attempts to quantify collection quantities as they depend on type of collector and socio-economic characteristics of the urban environment. Literature based summaries assume that collectors handle 17–38% of generated MSW in urban China (Linzner and Salhofer 2014). Findings by the author and colleagues in a subdistrict of Beijing's Haidian district on the other hand showed that informal collectors reclaim around 90% of waste recyclables generated by households (Steuer, Ramusch, and Salhofer 2018a). In e-waste collection, estimations for different Chinese cities record informal collection rates between 50–90% (Steuer, Ramusch, and Salhofer 2018b). Respective interviews with e-waste recyclers on their supply channels in turn shows that between 90–100% of what they process has been delivered by the informal sector (ibidem). Individual daily collection quantities equally vary, depending on the means of transport. In two separate surveys during 2013 and 2016 in Beijing the author and colleagues documented median collection quantities for collectors on foot, collectors with a tricycle and stakeholders with a van to be as diverse as 14–16 kg/cap/day, 80–311 kg/cap/day, and 116–890 kg/cap/day, respectively (Steuer, Ramusch, and Salhofer 2018b).

Acknowledging the various weaknesses of data collection methods, insights on the IRS nevertheless show the segment's substantial capacity to recover recyclables. More significantly, however, are the aforementioned informal rule structures, which build the foundation for the sector's effectiveness. In this regard, the overarching value and interest focus of waste collectors is centred on economic aspects. Having defied the harshness of past political rigidity in rural areas and with a capacity to adjust quickly to new circumstances, labour migrants have acquired particular entrepreneurial-organizational skill sets (Rawski 2008). This particularity often manifested in conversations with the author, when the interviewees mentioned the "need to make a living", "potentially get rich" and "being one's own boss" as key motivators. It must be assumed that out of these shared value visions emerged a sophisticated network structure that spans over entire cities and beyond. Herein, pivotal informal rule-systems evolved relating to a high degree in division of labor, strategies of self-marketing and public relation, techniques of material valorization, as well as information and material exchange systems. The organization of work duties implies different collector, intermediary, transfer, and pre-processing roles depending on spatial and individual coverage specifics. Communication is often made via cardboard signs indicating collectable materials and telephone numbers for doorstep pick-up. Through frequent interaction trust has been established between members of the IRS and their customers, i.e., shop owners and households, which enhances economic certainty. Particularly crucial is the valorisation of materials. Residents in urban China hold clear ideas regarding the value of cardboard, paper, plastics, metals, and other recyclable fractions. The initial transfer between residents and collectors lays the foundation for a profitable and unique value chain: Herein, the marginal material value rises with every transaction further along the IRS hierarchy until the product enters formal or informal processing. All such activities are guided by informal norms and conventions, in which trust through repeated interaction forms relationships that in turn give rise to physical structures: Waste markets (Tong and Tao 2016) or smaller-scaled Trading Points[1] (Steuer et al. 2017) epitomize these networks in the form of physical exchange nodes, where steady customers exchange information, deliver recyclables to specialized intermediaries, and where prices are set according to the market. It is these rule-based techniques of the IRS that not only render urban household waste recyclables into treasures, but also facilitate planning and executing field research on the segment.

A framework for planning, conducting, and assessing field surveys on China's IRS

Having outlined the specific characteristics of China's IRS and its surrounding socio–political context, the subsequent paragraphs aim to provide a guide for field surveys on urban informal waste collectors. The decision to focus on the collection segment of the IRS stems from its crucial role for the entire WM system because informal collectors not only reclaim secondary resources directly from the major generators but their high collection rates bode well for achieving high recycling rates (i.e., the fraction-specific quantity recovered from the entirety of the fraction contained in the waste stream).

It goes without saying that the first steps in preparation of survey activities on the IRS need to be centred on a set of objectives. Based on past experience, key goals are recording collection quantities and achieved turnovers. Besides these more standard aims in WM-oriented research, it is useful to take the aforementioned institutional structures into account. Research by the author and others (Li 2002; Tong and Tao 2016; Schulz 2019) has demonstrated the embeddedness of IRS practices in their own rule-environment. An analytical perspective on this subject matter essentially helps to understand and interpret the two preceding, more quantitative objectives. While these factors need to be kept in mind for the design of the main survey, the initial approach to assess IRS activities in a selected city is relatively straightforward: Reviewing articles on the subject matter as well as conversations with local colleagues provide first valuable insights on size, dimension, and potential locations. Such preparations should be followed by preliminary field observations, in which strolling through streets and side alleys can be very rewarding. Not only major cities such as Beijing, Guangzhou, or Shenzhen, but also smaller ones like Changchun or Jingmen boast a multitude of seemingly non-registered collectors using tricycles (Figure 11.1) or open vans (Figure 11.2) for material recovery (Figure 11.3).

Stakeholders encountered in such a casual manner are ideal sources of information on the local context of WM and the IRS. It must be kept in mind that their work is informal in nature and thus their presence alone may raise sensitivity among residents and potential competitors. Therefore, every approach should adopt the necessary level of politeness and respect, but also understanding if a collector refuses to talk. In the experience of the author, expressing genuine interest in and curiosity about their activity, paired with Chinese language skills, are appreciated and makes them more open to a short conversation. Once a level of understanding is achieved, the most fruitful way is to follow a semi-structured interview.

(a)

(b)

Figure 11.2   Waste collectors on tricycles in Jingmen (Hubei Province) (left) and Changchun (Jilin Province) (right) (© the author).

Figure 11.3   A middleman's truck in Changchun (Jilin Province) (© the author).

It enables the interviewee to provide a good account on the situation, while being guided in the desired direction on the basis of key questions:

• Engagement in WM: ability to make a living?
• Collection quantities and turnover?
• Collected recyclables: specialization?

- Transfer path: sources and destination of waste recyclables?
- IRS network in the area: size, exchange points, routines?
- Local government response to IRS?

Ideally, such preliminary interviews should be conducted with two goals in mind: First, to help verify the size and relevance of the IRS in the selected area and secondly the identification of larger waste-exchange structures. The latter goal is particularly important as such material exchange hubs constitute central, relatively stable nodes, where most of informal material transactions occur. While these take different forms in different Chinese cities, such as small street stores (Li 2002; Steuer 2018), street-market-like Trading Points (Steuer et al. 2017), or even waste villages (Tong and Tao 2016), the functional purpose is generally the same, that is, to reclaim materials from MSW (households) and transfer these to formal or informal recycling.

Based on these insights, subsequent survey preparations need to primarily focus on the designated spatial coverage of the field survey. While essentially being subject to discretionary research budgets, it is highly useful to decide upon several comparable spatial samples. For a project on Beijing, a field survey by the author and colleagues centred on a relatively heterogenous district in Beijing, which itself is composed of several subdistricts. Due to a lack of economic data at the time, the study adopted a sample stratification approach centring on population density, which to some extent served as proxy for household affluence: At the time, higher population density reflected areas with higher housing demand. Accordingly, the subdistricts were categorized along three quartiles of population density—upper (25%), middle (50%), lower (25%)—and four (4) respectively ranked pairs were selected for the field survey. By virtue of this arrangement, it was possible to investigate IRS activities within a sufficiently diverse, yet comparable socio-economic environment.

As outlined previously, the first objective of the main survey was to identify those structural nodes of IRS activities, which could guarantee a certain degree of continuity. Therefore, the initial step was to verify all positions of material exchange hubs in the designated area of the project. Following their identification, fieldwork facilitated the assessment of approximate material volumes flowing through the hubs and the identification of the key actors and groups. Verifying the frequency of Trading Point visits by different groups helped to ascertain the composition of the IRS in each subdistrict as well as the importance of the hub. For later analysis, this assessment enables mapping a hierarchy of significance among the many nodes in a subdistrict as well as extrapolating actor composition and respective collection quantities on a larger scale. Equally, these points were pivotal for administering the main survey. Individually designed for collecting actors as well as purchasing actors, the questionnaires were composed of six sections (Table 11.1).

Most of the questions warrant few further explanations with the exception of Sections 4 and 6. The former essentially serves two objectives. One is assessing secondary resource prices and how they increase in marginal value with every exchange from household, to collector, to middlemen and to formal/ informal recycler. The other is to calculate the profit margin, i.e., selling price minus buying price. By multiplying the per unit profit with the transacted fraction-specific quantities it is possible to extrapolate the turnover of each collector. This outcome can subsequently be verified with the figure provided by the interviewee in Section 5. Interestingly our validations had shown that figures provided by collecting parties were more precise than those given by intermediaries (Steuer et al. 2017).

Section 6 on the other hand helps discover the institutional setting which the IRS employs. These qualitative data allow for the identification of various preferences and the value ideals that affect informal collectors' choices. When it comes to the matter of formal rule structures and

Table 11.1 Survey questionnaire design (based on Steuer et al. 2017a and Steuer, Ramusch, and Salhofer 2018b)

Section	Content	Purpose
1	• Introduction to survey • Ethical standards	• Familiarizing the interviewee with the subject • Expressing interest in interviewee's work • Assuring personal anonymity and safety
2	• Socio-economic background • Activity in WM • Means of transport	• Identification of different groups according to background and activity
3	• Collected quantities (kg/cap/day) • Collected material types (tick-box selection)	• Data on material quantities
4	• Purchasing and selling prices for each material (CNY/kg) or (CNY/unit)	• Data on material value
5	• Turnover (CNY/day) or (CNY/month) • Expenditures (CNY/month)	• Data on economic performance
6	• Sources of transfer (household vs. shops vs. colleagues) • Destination of transfer (location, frequency, diversity & reasons for each item) • Pre-processing of materials (y/n & why) • Reason for engagement within IRS • Impact of governmental regulations (y/n & how) • Changes in practices, routines over time (y/n & why & how)	• Data on institutionalized WM practices

particularly when formal-private or public collectors exist and compete with the IRS, additional attention in questionnaire design could be given to overhead costs and expenditures. Should the survey be extended to include such formal systems, findings could then permit a direct comparison of formal and informal systems on the basis of various factors. Another potential extension relates to including household waste generation in the survey so as to assess the proportion of household recyclables reclaimed by the IRS. In this case, the use of a waste diary has proven effective (Steuer, Ramusch, and Salhofer 2018b). Essentially, the idea is to document daily household waste generation on the basis of two streams—one composed by the entirety of all recyclables (generally collected by the IRS) and one of the remainder. A leaflet handed out to participating household would identify recyclables to ensure consistency. The actual surveying process is done over a period of three to four weeks, via previously disseminated portable scales and waste bags. As for the questionnaire of collectors, a key requirement is to confirm that interviewees generally restrict their collection activities to households and only operate within the same delineated area of household survey.

The concluding step of the survey pertains to the questionnaire's data analysis. After having validated the samples, the standard data analysis in waste management studies covers daily recovery quantities, monthly turnover (i.e., gross income before deduction of overhead costs), composition of recovered materials, and distribution of particular material specialisations among individuals of the different groups. In order to obtain a more refined picture on the economic dimension of IRS operations, it is useful to assess the average marginal value increases of recyclable exchanges along the IRS value chain. The idea is to see how the value of all surveyed recyclable fractions increase along the exchange chain from households to final processing. While the profit is calculated via deducting the buying price from the selling price, data from individual collectors and intermediaries will exhibit substantial variations for the same fraction. To arrive at a more balanced insight, it is first necessary to order the questionnaire samples according to different locations, e.g., either by subdistricts or Trading Points, and in second instance by different IRS group, i.e., collector vis-à-vis intermediary. As a result, data will show that particular fractions are exchanged at different prices in different locations. These findings can be further analyzed in the context of different Trading Points, which in our case provided insight on a hierarchy of Trading Points defined by material values. By implication, it was found that certain larger Trading Points developed into hubs specialised in the trade of a particular waste recyclable stream (Steuer et al. 2017). Another option for using fraction profits is to individually assess the reliability of statements on collector/intermediary turnovers. In this context, statements regarding collected/purchased quantities (kg/cap/day) are multiplied by their profit (CNY/kg) and are then extrapolated to achieve at monthly turnover. For the early study in Beijing the results showed that tricycle collectors' turnover statements were very precise, whereas those by intermediaries showed large discrepancies, possibly due to overhead costs that were not covered in the survey questionnaire. Finally, extrapolations can also be used to assess the dimension of the IRS for the larger context of a city. In our Beijing study, we used the composition of IRS actors at Trading Points on the basis of the initial observational surveys. Taking the median per hour observed actor compositions extrapolations can be done on the basis of TPs/inhabitants in a district or city. Doing so provides more solid, yet not perfect, estimative insights on actual stakeholder size, transacted volume, and network density of the IRS in a surveyed city. This again can help to shed light on the actual impact of informal waste recovery on managing waste quantities in Chinese cities.

Reviewing the significance of institutions for informal waste collection

The previously discussed methods exemplify one feasible approach to better understand the dynamics of recovery, turnover, organization, and exchange activities practiced by the informal domain in urban China's WM. Whatever the chosen approach, it would seem that quantitative assessments alone do little to explain the phenomenon of the IRS. Rather, what should constitute the cornerstone for survey activities and analysis are the group's internal mechanisms of organization and exchange—the rules of the game—of this social group. By extension, observing, studying, and accounting for these informal institutions of the IRS it becomes possible to translate the here presented methodology to areas outside of China. The key lies in identifying informal patterns of material recovery and transaction, which as part and parcel of private economics in waste recovery are inherent to the IRS worldwide. Hence, an adjusted survey approach would follow the baselines of ascertaining patterns of collection and transac-tion, exchange structures and stakeholder roles so as to discern the qualitative underpinnings that explain the quantitative performance of the segment.

Beyond the immediate benefit for managing recyclable waste flows, informal rule systems of the IRS exert an even more encompassing effect. With practically no safety provided by the formal regulatory environment, actors of the IRS crucially depend on strategic routines and norms of behaviour that reduce uncertainty and that provide the means of subsistence. Informal rule systems shared and adhered to within the IRS do essentially constitute a highly resilient system that provides a safety net for its members when engaging in secondary resource man-agement. Coming from outside with the perspective on the latter, understanding the IRS's operations can only be successful if the IRS's rules and the interests behind it are understood.

Engaging in this type of research has its merits as informal systems offer an alternative to the formal public and private systems active in WM. By implication, tracing the roots of informal effectiveness, which in collection are clearly evident, may in turn help to excavate rule-related components that can potentially enhance public systems. The potential for systemic innovation offered by the IRS pertains less to matters of technology, but rather to organizational structures that effectively serve the interests of involved stakeholders (Steuer 2020b). Strengthening the work on this subject matter may by implication discover new sources of systemic innovation and reduce the frequently encountered stigmatization of what is a highly valuable, socio-environmental system.

Acknowledgments

The author wants to express particular gratitude to Professor Stefan Petrus Salhofer (BOKU University Vienna) and Dr. Roland Ramusch (European Bank for Reconstruction and Development), who have over several years provided substantial training and shared essential insights. Without their efforts, the development of research models on the IRS as applied to the context Beijing would not have been possible. Additionally, I want to thank Dr. Li Huijie, who has invested valuable efforts into more recent field research activities in Changchun.

Note

1 The term Trading Point is a direct translation of the Chinese term *jiaoyidian* 交易点, which was developed by the author and colleagues during an earlier research project (Steuer et al. 2017a).

References

Christensen, T. H. 2011. "Introduction to Waste Management." In T. H. Christensen (Ed.) *Solid Waste Technology & Management*. Blackwell Publishing. 3–16.

Edmonds, R. L. 1999. "The Environment in the People's Republic of China 50 Years On." *The China Quarterly* 159: 640–649.

Eurostat. 2020. "Municipal Solid Waste Statistics." https://ec.europa.eu/eurostat/statistics-explained/images/6/6d/Municipal_waste_statistics_10-07-2020.xlsx.

Fei, F., L. Qu, Z. Wen, Y. Xue, and H. Zhang. 2016. "How to Integrate the Informal Recycling System Into Municipal Solid Waste Management in Developing Countries: Based on a China's Case in Suzhou Urban Area." *Resources, Conservation and Recycling* 110: 74–86.

Goldstein, J. 2020. *Remains of the Everyday—A century of Recycling in Beijing*. University of California Press.

Goorhuis, M. 2014. "Developments in Collection of Municipal Solid Waste." In E. Worrell, and M. Reuter (Eds.) *Handbook of recycling–State-of-the-art for Practitioners, Analysts, and Scientists*. doi:10.1016/B978-0-12-396459-5.00026-X.

Gu, B., H. Wang, Z. Chen, S. Jiang, W. Zhu, M. Liu, Y. Chen, Y. Wu, S. He, R. Cheng, J. Yang, and J. Bi. 2016a. "Characterization, Quantification and Management of Household Solid Waste: A Case Study in China." *Resources, Conservation and Recycling* 98: 67–75.

Gu, Y., Y. Wu, M. Xu, H. Wang, and T. Zuo 2016b. "The Stability and Profitability of the Informal WEEE Collector Indeveloping Countries: A Case Study of China." *Resources, Conservation and Recycling* 107: 18–26.

Huang, P. C. C. 2009. "China's Neglected Informal Economy: Reality and Theory." *Modern China* 35, no. 4: 405–438.

Li, S. 2002. "Junk-Buyers as the Linkage Between Waste Sources and Redemption Depots in Urban China: The Case of Wuhan." *Resources, Conservation and Recycling* 36: 319–335.

Linzner, R. and S. Salhofer. 2014. "Municipal Solid Waste Recycling and the Significance of Informal Sector in Urban China." *Waste Management & Research* 32, no. 9: 896–907.

Liu., T., Y. Wu, X. Tian, Y. Gong, and Y. Li 2017. "The Statistics of Recyclable Resources in Beijing: Satus, Problems and Countermeasures." *IOP Conf. Series: Earth and Environmental Science* 178: 1–9.

MacBride, S. forthcoming. "Waste Metrics from the Ground Up." In S. Gille, and J. Lepawsky (Eds.) *Routledge Waste Handbook.* Routledge.

McDowall, W., Y. Geng, B. Huang, E. Bartekova, R. Bleischwitz, S. Turkeli, R. Kemp, and T. Domenech. 2017. "Circular Economy Policies in China and Europe." *Journal of Industrial Ecology* 21, no. 3: 651–661.

Ministry of Commerce of the PRC. 2012–2019. *Annual Report on the Development of China's Renewable Resource Recovery Industry (in Chinese).* http://ltfzs.mofcom.gov.cn/article/ztzzn/an/201605/20160501325666.shtml.

Ministry of Commerce of the PRC. and others. 1985. *Interim Regulation on Urban-Rural Individual Businesses Managing Discarded Material Resources (in Chinese).* No. 5/1985. http://www.12348cn.com/bwgz/gongan/news/bencandy.php?fid=49&id=3247.

Ministry of Commerce of the PRC. and others. 2016. *Opinions on Advancing the Transformation and Upgrading of the Renewable Resource Recovery Industry.* http://www.mofcom.gov.cn/article/h/redht/201605/20160501314905.shtml.

Ministry of Housing and Urban Rural Construction of the PRC. 2003–2019. *Household Waste Statistics (in Chinese).* http://www.mohurd.gov.cn/xytj/tjzljsxytjgb/index.html.

Mu, Z., S. Bu, and B. Xue. 2014. "Environmental Legislation in China: Achievements, Challenges and Trends." *Sustainability* 2014: 8967–8979.

Rawski, T. G. 2008. "Can China Sustain Rapid Growth Despite Flawed Institutions? The Sixth International Symposium of the Center For China-Us Cooperation University of Denver." 1–27. doi:10.4324/9780203834817.

SC (State Council of the PRC). 1991. *Notice on Strengthening the Management of Renewable Resource Recovery and Utilisation (in Chinese).* No. 73/1991. http://www.gov.cn/zhengce/content/2010-12/31/content_5041.htm.

State Council of the PRC. 2011. *Opinions on the Construction and Completion of An Advanced Recovery System for Discarded Products (in Chinese).* No. 49/2011. http://www.gov.cn/zwgk/2011-11/04/content_1986158.htm.

State Council of the PRC. 2013. *Notice on the Development Strategy and Short Term Action Plan for the Circular Economy.* No. 5/2013. Online: http://www.gov.cn/zwgk/2013-02/05/content_2327562.htm.

Schulz, Y. 2019. "Scrapping 'Irregulars': China's Recycling Policies, Development Ethos and Peasants Turned Entrepreneurs." *Journal für Entwicklungspolitik* XXXV 2/3-2019: 33–59.

Schulz, Y. and B. Steuer. 2017. "Dealing with Discarded e-Devices." In E. Sternfeld (Ed.) *Routledge Handbook China's Environmental Policy.* Routledge. 314–329.

Steuer, B. et al. 2017. "Is China's Regulatory System on Urban Household Waste Collection Effective? An Evidence Based Analysis on the Evolution of Formal Rules and Contravening Informal Practices." *Journal of Chinese Governance* 2, no. 4: 411–436.

Steuer, B. 2018. *The Development of the Circular Economy in the People's Republic of China -Institutional Evolution with Effective Outcomes?* PhD diss., University of Vienna.

Steuer, B. 2020a. "Governing China's Informal Waste Collectors under Xi Jinping: Aligning Interests to Yield Effective Outcomes." *Journal für Entwicklungspolitik* XXXVI, nomac_mac 1-2020: 61–87.

Steuer, B. 2020b. "Identifying Effective Institutions for China's Circular Economy: Bottom-up Evidence From Waste Management." *Waste Management and Research.* doi:10.1177/0734242X20972796.

Steuer, B., R. Ramusch, F. Part, and S. Salhofer 2017a. "Analysis of the Value Chain and Network Structure in Informal Waste Recycling in Beijing, China." *Resources, Conservation and Recycling* 128, no. 117B: 137–150.

Steuer, B., R. Ramusch, and S. Salhofer 2017b. "Invisible Markets—The role of households and waste collectors for resource management in urban China, Beijing. Research Project for the Jubilee Fund of the Austrian Central Bank." *Anniversary Fund, Project Number: 16763.* Austrian Central Bank.

Steuer, B., R. Ramusch, and S. Salhofer. 2018a. "Is There a Future for the Informal Recycling Sector in Urban China?" *Detritus Multidisciplinary Journal for Waste Sources & Residues* 4: 189–200.

Steuer, B., R. Ramusch, and S. Salhofer 2018b. "Can Beijing's Informal Waste Recycling Sector Survive Amidst Worsening Circumstances?" *Resources, Conservation and Recycling* 128: 59–68.

Tong, X. and D. Tao. 2016. "The Rise and Fall of a 'Waste City' in the Construction of an 'Urban Circular Economic System': The Changing Landscape of Waste in Beijing." *Resources, Conservation, Recycling* 10: 10–17.

Wang, F., R. Kuehr, D. Ahlquist, and J. Li. 2013. "E-Waste in China: A Country Report. StEP Green Paper Series, United Nations University." https://collections.unu.edu/eserv/UNU:1624/ewaste-in-china.pdf.

Xiao, S., H. Dong, Y. Geng, and M. Brander. 2018. "An Overview of China's Recyclable Waste Recycling and Recommendations for Integrated Solutions." *Resources, Conservation, Recycling* 134: 112–120.

Zhu, J., Y. Yan, C. He, and C. Wang. 2015. "China's Environment. Big Issues, Accelerating Effort, Ample Opportunities." *Goldman Sachs Equity Research*. http://www.goldmansachs.com/our-thinking/pages/interconnected-markets-folder/chinas-environment/report.pdf.

12

WASTE METRICS FROM THE GROUND UP

Samantha MacBride

Introduction

Scholars in the field of waste studies frequently seek out data on waste quantities, trajectories, and material characteristics (Offenhuber and Ratti 2017; Lepawsky 2018). Some are shocked or confounded to find that seemingly simple questions about "how much", "where", and "to what end?" resist straightforward answers. These challenges are not going to go away entirely. Nonetheless, it is possible to contend with waste indeterminacy from the ground up (Gille 2013). Everywhere in the world, direct, continuous measurements are taken as part of routine operations in the sector of commerce and governance known as "waste management" (Watson and Bulkeley 2005; Gregson and Crang 2010). The more we understand these practices, the more powerful we become in knowing, and acting on, waste.

This chapter describes operations, conventions, and practices of measurement within the waste management sector. It is based on my own experience of over 20 years as a government manager and researcher in this field. I draw from cases I am most familiar with in North America, but my points generally hold for formal, and even some informal, waste management systems throughout the world. I present here a primer on how institutions that provide waste management services like collection, transfer, tipping, acceptance, processing, and disposal keep track of wastes within cities, counties, districts, and other geographies that are used to organize these functions.

To this end, I define and discuss six major categories of waste metrics used in waste management today: generation, composition, disposal, diversion, capture, and contamination. As I will explain, these terms apply to rates, as well as tonnages that inform the calculation of rates. I also introduce some new terms. I use the term "disposition" to indicate a crux of decisions and practices that represent a fork in the road for waste travels. A disposition point marks a choice to send discards toward disposal, or alternatively toward some kind of utilization via recycling, composting, or other materials reuse. Toward the end of the chapter, I propose two additional neologisms: defilation (a way to measure harm), and dispossession (to measure displacement of harm and opportunity), which can also be understood through tonnages and other metrics.

For purposes of discussion, I use the term "discards" interchangeably with the term "waste". I acknowledge that these terms have different, though related meanings in the field of discard studies (Liboiron and Lepawsky 2018), but here I employ both to refer to solid materials arising as by-products of extraction, manufacturing, commercial activity, public space usage, and home life.

DOI: 10.4324/9781003019077-12

I use "jurisdiction" in a specific way in this chapter to refer to geographies that have their own operational governance arrangements for waste management (Figure 12.1). Making use of legal scholar Mariana Valverde's concept of "the work of jurisdiction" (Valverde 2009, 141), Josh Lepawsky discusses the importance of such work in the "legal ordering", of discards to achieve contingent "spatial, temporal, and moral patterns of social order", that affect not only waste management, but also production and consumption practices (Lepawsky 2018, 12).

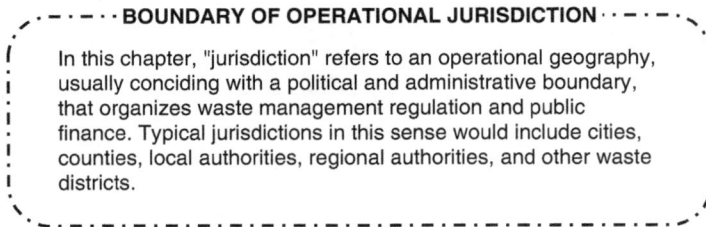

BOUNDARY OF OPERATIONAL JURISDICTION

In this chapter, "jurisdiction" refers to an operational geography, usually conciding with a political and administrative boundary, that organizes waste management regulation and public finance. Typical jurisdictions in this sense would include cities, counties, local authorities, regional authorities, and other waste districts.

Figure 12.1 Boundary of operational jurisdiction.

Here, I extend this notion of work and space to refer to services, rules, and economic arrangements organized by public and private institutions that serve inhabitants of a city, county, or other administrative boundary. Although there may be some broader scale planning involved, operational governance is focused primarily on picking up and moving discards that are generated within jurisdictional boundaries away from points of origin and towards a first point of disposition. This first aspect of disposition is important to distinguish from subsequent stages. As many scholars have chronicled, discards move in complex ways beyond this first point, and they become tough to measure after that. A simple schematic of these concepts is shown in Figure 12.2.

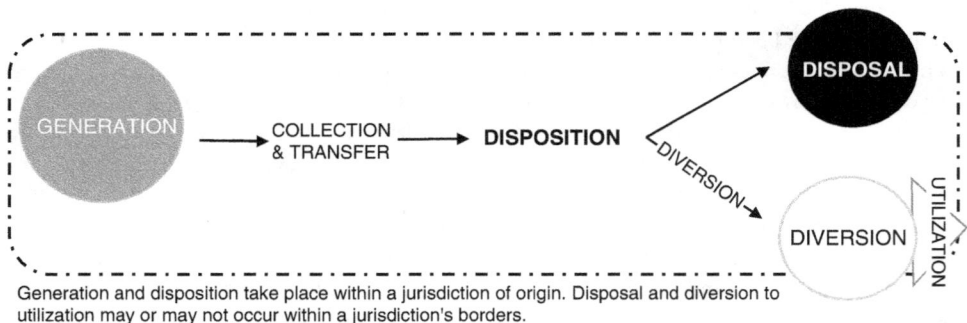

Generation and disposition take place within a jurisdiction of origin. Disposal and diversion to utilization may or may not occur within a jurisdiction's borders.

Figure 12.2 The flow of generation to a first disposition point.

Many wastes continue on, traversing into, through, and out of, other jurisdictions. As they travel, they encounter different structures of organizational governance, each with its own arrangements for waste management. Second, third, and further points of disposition may ensue. These differences multiply as wastes move across multiple jurisdictions, often crossing regional and national boundaries. As I will argue, heuristically limiting attention to the jurisdiction of origin, and the first point of disposition, has the benefit of keeping waste measurements closer to the terrain in which the waste is actually being generated, transferred, and transformed in the physical world. This perspective presents a complement to the study of waste flows, and one that I hope will be useful in people's struggles to improve their living environment and livelihoods in the present day.

Metrics used in the waste management field

Generation

Generation reflects the quantity of discards that arise from any source, regardless of whether these discards end up disposal, or go on to be beneficially used through reuse, recycling, or other methods.

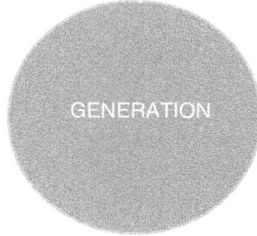

Figure 12.3 Generation.

A common inquiry I hear is, "How much waste do we make?"[1] The term "generation" offers an answer (Figure 12.3). A waste generator can be any person, household, building, or place of work that yields up discards for some sort of removal. Conventionally, generators are classified as (1) residential, (2) institutional, (3) commercial, and (4) industrial (Lagerkvist Meyers 2012). The first three generator types are related to what people consume as part of home and work life. The fourth, far more vast, spans extractive, agricultural, and manufacturing sectors.

Often, an undefined "we" implicitly refers to householders, public/NGO institutions, and commercial enterprises like offices, shops, and restaurants. Together, these sources generate Municipal Solid Waste (MSW), a term that not only connotes a jurisdiction (semi/sub/urban) and phase (solid) but also implies putrescible content. In most jurisdictions, there are different standards of containment, removal frequency, and disposal of putrescible wastes than for discards without putrescible content, such as sorted recyclables (metal, glass, paper, wood, plastic, textiles, or electronics) or inert materials such as sand, dirt, rock, brick, or concrete (Yessiler et al., 2011; Clark, Jambeck and Townsend 2006). Construction and demolition (C&D) debris, which bridges urban and industrial worlds, is generated through building, and consists largely of such inerts, as does fill from excavation. Usually, C&D and fill are considered distinct from MSW, although some jurisdictions report these waste streams combined with residential, institutional, and commercial generation as "municipal waste" (MacBride 2013). Medical discards, as well as consumer wastes with some degree of toxicity (household and commercial hazardous wastes), are covered by distinct regulations and accounting structures (Windfeld and Brooks 2015).

Industrial wastes, whether classified as hazardous or not, arise because of industrial consumption for commodity production. Mining wastes include displaced materials that are brought to the surface and dumped into nearby waterways or back into spent underground mines (Lagerkvist and Dahlén 2012). Agricultural wastes include residues, the excrement and bodies of animals, and plastic sheeting (Ramírez-García, Gohil and Singh 2019). In the realms of oil and gas drilling, as well as intermediate and finished product manufacturing, wastes include cuttings, sludges, slags, and residues that are specific to industrial processes. Power plants generate bottom and fly ash, along with residues from spent air pollution control systems (Dong et al. 2018).

Waste generation is sometimes confused with disposal, which may well be the fate of much or even all of generation, but is not necessarily so. Instead, generation reflects the total weight of what is thrown out, regardless of what happens next (Figure 12.4). Generation is always presented as a total tonnage over a set time period, often normalized on a per capita, per employee, or other basis (Mersky 2012).

Figure 12.4   Calculating generation.

An example of the presentation of waste generation statistics is shown in Figure 12.5, reproduced from the World Bank's *What a Waste 2.0: A Global Snapshot of Solid Waste Management to 2050* (Kaza et al. 2018, 22) (Figure 12.5). Generation rates worldwide have been increasing since they were first measured in the early 1960s, reflecting industrial growth in output, population growth, and increasing GDP in developing nations. On national scales, generation and income are highly correlated; within urban areas, this is generally the case as well (Kaza et al. 2018).

kg/capita/day

	2016 Average	Min	25th Percentile	75th Percentile	Max
Sub-Saharan Africa	0.46	0.11	0.35	0.55	1.57
East Asia and Pacific	0.56	0.14	0.45	1.36	3.72
South Asia	0.52	0.17	0.32	0.54	1.44
Middle East and North Africa	0.81	0.44	0.66	1.40	1.83
Latin America and Caribbean	0.99	0.41	0.76	1.39	4.46
Europe and Central Asia	1.18	0.27	0.94	1.53	4.45
North America	2.21	1.94	2.09	3.39	4.54

Note: kg = kilogram.

Figure 12.5   Ranges of average national waste generation by region.

In the text of the World Bank report referenced previously, the authors clarify that "National Waste Generation" actually means MSW generation, which they estimate at close to 2 billion metric tons total globally. The elision of MSW with total waste is common in aggregated reporting such as this. In fact, when industrial wastes are counted, other sources estimate the global solid wastes at 32 billion tons annually (Dewitt et al. 2018). There is wide variation in what goes into calculating global, national, and regional waste statistics, which are in turn aggregated up from reports made by millions of operational jurisdictions, each making idiosyncratic decisions about what to count as MSW or total waste. Different modelling methods can be used to confront such variations, but only by digging down to the jurisdiction of origin can one really witness the amount of waste being generated, disposed, and diverted. For reasons that this chapter will explore, direct measurement is essential to ensure that waste is managed and dealt with safely and fairly.

Disposition

DISPOSITION

DISPOSITION (in a waste management sense) indicates as set of decisions about whether to send generated wastes to disposal, or to divert them towards utilization.

DISPOSITION is institutionalized at the level of policy, service provisions, and contracting; but disposition decisions are enacted every day when when generators sort materials, and collectors decide where to tip their loads.

Figure 12.6   Disposition.

The word "disposition" (Figure 12.6) is an archaic term of art within the field of waste management (Fetherston 1915). I base this term on the act of "putting down", that follows "picking up" (Nagle 2017). Disposition describes a jurisdictional decision point in which practices and infrastructures split generated materials toward two general ends: semi-permanent containment (disposal) or utilization (diversion). Disposition is inscribed in the design and enforcement of laws and programs requiring generators to separate discards at home and at work, the organization of collection routes, contracts with receivers of sorted and unsorted discards, and technologies that process and/or attempt to prevent discards from migrating into the environment as pollution (Kaza et al. 2018). Disposition decisions are made through municipal political processes, funded with public budgets, and reflect a weighing of economic costs and benefits against political pressures to route discards toward different trajectories. Because disposition arrangements are extensively inscribed in operations, infrastructures, and legal constructs, they are tough to alter without comprehensive, long-term planning.

Disposal

Disposal refers to burying, burning, or submerging discards for a prolonged period of time.Such containment provides the illusion that discards have gone away.

DISPOSAL

Figure 12.7   Disposal.

The most prevalent disposition outcome throughout the world today is disposal (Kaza et al. 2018), (Figure 12.7). Disposal occurs when generated discards are collected, transferred, and consolidated for controlled operations including: sanitary or open landfilling, combustion, or (less frequently today) consolidated ocean dumping. An economic exchange takes place when a generator, or the jurisdiction that operationally governs generators, pays a disposer to accept material at a land-based facility. This service, almost always valued on a per ton basis, provides the appearance and actual short-term function of making materials disappear. This per ton valuation is priced via a "tip fee", that reflects the cost of acceptance at a facility for each delivered ton (*Waste Today* 2019).

Littering, or the uncontrolled dispersal of discards on land and in water, may or may not fall under the definition of disposal. Where there is insufficient investment in disposal infrastructure, littering into waterways or on land serves a disposal function—it is the means by which people keep their living spaces free of wastes that would physically and biologically harm them (Jambeck et al. 2015). In these cases, littering is a rational response to the failure of governance to provide collective disposition service. In zones where there are established systems of refuse collection, littering can instead be understood as an intermediate step before disposition, when litter collected for disposal via street cleaning/sweeping operations (Waring 1897).

Diversion

Discards that are not routed towards disposal are considered "diverted" from disposal, reinforcing disposal as a default condition.

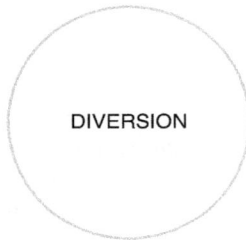

DIVERSION

Figure 12.8   Diversion.

In contrast to the terms "refuse" and "disposal", the term "diversion" is of 20th-century origin when applied to waste management, and it tends to remain within institutionalized discourses still.[2] The term refers both to a direction (towards utilization) and to quantities that do not end in disposal. The term diversion implicitly references a default status of disposal. What is "not disposed of" is "diverted from disposal" (Figure 12.8). This usage reinforces a still common view of waste utilization as exceptional, a view that unfortunately still pervades the waste management industry globally.

What counts as diversion? Among the most rudimentary forms of diversion are filling holes and trenches; among the most sophisticated, the transformation of sorted, cleaned, size-standardized wastes into durable objects of high economic value (Scarlett 2000). All can count equally as diversion in waste metrics, although it is up to each jurisdiction—in dialog with regional or national guidelines—to choose what counts. Among industrial generators, much diversion takes place on the shop floor, with clean industrial cuttings collected and fed back into production. Other forms of industrial diversion include the beneficial use of large-quantity residuals, such as crushed concrete, bottom or fly ash, manures, or crop residues for land reclamation or construction projects (Lyons, Rice, and Wachal 2009). Among residential, institutional, and commercial spheres of generation, most diversion takes place through collections of discards that generators have separated in accordance with local laws and programs. Public- and private-sector collectors drop off or "tip" diverted collections at a facility that will sort, clean, and aggregate materials. Common recyclables include cans, bottles, cardboard, paper, and textiles. Food scraps, garden trimmings, and other putrescibles may be transformed into compost, fuel gases, and/or chemicals.

Decisions on which materials to divert; as well as how to organize, encourage, enforce, and monetize diversion processes, are subject to political as well as economic considerations. Much of contemporary recycling in the urban sphere has been driven by political groups who argue for the diversion of materials in the absence of profit, for example with many types of plastics and glass recyclables. Such groups include environmental advocacy organizations, as well as manufacturing and consumer product sectors that gain from their spent products being deemed appropriate for inclusion in government-subsidized recycling programs (MacBride 2012).

With recycling and composting, jurisdictions pay firms to accept and process recyclables or compostables. Processing costs may or may not exceed the cost of disposal on a per ton basis, depending on geography and demand.[3] Generally, jurisdictions and recycling/compost firms negotiate a per-ton fee that covers processing costs, and shares in some of the revenue from the sale of commodities on secondary materials markets. Depending on what is being processed, as well as market conditions specific to time and place, this may be a negative value (when the jurisdiction pays the recycler or composter) or a positive value (when the recycler pays the jurisdiction).

A logical question is why any jurisdiction would choose to dispose of discards for a fee, as opposed to being paid a fee in a diversion scenario. There are three major reasons. The first is that refuse generally costs less to collect on a per ton basis in cases where it outweighs diversion because of the efficiencies in labor and equipment costs that occur with the clustering of refuse setouts (Bohm et al. 2010). The second is that markets only exist for certain materials in discards (finished compost, metals, paper/cardboard, textiles, electronics, and certain types of plastics), and not others (many other types of plastics and most mixed material products). Third, even for discards with markets, values fluctuate greatly in response to shifts in industrial demand which, in turn, reflect general economic conditions and the supply of raw material alternatives (Chang et al. 2019).

The diversion rate (Figure 12.9) is a fraction that has generation tonnage in the denominator and diversion tonnage in the numerator. This fraction, expressed as a percentage, is arguably the most important current metric by which North American jurisdictions are judged on their commitment and progress towards "sustainability", however defined (Wilson et al. 2012; Tufano 2015). Reliance on the diversion rate to publicize sustainability has pluses and minuses. The diversion rate does indicate a reduction in quantity of material going to disposal. However, it is agnostic on the categories included in the calculation. For example, in Europe, tonnage sent to combustion plants that generate heat and electricity is counted as diversion and placed in the numerator of the rate. In North America, convention designates incinerated tonnages as disposal in most places Figure 12.9.[4]

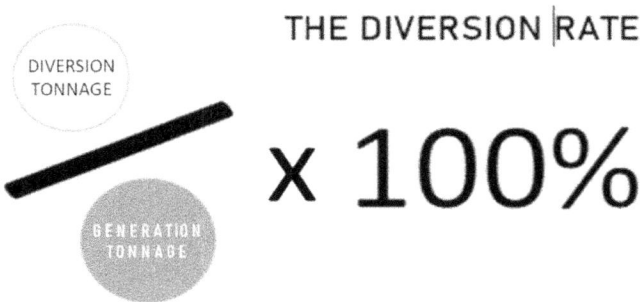

THE DIVERSION RATE

$$\frac{\text{DIVERSION TONNAGE}}{\text{GENERATION TONNAGE}} \times 100\%$$

Figure 12.9 The diversion rate.

The tip scale as a measurement tool

How are disposal and diversion tonnages measured? The most common, direct, and reliable method is to weigh discards as part of normal collection and transfer operations (Figure 12.10). In formal waste management systems, this almost always happens via the tip scale. As collection trucks hauling loads of mixed refuse or separated materials arrive at disposition points—including landfills, recycling plants, compost sites, transfer stations, and incinerators—they greet a gatekeeper and "scale in". Personnel and in some cases IT systems at the gate, called a scale house, record the weight of the full truck, which proceeds on to discharge its or "tip" its load.[5] A second weight is taken as the empty truck departs. The quantification of each delivery is an essential data point that no accepting operation can do without, for the simple reason that each transaction is the basis of a monetary exchange.[6]

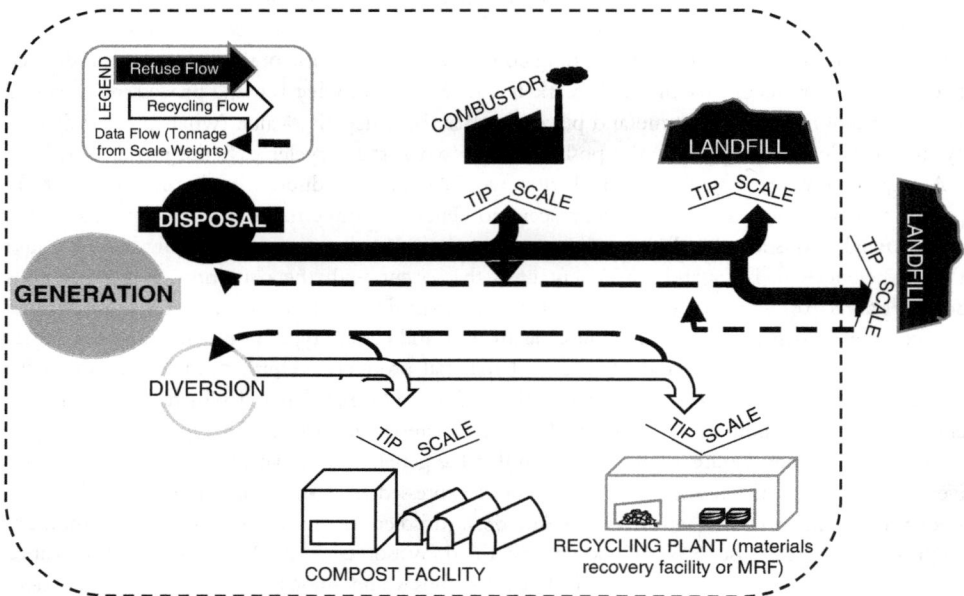

Figure 12.10 Materials and data flows.

Unless figures remain in the head of old-style facility operators (a dwindling informal practice), delivery tonnages will without fail be compiled in electronic or paper ledgers, because discharging or dumping is an economic transaction upon which the fortunes of carting, disposal, and recycling industries depend (Tonjes and Greene 2012; Gregson et al. 2013). The delivered truckload is the unit of measurement, and from this point it is possible to sum up tonnages for a jurisdiction. For this reason, relatively reliable, accurate, and timely data sets exist wherever waste is collected and moved by truck. Loosely similar informational requirements apply to waste management systems using pedestrian, animal cart, bicycle, or van-based collection.[7] In one fashion or another, the moment of tipping involves a record of weight or, less commonly, another metric (unit count, volume) that enables quantification for payment. These records are being kept continuously, day in and day out, and are specific to collection routes that serve groups of generators by category, location, or both.

Tracking weights of disposed and diverted collections as part of routine operations has great advantages over *ad hoc* projects that require special efforts. The literature is full of "waste audits"—research projects that entail weighing discards at the site of generation, *before* they are picked up. Here, workers place discards ready for pickup on a floor scale and record weights themselves (Owojori et al. 2020). Such inquiries have the benefit of being specific to the generator, as opposed to truck scale weights which, by definition, measure entire truckloads corresponding to routes. But waste audits cost money and interrupt other onsite operations, and as such are occasional rather than routine methods of measurement.

The tip scale method of measurement breaks down when wastes are disposed or recycled "on site", which mostly occurs with industrial wastes and may apply to certain municipal phenomena such as backyard composting or barrel burning (still a common practice in many parts of the United States) (Gullett et al. 2001). In these cases, there is no consolidation into trucks, nor a tip scale to impose a moment of measurement. This may be a reason why quantification of industrial generation, disposal, and diversion is more challenging than for MSW, which in turn is one driver of the disproportionate level of attention to MSW as compared to much larger industrial streams (MacBride 2012; Krones 2016). Another gap concerns discards which, as in the case of used goods sale and donation, may never enter the realm of waste management (Lepawsky and Mather 2011). As will be discussed, these gaps may be filled in with modelled estimates based on proxy factors such as sales volume or average generation, or disposal or diversion rate factors derived from other studies (Reynolds et al. 2016).

Composition

Part of a critical assessment of generation and diversion rates, as well as disposal risk and safety, requires understanding the composition of tonnages generated, diverted, and disposed (Figure 12.11). Waste composition is expressed as a set of percentages, and can be applied to generation, diversion, or disposal. Percentages correspond to material and/or product categories of different groups of discards, with some characterizations nearing 100 or more distinctions (New York City Department of Sanitation 2017).

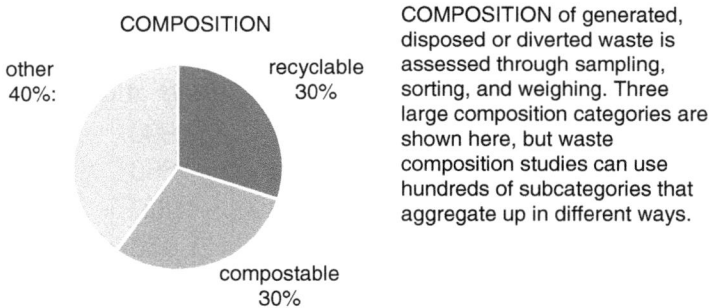

COMPOSITION

other 40%:

recyclable 30%

compostable 30%

COMPOSITION of generated, disposed or diverted waste is assessed through sampling, sorting, and weighing. Three large composition categories are shown here, but waste composition studies can use hundreds of subcategories that aggregate up in different ways.

Figure 12.11 Composition.

The composition of MSW generation in most cities in the developed world follows broadly consistent patterns. In jurisdictions with formal waste management systems, generation is 30–40% putrescible (compostable) materials; 30–40% familiar recyclable items like paper, cardboard, metal, glass containers, e-wastes, and marketable plastics. The remaining 20–40% is of widely varying, hard-to-recover materials including soft, foamed, or unmarketable rigid plastics; textile and rubber items; diapers and bathroom discards; inert constituents; items made of a mix of plastic with wood, glass, or metal parts; and many other products and substances lacking established diversion systems that accumulate in refuse (Kaza et al. 2018).

An understanding of composition is important to any critical evaluation of generation, diversion, or disposal tonnage metrics. Composition is also required to calculate two additional metrics frequently used in the waste management field: capture and contamination. Before continuing, a word is needed on how composition is studied.

Waste characterization

Mixture, or heterogeneity, is a core feature of waste streams. It is usually the case that mingling substances together—in a can, basket, bin, bag, dumpster, storage lagoon, or truck—is the first step in marking them as discards, regardless of whether future disentanglement is planned. Waste characterization is the act of literally "sorting things out" (Bowker and Star 2000) to gain an understanding of the range of substances that make up generated discards as they move from source to destination.

Waste characterization is specific to place, but is not continuous as is tip scale data. Instead, the activity uses statistical sampling to develop a snapshot composite breakdown of what is in disposed and diverted collections, which can in turn be used to estimate generation percentage (Figure 12.12).

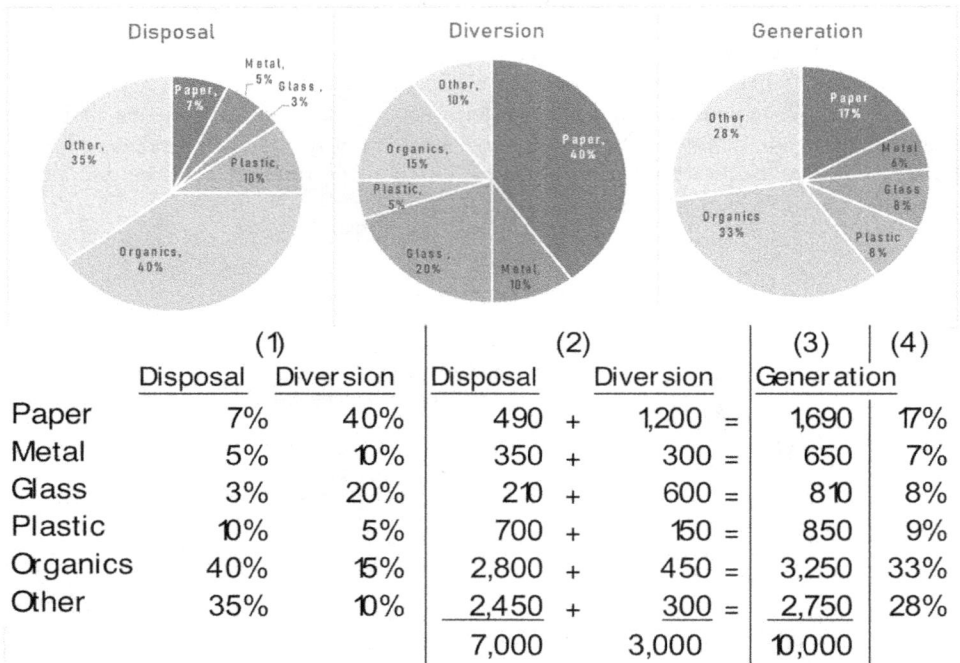

	(1)		(2)		(3)	(4)
	Disposal	Diversion	Disposal	Diversion	Generation	
Paper	7%	40%	490 +	1,200 =	1,690	17%
Metal	5%	10%	350 +	300 =	650	7%
Glass	3%	20%	210 +	600 =	810	8%
Plastic	10%	5%	700 +	150 =	850	9%
Organics	40%	15%	2,800 +	450 =	3,250	33%
Other	35%	10%	2,450 +	300 =	2,750	28%
			7,000	3,000	10,000	

Waste composition percentages are developed by taking random samples of disposal and diversion tonnages at tip scale(s) (before further processing)(1). Applied to the total disposal and diversion tonnage of the jurisdiction, we can calculate estimated tonnages for each material (2). We add disposal and diversion tonnages to get estimated generation by material group (3). From there, the percent composition of total generation can be calculated (4)

sample data for demonstration purposes

Figure 12.12 Waste composition percentages. Totals may not add to 100 due to rounding.

Such studies are costly and usually conducted by jurisdictions every few years. Academic researchers, public-sector officials, and/or consulting firms determine waste characterization categories to answer particular research questions relating to the economic value of collections first and foremost and, to a much lesser extent, the degree and conditions of health and ecological risk.

Classifications build on prior research and are often organized to allow some comparability across studies. Substances may be classified by gross material characteristic (metals, aluminum, plastics), or more fine-grained distinctions that fluidly engage product types and molecular structures—again driven, first and foremost, by economic considerations. In this regard, plans for waste incineration have long characterized the caloric value of discards. More recently, characterizations related to combustion, as well as gasification and pyrolysis processing methods, have divided calorie-laden fractions into biogenic and non-biogenic origins in order to argue that combustion of biogenic waste is a form of renewable energy (Millward-Hopkins and Purnell 2019).

Substantial changes in waste composition take place slowly over time. Looking at waste characterization data longitudinally, what emerges is a gradual but steady influence of production trends on composition. An examination of characterizations of NYC waste over the 20th century, for instance, shows the disappearance of coal ash from home cooking and heating by the 1930s and the appearance of plastics in the 1970s (Walsh 2002). Composition categories and percentages may increase over years instead of decades when, as with electronic waste, new forms of discards enter the mix of collections.

Composition assessed through regular operations

There is an additional source of information on the composition of diverted discards that *is* built into routine operational practice. It is available, if not promulgated publicly, on an ongoing basis and consists of weights of wanted materials that emerge from recycling, scrap, and composting facilities. While waste characterization studies hand-sort refuse and sometimes recycling streams into hundreds of specific categories, materials recovery facilities handling recyclables use automated and manual methods to sort and eject a smaller set of saleable commodities, as well as residue requiring disposal (Figure 12.13). These commodities emerge as compressed, cubed bales of a specific material, such as newsprint, corrugated cardboard, steel

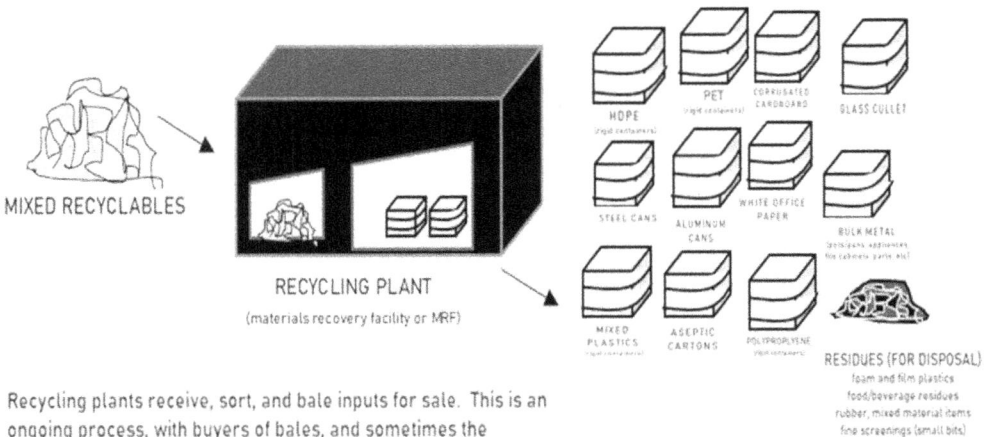

Recycling plants receive, sort, and bale inputs for sale. This is an ongoing process, with buyers of bales, and sometimes the materials mix in bales, changing often in response to market conditions.

Figure 12.13 Recycling plant diagram.

cans, PET bottles, HDPE rigid containers, and the like. Composting facilities use simpler sorting processes. Discards arrive as mixed putrescibles and often go through manual or automated storing stages to remove plastics and large woody items. After moving through biological processes, transformed organics emerge as tons of compost, mulch, or soil amendment, which may again be sifted to remove small pieces of plastic, glass, or metal.

The weight of sorted commodity outputs, as well as ejected residues requiring disposal, are always part of the ongoing recordkeeping of recycling, scrap, and composting facilities because these weights are the basis of buy-sell transactions. As noted above, jurisdictions routinely contract with, and pay, recycling and private waste management firms to accept, process, and market their collections, at times receiving a share of the revenues from sale of marketed commodities. It is the exception rather than the rule, however, that these firms report ongoing composition information is back to jurisdictional governance, especially on an ongoing basis. Recycling firms' reluctance to report stems from intense competition for buyers of separated discards. As compared to disposal and diversion rates, which are reported by jurisdictions as part of governance functions, details on marketed tonnages of recycled commodities may be considered proprietary business information, which can be kept from public disclosure on the grounds that revealing such information would compromise the firm's ability to compete with its rivals. Although reporting regulations differ from place to place, recycling facilities rarely publicize how much they are sending to market and where those markets are, especially when market arrangements frequently change in response to shifting demand.[8]

Managers of recycling facilities, however, possess this information in detail. They monitor the weight of sorted commodities and residues daily. They know the geographic regions of buyers of sorted materials, or at least of firms that broker purchase, and how their willingness to pay varies monthly. This fact, little discussed in the waste management literature beyond a respectful nod to "proprietary information", is one of the reasons that following recyclables, including e-waste, is notoriously difficult (Gille 2013; Offenhuber and Ratti 2017; Lepawsky 2018). Pressing firms to publicly disclose monthly or weekly data on outputs of sorted commodities and residues is often low priority for jurisdictional governance. It is handier to calculate diversion rates using tonnages tipped at the recycling plant. A city, county, or district can then proudly publicize its diversion rate as 60%, 70%, or 80% based on a calculation taken at the first point of disposition. Following where, what, and how much is then moved on to a series of second, third, and multiple points thereafter is not only logistically difficult because of the breadth of players involved, transnational aspects of shipments, and lack of harmonized metrics across jurisdictional borders, it is hobbled from the outset by what can only be described as commercially rational secrecy among recycling firms about where they market the materials that jurisdictions pay them to take.

Capture

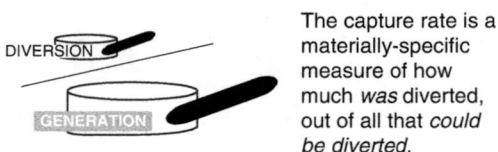

DIVERSION

GENERATION

The capture rate is a materially-specific measure of how much *was* diverted, out of all that *could be diverted.*

Figure 12.14 The capture rate.

The diversion rate measures how much, in total material, is being diverted from disposal. A capture rate, in contrast, measures how much specific material or product, or set of materials and products, is being recovered for beneficial use compared to how much of that same material is generated (Figure 12.14). The capture rate is materially specific; the diversion rate is not. To calculate a capture rate requires knowledge of composition; to calculate a diversion rate does not.

In a simplified example, we start with both a knowledge of summed tonnages from tip scales, and of composition from a fictional waste characterization study (Figure 12.15). Let's say metal pots and pans are 10% of all waste generation in a jurisdiction (which totals 10,000 tons), but 80% of all recycling collections (which total 1,000 tons), for a given month.[9] With these measurements of tonnage and composition, we can then calculate a capture rate for metal pots and pans of 80%.

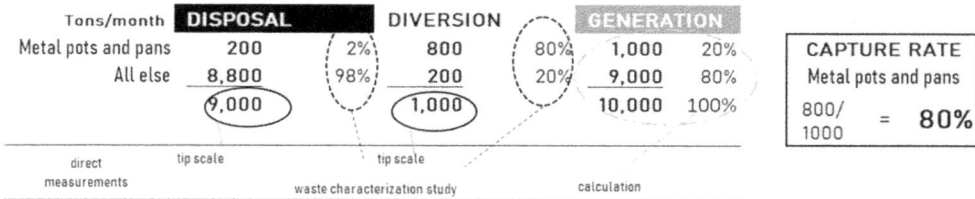

Figure 12.15 Calculating the capture rate.

These calculations show a simple case in which a jurisdiction with a low diversion rate (10%) can also have a high capture rate for a particular material (80%). Diversion and capture rates are related, but not equivalent; and neither the diversion nor the capture rate reflect how much is being generated to begin with overall. Complicating matters is the fact that the boundaries between what is "designated for recycling" or composting at any place and time are not fixed. These decisions, which are formalized in disposition arrangements, alter the fundamentals of diversion and capture rate calculations. One city may include, for example, small scrap metal items like pots and pans in its curbside recycling program, while another may instruct residents to put them in the trash.[10] In the first city, scrap metal collected will be in the numerator and denominator of the diversion rate; in the second, only the denominator. In the first city, we can assess a capture rate for metal pots and pans; in the second, the capture rate is zero because there is no recycling program for that material. Moreover, in the first city, metal pots and pans will count with other forms of formally organized recycling as "clean", while in the second, they will be considered "contamination" of recyclables. This brings us to a sixth concept needed to round out an understanding of waste metrics: contamination.

Contamination

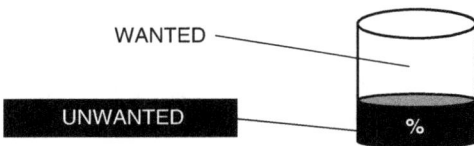

Figure 12.16 Contamination.

In the waste management field, contamination refers to unwanted discards mixed in with wanted discards (Figure 12.16). A contamination rate is shown as a percentage, with the numerator the weight of unwanted material, and the denominator the sum total of both wanted and unwanted stuff. Toxic contaminants like paints, solvents and other chemicals mixed with empty cans and jars create hazard. Long, flexible items like hoses, straps, or bags jam sorting equipment. Traces of food and beverages rot amid recycling—heating piles, drawing vermin, and adding to washing costs. Non-putrescible plastics, metals, and glass move through digestion and composting processes, sometimes releasing substances that mingle with biomass in risky ways. But the main problem for recycling and composting operations is that contaminants must be sorted out and disposed of—and this costs money (Marshall and Bandhauer 2017). The immediate effect on recycling or composting economics may be one reason why contamination is frequently framed by waste management industry voices as a "people problem", reflecting both careless behavior and "wishcycling" (Robinson 2018; Miller 2020). This perspective locates the source of contamination in the individual generator, but there are other drivers of contamination that occur before and after the moment of generation.

The first driver is the heterogeneous makeup of products and packaging. There is enormous variation in types and forms that commodities take, especially in the realm of plastic items. Variation, in turn, drives up costs of sorting for recycling. Only certain types and forms have value on the secondary materials market. Homogeneous, large volume, easy-to-sort discards command good prices because a lot of them are in production. Heterogeneous, hard to separate, and/or small-volume discards lack markets, because they are relatively scarce and the time it would take to accumulate enough to feed a production process costs recycling firms too much (Selwood 2017; The Association of Plastics Recyclers 2020). Much post-processing residue consists of such exceptional items: layers of paper and film plastic, often combined with metal parts; small-volume, dirt-attracting foamed plastics; relatively rare PVC containers that no buyer wants, and so on. Such items seem, to the average person, to be made of plastic or paper or metal and thus appear to logically belong in the recycling bin. Here, the design of products and packaging is an active agent in generating confusion among generators and higher costs at the recycling plant.

A second driver is the mutable definition of contamination, depending on markets. Following the above example, the individual who throws an old pan into the recycling bin is properly recycling in the first city, but "contaminating" recycling in the second. There is nothing prohibiting the firm contracting with the second city to set aside the old pan, along with other valuable ferrous scrap, and sell it when markets are favorable or to refrain from doing so if markets are poor. In such cases, deeming occasionally valuable recyclables as "contamination" gives recycling firms more flexibility in marketing than for officially "designated" materials. Most jurisdictions that contract with recycling firms require them to beneficially use, if not sell, designated recyclables. This does not apply to contaminants.

The flip side of this situation can be seen when new, yet basically unmarketable, items are added to a jurisdiction's list as "designated for recycling". This is often done in the name of "simplicity" of recycling rules, or because it is in the political interest of a producer to have the jurisdiction collect and attempt to market their spent products. In 2012, the plastics industry and mainstream environmental lobbying forced my city, New York, to expand the types of plastics accepted for recycling. Pre-2013, plastic tubs, trays, and other rigid plastics would have been classified as contamination. After 2013, these moved into the category of correct recycling (New York City Department of Sanitation 2017). The extent to which these additions made it to market is simply not a matter of public record. Establishing homogeneity standards for products, that is, reducing the range of different varieties of packaging and products to just a few economy-wide, cross-product versions, would go a long way to reducing generator-driven contamination because instructions on "what to recycle" would be simpler and more stable. Homogenization of packaging types across producers and brands would also help to build markets by driving the accumulation of specific, easy to recycle, and valuable material types (Sipper 2019). But because such standards, if more than voluntarily imposed, impinge on producer choice in selecting the most flexible, lightweight, innovative, or attractive qualities for each product, they tend to be opposed politically by powerful trade groups (Deloitte Sustainability 2017).

Bad generator behavior, heterogeneity in product make up, and shifting definitions of what is recyclable all contribute to contamination. Unless very costly processing steps are taken, most marketed bales emerging from recycling plants will contain some residues, including non-marketable recyclables, remainders of foods or liquids, and the like. Buyers naturally want to limit residue in bales they plan to purchase. It is not feasible to test the residue rate in every bale offered for sale, but it is standard industry practice to do so periodically. International trade in scrap relies on published "bale specifications", which set limits on the level of allowable residue that a bale marketed as a specific commodity may have. Bale specifications are updated annually and are internationally accepted as standards that sellers agree to respect. For example, 2020 bale specifications for PET bottles set a limit of 6% residue for "Grade A" bales and increasing allowances for lower grade bales (ISRI 2020). The trade restrictions that China has placed on the import of scrap have, in fact, been accomplished through the imposition of strict residue minimums for incoming bales, as opposed to, as is often framed in the media, refusal to accept "garbage" as imports (Resource Recycling 2018; Crawford and Warren 2020).

Other perspectives: materials flow analysis and lifecycle analysis

So far, I have described a bottom–up, continuous measurement approach that is integrated into the ongoing waste management operations. This perspective brackets inputs to consumption prior to generation, nor can it effectively follow outputs of acceptance and processing. As I have noted, bottom-up categories and information sources vary greatly from one jurisdiction to another—a common finding among discards researchers (Zeller et al. 2019). In many cases, especially in transnational situations, incomparable categories or the exclusion of certain sectors from regulation leave large knowledge gaps. Here, analysts turn to a quite different methodology, called Materials Flow Analysis or MFA, to measure waste.[11]

MFA tracks inputs, throughputs, and outputs of material systems using the tools of mass balance (Hodson et al. 2012; Fragkou et al. 2014). When used to estimate waste generation, diversion, disposal, composition, capture, or contamination, MFA may incorporate direct measurements taken at the jurisdictional scale, but seeks to expand the range of estimate through modeling (Haberl et al. 2019). Modeling entails the application of factors to predict waste quantities based on other units that are measured—such as production and sales volumes, population, or growth projections. In such cases, an entire economy, or an entire jurisdiction, is taken as a systems boundary, and information is gathered about material flows into, within, and from that economy, including wastes (Newell and Cousins 2015; Rochat et al. 2013).

MFA inquiries use well-documented methodologies and are important tools for conceiving of waste flows in aggregate. However, as all MFA practitioners will tell you, they confront immense problems in the availability and comparability of bottom-up discards data.[12] In addition, much MFA relies on industry-provided statistics on production and wasting activities, which puts potentially regulated entities in the position of supplying data for analyses that may well buttress arguments for or against regulation. MFA is an essential tool for getting a handle on the magnitude and breadth of waste flows, but it works at a relatively high level of abstraction. As a waste scholar, it is important for you to be aware of whether the waste metrics you are interested in are directly measured at a known place in recent time or are, instead, modeled using MFA.

Context and next steps

To recap: a jurisdiction's generation rate gives a sense of the size and scope of discards that must be operationally governed on an ongoing basis. Its diversion rate quantifies the portion of generation that is being routed away from disposal towards utilization. Due to its relative ease in measurement and calculation, the diversion rate offers a widely used metric of urban sustainability, often used in public-facing reports to position jurisdictions as leaders in greening (Tufano 2015). The capture rate provides material-specific information about the efficiency with which different substances and/or products are recovered for use, as opposed to "wasted" in the sense of lost economic opportunity. A contamination rate quantifies the adulteration of potentially valuable substances with unwanted fractions at many stages along the diversion path.

Why care about these metrics, though? Why should, for example, lower generation rate or a higher diversion or capture rate or a low contamination rate be pursued? I have reviewed the obvious commercial reasons: higher capture and lower contamination rates certainly make for more cost-efficient operations and viable secondary markets. Beyond the business case, it seems obvious that a higher diversion and/or capture rate means less trash going to disposal; as does a lower generation rate. Low contamination rates are also taken to mean less disposal of unwanted materials that come in with wanted ones. Less material going to landfills or incinerators seems, in this reflexive calculus, to be always a good thing.

But do these metrics really reflect good things? Not necessarily. It is incorrect to presume that reductions in quantities going to disposal will translate into closure of landfills or incinerators. Landfills normally close because they reach maximum capacity and/or cannot operate in accordance with air and water pollution regulations (Zimlich 2015). The success of recycling diversion in much of northern Europe has, in fact, triggered import of refuse from other nations to keep waste-to-energy plants fed (Yee 2018). What is more, increased recycling or composting can easily lead to greater, not reduced, truck traffic, as collected loads must travel to multiple points for disposition. Finally, more material flowing into spheres of commerce may well enrich the fortunes of private firms, but these firms may be disconnected from local livelihoods, and more allied with extractive and manufacturing industries than green or sustainable enterprises (Huber 2012; MacBride 2012). Lower generation rates, higher diversion rates, and so on, on their own do not indicate reduction in risk and harm from waste management practices anywhere, much less within the jurisdiction that measures and sometimes celebrates these rates.

Environmental justice requires a different perspective. I argue that what is first and foremost wrong with waste is what is *materially* experienced by people and other living things as a consequence of what happens during and after disposition (Fragkou et al. 2014; Dillon et al. 2019). Each and every such material experience happens in a specific time and place, and is set into motion by the output, or release, of substances that cause harm. So, while it is important to think in terms of global waste flows, the extraction and production processes that drive them, and the global effects on resource economies, it is also important to implement waste metrics that are maximally relevant to people's daily lives. From a waste management perspective, this occurs at the level of each waste facility within a jurisdiction.[13]

To this end, I recommend (1) an improvement in the frequency and transparency of re-porting tonnages and rates associated with generation, diversion, disposal, composition, capture, and contamination and (2) organizing these statistics at the level of individual waste facilities within a jurisdiction, as well as the jurisdiction in toto. I also propose two additional sets of direct metrics. One I call defilation, which refers to a rate of both air/water/ground pollution and solid refuse emanating from facilities that, sooner or later, accumulates to defile bodies, foodwebs, and ecosystems. The second is dispossession, a term I use to indicate the degree to which solid outputs are exported from the jurisdiction of origin. No wheel reinvention is proposed here. Systems of data gathering for these metrics are in large part already integrated into routine waste management operations, annual reporting requirements to regulatory agencies, and marketing/expenditure accounting within firms. The point is to make them public in a timely and user-friendly manner—a condition decidedly not in place currently.

Defilation: a measure of waste's localized, ongoing harm

There is widespread consensus that transfer and disposal facilities and the trucks that deliver to them pollute air, water, and land. As the disposal industry loves to point out, however, landfill and combustion technologies are far less polluting than in past generations (Blight 2011). Both perspectives are accurate. Today's sanitary landfills are designed to hold and treat leachate ef-fluent, monitor and mitigate methane and other air toxics and GHG emissions, and promote long-term stabilization of contents (Cossu and Stegmann 2018). They may do this in ac-cordance with or even exceedance of all applicable regulations, but emissions are still a fact of operation (Ehrig and Robinson 2010; Shu, Zhu, and Shi 2019). Even the most efficient landfill gas collection systems control only a part of air releases, with fierce debate as to ratio of fugitive to captured emissions (Karidis 2016). Likewise, today's mass burn combustion facilities feature emission control technologies that remove and consolidate hazards from flue gases and waste effluents, concentrating them as hazardous solid wastes requiring special disposal. These facilities ship fly and bottom ash to designated ash fills with possibly more stringent design requirements than MSW landfills (Mukherjee et al. 2020). They may even sift their ash for metals which are then sold on scrap markets and recycled (Šyc et al. 2020). The disposition of these solid outputs may well take place in accordance with all applicable laws. In all cases, however there is still (1) the liberation and migration of elements and compounds bound up in manufactured commodities into air, water and ground; and (2) the release of carbon and nitrogen in greenhouse gas forms.

These unwanted outputs happen for recycling, reuse, composting, anaerobic digestion fa-cilities, too. For a period in the 1990s, it was fashionable for free-market libertarian critics of recycling to argue this point. With delight, these voices highlighted instances in which life cycle analyses (LCAs) revealed a greater ecosystemic burden for recycling than, for example, com-bustion with energy recovery. Such claims led to a spate of improved LCA research and a concerted defense of recycling against its critics (Denison 1996; Hershkowitz 1998). Today, many of the same dialogs are reprised with regard to, for example, the environmental harms or benefits of bans on plastic bags (Morris and Christensen 2014). But the fact remains that any operation that receives and handles material inputs will generate emissions of pollutants and solid wastes, along with marketable outputs such as recycled paper, metals, or compost.

As with facility-level diversion, capture, and contamination metrics, a defilation metric would require that facilities compile and report inputs by generator category and/or broad material group, as well as outputs to air, water, land deposit, and sale—using classification systems specific to each jurisdiction (as presently). It would apply to transfer stations, landfills, waste incinerators, gasification and pyrolysis plants, compost sites, anaerobic digestion facilities, scrap yards, refurbishment enterprises, used goods exchanges, and industrial generators of pre-consumer scrap and residues, whether sent to on-site or off-site disposal. For solids, all metrics would be informed by incoming scale and/or outgoing bale/residue weights. A defilation metric would measure the weight of unwanted output tonnages, including residues, ash, spent pollution control equipment, as well as sludges, hazardous wastes, and refuse deposited in/on land, or exported for disposal elsewhere. Such measures should be reported on at least monthly, or at a regular timeframe that would correspond to the actual frequency of collection and disposition of such tonnages.

For releases to air and water, continuous environmental monitoring (CEM) is the only responsible way to publicly share pollution information. CEM systems, which use sensors to detect and report releases in real time, exist widely in industrial uses and have, as a result of justice-oriented activism, been implemented to monitor air releases in some waste-to-energy plants in North America and Europe. The implementation of CEMs has been hard fought by justice coalitions and is scarcely complete as a project of protection from harm (Brugh, n.d.; Gilligan 2013). At the same time, many aspects of CEMs—including testing protocols, sensors, and communication systems, are technologically mature and are increasingly used in industrial process control (Docquier and Candel 2002).

As with any of the existing metrics I have described previously, specific questions would require sub-calculations. There could be, for example, specific defilation metrics for hazardous, inert, and all other solid outputs to the environment. Air and water emissions, which are measured on vastly different scales than solid wastes and feed differently into various modelled estimates, would be reported separately, but comparably, across facilities following existing or amended models of regulatory reporting. When combined with a dispossession metric (discussed below), defilation could also be broken into categories of harm staying within the jurisdiction, as opposed to being passed along to another jurisdiction through export. With little difficulty, truck trips could also be quantified as arriving at and leaving the facility to paint a picture of transportation-related air emissions for any waste facility.

Possession and dispossession

It may seem odd to talk about possessing discards, when so much effort is put into sending them away. But, as we know well, discards are liminal and straddle multiple dichotomies (Gille 2007; Hird 2012). Demands of possession and dispossession reveal one such duality. Unwanted solid discards, if sent away, displace risk along lines of entrenched power. Such dispossession reflects and reinforces patterns of historical and current violence against Black and Indigenous peoples, and people of color; and reinforces geographies of exclusion based on gender, race, ethnicity, class, and ability (Taylor 2014; Liboiron 2021). This form of dispossession renders conditions from which waste was sent more livable, and more profitable, for the people-in-power and firms that remain (Harvey 2005; Samson 2015). But other discards—if managed well *and* equitably—offer opportunity as substrates upon which to add labor and build value. And beyond economic value, some may well be sources of utility and beauty to those who need and want them. When these materials—via disposition arrangements inscribed in institutions—are blocked from groups within a jurisdiction that might seek to utilize them, this also constitutes dispossession.

Disposal, recycling, and composting firms are well aware of the dynamics of dispossession from a business calculus. Such dynamics make export of goods *and* risk economically rational. Recycling firms also know the flip side: the headaches of possession of discards that resist valorization due to their material composition. Even when sorted, these materials are hard or impossible to sell, and it falls to jurisdictional governance to fiscally prop up their diversion. This approach continues today each time a jurisdiction pays more to collect and process recyclables than it does to dispose of them (which is common). Combustion facilities producing electricity and heat are in a similar situation in that they cannot sustainably operate without revenues from tip fees. This particular form of waste management seeks to possess discards both as a conduit for revenue, as well as material feedstock. Finally, all polluting facilities entail social costs of externalized harm that will be reckoned with both morally and financially in human, nonhuman, and ecosystemic health for centuries to come. In all cases, transfers of present or future public funds and spaces to the private enterprises that run waste facilities constitute a subsidy from which it is reasonable to expect recompense.

The recompense can be paid in part in transparent, real-time facility level reporting. But it should also include sharing wanted discards within the jurisdiction, and counter-subsidizing the development of local economies that receive and transform them. Diverse social movements organized around alternative economies call for access to wanted discards as part of their demands for positive environmental justice (Gutberlet 2011, 2012). Community composters and allied movements uniting around urban farming and progressive agriculture, thirst for clean, high-quality input material for transformation into compost at small to medium scales. In the Zero Waste movement, which has historically focused on reuse, repair, and recycling (while fully endorsing composting), there is ongoing support for local NGOs and for-profit firms to start and run smaller scale, renewable enterprises built on the reuse of textiles, furniture, and other durables; as well as to support small- to medium-scale recycling firms not integrated into larger-scale waste management corporations. These social movements hold that materials are more likely to be transformed safely into items of need, creating good jobs and building self-reliant enterprises, when the materials are located within or near to the jurisdiction of origin.[14]

A strong case can be made that offering local incentives, access to land, and a guaranteed supply of input materials should be a requirement for any larger-scale recycling or disposal operation that profits from jurisdictional disposal or diversion contracts. If a multinational firm is, for example, awarded a contract to accept and process residential recyclables, its contract should ensure a transfer of some materials and seed funding to cooperatives, NGOs, and community-based enterprises that accept and valorize portions of its output. If a waste-to-energy conglomerate is paid to accept and combust refuse, then it owes its neighbors not only full transparency about its role in defilation, but also monetary and logistical support for local activities that retrieve and utilize what would otherwise be combusted as refuse.

Elsewhere I have cautioned against waste management planning based on the expectation that small- or even medium-sized diversion enterprises, if they proliferate and network, can manage urban wastes on their own, obviating the need for disposal and larger-scale composting or recycling facilities (MacBride 2012). I stand by this argument. In cities, especially large cities, the concentration of materials—in particular unsorted and unmarketable portions of refuse and commodities like plastics that are of little use to buyers other than large-scale producers—must presently go through large industrial facilities for management. This condition will hold as long as heterogeneous, industrial, and/or hard-to-recover materials dominate consumption. Stemming the source of these materials is a radical political project that will take decades of fight to implement. As of today, product bans, consumption-fees, and moral suasion have yet to more than minutely reduce discards generation in most areas of the world—with the possible exception of food wastes, which occupy a far different status as discards than do products and packaging.[15] In other material realms, "producing less" and "consuming less" are at once vitally important, and structurally incompatible with the current systems of production and consumption that dominate national and global economies. Therefore, while it may be noble to call for "disruptive" economic restructurings to usher in the blessings of a future circular economy, it is even more important to be open about what happens on the ground, in place, now.

A measure of dispossession would quantify outputs from a facility that remain within the jurisdiction versus those that are exported to regional, national, or international buyers. In combination with specificity as to generator category and composition, we would see some tonnages remaining and being beneficially used within a jurisdiction, and many of others exiting the boundaries to travel a range of distances to their next facility landing site. Staying closest would be heavy, inert materials such as fill and rubble. Other materials, such as glass and compost (neither of which have global markets) would tend to land in jurisdictions relatively near to those of origin. Heavily traded scrap commodities such as metals, certain plastics, paper, and cardboard would move outward toward domestic buyers, and would also cross national boundaries extensively. Items such as textiles would show a complicated profile, with high-demand clothing staying local but much other clothing moving across borders.

A dispossession measure would not be meant to take the place of the maddening challenge of following flows, especially on their transnational peregrinations. In an ideal world, there would be enough transparency in flow data to be able, at any moment, to respond to the question, "Where does my recycling (or trash, or food waste) go?" It is doubtful we will ever reach this ideal, which would require intricate degrees of surveillance on a continuous basis using comparable metrics across national and subnational borders (Lepawsky and Mather 2011; Offenhuber and Ratti 2017; Lepawsky 2018). In the meantime, the current method of following things, especially discard tonnages, is not only tough, it only tells a story that is particular to time, a set of geographies, and circumstances. It is an important analytic perspective, but is not strictly necessary to address the immediate concerns of people living and working in jurisdictions.

Another dimension of dispossession, built into a facility-jurisdictional approach through geographic specificity, has to do with "Land" and its relationship to waste colonialism. Discard scholar Max Liboiron describes colonialism as "a system of domination that grants settler access to Land for settler goals" (Liboiron 2018, n.p.). Citing Kimmerer, Liboiron distinguishes land in the common usage from its capitalized alternative: "Land" which is, "about relations between the material aspects we might think of as landscapes—water, soil, air, plants, stars—as well as histories, spirits, events, feelings, and other more-than-humans" (Kimmerer 2013, 13). Historically, a defining feature of colonialism entailed the extraction of raw materials from colonized Lands, and their export to colonizer spaces of production and profit (Barham and Bunker 1995; Bunker 2005). The use of Land to site facilities engaged in disposal and diversion can be understood as spatial and material manifestation of waste colonialism. These facilities, which operate under varying degrees of secrecy behind factory gates, occupy Land originally taken from others. Now owned by settlers under settler law, this Land is continually used, with varying degrees of defilation and dispossession, to absorb and redirect discards. At the very least, firms that have settled in places (and who have been formally identified and regulated through the imposition of permits, also granted under settler law) owe complete transparency, at a frequency that rivals that of other spheres of commerce, on the nature and fate of waste inputs and outputs coursing through those Lands.

Conclusion

In this chapter, I have presented features of present-day waste measurements that lead to widely accepted metrics within the waste management field. I have discussed the benefits of an on-the-ground, continuous measurement approach in the context of current harms from unwanted outputs of waste operations, and trends of dispossession of resources from people who live and work near these same operations. I have stressed the importance of each jurisdiction requiring frequent transparency for all waste facilities located in its borders as to the quantity, nature, and general direction (if not precise destination) of wanted and unwanted outputs.

Through my own experience, I am painfully aware that the establishment of reporting systems I am suggesting will be costly and onerous, requiring increased budgets within operational governance, and increased costs of compliance to firms. At the same time, what I am suggesting will not require the invention of new systems of measurement, accounting, or data sharing. It is, furthermore, owed to the public. Obviously, the details and requirements of such reporting would need to be built using extensive engagement with the very people who seek reduction in defilation and dispossession, and who organize in grassroots organizations often kept at arm's length from institutions of jurisdictional governance. Part of the strategy of distancing is built on specialized knowledge of how waste measurements are taken, how data are stored and protected, and how a generic, timeless, and vague story of "what happens to my recycling (or trash, or organics)" can be put forward, often in conjunction with a reported high diversion rate, to stop further questions, polish political careers, and delight journalists. My goal is to inform new and continuing waste scholars about the workings of these practices and to help develop tools and frameworks needed to institutionalize waste metrics that materially matter for human and non-human well-being in the here and now.

Notes

1 This common beginner's question is a good one, although asking it in these terms is a problem. An abstract, collective subject ("we") has, in fact, never corresponded to the range of people, institutions, and processes that set conditions for waste generation (see Liboiron 2020). The statement also suggests that waste is "made", connoting creation or fabrication by the waster—which is again not what happens.

2 The term appeared in the early 1970s as waste management was overhauled in many nations through air, water, and ground regulations to protect health. For an early example, see United States Environmental Protection Agency, "Recycling: Assessment & Prospects for Success" (Washington, DC: U.S. Government Printing Office, 1972).

3 Scrap or secondary materials prices are published in trade data sources such as Secondary Materials Pricing, accessed 10/11/20, https://www.recyclingmarkets.net/secondarymaterials/index.html. Prevailing prices being paid for different scrap materials are collected from recycling firms on condition of anonymity.

4 See "Diversion of Waste from Landfill—European Environment Agency", Indicator Assessment, accessed October 11, 2020, https://www.eea.europa.eu/data-and-maps/indicators/diversion-from-landfill/assessment. Florida, a swampy state lacking landfills is one exception in North America. Many people in the waste combustion industry in North America would like to see that convention changed to be more like Europe.

5 For a good review of scale technology and its relation to accounting systems, see "AMCS Scale House", accessed October 12, 2020, https://us.amcsgroup.com/solutions/enterprise-management/scale-house/.

6 For a discussion of instances in which MRFs evade reporting requirements at tip scales, see David J. Tonjes and Krista L. Greene, "A Review of National Municipal Solid Waste Generation Assessments in the USA", *Waste Management & Research* 30, no. 8 (2012): 758–771.

7 For a good review of animal cart collection operations, see Hasan Muhammad Umer, "Solid & Municipal Waste Management of Local Area Of Karachi", May 28, 2012, http://www.slideshare.net/ umer87/solid-municipal-waste-management-of-local-area-of-karachi.

8 Work has been done in the UK to improve transparency under an initiative led by the Waste & Resources Action Programme that requires quarterly reporting by MRFs. See Tom Goulding, "First Quarterly MRF Reporting Results Published", July 16, 2015, https://www.letsrecycle.com/news/ latest-news/first-quarterly-mrf-reporting-results-published/.

9 Fictitious tonnages are used here to demonstrate calculations.

10 For example, New York City includes these materials (https://www1.nyc.gov/assets/dsny/site/ services/recycling/what-to-recycle); San Francisco excludes them from recycling (https://www. recology.com/recology-san-francisco/what-bin/, accessed October 11, 2020).

11 For an excellent overview of MFA applied to urban settings, see Mike Hodson, Simon Marvin, Blake Robinson, and Mark Swilling, "Reshaping Urban Infrastructure: Material Flow Analysis and Transitions Analysis in an Urban Context", *Journal of Industrial Ecology* 16, no. 6 (2012): 789–800; for a more critical approach to urban MFA and waste flow modeling, see David J. Tonjes and Krista L. Greene, "A Review of National Municipal Solid Waste Generation Assessments in the USA", *Waste Management & Research* 30, no. 8 (2012): 758–771.

12 For reference to this task as a "methodological dark art", see Christian Reynolds, Arne Geschke, Julia Piantadoxi, and John Boland, "Estimating Industrial Solid Waste and Municipal Solid Waste Data at High Resolution Using Economic Accounts: An Input–Output Approach with Australian Case Study", *Journal of Material Cycles and Waste Management* 18, no. 4 (September 2016): 678, https:// doi.org/10.1007/s10163-015-0363-1.

13 Other areas of direct material effect on people's lives are experienced as waste-related pollution. Such pollution enters foodwebs and organismal biologies (including human ones) at a micro-scale; and makes uncontrolled incursion on to lands at a macro-scale through the action of wind, tides, and erosion. Consistent with this chapter's focus formal waste management systems, I single out sited diversion and disposal facilities here.

14 A leading voice for this position in the United States is the Institute for Local Self-Reliance and its "Waste to Wealth" resources, accessed 11/27/20, https://ilsr.org/waste-to-wealth/.

15 A full discussion of this topic is beyond the scope of this chapter. See Jayne Cox, Sara Giorgi, Veronica Sharp, Kit Strange, David C. Wilson, and Nick Blakey, "Household Waste Prevention—a Review of Evidence", *Waste Management & Research* 28, no. 3 (March 1, 2010): 193–219, https://doi.org/10. 1177/0734242X10361506; Veronica Sharp, Sara Giorgi, and David C. Wilson, "Delivery and Impact of Household Waste Prevention Intervention Campaigns (at the Local Level)", *Waste Management & Research* 28, no. 3 (March 1, 2010): 256–68, https://doi.org/10.1177/0734242X10361507; Antonis A. Zorpas and Katia Lasaridi, "Measuring Waste Prevention", *Waste Management* 33, no. 5 (May 2013): 1047–56, https://doi.org/10.1016/j.wasman.2012.12.017.

References

AMC Group. n.d. "AMCS Scale House." Accessed October12, 2020. https://us.amcsgroup.com/ solutions/enterprise-management/scale-house/.

Barham, Brad, and Stephen G. Bunker. 1995. *States, Firms, And Raw Materials: The World Economy And Ecology of Aluminum.* University of Wisconsin Press.

Blight, Geoffrey. 2011. "Landfills: Yesterday, Today, and Tomorrow." In Trevor M. Letcher and Daniel A. Vallero (Eds.) *Waste: A Handbook for Management.* Elsevier Science & Technology. http:// ebookcentral.proquest.com/lib/baruch/detail.action?docID=667711.

Bohm, Robert A., David H. Folz, Thomas C. Kinnaman, and Michael J. Podolsky. 2010. "The Costs of Municipal Waste and Recycling Programs." *Resources, Conservation and Recycling* 54, no. 11: 864–871. doi:10.1016/j.resconrec.2010.01.005.

Bowker, Geoffrey C., and Susan Leigh Star. 2000. *Sorting Things Out: Classification and Its Consequences.* MIT Press.

Brugh, Mercedes. n.d. "Garbage In, Garbage Out." *Energy Justice Network* (blog) http://www. energyjustice.net/content/garbage-garbage-out.

Bunker, Stephen G. 2005. "How Ecologically Uneven Developments Put the Spin on the Treadmill of Production." *Organization & Environment* 18, no. 1: 38–54. doi:10.1177/1086026604270434.

Chang, Jiyoun C., Robert H. Beach, and Elsa A. Olivetti. 2019. "Consequential Effects of Increased Use of Recycled Fiber in the United States Pulp and Paper Industry." *Journal of Cleaner Production* 241. doi: 10.1016/j.jclepro.2019.118133.

Clark, Corrie, Jenna Jambeck, Timothy Townsend. 2006. "A Review of Construction and Demolition Debris Regulations in the United States." *Critical Reviews in Environmental Science and Technology* 36, no. 2: 141–186.

Cossu, Raffaello, and Rainer Stegmann. 2018. *Solid Waste Landfilling: Concepts, Processes, Technologies.* Elsevier. doi:10.1016/C2012-0-02435-0.

Cox, Jayne, Sara Giorgi, Veronica Sharp, Kit Strange, David C. Wilson, and Nick Blakey. 2010. "Household Waste Prevention—a Review of Evidence." *Waste Management & Research* 28, no. 3: 193–219. doi:10.1177/0734242X10361506.

Crawford, Alan, and Hayley Warren. 2020. "China Upended the Politics of Plastic and the World Is Still Reeling." *Bloomberg Green*, January 21, 2020. https://www.bloomberg.com/graphics/2020-world-plastic-waste/.

Deloitte Sustainability. 2017. "Blueprint for Plastics Packaging Waste: Quality Sorting and Recycling." *Plastics Recyclers Europe.* https://www2.deloitte.com/content/dam/Deloitte/my/Documents/risk/my-risk-blueprint-plastics-packaging-waste-2017.pdf.

Denison, Richard A. 1996. "Environmental Life-Cycle Comparisons of Recycling, Landfilling, and Incineration: A Review of Recent Studies." *Annual Review of Energy and the Environment* 21, no. 1: 191–237. doi:10.1146/annurev.energy.21.1.191.

Dewitt, Marc, Jelmer Hoogzaad, Shyaam Ramkumar, Harald Friedl, and Annerieke Douma. 2018. *The Circularity Gap Report: An Analysis of the Circular State of the Global Economy.* Circle Economy.

Dillon, Lindsey, Rebecca Lave, Becky Mansfield, Sara Wylie, Nicholas Shapiro, Anita Say Chan, and Michelle Murphy. 2019. "Situating Data in a Trumpian Era: The Environmental Data and Governance Initiative." *Annals of the American Association of Geographers* 109, no. 2: 545–555. doi:10.1080/24694452.2018.1511410.

Docquier, Nicolas, and Sébastien Candel. 2002. "Combustion Control and Sensors: A Review." *Progress in Energy and Combustion Science* 28, no. 2: 107–150. doi:10.1016/S0360-1285(01)00009-0.

Dong, Jun, Yuanjun Tang, Ange Nzihou, Yong Chi, Elsa Weiss-Hortala, and Mingjiang Ni. 2018. "Life Cycle Assessment of Pyrolysis, Gasification and Incineration Waste-to-Energy Technologies: Theoretical Analysis and Case Study of Commercial Plants." *Science of The Total Environment* 626: 744–753. doi:10.1016/j.scitotenv.2018.01.151.

Ehrig, Hans-Jürgen, and Howard Robinson. 2010. "Landfilling: Leachate Treatment." In Thomas H. Christensen (Ed.) *Solid Waste Technology & Management.* John Wiley & Sons, Ltd. 858–897. doi:10.1002/9780470666883.ch54.

European Environment Agency. 2019. "Diversion of Waste from Landfill." *Indicator Assessment.* European Environment Agency. https://www.eea.europa.eu/data-and-maps/indicators/diversion-from-landfill/assessment.

Fetherston, J. T. 1915. *Annual Report of the Department of Street Cleaning.* New York City Department of Sanitation.

Fragkou, Maria Christina, Luis, Salinas Roca, Josep Espluga, and Xavier Gabarrell. 2014. "Metabolisms of Injustice: Municipal Solid-Waste Management and Environmental Equity in Barcelona's Metropolitan Region." *Local Environment: The International Journal of Justice and Sustainability* 19, no. 7: 731–747. http://repositorio.uchile.cl/bitstream/handle/2250/132602/Metabolisms-of-injustice.pdf?sequence=1.

Gille, Zsuzsa. 2007. *From the Cult of Waste to the Trash Heap of History: The Politics of Waste in Socialist and Postsocialist Hungary.* Indiana University Press.

Gille, Zsuzsa. 2013. "Is There an Emancipatory Ontology of Matter?" *Social Epistemology Review and Reply Collective* 2, no. 4: 1–6. http://wp.me/p1Bfg0-Ig.

Gilligan, Keith. 2013. "Durham Touts Incinerator Monitoring Program as 'Best of the Best.'" DurhamRegion.com, October 3, 2013. https://www.durhamregion.com/news-story/4137550-durham-touts-incinerator-monitoring-program-as-best-of-the-best-/.

Goulding, Tom. 2015. "First Quarterly MRF Reporting Results Published." Letsrecycle.Com, July 16, 2015. https://www.letsrecycle.com/news/latest-news/first-quarterly-mrf-reporting-results-published/.

Gregson, Nicky, and Mike Crang. 2010. "Materiality and Waste: Inorganic Vitality in a Networked World." *Environment and Planning A* 42, no. 5: 1026–1032. doi:10.1068/a43176.

Gregson, Nicky, Helen Watkins, and Melania Calestani. 2013. "Political Markets: Recycling, Economization and Marketization." *Economy and Society* 42, no. 1: 1–25. doi:10.1080/03085147.2012.661625.

Gullett, Brian K, Paul M Lemieux, Christopher C Lutes, Chris K Winterrowd, and Dwain L Winters. 2001. "Emissions of PCDD/F from Uncontrolled, Domestic Waste Burning." *Chemosphere* 43, no. 4–7: 721–725. doi:10.1016/S0045-6535(00)00425-2.

Gutberlet, Jutta. 2011. "Waste to Energy, Wasting Resources and Livelihoods." *Integrated Waste Management* 1: 219–236. doi:10.5772/17195.

Gutberlet, Jutta. 2012. "Informal and Cooperative Recycling as a Poverty Eradication Strategy." *Geography Compass* 6, no. 1: 19–34. doi:10.1111/j.1749-8198.2011.00468.x.

Haberl, Helmut, Dominik Wiedenhofer, Stefan Pauliuk, Fridolin Krausmann, Daniel B. Müller, and Marina Fischer-Kowalski. 2019. "Contributions of Sociometabolic Research to Sustainability Science." *Nature Sustainability* 2: 173–184. https://www.nature.com/articles/s41893-019-0225-2.

Harvey, David. 2005. *The New Imperialism*. Oxford University Press.

Hasan, Muhammad Umer. 2012. "Solid & Municipal Waste Management of Local Area Of Karachi." May 28, 2012. http://www.slideshare.net/umer87/solid-municipal-waste-management-of-local-area-of-karachi.

Hershkowitz, Allen. 1998. "In Defense of Recycling." *Social Research* 65, no. 1 (Spring): 141–218. www.jstor.org/stable/40971254.

Hird, Myra J. 2012. "Knowing Waste: Towards an Inhuman Epistemology." *Social Epistemology* 26, no. 3–4: 453–469. doi:10.1080/02691728.2012.727195.

Hodson, Mike, Simon Marvin, Blake Robinson, and Mark Swilling. 2012. "Reshaping Urban Infrastructure: Material Flow Analysis and Transitions Analysis in an Urban Context." *Journal of Industrial Ecology* 16, no. 6: 789–800. doi:10.1111/j.1530-9290.2012.00559.x.

Huber, Matthew. 2012. "Refined Politics: Petroleum Products, Neoliberalism, and the Ecology of Entrepreneurial Life." *Journal of American Studies* 46, no. 2: 295–312. https://www.jstor.org/stable/23259138.

ILSR. 2012. "Waste to Wealth Overview." Institute for Local Self-Reliance. http://www.ilsr.org/overview-waste-to-wealt/.

ISRI. 2020. "ISRI Scrap Specifications Circular." https://www.isri.org/recycling-commodities/scrap-specifications-circular.

Jambeck, J. R., R. Geyer, C. Wilcox, T. R. Siegler, M. Perryman, A. Andrady, R. Narayan, and K. L. Law. 2015. "Plastic Waste Inputs from Land into the Ocean." *Science* 347, no. 6223: 768–771. doi:10.1126/science.1260352.

Karidis, Arlene. 2016. "What Are the 'real' Numbers behind Landfilled Waste Tonnage and Landfill Gas Emissions?" *Waste Dive*, April 11, 2016. https://www.wastedive.com/news/what-are-the-real-numbers-behind-landfilled-waste-tonnage-and-landfill-ga-1/417194/.

Kaza, Silpa, Lisa Yao, Perinaz Bhada-Tata, and Frank Van Woerden. 2018. *What a Waste 2.0: A Global Snapshot of Solid Waste Management to 2050*. The World Bank Open Knowledge Repository. http://hdl.handle.net/10986/30317.

Kimmerer, Robin Wall. 2013. *Braiding Sweetgrass: Indigenous Wisdom, Scientific Knowledge and the Teachings of Plants*. Milkweed Editions.

Krones, Jonathan Seth. 2016. *Accounting for Non-Hazardous Industrial Waste in the United States*. PhD diss., Massachusetts Institute of Technology.

Lagerkvist, Anders, and Lisa Dahlén. 2012. "Solid Wastes Generation and Characterization." In Robert A. Meyers (Ed.) *Encyclopedia of Sustainability Science and Technology*. Springer. doi:10.1007/978-1-4419-0851-3_110.

Lepawsky, Josh. 2018. *Reassembling Rubbish: Worlding Electronic Waste*. MIT Press.

Lepawsky, Josh, and Charles Mather. 2011. "From Beginnings and Endings to Boundaries and Edges: Rethinking Circulation and Exchange through Electronic Waste." *Area* 43, no. 3: 242–249. doi:10.1111/j.1475-4762.2011.01018.x.

Liboiron, Max. 2018. "Waste Colonialism." *Discard Studies* (blog), November 1, 2018. https://discardstudies.com/2018/11/01/waste-colonialism/.

Liboiron, Max. 2020. "There's No Such Thing as 'We.'" *Discard Studies* (blog), October 12, 2020. https://discardstudies.com/2020/10/12/theres-no-such-thing-as-we/.

Liboiron, Max. 2021. *Pollution is Colonialism*. Duke University Press.

Liboiron, Max and Josh Lepawsky. 2018. "The What and the Why of Discard Studies." *Discard Studies* (blog), September 1, 2018. https://discardstudies.com/2018/09/01/the-what-and-the-why-of-discard-studies/

Lyons, Donald, Murray Rice, and Robert Wachal. 2009. "Circuits of Scrap: Closed Loop Industrial Ecosystems and the Geography of US International Recyclable Material Flows 1995–2005." *Geographical Journal* 175, no. 4: 286–300. https://www.jstor.org/stable/25621840

MacBride, Samantha. 2012. *Recycling Reconsidered: The Present Failure and Future Promise of Environmental Action in the United States*. MIT Press.

MacBride, Samantha. 2013. "San Francisco's Famous 80% Waste Diversion Rate: Anatomy of an Exemplar." *Discard Studies* (blog). November 1, 2013. https://discardstudies.com/2013/12/06/san-franciscos-famous-80-waste-diversion-rate-anatomy-of-an-exemplar/.

Marshall, Cody, and Karen Bandhauer. 2017. "The Heavy Toll of Contamination." *Recycling Today*, April 19, 2017. https://www.recyclingtoday.com/article/the-heavy-toll-of-contamination/.

Mersky, Ronald L. 2012. "Solid Waste Disposal and Recycling." In Robert A. Meyers (Ed.) *Encyclopedia of Sustainability Science and Technology*. Springer. doi:10.1007/978-1-4419-0851-3_109.

Miller, Chaz. 2020. "Recycling Is Not Rocket Science. It's People Science." *Waste360*, July 29, 2020. https://www.waste360.com/recycling/recycling-not-rocket-science-its-people-science.

Millward-Hopkins, Joel, and Phil Purnell. 2019. "Circulating Blame in the Circular Economy: The Case of Wood-Waste Biofuels and Coal Ash." *Energy Policy* 129: 168–172. doi:10.1016/j.enpol.2019.02.019.

Morris, Julian, and Lance Christensen. 2014. *An Evaluation of the Effects of California's Proposed Plastic Bag Ban*. Reason Foundation, June 18, 2014. https://reason.org/policy-brief/an-evaluation-of-the-effects-of-californias-proposed-plastic-bag-ban-2/.

Mukherjee, C., J. Denney, E. G. Mbonimpa, J. Slagley, and R. Bhowmik. 2020. "A Review on Municipal Solid Waste-to-Energy Trends in the USA." *Renewable and Sustainable Energy Reviews* 119. doi:10.1016/j.rser.2019.109512.

Nagle, Robin. 2017. "New York's City's Dis-Contents." Presentation at the Getting to Zero: Open House New York, May 1, 2017.

New York City Department of Sanitation. 2017. *NYC Residential, School, and NYCHA Waste Characterization Study*. New York City Department of Sanitation. https://dsny.cityofnewyork.us/wp-content/uploads/2018/04/2017-Waste-Characterization-Study.pdf.

New York City Department of Sanitation. n.d. "What to Recycle for Residents and Apartment Managers." Accessed December 5, 2020, https://www1.nyc.gov/assets/dsny/site/services/recycling/what-to-recycle.

Newell, Joshua P., and Joshua J. Cousins. 2015. "The Boundaries of Urban Metabolism: Towards a Political–Industrial Ecology." *Progress in Human Geography* 39, no. 6: 702–728.

Offenhuber, Dietmar, and Carlo Ratti. 2017. *Waste Is Information: Infrastructure Legibility and Governance*. MIT Press.

Owojori, Oluwatobi, Joshua N. Edokpayi, Ratshalingwa Mulaudzi, and John O. Odiyo. 2020. "Characterisation, Recovery and Recycling Potential of Solid Waste in a University of a Developing Economy." *Sustainability* 12, no. 12: 5111. doi:10.3390/su12125111.

Ramírez-García, Robert, Nisarg Gohil, and Vijai Singh. 2019. "Recent Advances, Challenges, and Opportunities in Bioremediation of Hazardous Materials." In Vimal Chandra Pandey and Kuldeep Bauddh (Eds.) *Phytomanagement of Polluted Sites: Market Opportunities in Sustainable Phytoremediation*. Elsevier. 517–568.

Recology. n.d. "." *Recology*. Accessed December 5, 2020, https://www.recology.com/recology-san-francisco/what-bin/.

Recycling Markets Limited. n.d. "Secondary Materials PriciWhatBin Search Toolng." Accessed December 5, 2020. https://www.recyclingmarkets.net/secondarymaterials/index.html.

Resource Recycling. 2018. "From Green Fence to Red Alert: A China Timeline." *Resource Recycling*, February 13, 2018. https://resource-recycling.com/recycling/2018/02/13/green-fence-red-alert-china-timeline/.

Reynolds, Christian, Arne Geschke, Julia Piantadosi, and John Boland. 2016. "Estimating Industrial Solid Waste and Municipal Solid Waste Data at High Resolution Using Economic Accounts: An

Input–Output Approach with Australian Case Study." *Journal of Material Cycles and Waste Management* 18, no. 4: 677–686. doi:10.1007/s10163-015-0363-1.

Robinson, Susan. 2018. "The Dangers of 'Wishcycling.'" *Waste Management*, April 24, 2018. http://wm-mediaroom-alb-141381419.us-east-1.elb.amazonaws.com/the-dangers-of-wishcycling/.

Rochat, David, Claudia R. Binder, Jaime Diaz, and Olivier Jolliet. 2013. "Combining Material Flow Analysis, Life Cycle Assessment, and Multiattribute Utility Theory." *Journal of Industrial Ecology* 17, no. 5: 642–655. doi:10.1111/jiec.12025.

Samson, Melanie. 2015. "Accumulation by Dispossession and the Informal Economy – Struggles over Knowledge, Being and Waste at a Soweto Garbage Dump." *Environment and Planning D: Society and Space* 33, no. 5: 813–830. doi:10.1177/0263775815600058.

Scarlett, Lynne. 2000. "The Waste Diversion Tango." *MSW Management*, January 1, 2000. https://www.mswmanagement.com/collection/article/13000301/the-waste-diversion-tango.

Schmoldt, A., H. F. Benthe, and G. Haberland. 1975. "Digitoxin Metabolism by Rat Liver Microsomes." *Biochemical Pharmacology* 24, no. 17: 1639–1641. https://www.sciencedirect.com/science/article/abs/pii/0006295275900945.

Selwood, Daniel. 2017. "Mixed Materials Just Rubbish When It Comes to Recycling." *The Grocer*, May 26, 2017. https://www.thegrocer.co.uk/environment/mixed-materials-just-rubbish-when-it-comes-to-recycling/553284.article.

Sharp, Veronica, Sara Giorgi, and David C. Wilson. 2010. "Delivery and Impact of Household Waste Prevention Intervention Campaigns (at the Local Level)." *Waste Management & Research* 28, no. 3: 256–268. https://www.researchgate.net/publication/41848539_Delivery_and_impact_of_household_waste_prevention_campaigns_at_the_local_level.

Shu, Shi, Wei Zhu, and Jiangwei Shi. 2019. "A New Simplified Method to Calculate Breakthrough Time of Municipal Solid Waste Landfill Liners." *Journal of Cleaner Production* 219: 649–654. doi:10.1016/j.jclepro.2019.02.050.

Sipper, Bill. 2019. "It's Time for Glass Again—Can We End Beverage Industry's Use of Plastic?" *Waste Advantage Magazine*, August 13, 2019. https://wasteadvantagemag.com/its-time-for-glass-again-can-we-end-beverage-industrys-use-of-plastic/.

Šyc, Michal, Franz Georg Simon, Jiri Hykš, Roberto Braga, Laura Biganzoli, Giulia Costa, Valerio Funari, and Mario Grosso. 2020. "Metal Recovery from Incineration Bottom Ash: State-of-the-Art and Recent Developments." *Journal of Hazardous Materials* 393: 122433. doi:10.1016/j.jhazmat.2020.122433.

Taylor, Dorceta. 2014. *Toxic Communities: Environmental Racism, Industrial Pollution, and Residential Mobility.* NYU Press.

The Association of Plastics Recyclers (APR). n.d. "PVC Design Guidance." Accessed December 5, 2020. https://plasticsrecycling.org/pvc-design-guidance.

Tonjes, David J., and Krista L. Greene. 2012. "A Review of National Municipal Solid Waste Generation Assessments in the USA." *Waste Management & Research* 30, no. 8: 758–771.

Tufano, Linda. 2015. "Which U.S. Cities Are Recycling Champions?" *Waste Dive*, August 4, 2015. https://www.wastedive.com/news/which-us-cities-are-recycling-champions/403347/.

United States Environmental Protection Agency. 1972. "Recycling: Assessment & Prospects for Success." SW-81. U.S. Government Printing Office.

Valverde, Mariana. 2009. "Jurisdiction and Scale: Legal 'Technicalities' as Resources for Theory". *Social and Legal Studies* 18, no. 2: 139–157. https://journals.sagepub.com/doi/10.1177/0964663909103622.

Walsh, Daniel C. 2002. "Urban Residential Refuse Composition and Generation Rates for the 20th Century." *Environmental Science & Technology* 36, no. 22: 4936–4942. doi:10.1021/es011074t.

Waring, George Edwin. 1897. *Street-Cleaning and the Disposal of a City's Wastes: Methods and Results and the Effect Upon Public Health, Public Morals, and Municipal Property.* Doubleday & McClure.

Waste Today. 2019 "EREF Releases Analysis on National Landfill Tipping Fees." October 29, https://www.wastetodaymagazine.com/article/eref-releases-analysis-national-msw-landfill-tipping-fees/.

Watson, Matt, and Harriet Bulkeley. 2005. "Just Waste? Municipal Waste Management and the Politics of Environmental Justice." *Local Environment* 10, no. 4: 411–426. doi:10.1080/13549830500160966.

Wilson, David C., Ljiljana Rodic, Anne Scheinberg, Costas A. Velis, and Graham Alabaster. 2012. "Comparative Analysis of Solid Waste Management in 20 Cities." *Waste Management & Research* 30, no. 3: 237–254. doi:10.1177/0734242X12437569.

Windfeld, Elliott Steen, and Marianne Su-Ling Brooks. 2015. "Medical Waste Management—A Review." *Journal of Environmental Management* 163: 98–108. doi:10.1016/j.jenvman.2015.08.013.

Yee, Amy. 2018. "In Sweden, Trash Heats Homes, Powers Buses and Fuels Taxi Fleets." *The New York Times*, September 21, 2018. https://www.nytimes.com/2018/09/21/climate/sweden-garbage-used-for-fuel.html.

Yesiller, Nazli, Samuel A. Vigil, and James L. Hanson. 2011. "Assessment of State Composting Regulations in the United States." In *Proceedings, 26th International Conference on Solid Waste Technology and Management*, 1281–1291. https://www.researchgate.net/publication/265119930_Assessment_of_State_Composting_Regulations_in_the_United_States.

Zeller, Vanessa, Edgar Towa, Marc Degrez, and Wouter M.J. Achten. 2019. "Urban Waste Flows and Their Potential for a Circular Economy Model at City-Region Level." *Waste Management* 83: 83–94. doi:10.1016/j.wasman.2018.10.034.

Zimlich, Rachel. 2015. "Regional Landfill Capacity Problems Do Not Equate to a National Shortage." *Waste360*, August 18, 2015. https://www.waste360.com/operations/regional-landfill-capacity-problems-do-not-equate-national-shortage.

Zorpas, Antonis A., and Katia Lasaridi. 2013. "Measuring Waste Prevention." *Waste Management* 33, no. 5: 1047–1056. doi:10.1016/j.wasman.2012.12.017.

13

THE POTENTIAL ROLE OF GAMIFICATION: AN INNOVATIVE INTERVENTION METHOD IN WASTE STUDIES

Tammara Soma, Belinda Li, and Virginia Maclaren

Introduction

In the field of waste studies, a dominant approach to engage communities in waste reduction has been through education, whether in school, through social media, or through municipal awareness campaigns. Awareness campaigns are viewed as a way to raise people's awareness about a particular problem and therefore motivate people towards waste reduction and or more sustainable waste management, which depending on the context of the waste may include reduction, reuse, recycling, or recovery (Suttibak and Nitivattananon 2008). Beyond raising awareness about an issue, at the core of many awareness campaigns is an educational approach that will also provide individuals with the "how to" knowledge for implementing a particular solution. A vast array of tools such as flyers, brochures, social media ads, waste calendars, and fridge magnets have been utilized to disseminate information (Soma et al. 2020; van der Werf, Seabrook, and Gilliland, 2019). These tools may contain information such as statistics around the environmental, economic, and social impacts of waste, as well as information that explains how to address the issue. As studies have discovered, however, traditional communication and educational campaigns via posters and printed materials (McKenzie-Mohr 2000; Low, Mohammed and Choong 2013), often have little to no impact on behavior change.

Recently, a new approach to education called gamification has garnered increasing attention in the field of environmental communication. Gamification is the use of game design approaches and elements in non-game contexts (Deterding et al. 2011). With a growing interest in employing gamification to promote environmental-related behaviors, scholars have begun to explore its potential role in the field of waste studies. An example of gamification is making pro-environmental behavior, such as recycling, fun through game-type challenges on sorting waste. Gamification can range from low-tech approaches such as board games, or personal challenges, to highly complex technologies such as robotic and online video games. It has become an increasingly popular tool to motivate behavior change particularly in educational, health, and environmental interventions (Tobon, Ruiz-Alba, and García-Madariaga 2020). This chapter provides an overview of the gamification approach in promoting pro-environmental behaviors and highlights the potential role for applying gamification in the field

DOI: 10.4324/9781003019077-13

of waste studies. This chapter will also present a number of waste-related case studies in gamification that represent simple to more complex gamification designs.

The authors acknowledge criticisms of focusing on individual awareness, individual responsibility, and information delivery to address the issue of waste (e.g., Liboiron 2014; Shove 2010) rather than infrastructure, the state, and other actors that shape daily life. For example, it is well documented that the individual awareness approach to behavior change is problematic because there often exists a value action gap between individual attitudes and individual behavior (Shove 2010). The behaviors of individuals regardless of attitudes are also influenced by a range of internal and external factors that are in some cases out of their control (Shove 2010; Evans 2014). At the same time, we argue that gamification is a new instrument in the waste solutions toolbox that should be considered along with the need for broader systemic and structural changes.

Literature review

A brief overview of gamification

At first glance, the term *gamification* may bring to mind ideas of complex, new, and innovative gaming technologies. Indeed, the term *gamification* has become what some might say a "buzzword" (Richter, Raban and Rafaeli 2015). It is important to distinguish that there is a slight difference between play and game, as game is a form of play but with structure (Maroney 2001). Gamification as noted previously is "the use of game design elements in non-game contexts" (Deterding et al. 2011). Huotari and Hamari (2012) define gamification as an interactive feedback mechanism to the user of the game with the ultimate goal of supporting value creation for the users. An important element of gamification is that a user is having fun (Schiele 2018). According to Dyer (2015), the emergence of gamification roughly dates back to 2008, with the concept becoming popular around 2010. The gamification of systems is a method ripe for interdisciplinary collaboration, between computer engineers, sociologists, psychologists, game designers, and more (Mora et al. 2015). Gamification applications are also diverse and business scholars have identified the exponential growth potential of using this tool for marketing or consumer retention purposes. Kim and Werbach (2016) argue that gamification is one of the fastest dispersing behavioral tools in the business sector.

According to Nicholson (2015), gamification is almost synonymous with rewards, ranging from points, to badges, and leaderboards (recognition), which in a real-world setting will entice people to engage in the games to earn rewards. While the argument can be made that a system based on rewards can only work if the rewards keep on coming, there are many ways within a gamified system to disrupt the need for continued reward. This can be done through intrinsic motivation (in the case of educational games) (Deci and Ryan 2004), and in other cases, the need for reward can be disrupted through what is called "operant conditioning" or the idea in the mind of the participant that perhaps the next time or this time they will get a reward (Skinner 1938 as cited in Nicholson 2015). The idea behind operant conditioning is to get players to continue engagement without the need to continue supplying rewards, something that has been harnessed by casinos to keep gamblers playing, and also a common tool used by video game developers to continue players' interest to keep on playing despite no guarantee of a reward (Nicholson 2015). Another external reward system now ubiquitous in grocery stores, hotels, and other companies is loyalty programs, which is a type of gamification approach called the BLAP gamification system (Badges, Levels/Leaders, Achievements, Points) (Nicholson 2012).

However, there are critiques of the reward system. Scholars have argued that shaping individual's intrinsic motivation to engage without needing an external reward may result in longer-lasting engagement (Deci and Ryan 2004; Kohn 1999). In fact, Deci and Ryan (2004) argue that extrinsic rewards can negatively impact intrinsic motivations. Other concerns by critics also include the possibilities of user fatigue (Richter et al. 2015), and also the existence of many gamification approaches developed with poor design strategies (Mora et al. 2015). However, there are numerous strategies that scholars have explored as an alternative to providing continued rewards, with rewards being replaced by what Nicholson (2015) noted as more meaningful rewards, such as opportunities to join like-minded communities, playful activities, and opportunities for reflection.

From an educational perspective, gamification is a useful tool to teach skills in a non-traditional way, and as the skill is mastered, there may be a reduced need for rewards. Moreover, psychological studies have noted the importance of social comparison (Wood, 1989) in nudging people towards a particular behavior by comparing themselves with others. Rewards such as badges or leadership board allows gamification to address these particular theories and approaches. While gamification has been harnessed to encourage consumption through tools such as shopping loyalty points (Nicholson 2015), or to promote mobile marketing and consumption (Hofacker et al. 2016), the question is whether gamification can be harnessed for the opposite, that is to reduce waste and consumption. As the practice of disposing of waste itself is learned, it is worth exploring and identifying the types of gamification approach and rewards (external, internal, skill development) that can be developed in order to shift social values and practices towards prevention, reduction, or more sustainable waste management.

Gamification, pro-environmental behaviors, and waste: a nascent area of study

There are many reasons for the growing interest in gamification, particularly in using this tool to encourage pro-environmental behaviors and in the case of this chapter, waste-related behaviors. Based on findings from gamified interventions for health and well-being, 59% of such interventions had a positive effect, 41% had a neutral or mixed effect, and none had a negative effect (Johnson et al. 2016). However, very few of the studies explored the long-term or sustained effects of gamification (Johnson et al. 2016). In a systemic review of gamification studies with Human Computer Interaction [HCI], Seaborn and Fels (2015) found a generally positive but mixed picture of the effectiveness of gamification. It is important to note that while this field is still nascent, there are a growing number of studies harnessing gamification to encourage more sustainable practices in various fields including water use and water saving (Novak et al. 2018), energy conservation (Oppong-Tawiah et al. 2020; Wee and Choong 2019), and sustainable tourism (Negruşa et al. 2015).

To promote environmental behaviors, the concept of "Games with a Purpose" [GWAPs] allows people to address complex problems in an enjoyable manner (Albertarelli et al. 2018). Addressing environmental problems such as climate change, water scarcity, energy instability through reduction in travel or water use, or using alternative energy such as biogas may be overwhelming, particularly for individuals. As such, gamification can make the process of learning or acting differently more engaging, more fun, and less intimidating. For example, in the tourism and hospitality sector, numerous gamification platforms have been developed to motivate tourists to choose more sustainable options. For example, the City of Eindhoven provides incentive points which reward tourists who take public transportation, and the points can then be used in restaurants, at museums, or to be used for a reduction in tourist tax (Negruşa et al. 2015). Many tourism games go beyond external rewards and nudge tourists' intrinsic motivation through games that build communities and promote charities in the tourist destination. In the case of promoting sustainable energy use in the office, Oppong-Tawiah et al. (2020) developed a gamified mobile app that tracks employees' electricity usage and then engages them to reduce their consumption by using a gardening game where those using less electricity were rewarded online with a thriving virtual garden, whereas those who use a lot of energy will see their garden start to wither. There were no external rewards in this game, and it solely relied on the participants' intrinsic motivation (Oppong-Tawiah et al. 2020). The study found that the app was successful in reducing electricity consumption and there was a statistically significant difference post intervention (Oppong-Tawiah et al. 2020). While there were limitations, for example, when participants noted that they needed to use two computer monitors to be productive regardless of their electricity use, or the small number of participants (n = 12), the participants noted that the game helped remind them to conserve energy, and also motivated them to educate and remind others (Oppong-Tawiah et al. 2020). In a review of gamification interventions in energy efficiency, Morganti et al. (2017) found that gamification can help foster energy saving behaviors. The studies reviewed in the paper used gamification to target three key themes, the first being general environmental education, the second being awareness of consumption (personally, or in a community), and the third focusing on nudging/encouraging energy efficiency behaviors (Morganti et al. 2017). Some of the interventions, however, were comprehensive and tackled all the themes.

Gamification may play an important role in addressing what Princen (2002) has identified as a world of distancing. According to Princen (2002), distancing is the separation of the source of primary production or resource extraction with final consumption decisions, and it is through this process of distancing that it becomes easier to exploit, waste, and overuse resources. In a nutshell, gamification has the potential to enable users to visualize the impact of everyday practices and behaviors that the industrial and global production, distribution, consumption, and waste complex make opaque. Nowhere is this distancing phenomenon more relevant than the field of waste studies where the mantra of "out of sight, out of mind" often comes up (Clapp and Princen 2003), or where even at the local level, the public finds the process of waste management such as recycling and sorting confusing (Hopewell, Dvorak, and Kosior 2009).

Case Studies

The five case studies presented here illustrate a range of objectives for waste-related gamifiation and a range of rewards or incentives for participation (Table 13.1). Three of the case studies undertook an evaluation of the effectiveness of gamification for improving waste diversion outcomes using self-reported and/or measured (audited) waste data, and two of those compared the results to a a control group. Only one focused on whether contamination of the waste diversion streams was reduced.

The first case study is the web-based OpenLitterMap, which uses gamification to raise awareness about litter (Lynch 2018). It rewards players with a blockchain token known as a "Littercoin" when they upload photos from their smart phones of the geo-coded locations of more than 100 types of litter. As of 2018, participants had uploaded 8,200 photos of about 28,000 items of litter but no assessment of the effectiveness of the approach on raising awareness had yet been conducted. Although Littercoins had no monetary value at the time of the study, Lynch (2018) offered several suggestions for developing a sustainable funding strategy for the tokens. For example, a business could offer a discount or a product (e.g. coffee) in return for some Littercoins and then publicize its accumulation of the Littercoins as

Table 13.1 Characteristics of the gamification case studies

Authors	Objective	# of Participants	Intervention Length	Rewards and Incentives	Participant Recruitment	Control	Waste Change
Lynch 2018	Raise awareness about litter	150	Not specified	Blockchain tokens ("Littercoins")	Not specified	No	Not assessed
Aguiar-Castillo et al. 2019	Increase visitor recycling in tourism destinations	141	Visit length not specified	Points exchanged for rewards offered by tourism destinations	Convenience sample	No	Not assessed
Luo, Zelenika, and Zhao 2019	Improve waste sorting	308	6 weeks	$25 gift card draw	Tabling at university residence buildings	Yes	Measured quantity and contamination
Comber and Thieme 2013, Thieme et al. 2012	Increase diversion of food waste and recyclables from garbage	22 in 4 households	5 weeks	Leaderboards	Posters, flyers and personal contacts	No	Self-reported behavior
Soma et al. 2020	Reduce food waste	62	12 weeks	$10 gift card for playing at least 6 weeks, $20 gift card for 12 weeks	Cluster sample for single family homes; tabling at multi-family buildings	Yes	Self-reported behavior and measured quantity

a corporate social responsibility marketing strategy. Local governments could offer free recycling or composting bins.

The second case study is an application for mobile devices (WasteApp) that tourists can use to find the locations of recycling bins in cities that they visit as well as information about what is recyclable (Aguiar-Castillo et al. 2019). Participants gain points by reading QR codes on the waste bins that they use and posting comments about them using the project hashtag. The points could be exchanged for prizes offered by the participating cities. There was no assessment of whether the app changed the quality or quantity of recycling undertaken by participants or self-reported recycling behavior. The assessment was primarily about the quality of the app itself. Among other findings, the study discovered a significant relationship between participants' perceived ease of use of the app and its perceived usefulness, measured as its usefulness as a tool for travelling, for finding recycling bins, and for encouraging recycling behavior (Aguiar-Castillo et al. 2019).

There are very few studies that have attempted to measure the effectiveness of a gamification intervention on waste reduction or recycling. One of them is the digital sorting game (http://yuluo.psych.ubc.ca/studies/Sorting_MD) developed by Luo, Zelenika and Zhao (2019). Students in two university residences were asked to sort 28 waste items into one of four online bins (food scraps, recyclable container, paper, or garbage) and given immediate feedback on whether their sorting choices were correct. The expectation was that the game would lead to improved real-world sorting performance. By the end of the intervention period, the weight of source-separated food waste scraps collected had increased significantly and contamination decreased marginally. The increase in food scrap quantity also continued during the three-week period after the intervention concluded. Contamination of the recycling and paper streams decreased marginally between the intervention period and post-intervention periods but there was no change in quantity sorted. The authors speculated that the gamification effect on sorting outcomes could have been even higher with higher participation in the game. Only 44% of the residents of the two buildings participated.

Another waste-related gamification study that examined effectiveness was Comber and Thieme's (2013) study, using a BinCam system that had the ability to capture and then share images of waste disposed by participants in the household waste bin on an online platform. All households participating in the game became part of an online BinLeague in which households could track on leaderboards how well they were doing in reducing food waste and increasing recycling relative to other households. The researchers recognized the potential for privacy concerns from posting such information on a public platform, but felt that choosing only larger, shared households to participate in the study was sufficient to reduce the concern because it would be difficult to identify an individual's waste (Thieme et al. 2012).

Pre- and post-intervention surveys found that self-reported attitudes towards recycling and recycling behaviors for the group as a whole did not change significantly (Thieme et al. 2012). The study team did not measure differences in levels of recycling contamination among the four households or changes in recycling quantities producing during the intervention. Although gamification and the HCI approach did not change group attitudes or behavior significantly, focus groups conducted by the researchers found that participants felt that the technology was useful and fun to use. Using the BinCam raised awareness among household members, leading to participant self-reflection about their recycling knowledge and re-evaluation of their recycling practices. In some cases it also resulted in shame for what the participants saw as a gap between their values and their actual actions (Comber and Thieme 2013). Some participants felt both social pressure and feelings of guilt if they "let their house down" by not recycling better when their household waste behavior became visible to others by sharing photos online. Self-reflection during the intervention also led to action, with some participants starting to educate themselves about the recyclability of their waste items, exchanging recycling expertise with others, and improving their food planning practices (Comber and Thieme 2013). These changes occurred despite the fact that participants only looked at the photos from their own bin a few times, mainly at the beginning of the study, and even less at photos from other households (Thieme et al. 2012). Although the photos may not have been as influential as anticipated, the researchers found an unexpected influence of the BinCam, namely that when the bin closed, the camera made a small noise when taking a photo and this noise acted as an auditory trigger for participant self-reflection about frequency of using the bin and whether they had sorted correctly (Comber and Thieme 2013).

Another important finding of the study was that beyond gaps in actual knowledge of how to sort recyclables, several participants noted inadequacies in their waste management facilities that affected their ability to recycle. These included recycling bins that could not accommodate large amounts of bulky cardboard or bins that were not collected frequently enough (Comber and Thieme 2013). As such, while the game was able to help users better understand the correct procedures to sort and recycle, some felt constrained by infrastructure deficits that the game was not designed to address.

The fifth example of gamification is a simple online trivia-based game with food waste facts and tips on reducing food waste at home (Soma et al. 2020). This intervention was compared to results from an information-only food waste reduction intervention and a control group. Participants in both intervention groups received a pamphlet with tips to reduce food waste at home, a fridge magnet on how to organize a fridge for optimal food storage, and four newsletters with additional tips for food storage, meal planning, shopping, understanding best-before dates, and recipes for using leftovers or slightly spoiled food. The sample of households consisted of residents of single-family homes and multi-family buildings. All households were surveyed before and after the intervention period, asking about food waste awareness, practices, and attitudes. A waste composition analysis of the amount of edible food waste in the garbage, recycling, and organic streams was performed prior to and at the end of the intervention period but only for the single-family homes.

The gamified intervention lasted 12 weeks and those who participated could take as many attempts as they needed to select the correct response for each of five weekly multiple-choice questions (Figure 13.1). Participants earned 10 points for each week of the game that they played and in which that they had successfully answered all five of the trivia questions. Participants received a $10 gift card as a bonus for reaching 60 points (half of the weeks played) or a $20 gift card for 120 points (all weeks played). One surprise bonus points week was offered in the middle of the campaign. This bonus was an additional 10 points if a participant played that week, so the $20 gift card could have been obtained by playing only for 11 weeks. Bonuses are game elements that are typically provided on top of rewards after successful completion of a set of tasks, such as completing the game (Schöbel, Janson and Söllner 2020).

Based on survey results, the change in self-reported awareness of food waste and amount of food wasted at home was about the same between the information and gamification intervention groups at the end of the intervention period (Soma et al. 2020). In other words, gamification was no better than a traditional information campaign in raising awareness and influencing self-reported behavior. Compared to the control group, however, participants in the two intervention groups were more aware of food waste after the intervention and more reported that their food waste had decreased (Soma et al. 2020).

While participants in the information group perceived that they wasted less food after the intervention, results from the waste composition analysis demonstrated otherwise, showing no statistically significant change in food waste per capita. In contrast, households in the gamification group saw a statistically significant decrease in food waste per capita, but only marginally so. Soma et al. (2020) suggested that a possible reason for the fairly weak evidence of a decrease in food waste for the gamification group was that participation in the gamified intervention was greatly polarized. Participants appeared either to be highly engaged in the game, or did not engage in the game at all (Figure 13.2). Forty-two percent of participants obtained the \$20 gift

Question 2/5

How can I get more organized to reduce the amount of food that I intend to eat, but gets lost in the shuffle?

Buy on-sale grocery items at the store

Buy ugly fruits and veggies

Keep an 'eat-me-first' basket or shelf in your fridge

Eat out more

Figure 13.1 Example of multiple-choice question.

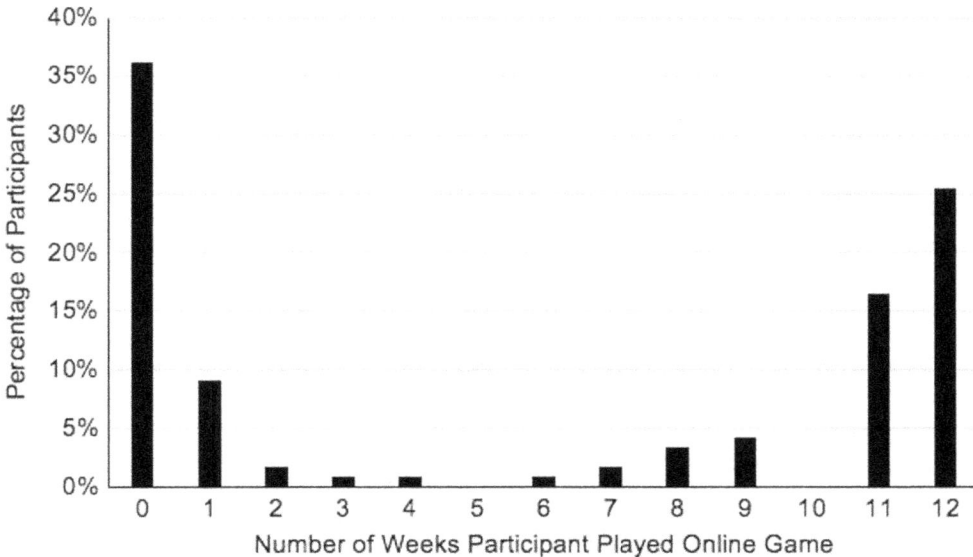

Figure 13.2 Participant engagement with online game.

card for playing 11–12 weeks but 45% did not play the game at all, and 9% only tried the game once. The researchers found multiple reasons why participants did not engage as much, such as not receiving the weekly reminders about participating that were sent out to members of the gamification group, being too busy, not being interested, or feeling that the game was too simple. The frequent gamers had a statistically significant reduction in their food waste whereas the less frequent gamers did not (Soma et al. 2020). In other words, high engagement in the gamified intervention seemed to correlate with lower food waste. Soma et al. (2020) defined frequent gamers as those who had played 11–12 weeks, which was the requirement for "completing the game" and receiving the bonus gift card of $20. They did not report whether other definitions of frequent gamers, such as including the 10% of participants who played 6–9 weeks (none played for 10 weeks), would have changed the results about less food waste being produced by frequent gamers. Those who participated in the game also expressed their appreciation of its playfulness, as is evident in this quote from a participant:

> I would get kind of upset when I got a wrong answer or something. Call my husband, say come here, what do you think of this?…So yeah, it was good. I enjoyed it very much. I was sorry when it ended, I really was. (quote from "Sally" in Soma et al. 2020, 13)

Conclusion

The multi-scalar and complex nature of waste issues calls for multifaceted and interdisciplinary solutions. This chapter highlights the potential role of gamification as one of the tools in the waste solution toolbox that may create more opportunities for reflective action, meaningful engagement, skills building, education, and enjoyment in participating in more sustainable practices. While gamification design is often tied to external rewards, the approach itself is broad and flexible and it is possible to apply alternative reward mechanism that use intrinsic motivations to nudge users towards a more fun approach to learning. A brief overview of gamification also distinguished the difference between play and gamification and clarified the purposeful end goal of gamification.

To demonstrate the application of gamification in a waste context, we showcased five case studies. The case studies illustrated a range of gamification approaches and outcomes. One gamification outcome was an increase in the weight of food waste source-separated by participants and a marginal decrease in food waste contamination (Luo, Zelenika, and Zhao 2019). A second gamification outcome was that participants engaged in self-reflection about their recycling knowledge and re-evaluated their recycling practices (Comber and Thieme 2013). A third outcome was that gamification performed no better than a traditional information campaign in effecting changes in self-reported awareness of food waste and self-reported food wasted at home (Soma et al. 2020). A fourth outcome was that people who were frequent gamers significantly reduced their food waste compared to less frequent gamers, as measured by waste audits (Soma et al. 2020).

The case studies suggest multiple future challenges that will need to be addressed if gamification is to be considered for wider use in waste education programs. One challenge is to develop a better understanding of the effectiveness and sustainability of rewards. A common element of gamification is a reward system. A reward system can entice participation, but despite the presence of rewards, Luo, Zelenika, and Zhao (2019) had a participation rate of only 44% in the sorting game and Soma et al. (2020) found that 45% of households that were part of the gamification group never participated in the game. In addition to understanding what types and values of rewards encourage participation, future research is needed on the long-term impact of rewards-based gamification. There are numerous examples from experiments in behavioral psychology of how once extrinsic motivations for pro-environmental behavior are removed, behavior returns to the pre-intervention state (Van Der Linden 2015). A possible reason for this is that when people are presented with an extrinsic reward, they undertake the behavior for the reward, and not because they are inherently interested in changing their behavior (Deci, Koestner, and Ryan 1999). While rewards are often key elements of game design in gamification, Ro et al. (2017) recommend using only small rewards to avoid this problem. This approach can also help with sustainability since the inclusion of rewards in gamification can be a challenge to fund, as noted by Lynch (2018).

A second challenge is to develop a better understanding of the usability of gamification by different socio-demographic groups. Soma et al. (2020) found that gamification participation rates were not significantly different across gender, age, household size, presence of children in the home, education, or income. However, there have been very few other studies on the influence of socio-demographics on adoption of gamification in general (Koivisto and Hamari 2019) and none that we know of in waste-related gamification.

A third challenge is the need for further research into the theorization of gamification in general and as it applies to waste-related gamification. Tobon, Ruiz-Alba, and García-Madariaga (2020) and Seaborn and Fels (2015) identify several theoretical frameworks that have been used in gamification studies to date but Tobon, Ruiz-Alba, and García-Madariaga (2020) find that theory is absent in many, and Seaborn and Fels (2015) state that there is little cohesion in the theoretical underpinnings of gamification.

A fourth challenge is the need for more assessments of the effectiveness of gamification on waste-related outcomes. Only three examples of effectiveness assessment for waste-related gamification were found in the literature. One used self-reporting, one used observable measurement and one used both. The last study (Soma et al. 2020) found a discrepancy between self-reported change in food waste in the home versus directly measured food waste from a waste composition analysis. This discrepancy is consistent with other research on self-reported versus measured waste (Quested et al. 2020; van der Werf, Seabrook, and Gilliland 2020; Elimelech, Ert, and Ayalon 2019; Giordano et al. 2019). Direct measurement of waste is generally considered the most accurate method for capturing changes in waste quantities and composition, but can be expensive and challenging to undertake.

A fifth challenge is the question of positive versus negative reinforcement in gamification. Negative reinforcement may be successful with some groups but not with others. For example, Wonneberger (2018) found that guilt arousal in an environmental campaign had a positive effect on the intention to donate to the campaign for those who already had a high level of environmental concern, but not for those with low environmental concern. The BinCam game employed a type of guilt arousal that Schiele (2018) refers to in a gamification context as "shamification". Shamification creates feelings of guilt or shame among participants for engaging in undesirable behavior. Hawkins (2006) points to a similar type of shamification found in a traditional waste education campaign meant to improve waste sorting practices. He contends that guilt, anxiety, and resentment are not productive. Instead, they inhibit responses and ways of being with waste that are only found in promoting enjoyment in behaviors. Gamification is normally designed to be playful, so would seem to be one way to address this concern.

A sixth challenge is that the long-term effects of gamification interventions have not been tested widely. Luo, Zelenika, and Zhao (2019) found that the increase in weight of food waste separated persisted three weeks after the end of the digital sorting game but they did not test for changes beyond that. Ro et al. (2017) found that pro-environmental behavior persisted among participants in their gamification study 12 months after the end of the gamification intervention, but those behaviors were based on self-reports rather than measured change. In a systematic review of gamification in the field of energy efficiency, Morganti et al. (2017) highlight the lack of research on the longevity of gamification outcomes. Gamification is particularly susceptible to concerns about longevity because of its use of rewards, as noted above, and the removal of those rewards at the end of the gamification period.

To conclude, by no means is gamification a panacea or an appropriate solution for all types of waste issues. However, there is much to explore in gamification's potential contribution to engage in capacity building and to align values in the public, political, and community sectors that may eventually lead to broader systemic and structural changes in production, consumption, and eventually waste generation.

References

Aguiar-Castillo, L., J. Rufo Torres, P. De Saa Pérez, and R. Pérez Jiménez. 2018. "How to Encourage Recycling Behaviour? The Case of WasteApp: A Gamified Mobile Application." *Sustainability*. doi: 10.3390/su10051544

Aguiar-Castillo, L., A. Clavijo-Rodriguez, D. Saa-Perez, and R. Perez-Jimenez. 2019. "Gamification as an Approach to Promote Tourist Recycling Behavior". *Sustainability* 11, no. 8: 2201.

Albertarelli, S., P. Fraternali, S. Herrera, M. Melenhorst, J. Novak, C. Pasini, A.-E. Rizzoli, and C. Rottondi. 2018. "A Survey on the Design of Gamified Systems for Energy and Water Sustainability". *Games* 9, 38. doi: 10.3390/g9030038.

Aldrich, C. 2009. *The Complete Guide to Simulations and Serious Games*. Pfeiffer.

Buchan, R., D. S. Cloutier, and A. Friedman. 2019. "Transformative Incrementalism: Planning for Transformative Change in Local Food Systems". *Progress in Planning* 134, 100424.

Clapp, J., and T. Princen. 2003. "Out of Sight, Out of Mind: Cross-Border Traffic in Waste Obscures the Problem of Consumption". *Alternatives Journal* 29, no. 3: 39–41.

Comber, R., and A. Thieme. 2013. "Designing Beyond Habit: Opening Space for Improved Recycling and Food Waste Behaviors Through Processes of Persuasion, Social Influence and Aversive Affect". *Personal and Ubiquitous Computing* 17, no. 6: 1197–1210.

Deci, E., and R. Ryan. 2004. *Handbook of Self-Determination Research*. University of Rochester Press.

Deterding, S., D. Dixon, R. Khaled, and L. E. Nacke. 2011. "From Game Design Elements to Gamefulness: Defining 'Gamification'". *MindTrek '11: Proceedings of the 15th International Academic MindTrek Conference: Envisioning Future Media Environments*, 9–15. doi:10.1145/2181037.2181040

Deci, E. L., R. Koestner, and R. M. Ryan. 1999. "A Meta-Analytic Review of Experiments Examining the Effects of Extrinsic Rewards on Intrinsic Motivation". *Psychological Bulletin* 125, no. 6: 627–668.

Dyer, R. 2015. "A Conceptual Framework for Gamification Measurement". In T. Reiners and L. C. Wood (Eds.) *Gamification in Education and Busines*. Springer International Publishing. 47–66.

Elimelech, E., E. Ert, and O. Ayalon. 2019. "Exploring the Drivers behind Self-Reported and Measured Food Wastage". *Sustainability* 11, no. 20: 5677. doi:10.3390/su11205677

Evans, D. M. 2014. *Food waste: Home Consumption, Material Culture and Everyday Life*. Bloomsbury publishing.

Farr-Wharton, G., J. H.-J. Choi, and M. Foth. 2014. "Technicolouring the Fridge: Reducing Food Waste through Uses of Colour-coding and Cameras". In *MUM '14: Proceedings of the 13th International Conference on Mobile and Ubiquitous Multimedia*, 48–57. doi:10.1145/2677972.2677990

Giordano, C., F. Alboni, C. Cicatiello, and L. Falasconi. 2019. "Quantities, Determinants, and Awareness of Households' Food Waste in Italy: A Comparison between Diary and Questionnaires Quantities." *Sustainability* 11, no. 12: 3381. doi:10.3390/su11123381

Hawkins, G. 2006. *The Ethics of Waste*. Rowan & Littlefield. 149 pp.

Helmefalk, M., and J. Rosenlund. 2020. "Hedonic Recycling: Using Gamification and Sensory Stimuli to Enhance the Recycling Experience." *EAI Endorsed Transactions on Serious Games* 18, no. 3.

Hofacker, C. F., K. De Ruyter, N. H. Lurie, P. Manchanda, and J. Donaldson. 2016. "Gamification and Mobile Marketing Effectiveness". *Journal of Interactive Marketing* 34, 25–36.

Hopewell, J., R. Dvorak, and E. Kosior. 2009. "Plastics Recycling: Challenges and Opportunities". *Philosophical Transactions of the Royal Society B: Biological Sciences* 364, no. 1526: 2115–2126.

Huotari, K., and J. Hamari. 2012. "Defining Gamification: A Service Marketing Perspective". In *MindTrek'12 Proceeding of the 16th International Academic MindTrek Conference*. ACM Digital Library. 17–22.

Johnson, D., S. Deterding, K.-A. Kuhn, A. Staneva, S. Stoyanov, and L. Hides. 2016. "Gamification for Health and Wellbeing: A Systematic Review of the Literature." *Internet Interview* 6, 89–106. doi:10.1016/j.invent.2016.10.002

Kim, T. W., and K. Werbach. 2016. "More Than Just a Game: Ethical Issues in Gamification". *Ethics and Information Technology* 18, no. 2: 157–173.

Kohn, A. 1999. *Punished by Rewards: The Trouble With Gold Stars, Incentive Plans, A's, Praise, and Other Bribes*. Houghton Mifflin.

Koivisto, J., and J. Hamari. 2019. "The Rise of Motivational Information Systems: A Review of Gamification Research." *International Journal of Information Management* 45, 191–210.

Konrad, A. 2011. "Inside the Gamification Gold Rush". Retrieved June 20th 2020, from https://fortune.com/2011/10/17/inside-the-gamification-gold-rush-2/

Liboiron, M. 2014. "Against Awareness, For Scale: Garbage Is Infrastructure, not Behavior". *Discard Studies* (blog). http://discardstudies.com/2014/01/23/against-awareness-for-scale-garbage-is-infrastructure-not-behavior/

Low, Sheau-Ting, A. H. Mohammed, Weng-Wai Choong. 2013. "What Is the Optimum Social Marketing Mix to Market Energy Conservation Behavior: An Empirical Study". *Journal of Environmental Management* 131: 196–205.

Luo, Y., I. Zelenika, and J. Zhao. 2019. "Providing Immediate Feedback Improves Recycling and Composting Accuracy." *Journal of Environmental Management* 232: 445–454.

Lynch, S. 2018. "OpenLitterMap. com–open Data on Plastic Pollution With Blockchain Rewards (Littercoin)". *Open Geospatial Data, Software and Standards* 3, no. 1: 1–10.

Maroney, K. 2001. "My Entire Waking Life". *The Games Journal*. http://www.thegamesjournal.com/articles/MyEntireWakingLife.shtml

McKenzie-Mohr, D. 2000. "New Ways to Promote Pro-Environmental Behavior: Promoting Sustainable Behavior: An Introduction to Community-Based Social Marketing". *Journal of Social Issues* 56, no. 3: 543–554.

Mora, A., D. Riera, C. Gonzalez, and J. Arnedo-Moreno. 2015, September. "A Literature Review of Gamification Design Frameworks." In 2015 *7th International Conference on Games and Virtual Worlds for Serious Applications (VS-Games)*. IEEE. 1–8.

Morganti, L., F. Pallavicini, E. Cadel, A. Candelieri, F. Archetti, and F. Mantovani. 2017. "Gaming for Earth: Serious games and Gamification to Engage Consumers in Pro-Environmental Behaviours for Energy Efficiency." *Energy Research & Social Science* 29: 95–102.

Negruşa, A. L., V. Toader, A. Sofică, M. F. Tutunea, and R. V. Rus. 2015. "Exploring Gamification Techniques and Applications for Sustainable Tourism." *Sustainability* 7, no. 8: 11160–11189.

Nicholson, S. 2012a, June. "A User-Centered Theoretical Framework for Meaningful Gamifi Cation." Paper Presented at Games+Learning+Society 8.0, Madison. Retrieved from online at http://scottnicholson.com/pubs/meaningfulframework.pdf

Nicholson, S. 2015. "A Recipe for Meaningful Gamification". In T. Reiners and L. C. Wood (Eds.) *Gamification in Education and Business*. Springer International Publishing.

Novak, J., M. Melenhorst, I. Micheel, C. Pasini, P. Fraternali, and A. E. Rizzoli. 2018. "Integrating Behavioural Change and Gamified Incentive Modelling for Stimulating Water Saving". *Environmental Modelling & Software* 102: 120–137.

Oppong-Tawiah, D., J. Webster, S. Staples, A. F. Cameron, A. O. de Guinea, and T. Y. Hung. 2020. "Developing a Gamified Mobile Application to Encourage Sustainable Energy Use in the Office." *Journal of Business Research* 106: 388–405.

Princen, T. 2002. "Distancing: Consumptino and the Severing of Feedback". In T. Princen, M. Maniates, and K. Conca (Eds.) *Confronting Consumption*. MIT Press. 103–131.

Quested, T. E., G. Palmer, L. C. Moreno, C. McDermott, and K. Schumacher. 2020. "Comparing Diaries and Waste Compositional Analysis for Measuring Food Waste in the Home". *Journal of Cleaner Production* 121263. doi:10.1016/j.jclepro.2020.121263

Reiners, T., and L. C. Wood. 2015. *Gamification in Education and Business*. Springer International Publishing.

Richter, G., D. Raban, and S. Rafaeli. 2015. "Studying Gamification: The Effect of Rewards and Incentives on Motivation". In T. Reiners and L. C. Wood (Eds.) *Gamification in Education and Business*. Springer International Publishing. 21–46.

Ro, M., M. Brauer, K. Kuntz, R. Shukla, and I. Bensch. 2017. "Making Cool Choices for Sustainability: Testing the Effectiveness of a Game-Based Approach to Promoting Pro-Environmental Behaviors." *Journal of Environmental Psychology* 53: 20–30.

Seaborn, K., and D. I. Fels. 2015. "Gamification in Theory and Action: A Survey." *International Journal of Human-Computer Studies* 74: 14–31.

Schiele K. 2018. "Utilizing Gamification to Promote Sustainable Practices". In S. Dhiman and J. Marques (Eds.) *Handbook of Engaged Sustainability*. Springer. 427–444. doi:10.1007/978-3-319-53121-2_16-1

Schöbel, S. M., A. Janson, and M. Söllner. 2020. "Capturing the Complexity of Gamification Elements: A Holistic Approach for Analysing Existing and Deriving Novel Gamification Designs". *European Journal of Information Systems* 29, no. 6: 1–28

Shove, E. 2010. "Beyond the ABC: Climate Change Policy and Theories of Social Change". *Environment and Planning A: Economy and Space* 42, 1273–1285. doi:10.1068/a42282

Skinner, B. F. 1938. *The Behavior of Organisms: An Experimental Analysis*. Appleton-Century.

Simon Fraser University [SFU]. 2019. "Oscar the Trash-Sorting Robot Reduces Waste at SFU's Surrey Campus". Retrieved June 14th 2020: https://www.sfu.ca/university-communications/media-releases/2019/07/oscar-the-trash-sorting-robot-reduces-waste-at-sfu-s-surrey-camp.html

Soma, T., B. Li, and V. Maclaren. 2020. "Food Waste Reduction: A Test of Three Consumer Awareness Interventions". *Sustainability* 12, no. 3: 907. 10.3390/su12030907

Suttibak, S., and V. Nitivattananon. 2008. "Assessment of Factors Influencing the Performance of Solid Waste Recycling Programs." *Resources, Conservation and Recycling* 53: no. 1–2: 45–56.

TrashTrack. 2011. "MIT Researchers Map the Flow of Urban Trash?" Last retrieved June 14th 2020, http://senseable.mit.edu/trashtrack/downloads/trash-track-nsf.pdf

Takahashi, D. 2018. "Game Market Expected to hit $180.1 Billion in Revenue". *Venture Beat*. Retrieven June 13th 2020 from: https://venturebeat.com/2017/11/28/newzoo-game-industry-growing-faster-than-expected-up-10-7-to-116-billion-2017/

Terlutter, R. and M. Capella. 2013. "The Gamification of Advertising: Analysis and Research Directions of in-Game Advertising, Advergames, and Advertising in Social Network Games". *Journal of Advertising* 42, no. 2/3: 95–112.

Thieme, A., R. Comber, J. Miebach, J. Weeden, N. Kraemer, S. Lawson, and P. Olivier. 2012. ""We've Bin Watching You": Designing for Reflection and Social Persuasion to Promote Sustainable Lifestyles". In *Proceedings of the SIGCHI Conference on Human Factors in Computing Systems*. 2337–2346.

Tobon, S., J. L. Ruiz-Alba, and J. García-Madariaga. 2020. "Gamification and Online Consumer Decisions: Is the Game Over?" *Decision Support Systems* 128: 113167. doi:10.1016/j.dss.2019.113167

Van Der Linden, S. 2015. "Intrinsic Motivation and Pro-Environmental Behaviour." *Nature Climate Change* 5, no. 7: 612–613.

van der Werf, P., J. A. Seabrook, and J. A. Gilliland. 2019. "Food for Naught: Using the Theory of Planned Behaviour to Better Understand Household Food Wasting Behaviour". *The Canadian Geographer / Le Géographe Canadien* 63: 478–493. doi: 10.1111/cag.12519

Van der Werf, P., J. A. Seabrook, and J. A. Gilliland. 2020. "Food for Thought: Comparing Self-Reported Versus Curbside Measurements of Household Food Wasting Behavior and the Predictive Capacity of Behavioral Determinants". *Waste Management* 101: 18–27. 10.1016/j.wasman.2019.09.032

Wee, S. C., and W. W. Choong. 2019. "Gamification: Predicting the Effectiveness of Variety Game Design Elements to Intrinsically Motivate Users' Energy Conservation Behaviour". *Journal of Environmental Management* 233: 97–106.

Wonneberger, A. 2018. "Environmentalism—A Question of Guilt? Testing a Model of Guilt Arousal and Effects for Environmental Campaigns". *Journal of Nonprofit & Public Sector Marketing* 30, no. 2: 168–186.

Wood, J. V. 1989. "Theory and Research Concerning Social Comparisons of Personal Attributes". *Psychological Bulletin* 106, no. 2: 231–248.

PART IV

Cases waste scholars investigate

14

THE EXPERIENCE OF NUCLEAR WASTE

Romain J. Garcier

Introduction

On March 11, 2011, the day of the Fukushima disaster, I was visiting the nuclear complex at Cap La Hague, in France, together with colleagues and a group of students. The complex reprocesses spent nuclear fuel to separate uranium, plutonium, and fission products (Zonabend 1993; Blowers 2016). Plutonium and uranium can be re-used but there is no redeeming fission products, which are high-level waste. The company that operates the plant, Orano (formerly Areva), mixes them in a special glass matrix and then pours the molten glass into a thick metal cask for prolonged storage. Regular visitors can't see the actual process but we passed through the dry storage area where high-level waste is kept. The storage room's floor was dotted with circles, each labeled with a code (Figure 14.1). Under each metal circle, there are caskets of high-activity radioactive products. Thick concrete and metal casings shield workers and visitors from radiation. As I walked over the surface, I noticed that the floor was warm from the heat of waste. "This is the closest I have ever been to high-level waste", I thought. Without the protective casing against radiation, the waste would kill me within minutes. It was a strange experience, sensing but not seeing waste.

My reflection resonates with radioactive waste's status in contemporary waste studies. They often invoke radioactive waste rather than engaging with it. Most studies put it in a class of its own, endowed with unique and threatening material properties. The fact that it originates in industrial, scientific, or military environments that are not easily accessible to enquiry certainly complicates research. The rare study, in turn, generally emphasizes the complete, radical exceptionality of radioactive waste (O'Brien 2008). Ask around: in any conversation on waste, at some point, somewhere, radioactive waste will surface as the great Other, waste so extraordinarily dangerous that it can never become a resource again, it can only be contained. Of course, other people are keen to downplay the significance of radioactive waste, generally insisting on the (debated) potential for recycling (Garcier 2012), on its low volume and its expert management.

Both attitudes signal, I suspect, a reality about radioactive waste: everybody knows of radioactive waste but few people have any experience of it. For waste studies, this presents a challenge because contemporary scholarship emphasizes how gaining a form of proximity, a direct experience of waste is epistemologically and politically important (Reno 2016). Waste

DOI: 10.4324/9781003019077-14

Figure 14.1 Storage room for high-level waste at Cap La Hague, France. February 2018
Credits: ©PHOTOPQR/OUEST FRANCE/Arnaud Le Gall

ethnographies, in particular, help document how social practices produce and animate waste objects and waste flows, intersecting with waste's labile materiality. Following waste throughout its transformations, as Arjun Appadurai would have us (Appadurai 1986), is essential to recover its meaning and political import, the complex value circuits along which waste circulates (Lepawsky and Mather 2011). For Gay Hawkins, waste ethics originates in our careful attentiveness to waste materials (Hawkins 2006). But how can we follow or be attentive to something that we cannot encounter? Because radioactive waste is so elusive and at the same time ever present, it is the locus of a tension between meaning and experience: the meaning we give to it is divorced from any actual experience we may have of it.

This chapter offers a reflexion on the experience of waste, with a specific focus on radio-active waste. I revisit the forms of waste experience that are accounted for in the literature, and argue that dominant approaches frame radioactive waste as a material beyond direct experience. Analyzing the variety of waste experiences and the framings that make them possible, I argue, is necessary to subject radioactive waste to the same type of critical scrutiny as other forms of waste and challenge its otherness.

Waste and experience

John Dewey noted that most philosophers speak of experience "at large". They refer to a fairly "inchoate" stream of experience: "Experience occurs continuously, because the interaction of live creature and environing conditions is involved in the very process of living" (Dewey 1980, 35). Experience, then, refers to the constant interactions between a living entity and its environment, whereby reciprocal changes are brought about. Experiencing something is being changed by it. Dewey, however, also notes that generally, we do not refer to experience in such a continuous fashion and that our approach to experience is discrete. We speak of "an experience": "an experience has a unity that gives it its name, that meal, that storm, that rupture of friendship. The existence of this unity is constituted by a single quality that pervades the entire experience in spite of the variation of its constituent parts" (Dewey 1980, 37). Sometimes then, experience coalesces into a discrete experience that "has pattern and structure, because it is not just doing and undergoing in alternation, but consists of them in relationship" (Dewey 1980, 44). In other words, a discrete experience does not emerge randomly nor mechanically out of an indiscriminate flow of events, but because the changes it brings about merge together practice and meaning.

These considerations are useful in the context of waste studies for they illuminate a tension about waste as an object of experience. Waste studies have noted how some materials are stricken by a form of selective invisibility (Alexander and O'Hare 2020). They disappear as meaningful objects: in other words, they lose the capacity to become the object of "an experience" because they are not seen anymore. Jane Bennett's famous opening in *Vibrant Matter* (Bennett 2009) illustrates precisely the reverse: encountering a bunch of dead things on a Baltimore sewer grid, Bennett has an experience of waste that marks the beginning of her ontological inquiry into the political ecology of things. Waste studies literature is replete with such examples where inchoate flows of waste and materials suddenly take on relief and contours and coalesce into discrete experiences that open up to new regimes of meaning and visibility. Such experiences of waste are, of course, dispersed: different people have different experiences but they can also be shared and become socialized. Dewey reminds us that a defining quality of "an experience" lies in its capacity to be told, circulated, and form the basis of new knowledge. Mathias Girel remarks that German has two words for experience: *Erlebnis*, which conveys the idea of a something that is privately lived (*Leben*) and felt; and *Erfahrung*, which refers to something that can be shared, capitalized, and narrated (Girel 2014).

It is those various threads that I would like to bring together in this chapter, by interrogating the very experiences of radioactive waste that we may have. The atomic age has created new forms of waste but it has also shaped how we may experience it. Radioactivity escapes direct sensory experience and is incommensurate with our understanding of time, making all experiences of it derivative and opening up to a variety of modes of knowing and experiencing. Moreover, a major concern in government and industry has precisely been to *prevent* the general public from gaining any actual experience of nuclear waste for safety reasons (Garcier 2014). Since the 1970s, nuclear waste management has relied, by and large, on containment strategies that remove radioactive waste from the public eye and segregate it from other waste flows. What then is the nature of the experience of waste we may have? In the following paragraphs, I introduce the idea that the presence of radioactivity in waste is not enough to produce an experience of "radioactive waste" and that we need to account for a greater variety of experiential forms.

Inventing (extra)ordinary materials

Radioactivity is a peculiar property of matter. Chemical elements have various isotopes, some of which are unstable: they are called radioactive. Their atoms spontaneously emit energy, in the form of ionizing rays (alpha, beta or gamma, neutrons). However, this release of energy is not directly observable: it requires specific equipment to become visible or detectable—Geiger counters, cloud chambers (Cloudylabs 2014), radioluminescent substances, particle detectors.

Without this equipment, radioactivity cannot be perceived. It will only manifest itself, belatedly, if it meets living tissues, which it will harm depending on the quantity of energy (absorbed dose) delivered by contamination or activation. Contamination refers to "radioactive substances on surfaces, or within solids, liquids or gases (including the human body), where their presence is unintended or undesirable, or the process giving rise to their presence in such places" (International Atomic Energy Agency 2018). Activation is the "process of inducing radioactivity in matter by irradiation [i.e., exposure to ionizing electromagnetic energy] of that matter" (International Atomic Energy Agency 2018). Depending on radiation type, energy, and how it touches or enters the body, detrimental effects will vary tremendously (Cram 2015). Those two properties of radioactivity—energy emissions invisible to human sensory perception and capacity to create harm at a distance—are not constant in time and space. Because of radioactive decay, radioelements change and transform, altering the type and intensity of their radiance until they reach an ultimate, steady state. Decay can be very quick. It may also take a long time, several billion years in the case of the transformation of uranium into lead.

It is tempting to think that radioactive waste is simply waste that contains significant amounts of radioactivity. That is not the case. As the International Commission on Radiological Protection notes:

> Radioactive waste is extremely difficult to specify. Everything is radioactive, and the specification of that part of waste that is to be treated as radioactive depends on the existence of definitions of what is to be excluded or exempt from the scope of the relevant documents. (ICRP-CIPR 1997, 3)

Accordingly, the question of whether waste will be considered and treated as "radioactive waste" is not predicated on radioactive contents alone: it is the result of a techno-political intervention (Garcier 2012). In this sense, we can say that radioactive waste has historically been "invented" (Sundqvist 2002; Hamblin 2008). The identification of what is of "regulatory concern" (and hence, what constitutes radioactive waste for law and policy) does not linearly depend on the nature or quantity of radionuclides in waste, nor on the origin of the waste. For example, uranium mine tailings have historically not been considered as "radioactive waste", although their radioactivity can reach very significant levels (International Atomic Energy Agency 2002). The financial cost and logistical problems of including the tailings in the nuclear waste category would have been phenomenal—even threatening the financial sustainability of the nuclear industry itself.

Such selectivity is an example of the work of what historian Gabrielle Hecht has termed "nuclearity" (Hecht 2006; 2007). Nuclearity is the techno-political construction of something as "nuclear", whether *or not* that thing is significantly radioactive. Radioactivity decays but nuclearity fluctuates depending on circumstances—e.g., the state of knowledge or political priorities. Historically, the emergence of "nuclearity" has been intimately associated with the development of atomic weapons and electronuclear energy, which have created large amounts of very radioactive waste. After World War 2, companies and governments mined uranium resources all over the world and developed new chemical and physical processes to extract, purify, and enrich useful isotopes for military and later, energy generation purposes. Starting with the Shinkolobwe uranium mine in the Congo (that provided some of the uranium for the Hiroshima bomb: Zoellner 2009), mining sites, ore processing and uranium enrichment facilities have become home to large stockpiles of radioactive waste. For uranium tailings, an IAEA study conservatively estimated the global amount at nearly 1 billion cubic meters (IAEA 2004, 9–10). Uranium mining and refining have left a trail of waste products, be it in Tajikistan (where the Soviets mined), in the U.S. Rockies or Southwest (Eichstaedt and Haynes 1994; Malin and Petrzelka 2010; Pasternak 2011; Malin 2015; Voyles 2015), in the central mountains of France (Brunet 2004), in Canada (van Wyck 2010), Australia (Marsh and Green 2020), in Madagascar, Niger, or South Africa (Hecht 2012). Some sites where uranium has been transformed for military or civilian purposes have become so intimately associated with radioactive waste that their cleanup is highly problematic (Krupar 2011; 2012; Cram 2016). Because the extraction and manufacturing of radioactive substances involve advanced chemistry and physical processes, the waste produced on these sites contains a panoplie of rare products, firmly setting this waste apart in a separate ontological realm: recent research on the environmental impact of the Sellafield site in the United Kingdom, for example, has shown how radioactive Cesium, Plutonium, and Americium from low-level liquid waste effluents have accumulated over the course of 50 years in an area of the Irish Sea known as the "Mudpatch" (Ray et al. 2020). Those elements are entirely anthropogenic: they do not occur naturally.

Atomic fission produces a second, even starker, range of waste. Chain reactions can be controlled (in nuclear energy generation) or not (in atomic explosions). They create new elements and new forms of waste. Spent nuclear fuel and waste derived from electronuclear energy are the best known and the most studied and I shall go back to them shortly (Rosa et al. 2010). But we should also remember that atomic tests have contaminated vast expanses of land in the U.S. deserts (Kirsch 2004; 2005; Masco 2006) and in the Kazakh steppe near the closed city of Kurchatov (Brown 2013; Alexander 2020). Tests have contaminated bodies and environments in the South Pacific, where the French tested their nuclear arsenal at Mururoa and Fangataufa atolls (Barthe 2017). Prior to the Partial Test Ban Treaty (PTBT) signed in 1963, nuclear powers conducted more than 500 atmospheric nuclear tests, disseminating large amounts of radioactive elements across the globe. Nuclear catastrophes—such as Chernobyl and Fukushima—have also contributed vast amounts of waste and pollution, creating the need for large-scale attempts at radioelement containment and "landscape decontamination" (Brown 2019; Yasunari et al. 2011; Cram 2016; Evrard, Laceby, and Nakao 2019). Nuclear fallout has become the sign of a new epoch, a putative global marker for the beginning of the Anthropocene (Waters et al. 2015; Masco 2015). Crucially, however, some of that waste and contamination never graduated to the "radioactive waste" category, thus naturalizing contamination and sheltering militaries and industry worldwide from the liabilities associated with proper radioactive waste management (Petryna 2002).

Frames of waste experience

The creation of waste that contains radioactivity and the workings of nuclearity have had a very large influence on our experience of radioactive waste. As Dewey noted, experience does not take place on a *tabula rasa* of the mind: it is predicated on and articulated by previous knowledge, expectations, selective perceptions...Experience does not derive from a series of chance encounters or a string of passive exposure to events: it proceeds from the circulation and capitalization of meanings and interpretations. In this sense, having *an* experience of radioactive waste involves materials but also registers of meaning attached to them. It is precisely those registers that I will try to outline here. The typology that follows does not claim to be exhaustive, but seeks to provide a rough categorization for a large body of literature and a variety of discrete experiences with and of waste.

Nuclear waste as an experience of the technological condition

Many scholars, activists, and artists have analyzed and elaborated on the historical significance of atomic events for the modern condition (Weart 1989; Beck 1992; Carpenter 2016). The bombings of Hiroshima and Nagasaki, with their lasting effects on generations of survivors and their offspring, have radically changed our visions of modernity, technology, and progress (Alexis-Martin 2019). They mark radioactivity's escape out of the laboratory and into the lives and political imaginaries of billions of people (Boyer 1994; Masco 2015), marking the "obsolescence of mankind" (Anders 2018). Radioactive waste cannot be divorced from these legacies of the past: our experience of it is fraught with collective memories and imaginaries that bring together technology, power, and natural forces.

Nuclear waste has undergone several waves of sociocultural framing and construction that have shaped our perceptions of it (Pajo 2016). In the early days of the nuclear industry, radioactive waste was purely framed as an object for management, a technological issue. In the 1960s and 1970s, Project Ploughshare in the United States even explored the feasibility of using nuclear explosives for "peaceful construction purposes", generating serious ground contamination without much concern (Kirsch 2005). A few years later, a new framing gradually appeared that transformed radioactive waste into a dangerous, highly political substance, and a predicament that could not be solved technologically.

Nuclear waste as an experience of risk and danger

Scientific accounts of encounters with radioactive waste generally have a very factual tone, but they also shed a rather sinister light on its ability to tear the mundane apart. The 1987 Goiania incident in Brazil is a case in point (M. Anjos et al. 2002). In September 1987, two scavengers stole a radiotherapy source from an abandoned hospital in the city of Goiania, hoping to resell it as scrap metal. After it was dismantled, steel and lead from the waste source entered various recycling streams in the city, finding their way into people's homes. Scrapyard owner Devair Alves Ferreira noticed the strange blue glow emanating from the caesium chloride capsule he had just bought, without knowing what it was. He brought it back home to show his friends and family. In two weeks, 97 grams of caesium eventually killed four people (including Devair Alves Ferreira's own daughter), contaminated more than 200 others, and created lasting environmental pollution. The Goiania incident has remained famous, because it exemplifies how the ominous agency of radioactive waste may contaminate everyday life.

This incident and several others (e.g., Associated Press 1981) have greatly contributed to the production of narratives and imaginaries around radioactive contamination in waste (Dunlap, Kraft and Rosa 1993; Freudenburg 2004). Radioactive waste is not waste as we know it and should not be taken as such: experiencing it is exposing oneself to death. What is more, radioactive waste is deceitful: it contradicts our experience of waste. Metal is not metal. Dust is not dust. Nowhere is the treacherous agency of radioactive waste more obvious that in the aftermath of atomic tests, uranium mines, or nuclear catastrophes (Brown 2019), where the landscape itself is replete with invisible corruption (Lepage 2012) but where the workings of nuclearity mask the danger. Writing on waste after Chernobyl, Nobel Prize Winner Svetlana Alexievitch conveys the voice of a man, Sergei Sobolev, deputy head of the Executive Committee of the Shield of Chernobyl Association, who warns of the threat of radioactive waste being passed off as conventional waste. The man describes the waste "graveyards" where "tons of metal and steel, small pipes, special clothing, concrete constructions", where "thousands of individual pieces of automotive and aviation machinery, fire trucks and ambulances" have been dumped—only to be stolen and resold to unknowing citizens (Alexievich 2017: "Monologue about lies and truths"). What can be done then? Should zones of exclusion be created, giving over space to radioactive waste (Alexis-Martin and Davies 2017)? Should attempts be made to contain radioactive waste's agency to prevent our experiencing it?

An experience of presence/absence

In their 2015 documentary *Containment*, Peter Galison and Rob Moss show how incredibly complex containment is (Galison and Moss 2015). Not only does containment require nuclear waste streams to be separated from other streams, to avoid the Goiana effect of radioactive waste passing for conventional waste; but it also rests on controlling radioactivity's movements for eons. The "boundary of control" is the shifting time period where such control can be achieved, first by institutional means, and then, by technical artifacts (Berkhout 1991). Berkhout shows how governments have tried to expand the "boundary of control" over radioactive waste by developing extraordinary engineering concepts—such as the geological repository, which functionally projects human agency into the distant future. For this, it is necessary to enroll natural forces and engineered materials—stable geological layers, dry environments, sturdy metals.

Geological repositories are the consensus solution for high-level waste storage (Barthe 2006; Madsen 2010). A few have been constructed—e.g., the Waste Isolation Pilot Plant or WIPP, in New Mexico—and their operation is far from simple (Ialenti 2018; 2021). Others are planned or being built in Finland, Sweden, Belgium and France (Nucci et al. 2015). Repositories are complex structures, designed to create physical barriers between high-level waste, the surface environment, and human bodies (Macfarlane and Ewing 2006). But to avoid the Chernobyl effect of people digging up waste to use it or sell it, containment also hinges on conveying a sense of danger across time, to future people who may not share any of our languages or cultural signs (Bloomfield and Vurdubakis 2005). How can we do that? What art forms are best attuned to warning future Earth dwellers about any possible encounter with high-level waste (Poirot-Delpech and Raineau 2016)? Some artists suggest that songs are the best art form to talk about radioactive waste for millenia. Haven't the chants of the Veda outlived most human-built structures? Can we write and sing songs about waste and our experience of it (Mars 2014)?

There is one interesting consequence of the emphasis, in the literature, on high-level waste and future geological repositories to be built (Nucci et al. 2015; Cotton 2017; Murdock, Leistritz and Hamm 2018). The experience of waste they analyze is based on institutional assessments of policies and large infrastructural projects (Lehtonen, Joly and Aparicio 2016) rather than on an intimate, physical connection with materials (Gregson 2012). Waste as "stuff" disappears: the focus is on potential sites, funding schemes, fierce local opposition and stakeholder engagement and participation (Di Nucci and Brunnengräber 2019; Ferraro 2019). The experience of nuclear waste, then, is an experience where waste itself is curiously absent (Bickerstaff 2012).

Encounters with nuclear waste

The three modalities of experience framing I have just detailed share a strange property: they all involve a form of waste euphemizing, where the materiality of radioactive waste disappears as such. Such absence removes the materiality of waste from consideration and transforms radioactive waste into an empty signifier, a virtual object. What then, would a concrete engagement with radioactive waste mean? What insights could it provide? In this last section, I would like to share a personal experience. A few years ago, I attended a four-day course on radioactive waste management at the national low-level radioactive waste repository in eastern France. The course was designed to train waste managers on the acceptance rules at the repository. Among the roughly 15 waste managers in attendance, most had encountered *problems* with waste—often, problems with understanding why the disposal rules are so strict and how they can fully abide by them. For a few days, I got to listen to and interact with very qualified waste managers. They were quite surprised to see an academic interested in what they said was the "mundane business" that occupied their days. But they were also very keen to talk, and passionate about their activities. It struck me then—and now—that what differentiated their approach to nuclear waste from mine was their thorough experience of day-to-day management. Their framing of radioactive waste was entirely different from mine. How they experienced it was strongly linked to how waste manifested itself in their organizations. In a way, the whole course was designed to draw out what is common in their individual experiences and to transform lived reality (*Erlebnis*) into shared knowledge (*Erfahrung*).

On the first morning we went around the circle of participants and explained why we were there. A man was the last to speak. His company had started work on decommissioning large pieces of military equipment that contain radioactive sources—old radars, for example. Such sources often contain tritium (radioactive hydrogen), the man explained, an element that is problematic in the context of waste disposal. "I know that when we speak of tritium, we are not very welcome here...", the man said cautiously. The course convener agreed, smiling: "It is true. We are not madly in love with tritium waste on this site". Everybody laughed. A woman from another company reacted: "For tritium they say 'put that waste in a corner and wait, its half-life is short'". The armament man agreed: "Yes, we don't have a problem now, we can store the waste. But in the long run, what should we do with materials that contain tritium? For example, I need to decommission a tank that has tritium contamination because radioluminescent plates provided the lighting inside the cockpit. We removed the plates but the tank structure is contaminated. So what do I do with this tank? How should we cut it? In how many pieces do you want it? And how do you want the pieces? In drums? In big bags?"

Such questions make profound sense. In the mid-nineties and early noughties, when the French low-level and very low-level waste repositories were created in Soulaines and Morvilliers, respectively, they were not simply conceived of as disposal "sites", concrete-lined pits where all sorts of low-activity radioactive waste would be indiscriminately sent. They were designed as complex waste infrastructures: the investment included considerable research into radioactive waste agency, which resulted in the creation of a waste acceptance "referential" for each site. The referential includes specifications on the chemical and radiological composition of the waste. Known as "the specs", these criteria are a powerful source of waste determination: they require waste producers to create detailed information on the waste they generate and send for disposal, and they form the basis of the contractual agreements between waste producers and the operator of the site, the National Radioactive Waste Agency (ANDRA). The waste referential also includes sets of standards on packaging types and physical parameters for waste packaging (e.g., resistance to compression, porosity to water...collectively known as "technical standards"). Altogether, respecting the three sets of standards in the referential (the specs, the packaging standards, and the technical standards) is a technical challenge that poses a serious risk. The risk is discovering too late that waste sent to a disposal site does not meet either the contractual agreement, or even worse, does not meet the "specs". Considerable resources and imagination go into making waste fit inside these tight standards.

These regulations strongly frame waste managers' own experience of waste. The participants did not deal with high-level waste but with casual radioactive stuff that was anything but virtual. They described the minute considerations that they had to devote to waste's materiality. Brigitte, for example, told me about the problems with vermiculite, a powder mineral used in a variety of nuclear applications. "Vermiculite is powdery, so it cannot be compacted. And it is so light that if you try to fix it using grout or mortar, it will float on the surface. We have tried many things—we still haven't found the perfect solution". A colleague concurred: "It is the same problem with dust, for example when a facility is decommissioned. We vacuum the dust but then what? What should I do with it? It's so light I cannot congeal it into a matrix". Another woman, Lydia, works for a small waste producer that uses radioactive reagents in its laboratories: "Our problem is small, exotic waste without an outlet. Every day we have different garbage. Animals. Worms. Rats. Every day is different". She recalls an episode that became famous in the nuclear community. Another laboratory sent a batch of 16 monkey cadavers for disposal. The corpses had been insufficiently quicklimed. A dreadful stench floated on the site for days.

The course convenor listed other problematic waste materials, a Borgesian inventory of reluctant garbage. For example, "Safety shoes are a pain in the ass. You can't compact them. In direct storage, it's problematic because they're full of empty spaces. If there is labile contamination, what we do is we put the shoes in bulk (not bagged) in a container to be injected. And if there isn't, the best thing to do is to put them in bulk in a rack (but with the hole on top) and to place a welded grid on the top of the rack. Then we will inject them with grout. The grid is necessary, because being so light, the shoes can move up if there is no grid". Someone asked "why can't we grind or cut them?" To which Sophie replied: "It would take too much time and cost too much. Just like mask cartridges. Plus the risks associated with handling them". As for Denis, he had "a problem with asbestos. We put a doctrine in place back in 2011, but I'm afraid of doing something stupid, so I don't touch it. Asbestos in the waste, it has been stored for 50 years. So it's very varied … Before doing anything, I want to know how to do it and where to send the waste". The experiences I hear about escape the framings of waste I am accustomed to: they show the persistent materiality of mundane radioactive waste, how waste has to be specifically treated and transformed precisely because its combined materiality and radioactivity create a problem.

The experiences emphasize the interaction between institutions and waste, an interaction mediated by the generation of information on waste. Every waste producer has to declare what waste it has in store, how much more will be generated and which waste will be sent off for disposal in a given year. Such information is required because the entire waste management chain is organized from downstream up: if waste is to be disposed of, it has to abide by the strict terms of the "referential", themselves the product of radioprotection rules and disposal site characteristics. An ANDRA engineer explained: "For example, company A once told us, 'we have a batch of waste that contains 44 tons of asbestos'. Now, the maximum amount of asbestos we can put in a single storage cell at the disposal facility in Morvilliers is 13 tons. This batch alone would have used four cells. We asked the company for details on their methods of characterization and they answered 'in fact, we only have 24 tons of asbestos in the waste'. Therefore, the problem of characterization is essential". "Maximum acceptability limits" (MAL) define the maximum quantities of radioactive content in any batch of waste. The MAL are determined by dose constraints, designed to protect site operators in case the packaging fails. I asked Stéphane how waste producers integrate this particular metric (MAL) into their daily practice. He explained: "For radionuclides, we think in terms of mass when we plan how we will fill the waste bags. Then we send the numbers to Sophie, who does a global verification and compares it with the rules. If it's consistent, then Sophie translates it back into MAL terms". Stéphane concluded: "Sophie's the one who speaks the language of ANDRA".

Conclusion

Stéphane's last sentence states quite clearly the connection between language and experience. We need a language to convey and share our experiences of waste. But which one? Stéphane does not get to choose his language: the disposal chain provides the words and the syntax that define waste and describe how it should be dealt with. Sophie and her colleagues permanently have to juggle between their material, concrete, immediate experiences of waste, and the intricacies of an acquired language. They seek to align their practices with the complex demands of the management system and turn waste into an object of experience mediated by technical procedures, information systems, and personal know-how. The challenge, then, is to create a fit between the words and the materials, between the normative requirements of management and the reluctant materiality of garbage. Of course, this is far from specific to radioactive waste and it happens in other waste strands. It may even sound familiar to those of us who have tried to sort domestic waste for recycling and need to decide whether this yogurt container should go in the yellow or in the black bin. In a sense, the accounts of Sophie and her colleagues illustrate how thoroughly mundane nuclear waste may be as an object of management, and experience. Paying attention to how experiences of waste emerge and are organized helps dispel the illusion that some forms of waste are intrinsically more exceptional than others. Waste experiences question the relations between materials, practices, consciousness, and institutions.

They also challenge the idea that materiality alone provides a key to understanding waste's status in society. Indeed, I have tried to show that experiences of waste go beyond the encounter of waste materials. Forms of experience can occur without encounter: most of us never meet and don't produce radioactive waste. We never experience it as such. How can we explain then why nuclear waste resonates so strongly in the general public who have never come across the actual residues? Providing an answer requires recognizing that not all experiences of waste are material. In waste, it is not merely materials we may encounter, but infrastructural complexity or uncertainty; corporate secrecy or violence; symbolic imaginations and a variety of other framings. These dimensions are integral to our experience because they provide the interpretative bricks with which we create meaning. I suspect that the waste managers I met would absolutely agree: they know that the reality and the breadth of their experience of waste do not lie in the materials only, but in the infrastructural and legal environment that give meaning to them and in the practices in which they are enmeshed. As such, the diversity of waste experiences opens up to a variety of research questions and political possibilities.

Acknowledgments

In Section 4, all names have been changed and companies anonymized. This part of the chapter has been read and commented upon by the national radioactive waste agency ANDRA. I thank Robert Mandoki for his insights and clarifications. All analyses and possible errors remain mine.

This research is part of the METROPOLITIN project, funded by the French Government under the Programme Investissements d'Avenir, with the technical assistance of ANDRA.

References

Alexander, Catherine. 2020. "A Chronotope of Expansion: Resisting Spatio-Temporal Limits in a Kazakh Nuclear Town." *Ethnos*: 1–24. doi:10.1080/00141844.2020.1796735.

Alexander, Catherine, and Patrick O'Hare. 2020. "Waste and Its Disguises: Technologies of (Un) Knowing." *Ethnos*: 1–25. doi:10.1080/00141844.2020.1796734.

Alexievich, Svetlana. 2017. *Chernobyl Prayer: A Chronicle of the Future*. Penguin Classic.

Alexis-Martin, Becky. 2019. *Disarming Doomsday: The Human Impact of Nuclear Weapons since Hiroshima*. Pluto Press.

Alexis-Martin, Becky, and Thom Davies. 2017. "Towards Nuclear Geography: Zones, Bodies, and Communities." *Geography Compass* 11, no. 9: e12325. doi:10.1111/gec3.12325.

Anders, Günther. 2018. *Die Antiquiertheit Des Menschen Bd. II: Über Die Zerstörung Des Lebens Im Zeitalter Der Dritten Industriellen Revolution*. C.H.Beck.

Appadurai, Arjun. 1986. "Introduction: Commodities and the Politics of Value." In Arjun Appadurai (Ed.) *The Social Life of Things: Commodities in Cultural Perspective*. Cambridge University Press. 3–63.

M. Anjos, R., N. K. Umisedo, A. Facure, E. M. Yoshimura, P. R. S. Gomes, and E. Okuno. 2002. "Goiânia: 12 Years after the 137Cs Radiological Accident". *Radiation Protection Dosimetry* 101: 201–204. doi: 10.1093/oxfordjournals.rpd.a005967.

Associated Press. 1981. "Radioactive Jewellery In Western New York." *The New York Times*, February 1981. http://www.nytimes.com/1981/02/07/nyregion/the-region-radioactive-jewelry-in-western-new-york.html.

Barthe, Yannick. 2006. *Le Pouvoir d'indécision. La Mise En Politique Des Déchets Nucléaires. Etudes Politiques*. Economica.

Barthe, Yannick. 2017. *Les retombées du passé—Le paradoxe de la victime*. Le Seuil.

Beck, Ulrich. 1992. *Risk Society: Towards a New Modernity*. Sage.

Bennett, Jane. 2009. *Vibrant Matter: A Political Ecology of Things*. Duke University Press.

Berkhout, Frans. 1991. *Radioactive Waste: Politics and Technology*. Routledge.

Bickerstaff, Karen. 2012. "'Because We've Got History Here': Nuclear Waste, Cooperative Siting, and the Relational Geography of a Complex Issue." *Environment and Planning A* 44, no. 11: 2611–2628.

Bloomfield, Brian P., and Theo Vurdubakis. 2005. "The Secret of Yucca Mountain: Reflections on an Object in Extremis." *Environment and Planning D: Society and Space* 23, no. 5: 735–756.

Blowers, Andrew. 2016. *The Legacy of Nuclear Power*. 1st Edition. Routledge.

Boyer, Paul. 1994. *By the Bomb's Early Light: American Thought and Culture at the Dawn of the Atomic Age*. 1st Edition. University of North Carolina Press.

Brown, Kate. 2013. *Plutopia: Nuclear Families, Atomic Cities, and the Great Soviet and American Plutonium Disasters*. Oxford University Press.

Brown, Kate. 2019. *Manual for Survival: A Chernobyl Guide to the Future*. 1st Edition. W. W. Norton & Company.

Brunet, P. 2004. *La Nature Dans Tous Ses États: Uranium, Nucléaire et Radioactivité En Limousin*. PULIM.

Carpenter, Ele (Ed.). 2016. *The Nuclear Culture Source Book*. Black Dog Press.

Cloudylabs. 2014. *Cloudylabs Cloud Chamber Working*. https://www.youtube.com/watch/ZiscokCGOhs.

Cotton, Matthew. 2017. *Nuclear Waste Politics: An Incrementalist Perspective*. Routledge.

Cram, Shannon. 2015. "Becoming Jane: The Making and Unmaking of Hanford's Nuclear Body." *Environment and Planning D: Society and Space* 33, no. 5: 796–812. doi:10.1177/0263775815599317.

Cram, Shannon. 2016. "Wild and Scenic Wasteland: Conservation Politics in the Nuclear Wilderness." *Environmental Humanities* 7, no. 1: 89–105. doi:10.1215/22011919-3616344.

Dewey, John. 1980. *Art as Experience*. Perigee Books.

Di Nucci, Maria Rosaria, and Achim Brunnengräber. 2019. "Making Nuclear Waste Problems Governable." In Achim Brunnengräber and Maria Rosaria Di Nucci (Eds.) *Conflicts, Participation and Acceptability in Nuclear Waste Governance: An International Comparison* (Vol. III). Springer Fachmedien. 3–19. Energiepolitik Und Klimaschutz. Energy Policy and Climate Protection. doi:10.1007/978-3-658-27107-7_1.

Dunlap, Riley E., Michael E. Kraft, and Eugene A. Rosa. 1993. *Public Reactions to Nuclear Waste: Citizens' Views of Repository Siting*. Duke University Press.

Eichstaedt, Peter H., and Murrae Haynes. 1994. *If You Poison Us: Uranium and Native Americans*. Red Crane Books.

Evrard, Olivier, J. Patrick Laceby, and Atsushi Nakao. 2019. "Effectiveness of Landscape Decontamination Following the Fukushima Nuclear Accident: A Review." *SOIL* 5, no. 2: 333–350. doi:10.5194/soil-5-333-2019.

Ferraro, Gianluca. 2019. *The Politics of Radioactive Waste Management: Public Involvement and Policy-Making in the European Union.* Routledge.

Freudenburg, William R. 2004. "Can We Learn from Failure? Examining US Experiences with Nuclear Repository Siting." *Journal of Risk Research* 7, no. 2: 153–169.

Galison, Peter, and Rob Moss. 2015. *Containment.* Redacted Pictures.

Garcier, Romain. 2012. "One Cycle to Bind Them All? Geographies of Nuclearity in the Nuclear Fuel Cycle." In Catherine Alexander and Josh Reno (Eds.) *Recycling Economies: Global Transformations of Materials, Values and Social Relations.* Zed Books.

Garcier, Romain. 2014. "Disperser, confiner ou recycler?Droit, modes de gestion et circulations spatiales des déchets faiblement radioactifs en France." *L'Espace géographique* 43, no. 3: 265–283. doi:10.3917/eg.433.0265.

Girel, Mathias. 2014. "L'Expérience Comme Verbe?" *Education Permanente* 1: 23–34.

Gregson, Nicky. 2012. "Projected Futures: The Political Matter of UK Higher Activity Radioactive Waste." *Environment and Planning A* 44, no. 8: 2006–2022.

Hamblin, Jacob Darwin. 2008. *Poison in the Well: Radioactive Waste in the Oceans at the Dawn of the Nuclear Age.* Rutgers University Press.

Hawkins, Gay. 2006. *The Ethics of Waste: How We Relate to Rubbish.* Rowman & Littlefield. https://www.loc.gov/catdir/enhancements/fy1701/2005021924-b.html

Hecht, Gabrielle. 2006. "Nuclear Ontologies." *Constellations* 13, no. 3: 320–331.

Hecht, Gabrielle. 2007. "A Cosmogram for Nuclear Things." *Isis* 98, no. 1: 100–108.

Hecht, Gabrielle. 2012. *Being Nuclear: Africans and the Global Uranium Trade.* MIT Press.

IAEA. 2004. *The Long Term Stabilization of Uranium Mill Tailings.* TECDOC Series 1403. International Atomic Energy Agency.

Ialenti, Vincent. 2018. "Waste Makes Haste: How a Campaign to Speed up Nuclear Waste Shipments Shut down the WIPP Long-Term Repository." *Bulletin of the Atomic Scientists* 74, no. 4: 262–275. doi:10.1080/00963402.2018.1486616.

Ialenti, Vincent. 2021. "Drum Breach: Operational Temporalities, Error Politics and WIPP's Kitty Litter Nuclear Waste Accident." *Social Studies of Science.* doi:10.1177/0306312720986609.

ICRP-CIPR. 1997. *Radiological Protection Policy for the Disposal of Radioactive Waste.* Annals of the ICRP 77. International Commission on Radiological Protection.

International Atomic Energy Agency. 2002. "Management of Radioactive Waste from the Mining and Milling of Ores." *Safety Standards.* IAEA.

International Atomic Energy Agency. 2018. *IAEA Safety Glossary: Terminology Used in Nuclear Safety and Radiation Protection—2018 Edition.* International Atomic Energy Agency.

Kirsch, Scott. 2004. "Harold Knapp and the Geography of Normal Controversy: Radioiodine in the Historical Environment." *Osiris* 19: 167–181.

Kirsch, Scott. 2005. *Proving Grounds: Project Plowshare and the Unrealized Dream of Nuclear Earthmoving.* Rutgers University Press.

Krupar, Shiloh. 2011. "Alien Still Life: Distilling the Toxic Logics of the Rocky Flats National Wildlife Refuge." *Environment and Planning D: Society and Space* 29, no. 2: 268–290. doi:10.1068/d12809.

Krupar, Shiloh. 2012. "Transnatural Ethics: Revisiting the Nuclear Cleanup of Rocky Flats, CO, through the Queer Ecology of Nuclia Waste." *Cultural Geographies* 19, no. 3: 303–327.

Lehtonen, Markku, Pierre-Benoît Joly, and Luis Aparicio. 2016. *Socioeconomic Evaluation of Megaprojects: Dealing with Uncertainties.* Routledge.

Lepage, Emmanuel. 2012. *Un Printemps à Tchernobyl.* Futuropolis.

Lepawsky, Josh, and Charles Mather. 2011. "From Beginnings and Endings to Boundaries and Edges: Rethinking Circulation and Exchange through Electronic Waste." *Area* 43, no. 3: 242–249. doi:10.1111/j.1475-4762.2011.01018.x.

Macfarlane, Allison, and Rodney C. Ewing (Eds.). 2006. *Uncertainty Underground: Yucca Mountain and the Nation's High-Level Nuclear Waste.* MIT Press.

Madsen, Michael. 2010. *Into Eternity.* Chrysalis Films.

Malin, Stephanie A. 2015. *The Price of Nuclear Power: Uranium Communities and Environmental Justice.* Rutgers University Press.

Malin, Stephanie A., and Peggy Petrzelka. 2010. "Left in the Dust: Uranium's Legacy and Victims of Mill Tailings Exposure in Monticello, Utah." *Society & Natural Resources* 23, no. 12: 1187–1200. doi:10.1080/08941920903005795.

Mars, Roman. 2014. *Ten Thousand Years*. 99% Invisible. https://99percentinvisible.org/episode/ten-thousand-years/.

Marsh, Jillian K., and Jim Green. 2020. "First Nations Rights and Colonising Practices by the Nuclear Industry: An Australian Battleground for Environmental Justice." *The Extractive Industries and Society* 7, no. 3: 870–881. doi:10.1016/j.exis.2019.01.010.

Masco, Joseph. 2006. *The Nuclear Borderlands: The Manhattan Project in Post-Cold War New Mexico*. Princeton University Press.

Masco, Joseph. 2015. "The Age of Fallout." *History of the Present* 5, no. 2: 137–168. doi:10.5406/historypresent.5.2.0137.

Murdock, Steve H., F. Larry Leistritz, and Rita R. Hamm (Eds.). 2018. *Nuclear Waste: Socioeconomic Dimensions Of Long-Term Storage*. Routledge.

Nucci, Maria Rosaria Di, Achim Brunnengräber, Lutz Mez, and Miranda Schreurs. 2015. "Comparative Perspectives in Nuclear Waste Governance." In *Nuclear Waste Governance: An International Comparison*. Springer. 25–43. Energy Policy and Climate Protection.

O'Brien, Martin. 2008. *A Crisis of Waste? Understanding the Rubbish Society* (Vol. 33). Routledge Advances in Sociology. Routledge.

Pajo, Judi. 2016. "Two Paradigmatic Waves of Public Discourse on Nuclear Waste in the United States, 1945–2009: Understanding a Magnitudinal and Longitudinal Phenomenon in Anthropological Terms." *PLoS ONE* 11, no. 6. doi:10.1371/journal.pone.0157652.

Pasternak, Judy. 2011. *Yellow Dirt: A Poisoned Land and the Betrayal of the Navajos*. Free Press.

Petryna, A. 2002. *Life Exposed: Biological Citizens after Chernobyl*. Princeton University Press.

Poirot-Delpech, Sophie, and Laurence Raineau. 2016. "Nuclear Waste Facing the Test of Time: The Case of the French Deep Geological Repository Project." *Science and Engineering Ethics* 22, no. 6: 1813–1830. doi:10.1007/s11948-015-9739-9.

Ray, Daisy, Peter Leary, Francis Livens, Neil Gray, Katherine Morris, Kathleen A. Law, Adam J. Fuller, et al. 2020. "Controls on Anthropogenic Radionuclide Distribution in the Sellafield-Impacted Eastern Irish Sea." *Science of The Total Environment* 743: 140765. doi:10.1016/j.scitotenv.2020.140765.

Reno, Joshua. 2016. *Waste Away: Working and Living with a North American Landfill*. University of California Press.

Rosa, Eugene A., Seth P. Tuler, Baruch Fischhoff, Thomas Webler, Sharon M. Friedman, Richard E. Sclove, Kristin Shrader-Frechette, et al. 2010. "Nuclear Waste: Knowledge Waste?" *Science* 329, no. 5993: 762–763. doi:10.1126/science.1193205.

Sundqvist, Göran. 2002. "The Discovery of Nuclear Waste." In Göran Sundqvist (Ed.) *The Bedrock of Opinion: Science, Technology and Society in the Siting of High-Level Nuclear Waste*.49–73. Springer. Environment & Policy. doi:10.1007/978-94-015-9950-4_4.

Voyles, Traci Brynne. 2015. *Wastelanding: Legacies of Uranium Mining in Navajo Country*. 1st edition. University of Minnesota Press.

Waters, Colin N., James P. M. Syvitski, Agnieszka Gałuszka, Gary J. Hancock, Jan Zalasiewicz, Alejandro Cearreta, Jacques Grinevald, et al. 2015. "Can Nuclear Weapons Fallout Mark the Beginning of the Anthropocene Epoch?" *Bulletin of the Atomic Scientists* 71, no. 3: 46–57. doi:10.1177/0096340215581357.

Weart, Spencer R. 1989. *Nuclear Fear: A History of Images*. Harvard University Press.

Wyck, Peter van. 2010. *The Highway of the Atom*. McGill-Queen's University Press.

Yasunari, Teppei J., Andreas Stohl, Ryugo S. Hayano, John F. Burkhart, Sabine Eckhardt, and Tetsuzo Yasunari. 2011. "Cesium-137 Deposition and Contamination of Japanese Soils Due to the Fukushima Nuclear Accident." *Proceedings of the National Academy of Sciences of the United States of America* 108, no. 49: 19530–19534.

Zoellner, Tom. 2009. *Uranium: War, Energy, and the Rock That Shaped the World*. Penguin.

Zonabend, Françoise. 1993. *The Nuclear Peninsula*. Cambridge University Press.

15

URANIUM LEGACIES AND SETTLER-COLONIAL IMAGINARIES: NUCLEAR WASTE AS HISTORY, PROXIMITY, AND COLONIAL MATTER

Emily Potter

Introduction

Australia has a complex relationship with uranium, one of its most significant mineral exports. Since its extraction began in the early 20th century, uranium has offered a spectrum of material rewards and devastating destruction in a domestic context, resulting in dissonant community and state responses. These histories are local, but they are connected to the globalized nature of nuclear development, which unevenly connects countries across the Global North and South in a network of extraction, production, use, and disposal. While Australia's intimate history with uranium is not unique, its responses to nuclear activity are implicated in settler-colonial imaginaries that are inextricably linked to the country's colonial project itself. The spectre of nuclear waste has drawn divisive attention to this, as Australia has been forced to engage with the specific space-time of uranium and the fact that it never just "goes away". This comes into tension with the colonial project's future-focused logic, concerned with displacing First Nations' time and connection to Country, and the spatial imaginary of "remote" Australia as a "wasteland" (Voyles 2015), unpeopled and under-utilized. The "last continent found in Europe's long search for treasure" (Blainey 2), was also a site where, it was imagined, the externalities of this pursuit of wealth and power could be contained. The legacies of British atomic testing in Australia in the 1950s, legacies that will haunt for millennia, are a darker undercurrent to the contemporary debate around international nuclear waste storage in Australia, which move through the charged territory of national visions and extractivist imaginaries as much as they do ethical, economic, and political terrains. Waste, in this story, becomes a signifier of the impossibility of Australia excizing its violent and problematic histories. Renewing national imaginaries that acknowledge this suggests a way of coming to alternative, more potentially generative relations with the radiating resonance of nuclear waste.

DOI: 10.4324/9781003019077-15

Nuclear nation

Australia is one of the top producers of mineral resources globally, and mining—one of this settler colonial country's earliest commercial sectors—currently comprises its largest single export industry. The consequent role of mining in the nation's economic and social history is significant, and deeply imbricated in the logic and self-imagining of the nation. Australian uranium mining currently constitutes a little under a third of the global stockpile, and generates export revenue of roughly AU\$575 million per annum (McKay and Mietzitis 48; Senior et al. 65; Minerals Council of Australia 2020). While Australia prides itself on its bilateral agreements that stipulate non-military use of exported uranium (Senior et al. 66), the nation's relationship to nuclear activity, and its history as, what Gabrielle Hecht terms, a "nuclear state" is not straightforward (3). This complexity relates to Australia's role as both a producer of uranium, and also its participation as both a site of, and agent in, "nuclear colonialism" (Santayana 1). In this case, the colonizing capacities of nuclear activity relate to the nation's status as an historic subject of empire, yoked to Britain's post Second World War nuclear testing program, and beyond this, to the extractivist rationale at the heart of the settler colonial state, and the displacements and erasures that are endemic to the extractivist process.

Extractivism is an ultimate mode of territorial claim, "settler colonialism's specific, irreducible element" (Wolfe in Voyles 24). From this perspective, Traci Brynne Voyles contends, settler colonialism is always "deeply about resources" (23), and both environments and Indigenous bodies (representing both contested sovereignty and distinct ontologies of being in place) become expendable to this pursuit. Settler colonialism, as its theorists have pointed out (Veracini 2007; Huggan and Tiffin 2010) is a distinct form of imperial claim, involving an ongoing aspect of place-making for the settler colonial subject, as well as the correlate re-making of place for the colonized subject. Unlike lands formerly held by an occupying power, the claim to First Nations' lands is never rescinded by the settler-colonial state, requiring ongoing state efforts of possession, subjugation, and denial of enduring Indigenous connections to Country. The history of uranium mining and nuclearity in Australia must be understood in this context, in which the emergence of a uranium industry is at once a story of the drive to exhaustive consumption—where all available resources are used up in the service of the market—and colonial dominion that requires constant assertion. The imaginaries that enable this manifestation of the nuclear state still structure the nation's ongoing relationship to uranium as both a signifier and a material force of colonial power haunted by its own unstable, unsustainable claims.

Australia's export history of uranium dates to the 1950s, 60 years or so after ore was first identified at Carcoar in New South Wales in 1894. Exploration for uranium began in earnest following requests from the British and U.S. governments, with the Australian government offering tax incentives for its discovery. Four new key deposits were located between 1949 and 1956, and the opening of the Rum Jungle uranium mine in the Northern Territory in 1954 saw the export industry capacity rapidly grow, supplying uranium oxide to the US-UK Combined Development Authority (CDA) that (in contrast to today's agreements) enabled the development of nuclear weapons in both countries. Between 1954 and 1971 Rum Jungle, Radium Hill, and the Mary Kathleen mine in Queensland supplied over 7,700 tons of uranium to the CDA and the UK Atomic Energy Authority. All these mines were exhausted by 1971. The governmental intent behind uranium supply during the 1950s and into the 1960s was defense-focused rather than economic, and, as Hecht suggests, racialized. Reflecting the nation's alignments with other white Western forces, Australia's conversative Prime Minister Robert Menzies (1949–1966) advocated the supply of uranium for nuclear weapons for "our security…[and] the superiority of the Free World", assured by nuclear capability (in Harris 2). By securing the "white race's superweapons", Australia and its allies could defend against "their own colonization by superpowers" (Hecht 252). While a change of government brought economic imperatives to the fore in Australia's ongoing mining of uranium, an aquisatory, extractivist logic prevailed and the new Labour government continued to heavily invest in expanding uranium mining, with four additional mines established during the 1970s, all in the Northern Territory.

The introduction of laws to limit the harms done by uranium mining reflect the predicates of extractivist colonial logic as much as they do the attempt to mitigate these. Early examples of legislative protections such as the 1975 *National Parks and Wildlife Conservation Act* and the Northern Territory 1974 *Aboriginal Land Rights Act* focused on Indigenous land rights protection, place conditions on the production and location of uranium mining in Australia. But they also point to the key externalities of the extractivist agenda: the environment beyond instrumentalist, economic interests, and Indigenous lives and relationships with Country. These, once again, connect to the logic of settler colonialism itself, which requires resources to exploit and must therefore optimize the removal of impediments to their acquisition. Competing soveriegnty is key to this. It is therefore necessary, as Voyles contends, to deny the relationships that inform and enable Indigenous sovereign relations to place. The particular suitability of the Northern Territory to mining and other nuclear activites, as I will discuss, lies not just in its physical conditions and relatively small, geographically dispersed populations. It also lies in its amenability to, in Voyles terms, "wastelanding", a process of racial and spatial signification that "renders an environment and the bodies that inhabit to pollutable" (9). This reflexively feeds the project of setter colonial claim: through wastelanding settler colonial power denies that environments "could be sacred, could be claimed, could have a history, or could be thought of as home" (Voyles 26). Legal protections can also be altered over time, subjecting these environments and bodies always to the settler colonial state and its powers. Since the 1970s, the production and export of uranium has increased, enabled by legal changes that allow for the opening of new mines (for instance, Olympic Dam in 1988) and limit legal avenues for relevant communities to express opposition to the impact of mines (Harris 11; McKay and Mietzis 8; Panter 7–8).

Compounding this export-focussed mode of nuclear colonization are uranium's domestic, more shadowy legacies. Australia's nuclear identity has been dynamic and not simply a one way process of extraction and export. This reflects the nation's situation within global western chains of power, and also its status as the "wasteland" of empire, so distant from the "centre" and vast in its relative proportions of population and space, in Britain's imperial network. While Australia was supplying uranium to both the British and U.S. nuclear weapons programs, the British were bringing this back to Australia in the late 1950s in a series of 12 atomic tests that took place on the Montebello Islands off the coast of northwest Western Australia, and in Emu Field and Maralinga, in the north and west of South Australia. These tests were conducted "with the greatest of secrecy" (Royal Commission 9) at the time and remained so until they were exposed in a 1985 Royal Commission. Overall, the tests included three in the Montebello islands (1952 and 1956), two in Emu Field in 1952, and seven in Maralinga, from 1956–1957 (Martin 26). Also carried out were 550 "minor" tests in Maralinga until 1963, when the Partial Test Ban Treaty came into effect (Pockley 20). In all, more than 20 kg of plutonium, a by-product of uranium-238, was unleashed upon Emu Plains and Maralinga, and dispersed across hundreds of square kilometers of its surroundings (Mittmann 27).

While the remote Montebello islands were uninhabited at the time of these tests, Emu Plains and Maralinga had been continuously inhabited by their traditional owners, the Anangu Pitjantjatjara. Despite the British defending their choice of Maralinga and Emu Plains (200 km north) as "desolate and unpopulated", this was a fabrication in keeping with the *terra nullius* myth that discursively underscored the British colonization of the continent. This was never an empty land. At the time of the British tests, many traditional owners were residents at the Ooldea Aboriginal reserve—having been earlier moved there by the state—close to the Maralinga site. In 1955, ahead of the tests, most were forcibly moved to the newly established Yalata Lutheran mission, south of Maralinga, but still within the fall-out zone (Pockley 20). However, as the 1985 Royal Commission Report publicly acknowledged, the Anangu Pitjantjatjara "continued to move around and through the prohibited zone", living on Country, throughout—and subsequent to—the testing years. The report was explicit on this point": The attempts to ensure Aboriginal safety during the Buffalo series demonstrates ignorance, incompetence and cynicism on the part of those responsible" (Royal Commission 323).

The British atomic tests in Australia, while the most pernicious and devastating, are not the only spectre of Australia's uranium exports "coming home". The management of radioactive waste has been a national focus since the mid 1980s, when a proposal for a domestic "near-surface radioactive waste repository" for low-level waste (Commonwealth of Australia) kicked off a series of suggested locations, community opposition and government re-routes, that was potentially resolved in early 2020, with the announcement that a farming property in Napandee, near Kimba, southwestern South Australia, had been determined as the site for the National Radioactive Waste Management Facility. When the facility finally opens, it will take waste from over 100 sites around Australia, including the Lucas Heights reactor.

However, a "broader" vision has paralleled the hunt for a domestic "nuclear dump" (Boisvert): to find an Australian location for international high-level nuclear waste. Driven by the Australian arm of Pangea Resources, a global natural resources exploration and production company, this proposal first came to light in a corporate video leaked to environmental action group Friends of the Earth in 1998. It was to receive and store 75,000 tons of high-level radioactive material in Australia over 40 years. As Pangea's proposal went public, it activated a subtext of nuclear stewardship that latently spoke back to the nuclear fuel cycle and its chain of responsibility. This subtext had precedent in the 1976 Ranger Uranium Environmental Inquiry (established by the progressive Whitlam government to explore the environmental concerns around uranium mining as the industry expanded), and the more recent South Australian Nuclear Fuel Cycle Royal Commission (2015–2016), which both recommended the storage of international nuclear waste from Australian uranium on the grounds of responsible oversight. Pangea sought to capitalize on this earlier finding, and its latest iteration via the Royal Commission by suggesting Australia's storage of international radioactive waste would "provide the host country with the opportunity to play an unprecedented role in enhancing non-proliferation, encouraging nuclear weapon states to disarm" (Green 2016, 22). While two citizen juries established in 2016 rejected the idea of storing 138,000 tons of used fuel and 390,000 m^3 of intermediate waste in South Australia (Commission 98), the Royal Commission's recommendation to find ways of fostering community consent leaves the door open for this prospect.

Extractivist imaginaries

As I have indicated, Australia's history of uranium invites a reading through the country's imperial, colonial origins, which speaks to both its servicing of British and U.S. nuclear ambitions, but also more psychically to its founding logic as an extractivist site. This totalizing extractivist project, driven by an ambition to claim all "useful" resources for the acquisition of capital by the colonizing power, implicates dispossession and genocide as additional founding logics of the settler-colonial state because the acquisition of resources requires the "removal" of those who already hold claim to these (although with very different frames of relationality).[1] As colonizers moved in from the coast at their point of arrival, establishing pastoralism, systems of Western agriculture,[2] and eventually mining industries, they displaced and killed traditional owners (although not without resistance) on a scale that is still unacknowledged in broad public consciousness. A population of an estimated as upwards of 750,000 Indigenous inhabitants, spanning more than 250 different language groups and nations at the time of British arrival, was reduced by two thirds within the first 100 years of colonization (that is, by 1878) (AIATSIS).

The declaration of *terra nullius* in the British legal framing of their rightful claim to these lands spoke to the refusal to acknowledge their evident occupation. This willful and dissonant vision was perpetuated until 1992,[3] and coexisted with the state's constant obsession with regulating and controlling Indigenous lives (see, for instance, Lea et al. (2012) and Haebich (2011)), including the systematic removal of traditional owners from their lands, and the taking of Indigenous children from their families. The latter process, creating what is now known as the "Stolen Generations", took children into state- and church-run institutions, and also placed them into "service" roles, with minimal pay and no legal freedoms, within Australian households in a practice that ran from the 1920s–1940s (Austin). This additional form of extractivist logic echoes more well-known versions in the trade of human capital across the British Empire, and still remains little acknowledged in Australia.

This history of colonial violence and the centralized place of First Nations peoples as—at once—an imagined absence, an impediment to progress, and a resource to exploit, is a key aspect of what I term Australia's "extractivist imaginaries", a concept informed by Helen Verran's rendering of "imaginary" as a "conceptual resource" that underpins how we know and exist in the world (Verran 1998, 242). These resources are "constitutive of, and constituted by, ontic and epistemic commitments" (239), that register how things are made meaningful, and animating practices of living. As I have written elsewhere, "ascendent imaginaries represent designs on the world…shaping habit, practice and authority", and they underpin normative and dominant stories, how they are told and what they tell. This can singularize reality and "sustain entrenched structures of power" (Potter 2019, 5). Imaginaries thus underscore ways of thinking about, narrating, and acting that sustain a reality as entirely natural and normalized (for instance, the "reality" of uranium's "untapped potential" (Minerals Council of Australia 2020)). In the case of extraction in Australian history, these imaginaries have dominated to such an extent that the settler-colonial imaginary and the mining-capitalist imaginary are deeply imbricated in the other, and extraction is inextricable from the ongoing colonial project of erasure and (where this did not succeed) economic, political, and social disenfranchisement of First Nations peoples.

Particular renderings of space and time are key to these extractivist imaginaries. As indicated, the fantasy (and fallacy) of an empty land gave legal legitimacy to the colonial project in Australia, and it continued to be mobilized in visions of Australia's "vast and uninhabited" interior, supposedly empty of human life, or, at least, capital-claiming human life, and rich with untapped resources. Evocative of John Locke's "primordial waste" awaiting the intrepid settler in his *Second Treatise of Government* (1689) (Allred 2018), is a related narrative that sustains Australia's extractivist imaginaries: that of a bounty awaiting discovery in these lands which will reward entrepreneurship and hard labor. Like Locke's pioneers in the Americas laying claim to wasted lands with their toil and productive utilization, as if this were "the first peopling of the world" (Locke in Alldred 218), in Australia the trope of the "battler" hero has come to signify an autochthonous connection to land earned through toil. In the words of one of Australia's most iconic "battler" authors, Steele Rudd, whose novels and short stories were wildly popular in the early 20th century, these "pioneers…gave our country birth", with "giant enterprise [and] deeds of fortitude and daring" (epigraph to *On our Selection*, Lamond 2007).

The fantasy of "first peopling", or of a connection that displaces prior occupants, relies upon a revisionist conception of time that invests in linearity—a rolling tide of time—but one that re-inaugurates "the beginning". To install autochthony ("the native born"), and to re-animate and claim wasted lands, a "Year Zero" has to be imagined (Rose 1997). This temporal juncture, as Deborah Bird Rose explains, "cuts an ontological swathe between "timeless" land and historicized land". "Whatever happens within that Year Zero", she continues, "will be disjunctive with what follows as well as what existed before. This is a moment of transcendence" (28). A reset of time is thus embedded in the "battler" perspective in which labor of a particular kind generates resources from bare, fallow land. Australia is imagined as a young and industrious country, shaped through hard work—and what can be harder than digging through rock to get to the riches that lie below? At the same time, while wasted land can be put to "use" and claimed by industry, it can also be pillaged and re-wasted as part of the same moral economy. Although Australian governments have taken varying approaches to the uranium industry, both sides of politics have promoted a continued investment in mining that reflects the nation's strong attachment to extraction as a hallmark of economic development and nationalist identity, enrolled in a particular narrative of Australian life and values.

This imaginary underscores uranium histories in Australia, most strikingly around the presence, and fabricated absence, of First Nations people. The stages of colonialism, from violent dispossession to the paternalistic management of Indigenous lives over the course of the 19th and 20th centuries, worked through a paradigm of spatial and temporal order in which Indigenous bodies became matter out of place (Lea et al. 2012), consigned to anachronism or rendered invisible through the lens of *terra nullius*. The eastern seaboard-centric development of Australia furthered the association of mining locations with a Lockean vision of "empty", under-utilized spaces. The Northern Territory and South Australia particularly (once part of the same jurisdiction from 1863–1911), locations of all but one of Australia's uranium mines over time, are associated with underpopulation and occupy a lower status in the national hierarchy of internal jurisdictions. With over a quarter of the continent's land mass, the NT and SA currently host less than 10% of Australia's population.

The case of the Maralinga nuclear tests highlights the persistent framing of its location as "remote", which is a pejorative classification of the colonial mind. The map of Aboriginal Australia has no gaps—all the landmass of the continent is covered by traditional owner language groups and nations, including six which span the northern part of South Australia, most proximate to the Maralinga detonation zone. First-hand accounts of First Nations people who experienced and suffered from these tests refute a narrative of absence or erasure: for example, Edie Mulpiddie, who told the 1985 Royal Commission that she and her family had camped on the edge of one of the explosion site craters in 1957. She was pregnant at the time, and soon after miscarried; this was followed by a series of miscarriages and ill health over years (Tynan 28). More well known is Yankunytjatjara man Yami Lester, who became a community leader and campaigner for justice for traditional owners and the victims of atomic testing. Lester was blinded by the effects of radioactive fallout from Maralinga, and reported to the 1985 Royal Commission about the death and illness of many of his family members in the wake of the testing, after a "black mist"—a "black, greasy, unsettling miasma…covering hundreds of Aboriginal people in its path"—infiltrated waterways and all life forms (Tynan 27). Lallie Lennon also gave evidence to the Royal Commission, describing how the detonation "rumbled, the ground shook, it was frightening…We thought we were going to die. We reckon it was poison" (27).

The victims of the British atomic testing in Australia included many British servicemen—upwards of 35,000 who were stationed in West Australia and South Australia during the periods of these tests—and their claims for recognition and justice have also fought to be heard. These men were in repeated, dangerous proximity to the atomic tests, with little safety provision and no apparent understanding of the effects this exposure would wreak on them. At the same time, they were housed and provided with food from far away sources. They also wore shoes. Traditional owners, on the other hand, "still traversed the lands" usually barefoot, and sourced food and water directly from Country scattered with radioactive fallout (27). Subject unevenly to the impact of these detonations, the Anangu Pitjantjatjara and the British nuclear test "veterans" (as they are now know) were all exposed to extraordinary levels of radioactive material, including deadly plutonium—easily ingested and inhaled in small particles—as the result of this atomic testing. While we might consider this common experience productive of what Alexis-Martin and Davies call a "nuclear community", defined as "any group that is associated with ionising radiation", particularly in situated historical events, the Anangu Pitjantjatjara are erased from the story that Alexis-Martin and Davies tell about Maralinga. It is the British veterans alone who are considered in their definition of nuclear community (11–12).

While the extractivist imaginary at work in the British atomic tests in Australia reiterated the *terra nullius* narrative of colonization, the latter part of the 20th century saw formal agreements established with traditional owner groups around royalty payments for uranium mining. The predicate to this was the Aboriginal Land Rights Act (NT) in 1974, which vested title to land in traditional owners represented by newly established Land Councils. Property rights to minerals would be retained by the Commonwealth. Mining companies were required to negotiate consent and the terms on which exploration and mining could occur with the relevant Land Council and royalties from mining would be paid to the Commonwealth, the Northern Territory Government, and the traditional owners. The realities of this arrangement are complex, however (Wilson 1997). A 1997 report to Parliament on the impact of uranium mining on Indigenous communities in the NT indicated that, despite the access to money enabled through mining agreements, "there has not been any appreciable improvement in the standard of living in Aboriginal communities" (Wilson 2). This report also pointed to the (still ongoing issue) that "once consent has been given to exploration, Aboriginal people cannot withdraw consent to mining". Moreover, "pressure and tactics used to gain Aboriginal consent to exploration and mining" were commonly apparent (2).

While more recently, there has been an identified shift to a more "collaborative approach" between traditional owners and mining companies (Langton 2015, 23), resulting in stronger economic outcomes for traditional owners, including an increase in direct economic participation (7), the claim that minerals companies have grown in "respect for Aboriginal culture and history" (17) does not sit universally. In early 2020, one of the world's largest mining companies, Rio Tinto, deliberately destroyed a 46,000-year-old sacred rock shelter in the Juukan Gorge in Western Australia's Pilbara region to access iron ore beneath it, despite direct appeals from traditional owners to halt the process. As Indigenous Professor Marcia Langton, who has worked with mining companies to build social license in Indigenous communities, commented on the process, "Rio Tinto has done an about-face and thrown away 30 years of relationship-building [here]". "I want you to know that the history of this kind of treatment of Aboriginal sacred sites in Western Australia by the company goes back a long, long way in history" (Aston).

Waste as colonial matter

The extractivist imaginary in (post)colonial Australia, as we have seen, invests in erasure, linear time, entitlement earned through labor, and the overarching frame of colonial logic: the acquisition of wealth through the claim to resources, legitimized through a moral discourse of waste and industry. The land is without value if it is not put to use and extracted for riches. There is a further aspect to this imaginary, however, one which shadows the others, and connects into the Western tendency to regard the end point of production as something to be tactically separated from the self (Hawkins 2006 4). This is the psychic role of waste in the colonial imaginary. Australia has a strong legacy of fearing its waste that stems from the process of colonization as one that creates inherent externalities. A colonized nation such as Australia, established on dispossession, genocide, the appropriation of resources and dominant ideas of race and the white settler-colonial as the normative bedrock of national identity, will always create a shadow reality of what is excluded from, or hidden behind, its self-narration. Australia's continued investment in the idea of itself as industrious, egalitarian, and forward-looking—continuing a path of "common purpose" (Howard in Rundle 20), and a "remarkable success story" (Howard in Burke 221), as conservative Prime Minister John Howard declared during his term in the 1990s—models a disposition of waste management ubiquitous to modern waste practices. This is waste as "the shit end of capitalism" (Hawkins 2006 vii): the "ironic testimony to a desire to forget" (Hird 117).

As I have argued elsewhere (Potter 2019), the settler-colonial focus on *settlement*—on things being assured and in their place, in the claim to and acquisition of territory and resources—requires the practical imaginative work of modeling stable ground, of externalities remaining outside of the picture of dominant national identity. Waste, in this logic, is a potentially dispossessing force even when the underside of colonization is acknowledged, and its violent destruction put into the limelight. This was the case in Australia during the 1990s when a series of revelatory landmarks drew attention to the myths that the nation had perpetuated, and the histories that were little acknowledged. Most prominently, these landmarks were the 1992 Mabo decision, which overthrew the founding principle of *terra nullius*, and the 1997 *Bringing them Home* report on the Stolen Generations, which brought the genocidal intent of successive colonial governments into the spotlight. As these revelations induced an outpouring of white settler anxiety and guilt, they also coincided with an increasing public awareness of the environmental impacts of colonization, and with the imminent threat of climate change so associated with modernity and its relentless consumption of resources.

The recognition that the creation of modern Australia, and the project of non-indigenous belonging to this place, were implicated in destructive, exploitative, and inherently dispossessing forces (including, of course, climate change), induced an outpouring of non-indigenous disconcertion in which the spectre of waste, symbolic of colonial despoilation, often figured. This was waste in the classic Western trope as a signifier of what we do not want, and what we need to disconnect from ourselves. Our historical waste needed to be put out of sight and out of mind. "Cleaning up" the country becomes euphemistic for cleaning up the past. As non-indigenous public intellectual Germaine Greer exclaimed on this point in 2003: "The country I love has been crazily devastated by whitefellas…If we truly felt that this country was our home we would not despoil it in this manner. We are trashing it because it belongs to someone else" (Greer 117). Making a similar point in his address to the nation on Australia Day 2002, "Australian of the Year", environmental scientist Tim Flannery called on the nation's non-indigenous majority to clean up the "mess" of the preceding 200 plus years, and cast off the "shadows" of an "arrogant colonial vision" (2002, 1–2). Doing so would inaugurate a "yet-to-be-formed Australian culture" (3), premised on a "surrendering of [non-indigenous] otherness" (5). Waste is an impediment to this vision, and an eventual "right" belonging, into which Indigenous experience is also subsumed.

This discourse relates to the temporal structure mobilized by colonization. The "rolling" frontier inaugurates the beginning of history, and the settler-colonial becomes the agent of time. What disturbs this imaginary must be disconnected and left behind. Waste is necessarily out of colonial time. In contrast, waste occupies a very different status in accounts of First Nations' cultures and time, in which the abjection of waste, "out of sight, out of mind", is refused in a profoundly relational ontology. As Rose explains, the act of disconnecting waste amounts to "self-erasure" for Indigenous Australians on Country, which she extrapolates as "the performance of a lie, a refusal to witness" (Bird Rose quoted in Hird 117): "The remains of people's actions in Country tell an implicit story of knowledgeable action…Antisocial people who do not announce themselves, and use special techniques to avoid leaving tracks or traces, are up to no good" (Hird 117). Lucy Bell reiterates this point when she writes that waste in a Global South context signifies in vastly different ways: the very notion of "littering" is deeply rooted in a Western, dualist system of thought that opposes people and places/objects, humans, and non-humans. Through a non-western lens, "the notion of "dispersing" or "scattering" residues, so they are reintegrated into the environment, emerges from an interactionist view" of the world (115).

Taking these divergences into the context of radioactive waste is not straightforward: no one wants radioactive material lying about their home. However, the ways in which nuclear waste is physically and imaginatively treated in the Australian (post)colonial present speaks to both pervasive cultural tendencies amongst the settler-colonial majority, but also their possible limits and points of strain, where Indigenous modalities occupy a more sympathetic accord with the material realities of radioactivity, and what it is to live with history and its effects. Maralinga's toxic legacies reveal the logic of disposal on a large scale. By the end of the "minor trials" in 1963, 22.2 kg of deadly plutonium-239 was spread around the test site—this is what stayed behind after the wind picked up and dispersed plutonium oxide in plumes that were recorded as 150 km long and several meters wide (Tynan 2016, 28). These were carried hundreds of kilometers away: a radioactive cloud was reported passing the skies in Adelaide, and then Melbourne, ultimately almost 2,000 kilometers from the test site, in October 1956. Radioactive rain likely fell during the 1956 Melbourne Olympic Games (Roff 2017). When they withdrew from Australia at the end of the testing period, the British claimed that 20 kg of the plutonium-239 had been buried in deep pits on-site, while only 2 kg was scattered "evenly" around the testing range. According to Tynan, the British "almost certainly knew that was incorrect" (28). Yet they left, their waste in their wake, separated cleanly in an ultimate modern fantasy. Subsequently, examinations of the site indicated that around 3 million fragments (or sub-millimeter particles) of plutonium were loose, aboveground. Levels of contamination were estimated at around ten times higher than admitted by the British. The 1985 Royal Commission made it clear that the British had knowingly left the site polluted, and they resisted calls to pay for effective remediation into the mid-1990s.

The Maralinga Rehabilitation Project was finally completed in March 2000, with all debris buried on-site, and the most heavily contaminated parts of the site (roughly 3,000 km sq) were "returned" to traditional owners with no restricted access (native title had been granted in 1984). Now Indigenous-operated Maralinga Tours takes visitors on tours of the testing range (Tynan). The Montebello islands are also tourist destinations today, although radioactive warnings remain in place. Visitors are advised that they should limit their time to one hour per day in the worst affected of the islands (Parks and Wildlife 2020). At Maralinga, however, the cleanup, according to experts, was far from best practice, and the burial method, without concrete bunkering and relatively close to the surface, remains open to leakage. Moreover, the scope of the site has been unevenly cleaned through an under-resourced and piecemeal effort, meaning that 120 square kilometers are still actively contaminated. The force of dispossession here is ongoing. Despite traditional owner control, the 450 square kilometers of the testing range is "not suitable for permanent occupation" in its current state. The half-life of plutonium, after all, is 24,000 years (Parkinson 80).

While there may be no conclusively "safe" way to live with this waste, at least in its current state of treatment, the denial of waste's presence here enables harms both physical and psychic. It incorrectly affirms that pasts, still active and pressing, are resolved and—again—out of time, which is reset through the act of clearing up. A similar logic informs the contested question of nuclear waste storage in Australia. Australia's association with "remote, empty" regions (and perhaps its history, too, as Britain's site for its own externalities, the convicts first sent to its eastern states), as well as its relative geographic stability, make it a preferred candidate for off-shore nuclear waste storage in the Global North. In addition to Pangea's proposal to store nuclear waste in Australia, there have been other plans over the years, most initiated by Federal or State Governments. Despite persistent public opposition to these plans, the first shipment of "returned" uranium waste, processed in France, was received in 2015 and housed at the Lucas Heights reactor until a "long-term" storage option is realized (Agence France-Presse), as flagged earlier, close to the South Australian town of Kimba.

The Federal Government selected this as the location for an official disposal and storage facility for nuclear waste after a four-year consultation period in early 2020. Sixty-two percent of the town's residents endorsed the decision. In contrast, traditional owners over-whelming opposed it, and took their opposition to the Federal Court with the claim that the Racial Discrimination Act had been contravened by the ballot which excluded the Barngarla people from voting. They claimed that this exclusion was on the basis of race; the court found this not to be the case—rather it was on the "rateable" nature of the land to which their Native Title applied (that is, the status of the Traditional Owners as rate payers). While Native Title was recognized, the views of traditional owners were not, despite the Court citing a previous Native Title case on the significance of Indigenous connections to Country: "the loss of those rights to, and the relationship with, particular land...does not extinguish the powerful spiritual and cultural connections Aboriginal people have generally with the lands of Australia" (Edelman J cited in McKerracher, Rangiah and Charlesworth JJ). The paradox of this situation offers an indication of how the colonial mind, including its legal system, per-petuates its own imaginaries, and consequently the normative realities which constantly discipline Indigenous Australian life.

Living with radioactive pasts

The space-time of the colonial project necessitates that waste is cleared away, however this is a fantasy made tenuous by the "mutual implication" (Adam 1998, 148) that is always there with the resources that national subjects, unevenly of course, extract and consume. Waste will inevitably return in some way and we need to admit that this is so. Externality is a discursive rather than a material reality. The infrastructure set up to put waste aside, out of view and out of touch, as Hird elaborates, is "an ironic testimony to a desire to forget…[they are] a material enactment of forgetting" (106). The impossibility of disciplining radioactive waste and keeping it in place, "forgotten", is perhaps the best illustration of this fantasy. Barbara Adam elaborates on the particular temporalities of radioactivity, which counter colonial structures, and demonstrate graphically the incalculability of our relations with waste. Invisibly present, radiation eludes visual capture, and cannot be measured for its effects in an ordered, equalized quantification of damage. Exemplifying this with the Chernobyl nuclear explosion, she traces the unstable temporality of nuclear release, as radioactivity is at once irreversible but also elastic, operating on diverse timescales, some of which are unimaginable: "it materializes as symptom in un/predictable temporal and spatial positions" (150).

Presaging what Timothy Morton posited as the conditions of the "hyperobject" (2013), something so significantly distributed across time and space that it exceeds any systematic attempt to map and know it, Adam describes the dispersal of radiation from a nuclear centre as inequitable, unpredictable and as "presenting" in multiple ways. As a consequence, she concludes, radiation "poses problems for traditional ways of knowing and relating to the world" (138). It undoes the fantasy of separation and distinction from which to observe an event. The unknown and unknowable fallout of radioactive matter means that we all must consider it from within, with varying degrees of responsibility and subjection. The dispersal of radiation throughout the bodies of all life forms—entering waterways and seeping into the ground, or hanging in the air to come down as nuclear rain—creates a network of relations in which power is uneven, but tactical separation is impossible. "The science and engineering of landfills is all about making sure waste doesn't leak", Hird tells us (107). The impossibility of this promise in the life span of radioactive matter, however, forces us to consider our ongoing enrollment in complex nuclear histories, as well as geographies.

Australian uranium is part of dynamic global networks that present no easy picture of where "our" responsibility begins and ends. Uranium processed in one of Australia's earliest mines, Hunter's Hill, in NSW, fed the foundational research of Marie Curie and Ernest Rutherford (Mittmann 2016), while nuclear physicist William Penney, who oversaw British nuclear testing in Australia, was a witness to the bombing of Nagasaki from an observational plane. These entanglements need to be acknowledged rather than denied, as have the First Nations Dene people who inhibit the northern boreal and arctic regions of Canada, some of whom were employed to mine uranium in the 1940s that made its way into the atomic bombs dropped on Hiroshima and Nagasaki. As Hird relates, "in 1998, a group of Dene traveled to Japan and apologized to the Japanese people for the catastrophic destruction caused" (118). There are lines of connection and responsibility to follow, even if these never tell a whole story, or put a matter to rest. Despite their attempted disavowal by an imaginary in which time is junctural, power is radicaly uneven, and environments and non-white bodies are inherently "pollutable" (Voyles 24), these relations stay alive: they bring us close to the making of the capitalist-colonialist project that has shaped our current world so emphatically.

However, this imaginary is not the only one and, despite its violence, it doesn't erase the possibility of others. Alternative ways of living with waste, like those exemplified by First Nations peoples, indicate alternative imaginaries that extend in much more generative directions. Rather than appropriate these, we can acknowledge them, and look to the unsettlements within Western modernity's normative and dominant imaginaries that suggest other imaginaries already imminent to these. Toxic imaginaries are a case in point, where disconcerting proximities and radiating effects are an enactment of entangled lives and the impossibility of extracting ourselves from our shadows. This makes for transcorporeal realities in which "[o]ur relations with waste cannot be so easily severed" (Hawkins 10). Profound relationality is thus acknowledged as the condition of being which means that, despite efforts to the contrary, things can still "pollute for eternity" (Alaimo 2012, 487), and there is "no safe place to stand" in relation to colonial extractivist histories and colonial nuclearity (2012, 490). Such uncertainty and proximity are the signs of enmeshment, in which distances collapse, dispelling ontological security as a possibility in a world alive and remade with the energies of capitalist modernity. In this imaginary, "taking back our waste" means recognizing its ever-presence in the making of our world, and in the possibility for political futures that no longer rely upon cleaning up and moving on. This is a "nuclear community" to which we have all—with uneven impacts—been forced to belong, and a possible path for more just futures as Australia negotiates its status as a nuclear nation.

Notes

1   Without intending to conflate Indigenous ontology, it is generally understood that—as Indigenous relations to waste are elaborated further in this paper—First Nations' peoples in Australia evince an entirely different way of being and belonging from a paradigm of resource ownership, based instead on relationality and responsibility: living "of", and not "in" or "on" the land (Watson 257).
2   A sophisticated agricultural system was already in existence, as Bruce Pascoe has evidenced.
3   The Mabo decision successfully challenged and nullified the *terra nullius* doctrine and opened up the (restricted) possibility of native title claims for traditional owners. Famously, the judges who handed down the decision referred to this history of dispossession and violence against First Nations peoples in Australia remains "the darkest aspect of our history". The ruling declared that: "The nation as a whole must as diminished unless and until there is an acknowledgement of, and retreat from, those past injustices" (in Brennan 1995, x–xi).

References

Adam, B. 1998. "Radiated Identities: In Pursuit of the Temporal Complexity of Conceptual Cultural Practices." In M. Featherstone and S. Lash (Eds.) *Spaces of Culture: City-Nation-World*. Sage. 138–158.

AIATSIS (Australian Institute for Aboriginal Torres Strait Islander Studies). http://aiatisis.gov.au. Accessed January 24, 2021.

Alaimo, S. 2012. "States of Suspension: Trans-Corporeality at Sea." *ISLE: Interdisciplinary Studies in Literature and Environment* 19, no. 3: 476–493.

Alexis-Martin, B., and T. Davies. 2017. "Towards Nuclear Geography: Zones, Bodies, and Communities." *Geography Compass* 11, no. 9: 1–22.

Agence France-Presse. 2015. "Nuclear Waste Returned to Australia, Raising Concerns About Future Dump Site." *The Guardian Australia*, December 5. https://www.theguardian.com/environment/2015/dec/05/nuclear-waste-returned-to-australia-raising-concerns-about-future-dump-sit. Accessed October 12, 2020.

Allred, N. 2018. "Locke's American Wasteland." *The 18th Century Common*, April 16. https://www.18thcenturycommon.org/lockes-american-wasteland/. Accessed September 15, 2020.

Aston, J. 2020. "Marcia Langton eviscerates Rio Tinto." *Australian Financial Review*, July 22. https://www.afr.com/rear-window/marcia-langton-eviscerates-rio-tinto-20200722-p55ehg (Accessed October 10, 2020).

Austin, T. 1991. "A Chance to be Descent: Northern Territory "Half-Caste" Girls in Service in South Australia 1916—1939." *Labour History* 60: 51–65.

Bell, L. 2019. "Place, People and Processes in Waste Theory: A Global South Critique." *Cultural Studies* 33, no. 1: 98–121.

Blainey, G. 1993. *The Rush That Never Ended: A History of Australian Mining*. Melbourne University Press.

Boisvert,. "Federal Government Chooses Kimba Farm Napandee on the Eyre Peninsula for Nuclear Dump." *ABC News*, February 1, 2020. https://www.abc.net.au/news/2020-02-01/kimba-farm-eyre-peninsula-chosen-for-nuclear-dump/11920514. Accessed October 6, 2020.

Brennan, F. 1995. *One Land, One Nation: Mabo Towards 2001*. UQP.

Burke, A. 2001. *In Fear of Security: Australia's Invasion Anxiety*. Pluto.

Commonwealth of Australia Bureau of Resource Sciences. 1997. *A Radioactive Waste Repository for Australia: Site Selection Study—Phase 3: Regional Assessment: A Public Discussion Paper*, Canberra.

Flannery, T. 2002. "The Day, the Land, the People: Australia Day Address." January 23. https://www.australiaday.com.au/events/australia-day-address/dr-tim-flannery. Accessed October 29, 2016.

Green, J. 2016. "Radioactive Waste and the Nuclear War on Australia's Aboriginal People." *Chain Reaction* 127: 31–33.

Greer, G. 2003. *Whitefella Jump Up: The Shortest Way to Nationhood*. Quarterly Essay 11. Black Inc.

Haebich, A. (2011). "Forgetting Indigenous Histories: Cases From the History of Australia's Stolen Generations." *Journal of Social History* 44, no. 4: 1033–1046.

Harris, M. (2011). *The Origins of Australia's Uranium Export Policy. Background Note*. Parliamentary Library: Parliament of Australi. 1–13.

Hawkins, G. 2006. *The Ethics of Waste: How We Relate to Rubbish*. UNSW Press.

Hecht, G. (2014). *Being Nuclear: Africans and the Global Uranium Trade*. MIT Press.

Hird, M. (2013). "Waste, Landfills, and an Environmental Ethic of Vulnerability." *Ethics and the Environment* 18, no. 1: 105–124.

Holland, I. (2002). "Waste Not Want Not? Australia and the Politics of High-Level Nuclear Waste." *Australian Journal of Political Science* 37, no. 2: 283–301.

Huggan, G. and H. Tiffin (2010). *Postcolonial Ecocriticism: Literature, Animals, Environment*. Routledge.

Lamond, J. (2007). "The Ghost of Dad Rudd, on the Stump." *JASAL: Journal of the Association for the Study of Australian Literature* 6, no. 1: 19–32.

Langton, M. (2015). *From Conflict to Cooperation: Transformations and Challenges in the Engagement Between the Australian Minerals Industry and Australian Indigenous Peoples*. Minerals Council of Australia.

Lea, T., M. Young, F. Markham, C. Holmes, and B. Doran. 2012. "Being Moved (on): The Biopolitics of Walking in Australia's Frontier Towns." *Radical History Review* 114: 139–163.

McKay, A. and Y. Miezitis. 2001. *Australia's Uranium Resources, Geology and Development of Deposits: AGSO—Geoscience Australia Mineral Resource Report*. AGSO.

McKerracher and Charlesworth, J. J. 2020. *Love v Commonwealth of Australia*. HCA3.

Martin, B. 1982. "The Australian Anti-Uranium Movement." *Alternatives: Perspectives on Society and Environment* 10, no. 4: 26–25.

Minerals Council of Australia. "Uranium: Untapped Potential." https://minerals.org.au/minerals/uranium. Accessed September 15, 2020.

Mittmann, J. D. 2016. Ed. *Black Mist, Burnt Country: Testing the Bomb, Maralinga and Australian Art*. Burrina.

Morton, T. 2013. *Hyperobjects: Philosophy and Ecology After the End of the World*. University of Minnesota Press.

Panter, Rod. 1992. *Radioactive Waste Disposal in Australia*. Parliamentary Research Service. Issue Paper Number 6.

Parkinson, A. 2002. "'Maralinga: The Clean-Up of a Nuclear Test Site." *Medicine and Global Survival* 7, no. 2: 77–81.

Parks and Wildlife Service. 2020. "Montebello Islands Marine Park. Government of Western Australia." https://parks.dpaw/wa/gov.au/park/montebello-islands. Accessed October 6.

Pockley, P. 2000. "Legacy of Maralinga Lingers." *Australasian Science* 21, no. 3: 20–22.

Potter, E. 2019. *Writing Belonging at the Millennium: Notes From the Field on Settler Colonial Place*. Intellect.

Roff, S. R. 2017. "Australia's Nuclear Testing Before the 1956 Olympics in Melbourne Should be a Red Flag for Fukushima in 2020." *The Conversation*, October 19. https://theconversation.com/australias-nuclear-testing-before-the-1956-olympics-in-melbourne-should-be-a-red-flag-for-fukushima-in-2020-85787. Accessed October 7, 2020.

Rose, D. B. 1997. "The Year Zero and the North Australian Frontier." In D. B. Rose and A. Clarke (Eds.) *Tracking Knowledge in North Australian Landscapes: Studies in Indigenous and Settler Ecological Knowledge Systems*. Australian National University North Australia Research Unit. 19–39.

Royal Commission into British Nuclear Tests in Australia. 1985. *Mr Justice J. R. McLelland*. Parliament of the Commonwealth of Australia. December 5.

Rundle, G. 2001. *The Opportunist: John Howard and the Triumph of Reaction*. Quarterly Essay 3. Black Inc.

Santayana, V. 2020. "His Own Chernobyl: The Embodiment of Radiation and the Resistance to Nuclear Extractivism in Nadine Gordimer's *Get a Life*." *The Journal of Commonwealth Literature*. doi:10.1177/0021989420933987

Senior, A., A. Britt, D. Summerfield, A. Hughes, A. Hitchman, A. Cross, D. Champion, D. Huston, E. Bastrakov, M. Sexton, J. Moloney, J. Pheeney, and A. Schofield. 2020. *Australia's Identified Mineral Resources 2019*. Australian Government.

Tynan, E. (2016). "Thunder on the Plain." In J. D. Mittman (Ed.) *Black Mist, Burnt Country: Testing the Bomb, Maralinga and Australian Art*. Burrinjapp. 21–35.

Veracini, L. 2007. "Settler Colonialism and Decolonization." *Borderlands* 6, no. 2. http://www.borderlands.net.au/vol6no2_2007/veracini_settler.htm. Accessed May 13, 2017.

Verran H. 1998. "Re-Imagining Land Ownership in Australia." *Postcolonial Studies* 1, no. 2: 237–254.

Voyles, T. B. 2015. *Wastelanding: Legacies of Uranium Mining in Navajo Country*. University of Minnesota Press.

Watson, I. 2002. "Buried Alive." *Law and Critique* 13: 253–269.

Wilson, I. 1997. *Impact of Uranium Mining on Aboriginal Communities in the Northern Territory. Report to the Senate*. Parliament of Australia. April. https://www.aph.gov.au/Parliamentary_Business/Committees/Senate/Former_Committees/uranium/report/c11Witman. Accessed September 10, 2020.

16

BROWNFIELDS AS WASTE/RACE GOVERNANCE: U.S. CONTAMINATED PROPERTY REDEVELOPMENT AND RACIAL CAPITALISM

Shiloh Krupar

Introduction

Industrial practices have left widespread contamination impacting land, water, air, human, and nonhuman bodies. In the absence of effective tracking of historic hazardous substance releases, environmental politics in the United States has sought to address waste's material, economic, political, and cultural management. Dealing with contaminated land plays a significant part in the governance of waste, necessitating mechanisms of land conversion, from legislation aimed at holding parties liable for site remediation, to financial instruments that parlay land redevelopment risk between private and public agencies. Land conversions involve complex shifts in meanings and memory, investment structures, policy proposals, and physical arrangement. Within a capitalist property system, land's history of conversion involves temporal mixing and material contingencies where the real or perceived presence of waste devalues land yet provides the opportunity for future surplus value production. *Property* redevelopment of contaminated land treats the ground as receptacle of a history of deposits, many of them highly toxic and hazardous, with the goal of producing a surface of transfer, seizure, and exchange that allows for the burial, monitoring, and/or forgetting of waste (Crysler 2017). Land reuse is therefore a central case for waste studies.

Over the last few decades, local, state, and federal government policies promoted the redevelopment of contaminated land using voluntary and market-based policy instruments that depart from traditional regulatory strategies. In this environment, "brownfields" emerged as a widespread and prominent mechanism of land conversion and waste governance. Brownfields, in this context, are abandoned or underutilized industrial and commercial sites that are, or are perceived to be, physically, chemically, or biologically contaminated; they are reused land or property complicated by the (potential) presence of a hazardous substance, pollutant, or contaminant. Such land is often what remains of former industrial and/or military occupation, but the term is used more broadly to include other idle, abandoned, underused, derelict, damaged, vacant land or buildings. Brownfields therefore may refer to land known/suspected to be

DOI: 10.4324/9781003019077-16

polluted or contain hazardous waste, *or* to land that is not being used to its potential value. For advocates, brownfield policies offer not only environmental cleanup, but also locally driven land recycling and health improvement, "with the promise to transform distressed sites across the U.S. from blight to valuable economic and environmental resources" (Dull and Wernstedt 2010, 120).

Brownfield property redevelopment relies on modernist assumptions about waste as a quantifiable and measurable object that can be separated and removed from land, or kept in place and contained to levels of risk deemed "safe enough" for a particular post-cleanup end use. This way of thinking about waste maintains the illusion of a border between contamination and social life—a border upon which capitalism's constant expansion and revival depends. Brownfield land redevelopment essentially seeks to dis-embed the land market from the uncertain material conditions of waste, by converting waste's presence or excess into mere financial, legal, and technical matters. While brownfield redevelopers strive to limit and contain their liability for responding to lingering contamination and changing land uses, such property redevelopment schemes rely upon an underlying binary of waste/society to establish "the conditions for [the] revival of profitability", predicated on a racialized "logic of *cheapening*" and improvement (Moore 2011, 139; Conroy 2018, 210). Private developers are invited to invest in brownfields as an opportunity to turn a profit on converting economically "unproductive" land to "public use". Brownfields deepen private sector engagement through amenities that boost private investment while immunizing investors from financial risk and biopolitical consequences. This process builds on longstanding practices that made—and maintained—environmentally degraded, economically divested, and racially marked lands. Building on the entrenched relationship between waste, race, and space in the United States, brownfields operate as a form of racial governance by driving a land market that facilitates and obscures "conjoined processes of racialized property making and property taking" (Ranganathan 2016, 22).

The intimate proximity of environmental hazards and racialized bodies reveals that waste is one of the central modalities through which race was lived in the 20th century. In turn, race has been and is still pivotal to signifying the "waste-ability" of space (Dillon 2014, 1209). As environmental justice activists and scholars across a range of fields attuned to critical studies of race have shown, race and waste have been articulated in specific geographical ways, leading to racialized health inequities and other disparities (Dillon and Sze 2019; Pulido 2016). Unequal social formations reproduce racial difference through spatial de/valuation connected to the material presence of waste, as well as perceptions of "who" and "where" are unproductive and waste-able. Property value is not the neutral measure of the worth of a particular piece of real estate, but "better understood as the result of social and economic relations among places"; the property system—and the federal legislation, various authorities, and funding that support it—"builds racial differentiation, class, and normative gender and sexual relations into the landscape" (Loyd 2014, 28). Historically, for example, white single-family residential neighborhoods received low-interest loans for mortgages and high ratings in terms of their security value. Conversely, nonwhite land—often mixed-use and adjacent to industry—was frequently "redlined" and considered an investment risk, subject to the racialized labels of "nuisance" and "blight". Couched in the supposedly impartial language of public health and planning, exclusionary zoning sequestered nonwhite people to contaminated, vulnerable, and less desirable locations in an effort to protect white health and secure whiteness as property (Loyd 2014, 29–30; Freund 2007, 77–78 and 118; Harris 1993).

Brownfield redevelopments extend this legacy of property as an operation of racial capitalism. Brownfield programs expand practices that invested in public health for white communities and sanctioned contamination as part of historical racial segregation and disin-vestment; they facilitate site conversions and clean-up remedies that minimize investor liability and thus often ensure that contamination remains (Liboiron, Tironi and Calvillo 2018; Gille 2007; Dillon 2014). Scholars have warned that brownfield programs risk further entrenching "environmental apartheid" by implementing separate and unequal environmental standards across regions, particularly between inner cities and wealthier suburbs (Poindexter 1996, 5). Grassroots activists and national environmental organizations have both argued that "differential cleanup standards at brownfield sites could lead to a dangerous double standard and to a concentration of redeveloped sites in the inner cities where contamination has not been re-moved but rather contained on site" to buffer life elsewhere (Wernstedt and Hersh 1998, 160). Furthermore, brownfields can serve as a technocratic instrument to determine whose bodies and communities *can be wasted*. Brownfields facilitate contingent determinations of value about whose lives and whose health matter and whose racially marked bodies and futures do not. Due to the uncertain presence of waste, contaminated land always has *potential* in terms of economic viability. This potential for economic productivity is a settler-colonial way of understanding contaminated land as "wasteland". Physical contamination of land articulates with the racialized settler-colonial logics of improvement to create and delineate this new form of enclosure: Devalued and/or abandoned post-industrial land with (real or perceived) contamination that is "under-utilized" and thus available (Bhandar 2018). Extending from the historical colonial understanding of land or wilderness as "waste", because it had not yet been drawn into colonial or capitalist economic relations, brownfield conversions orchestrate a violent feedback loop of land *re*possession as dispossession, that reenacts settler-frontier waste logics (Gidwani 2008, 18–19; Goldstein 2013). With the entire property system built on stolen native land, this violent land dispossession animates a *brownfield frontier* (rural and urban) of renewed rounds of forced removal, occupation, and erasure of BIPOC inhabitants, amplifying the need to address brownfields as a deeply historical, spatial, and cross-generational racial justice issue (Berger 2006, 74).

This chapter, in response, draws on the justice orientation of much waste studies scholarship to suggest that *brownfield policy could be different*. Brownfields could operate as a *heuristic* of racial justice against the prevalent understanding of (white) property and health. They could open up discussion of the legacies of racist land policies—settler-colonial seizure, eviction, blight re-moval, redlining, predatory mortgages, highway constructions, pollution, hospital location/ segregation, urban renewal, business opportunity zones, and so forth—and their racially dis-parate harmful effects on bodies, lands, and futures. To that end, this chapter models how brownfields might serve to galvanize research on racial capitalism and land reuse within waste studies. The analysis addresses brownfields and a more recent iteration called "healthfields" as forms of racial governance. The following section further details how brownfield policy en-trenches a new waste frontier that expands the rationale for and application of eminent domain to the racialized logics of blight and industrial wasteland, intensifying the everyday dispossession experienced by BIPOC communities. Brownfield site conversions generate value through contingent determinations of "productivity" and "public use", implementing waste remedia-tion and land renewal of already racialized historical geographies of environmental degradation, disinvestment, displacement, and poverty. This chapter then turns to the ways that healthfields target contaminated land located in black and brown communities who suffer from the pre-sence of waste and who are underserved by health care providers. Originating in Florida, the initiative aims to convert such areas into hospitals, wellness centers, and grocery stores, offering tax breaks that invite "green health" projects to remedy toxic blight while lowering cleanup standards. These two cases—and their respective goals of economic redevelopment and public health improvement—expand on how blight designation can be used to gentrify inner city areas or arbitrarily secure more profitable site usage, and how hospital/health service installations may paradoxically entrench health inequities, economic injustices, and environmental hazards stemming from segregation. The chapter concludes by advocating the transdisciplinary method of *antiracist soil exegesis* within waste studies of land reuse, to document the sedimented power relations and physical geographies of the "color line" that allow for value creation through racialized property financialization and transfer (Du Bois 1996, 325).

Brownfield redevelopment: frontiers of unproductive land and blight

In the early 1990s, big-city mayors and legislators from urban industrial states pressured the U.S. Congress and U.S. Environmental Protection Agency to start a pilot program to redevelop land in highly desirable urban infill areas that was considered to be underutilized and/or damaged (Greenberg and Hollander 2006). Far cheaper than comparable non-polluted properties, due to the environmental and financial risks involved in their reuse, brownfields became sites of potential for rebranding and reusing a struggling city's vacant or depleted land—as a "green investment" effort to conserve greenfields (unused land) by redeveloping brownfields. Such properties could become new factories, businesses, housing, and other job- and tax revenue-creating endeavors. Brownfields may involve more risk, but offer higher rates of return potentially. They usually require special financing, as well as risk transfer mechanisms that enable developers to limit environmental and financial liability. The central premise of most brownfield redevelopment programs is that "regulatory flexibility is necessary to remediate contaminated properties and bring them back onto the tax rolls"—that is, brownfield strategies eschew traditional regulatory models of environmental protection to pursue "cooperative approaches" that encourage voluntary environmental improvements (Wernstedt and Hersh 1998, 157; Dull and Wernstedt 2010, 136). To that end, the U.S. Environmental Protection Agency (EPA) first launched a pilot Brownfields Program in 1995 to support the agency's land revitalization goals of reusing contaminated properties to reinvigorate communities and jumpstart local economies, preventing sprawl, preserving green space, and protecting the environment and health (Dull and Wernstedt 2010, 120). The program was given statutory footing by the 2002 Small Business Liability Relief and Brownfields Revitalization Act. This Act provided the EPA pilot with a congressional mandate, clarified liability issues to make redevelopment more attractive, and adopted new tools to promote land conversion and increase funding up to a level of $250 million per year (Dull and Wernstedt 2010, 119).

Brownfield legislation sought to remove barriers to redevelopment stipulated in existing environmental laws, such as the Comprehensive Environmental Response, Compensation, and Liability Act (CERCLA)—also known as "Superfund"—and the Resource Conservation and Recovery Act (RCRA). In contrast to federal programs that rely predominantly on liability and enforcement to initiate cleanups, brownfield voluntary programs permit site owners and developers to approach the state to identify potentially valuable sites, especially within inner cities (Wernstedt and Hersh 1998, 158). Numerous states have proposed and enacted their own brownfields programs, with different mixes of incentives, regulatory pressure, information provision, and public involvement. These programs typically limit cleanup costs and responsibility for adverse consequences of land conversion, while implying there will be job creation in economically distressed areas along with improvements in environmental health. Cities and other planning authorities draw on federal seed funding and state programs to apply advanced appraisal techniques to prospective brownfield site projects, including spatial demarcation and accounting in GIS databases and other parcel/property listings. This technological work subsidizes private industry's acquisition, remediation, and adaptive reuse of brownfield sites by optimizing data production for effective capital planning. Brownfield developers can qualify for a range of subsidies, including tax increment financing (TIF), revolving funds (loans), trust funds (tax or fee-based accounts), real estate trusts (private investments), tax credits and deferrals, state grants, and so forth. The 1997 Federal Taxpayer Relief Act enables developers to immediately reduce their taxable income by the cost of their eligible cleanup expenses: The law includes "allowances for the costs of environmental cleanup to become fully deductible in the year they are incurred", thus helping to offset short-term cleanup costs (Berger 2006, 70). To receive this incentive, the property earmarked for redevelopment must be located in a census tract area where more than 20% of the population resides below the mean poverty level, and where 75% or more of the area's land is zoned for commercial or industrial use. Because these areas are often located within cities experiencing industrial-manufacturing decline and devaluation, developers "reap the rewards of reusing inexpensive, urban, contaminated waste landscapes and enjoy limited liability and high resale value" (Berger 2006, 74).

Brownfield projects seek to reuse and redevelop property deemed distressed, hazardous, toxic, even ugly. This is justified as a form of sustainable development and public health improvement. The policy framework expands the potential to seize land based on "blight"—a longstanding methodology of removing poor people and communities of color from land. The government's power of eminent domain faces two limitations: The stipulation that the taking of property must be for public use, and it must involve just compensation. However, definitions of what constitutes public use have typically been left to each state, and the U.S. Supreme Court has made increasingly broader determinations over the years (McNiece 2006, 230–235). The status quo essentially upholds the sanctity of private property (which is already racialized) while allowing for the seizure of property racially marked as blight in the name of public use. As such, brownfield programs deepen the racialized operations of the private property system by expanding the opportunity for land dispossession to areas of actual or perceived contamination, uncertain environmental hazards, and risks to health. The power to take private property based on blight—amplified by the ambiguous/elastic definitions of public use found in brownfield programs—"could become the newest tool that local and state governments could use to accelerate the gentrification and displacement that is already affecting low-income black and brown communities" (Albright 2017). The symptoms of a neighborhood's systemic neglect, including degraded infrastructure, environmental hazards, and racialized health inequities, can serve as justification for exercising eminent domain based on the argument that it is a public necessity and/or for the public good. Any number of conditions resulting from the nexus of race, waste, and space—from sewer floodplain hazards and high quantities of lead in water systems to asthma rates linked to concentrations of industry and their pollution/effluvia—might provide the rationale for eminent domain to enhance land productivity to support racial capitalism. Even as eminent domain serves crucial state functions of creating and protecting public infrastructures and populations, cities and their favored developers can point to the legacies of structural-institutional-environmental racism as justification for further dispossessing low-income residents and communities of color from land, homes, and the ability to secure safe, stable living conditions under the proviso of public care and improvement.

Cities are increasingly designating property as blight not because the land exhibits conditions of toxicity or other hazards to public health, but because the city views the property as unproductive from a tax-revenue perspective or aesthetic one linked to tax revenues (Berger 2006, 75). In 2005, the Supreme Court ruled that cities may legally seize private property for "economic development" even if that property is not contaminated or blighted, essentially sanctioning any city seizure of private land that is believed will receive higher property-tax revenues with a new/different public use. Thus, municipalities and local governments can opportunistically define blight according to their own city or regional planning interests. Scenarios wherein an urban renewal authority or city agency attempts to declare eminent domain on a property in order to attract big box stores have grown rapidly in areas that are experiencing economic decline. The chain home-improvement store Home Depot actively seeks to develop store locations on urban brownfield sites, receiving tax breaks to lay vast parking lots over contaminated/unproductive land in Honolulu, HI, and East Palo Alto, CA, to Cleveland, OH, and Pittsburgh, PA. Walmart similarly pursues brownfield redevelopment: The Denver Urban Renewal Authority targeted the largest Asian grocery store in the city, along with a strip mall of popular Asian restaurants, to grant Walmart $10 million in tax subsidies to redevelop the site (Berger 2006, 75). The city's eviction and redevelopment plan was only thwarted after a lengthy community petition process.

Brownfield programs pose significant problems for public participation, intensifying already curtailed participation in governance by minority communities. Heightened reliance on private investments and private property controls to address residual site hazards and contamination means decisions about future land use and reuse rely more and more on proprietary information to which the public is not privy. Private investors who spur on brownfield redevelopments may seek to curtail or eliminate public input in order to limit liability and facilitate faster turnover of the site, even as citizen groups demand a say in the recycling of land and/or whether re-development would cause more harm to the community by creating new pathways for community exposure to existing waste. Deed restrictions enshrine private ownership of property, not environmental law, leaving the public at risk—even as brownfield projects aver goals of improving environmental health. The economic calculus of brownfield redevelopment restricts—even obstructs—a knowing and involved public. Brownfield developers receive state-certified liability releases for cleanups in the form of covenants not-to-sue after cleanup, no-further-action agreements, gag order clauses in property sales documents, and so forth. Freeing developers from any further responsibility for adverse environmental and biopolitical effects, the state agrees it will not require or impose additional cleanup requirements at a later date if criteria are followed and the accepted cleanup standards are implemented. Such standards are tied to a risk-based understanding of anticipated land use. Because different end uses of the site require distinct levels or tiers of remediation standards, land recycling efforts enlist remediation options that range from minor remediation needed for limited human contact, such as future use as a parking lot, to widespread waste removal from the site ("dig and dump" somewhere else) and/or on-site underground waste containment required to support future housing or premium infrastructure.[1] Developers typically remove the upper level of soil and replace it with clean soil, placing an impervious cap over it to prevent exposure to arsenic, lead, chromium, or any other remaining contamination capable of reaching the surface. In some cases, engineered systems may be constructed to pump out and/or monitor contaminated groundwater or to capture noxious odors. Governance of the remaining on-site contamination hinges on what are con-sidered safe levels of risk based on the type of land reuse and ongoing hazards. The process essentially locks communities into a future of permissible contamination tied to a specific site use with little-to-no public discussion. Moreover, local governments often have little incentive to restrict land use and impose controls, few resources and limited financial capacity to monitor or enforce controls, and in many cases face strong political pressure for unrestricted use of a site (Wernstedt and Hersh 1998, 172).

Brownfield programs ultimately wed environmental risk reduction to economic development, as part of the broader integration of economic priorities in federal government hazardous waste policy *and* the cycle of land dispossession/repossession that characterizes racial capitalism (Hula 2001). This raises equity issues: Many brownfield sites lie in minority communities, where lower-income and poor people have been relocated to devalued land or left under-served for decades. Yet scholars have found "negative correlations between the proportions of local populations that are nonwhite or low-income and the likelihood of receiving an award" for brownfield redevelopment (Dull and Wernstedt 2010, 119). Contrary to the EPA's explicit commitments to equity with respect to land revitalization, applicants from localities with higher concentrations of poverty and higher numbers of self-identified nonwhites have been historically less likely to receive an EPA Brownfields award (Dull and Wernstedt 2010, 134). Rather than widespread community applications, the brownfields framework instead assembles a new property frontier that aggregates and homogenizes a diverse array of land types under a new label of "underutilized potential"—a category that has existed in many national land systems since an earlier colonial era; renders such lands as commensurable and available for redevelopment projects with the potential to generate a high rate of return due to financial subsidies and liability caps; and facilitates targeting specific parcels of land by scaling down any necessary remediation/cleanup to discrete areas (Li 2014, 593–594). To revive conditions of profitability, a kind of "slum reasoning" prevails that treats each brownfield site as discrete territory, regardless of regional patterns of disinvestment and environmental decline (Loyd 2014, 30).

The spatial fetishism of the brownfield enclosure enables city officials and developers to target and intervene in delineated properties, and, in doing so, negate any site uses or meanings that may be important to marginalized communities. In the case of Emeryville, CA, this entailed taking control of a hallowed shellmound of the Ohlone/Lisjan people through eminent domain proceedings that considered the site merely as postindustrial urban wasteland and taxable city property (DelVecchio 2002). Dubbed a poster child for brownfield redevelopment, Emeryville conducted urban renewal of this sacred ground by digging into the massive human-made mound of shells, tools, bowls, animal bones, and human burials created over the course of 2500 years, to construct the Bay Street retail and entertainment complex. At one time standing more than 30 feet high and 300 feet long, the shellmound had been desecrated by earlier land conversions, including the occupation of the site by amusement park/dancehall, followed by heavy industry in the 1920s that left vats of toxic chemicals and polluted soils from a defunct pigment plant (Arias 2005). Emeryville's redevelopment agency subsequently stripped the ground of toxic dirt and hired a developer to create the "Main Street" commercial village that is now hailed nationally as a model of urban land reclamation. Every Black Friday Shellmound descendants and protestors converge near the intersection of Shellmound Street and Ohlone Way to honor the site's significance and remind shoppers that they are standing on a living cemetery, where reportedly 100 human burials were taken from the metered parking lot behind Victoria's Secret and several other hundred were reburied on-site in an unmarked grave anchoring the mixed-use development (Griner 2019).[2] While the EPA now acknowledges brownfields may be tribal lands, it remains unclear how brownfield programs will have the capacity to conduct land reparations for native communities and descendants: The productivity imperative of the racial-capitalist property system actively organizes and maintains *the social death of land* to the extent that, in Emeryville, a strip mall now contains ancestral burials and toxic waste in a still-active cycle of repudiation and erasure.

Healthfields: "greenwashing" austerity or community-driven health justice?

An environmentally inequitable and extractive logic potentially animates land redevelopment projects that plant hospitals and health services on brownfields for tax breaks and trickle-down benefits to local health-stricken communities. Both the EPA and the Agency for Toxic Substances and Disease Registry recognize another aspect of land reuse beyond economic development: Public health improvements (U.S. Environmental Protection Agency 2018). A new form of brownfield project called "healthfields" holds the promise of reframing land revitalization as an ongoing public health project involving community stewardship of bodies of land and human health. However, this new iteration of brownfield also reveals the continued prioritizing of property values and, even worse, the "greenwashing" of austerity, municipal bankruptcy, and poverty. Originating in Florida, the policy framework targets contaminated land located in predominantly minority communities that are underserved by the healthcare system. The initiative offers tax breaks and lowers cleanup standards and liability for "green health" projects that remedy toxic blight.

Healthfields are any brownfields used for health purposes, such as healthcare centers, grocery stores, farmer's markets, greenspaces, and in some cases affordable housing. The "EPA Brownfields to Healthfields" website defines healthfields as an "economic development strategy that has served lower income families living in environmentally overburdened neighborhoods" (U.S. Environmental Protection Agency n.d.). An EPA storymap detailing healthfields through visual media and textual narrative states, "Healthfields increase local access to healthcare and community clinics, parks and open space, food access and housing—all through cleanup and reuse of a former brownfield site" (U.S. Environmental Protection Agency 2019). The term *healthfield* first cropped up around 2014 in Florida and has been popularized by Miles Ballogg, the director for Brownfields and Economic Development for Cardno TBE, an engineering consulting firm self-described as a professional infrastructure and environmental services company (Cardno n.d.). Ballogg and Cardno have been leaders of healthfield projects and advocacy in Florida, but there are also prominent examples spanning McComb, Mississippi, to Los Angeles, California (Kaysen 2012). The EPA report "Improving Public Health in Brownfield Communities" details the benefits of healthfield redevelopment: "In addition to the restoration of blighted, idle land and the removal of contamination, residents now have improved access to health care, new jobs, and local economic engines that leverage additional improvements and enhance quality of life" (U.S. Environmental Protection Agency 2008). The literature on healthfields emphasizes that brownfield law provides local governments and brownfield communities with the opportunity to link land revitalization with public health through provisions that allow local governments to spend up to 10% of their grants to conduct health monitoring of populations in sites where people may be exposed to hazardous substances and "legacy contamination". Frequently cited healthfield projects range from asthma surveillance mapping of children linked to school-based health programs to the conversion of defunct gas stations into parks, farmer's markets, and health services centers.

The healthfield program debuted with the Willa Carson Health and Wellness Center in Clearwater, Florida. This laudable case was driven by Willa Carson herself, who had already been operating a community health care center and wanted to raise enough money to open a more permanent facility for a free clinic to service residents of the city's North Greenwood community. A derelict gas station (with an out-of-state property owner) was deemed an ideal place for the health center due to its central community location. Using EPA and state brownfield programs funds, four underground storage tanks and 450 tons of contaminated soil were removed from the property. The city then leased the property to the nonprofit clinic—which Carson had previously operated out of two refurbished apartments—for 30 years at the rate of $1 per year (U.S. Environmental Protection Agency 2008). Opening in 2001, the Willa Carson Health Resource Center provides free health care predominantly to the surrounding community and is operated on donations and grants with the help of a volunteer workforce (Ballogg 2013).

Another brownfield conversion that fulfills critical needs and greatly improved medical access for local residents is the Johnnie Ruth-Clarke Health Center at the historic Mercy Hospital in St. Petersburg, Florida. Beset with petroleum contamination from a former cab company as well as hazardous waste and contamination from the African American hospital that operated on the grounds for over forty years, this healthfield project sought to install a new community-oriented and -run health services center in the tradition of Jim Crow–era African American medical activism, providing residents with immediate access to health care and an economic anchor for further neighborhood redevelopment. Funded with a $3.75 million U.S. Department of Health and Human Services grant and $463,000 U.S. Housing and Urban Development (HUD) Community Redevelopment Block Grant funds, construction of the center began in 2003 and included the preservation of the 1923 historic hospital building and a new museum dedicated to the history of African American medicine in Pinellas County. The Johnnie Ruth-Clarke Health Center foregrounds the potential of healthfields to convert legacies of health disparities tied to racial segregation into geographies of justice based on community-driven health services and land revitalization. The site today is a brownfield due in part to the historic lack of infrastructure and medical waste removal—what an EPA Brownfields conference presentation described as "abandoned historic African American Hospital environmental issues" (Ballogg 2013). The development of the health center at this historic Jim Crow–era hospital brings more community health services to an under-served area, but it also does not remedy the spatialization of waste and racial inequity. The emphasis of brownfield programs—and by extension healthfields—on the turnover of land to maximize economic productivity and less stringent environmental cleanup standards associated with medical care facilities means that potentially significant but uncertain contamination remains in the land/area that may (continue to) endanger health. In this sense, healthfields represent a contradiction, in that such programs may bring badly needed health services and community-oriented land remedies, but they risk re-entrenching health disparities stemming from historic segregation. Healthfields may only surficially address the land's impairment and hazards, thus ensuring environmental exposures to contamination—a legacy of Jim Crow and decades of disinvestment—continues.

This negative outcome of healthfields may be further intensified by the type of healthfield developed: The broad definition includes corporate packaged-food box stores, for-profit health service chains, or pharmaceutical manufacturing—the latter being an especially well-known polluting industry. Healthfields may spur the growth of a nonprofit philanthro-capitalist complex of medical surveillance, as well as high-end hospital zones that exploit devalued land and the poor communities that live there while receiving tax breaks under the banner of environmental justice, sustainability, and humanitarian services. Even as healthfields create an opportunity to bring health care to underserved areas and rearrange historical geographies of hospital locations that had been based on segregation rather than epidemiological rationales and needs, they also potentially advance what I call "Jim Crow tax shelters" that thrive on blight renewal.[3] Medical administration and care in Camden, New Jersey, demonstrates the institutionalization of racial inequity in ways that reveal how some healthfield projects might intensify geographies of waste and race (Sicotte 2016). Following the collapse of its manufacturing base and decades of white flight and disinvestment, Camden has turned to large-scale waste processing—a regional sewage treatment plant, open-air sewage sludge composting facility, trash-to-steam incinerator, power-cogeneration facility, coke transfer station, chemical companies, cement-grinding plants, and more—as the means to reverse industrial decline within a deeply racialized region (Krupar 2020). The presence of these toxic industries, combined with poverty and violent crime, all have contributed to a dire public health problem in Camden, where the city's residents are majority black and/or Latinx. Camden now hosts a growing nonprofit medical complex that facilitates developmentalist NGO interventions to help poor people who have inadequate or no health care. Politicians have pushed an "Eds and Meds" approach to land redevelopment, with higher education and health care nonprofits as anchor institutions for the city's "rebirth", along with the state's designation of Camden as a "growth zone" that offers major tax breaks.

On one end of the spectrum, Camden has innovated a data-based approach to decreasing exorbitant health care spending on the medically indigent; known as "medical hot spotting", the strategy entails sharing medical metadata across hospitals, jails, and schools, and uses GIS technologies to locate and target "high utilizers" of intensive outpatient care, for the purposes of motivating them to do better self-care and cost the system less (Ehlers and Krupar 2019, 46–68; Dizikes 2020). Such "care" endeavors to integrate people within the surveillance of poverty, including jobs training for residents in medical data entry to support a medical intelligence industry that surveils and intensifies the division of labor of managing the poor and industrial remains—without reference to the actual health conditions of Camden, where widespread environmental hazards attenuate quality of life. Camden has also attracted high-end hospitals, such as the biomedical research facility and teaching-oriented Cooper University Hospital, that operate with little reference to the surrounding poverty and environmental health inequities experienced by Camden residents.[4] The affluence of these elite nonprofit institutions and other trophy companies purportedly "trickles down" to the local population. Yet the tax exemptions and tax breaks that drew them to Camden facilitate billions in private profits and/or lost property taxes, thus strip mining the city of its tax base (Solomon and Pillets 2019).

Similarly, the land development strategies of the Cleveland Clinic in Cleveland, Ohio underscore how healthfield tax exemptions may intensify legacies of Jim Crow segregation: The hospital's unrelenting expansion of operations—what local African American residents have called "the plantation"—thrives on the surrounding neighborhoods' dereliction and abandonment (Diamond 2017). The nonprofit clinic occupies a 17-block stretch of land, with smooth roads, paved bike lanes, and glassy white buildings interconnected by skyways; the campus has its own private police force, hosts a high-end Intercontinental Hotel, and offers a variety of amenities akin to an airport terminal or resort, such as live music, shopping, and a farmer's market. The second biggest employer in Ohio (just behind Walmart), pride of Cleveland, and one of the greatest hospitals in the world, the Cleveland Clinic brings billions to the state *and* reveals starkly uneven health care disparities linked to the treatment of the land and surrounding neighborhoods as an extraterritorial/special economic zone. One of the best-known global brands in health care essentially creates local conditions of medical apartheid, exacerbating wealth and health inequities. The surrounding neighborhood of Fairfax has an infant mortality rate of nearly three times the national average; more than one-third of residents in the census tract surrounding the clinic have diabetes; and the predominantly African American population experiences higher rates of cancer, chronic kidney disease, and coronary heart disease (Diamond 2017). While wealthy international patients receive heart transplants at the clinic, local people do not go there for an emergency; the hospital has in fact been sued for not providing enough emergency care. Like many high-end hospitals, attracting wealthy patients and expanding operations to cities like London and Abu Dhabi take precedence over serving poor patients for Medicaid rates or receiving the fractional payouts and charity write-offs for treating the uninsured.

As a tax-exempt organization, the Cleveland Clinic saves tens of millions in annual property taxes from its billion-dollar property value, with only a loosely defined commitment to reinvest in the local community. The clinic claims it improves the surrounding blighted area of vacant lots, unsafe homes, and bail-bond outfits, and that its world-renowned physicians and well-paying jobs lift up the community. Yet in a grotesque revival of urban renewal's draining of wealth from inner-city areas through investments in white suburbs and highways, the institution has pursued an over $330 million construction project called the "Opportunity Corridor"—a three-mile highway that requires ripping up streets and tearing down residential and commercial buildings—to expedite shuttling staff and patients to the hospital from interstate-490 (Litt 2018). The boulevard construction vacuums up what could have been the clinic's property taxes and local investments in schools and city services desperately needed in the area, such as out-of-hospital care and social support, long-term training, and acculturating local workers successfully into clinic jobs, neighborhood health monitoring and land/structural remediation to address lead exposure as equally as important as treating heart attacks. The Cleveland Clinic reveals how healthfields can serve to install extraterritorial enterprise zones, born from and perpetuating Jim Crow financial, infrastructural, and medical redlining across generations and transnational space.

Conclusion: antiracist soil exegesis

This chapter has explored land reuse—specifically brownfields—as an important case for waste studies. Brownfield programs create a framework for land reuse "investment, resignification, and value formation" that may support community-driven justice efforts, but more frequently, (re)produce inequitable living conditions and unequal social formations (Dillon 2014, 1206). Brownfield land redevelopments leverage environmental and financial liabilities to galvanize surplus value production, through a property system that maintains racial difference and ensures that waste remains. Case studies examined blight designation to gentrify inner-city areas in the name of environmental health or more productive use, and health service installations that may paradoxically further entrench health inequities, economic injustices, and environmental hazards stemming from segregation. The brownfields framework, however, can also serve as a heuristic for exploring "historical and contemporary articulations of race and toxic waste" (Dillon 2014, 1209). Scrutinizing how the private property system is racialized in terms of access and value shows that brownfield projects leave unfulfilled—and still open/possible—the promise to attenuate and repair systemic neighborhood neglect and health/wealth disparities. This is critical because estimates place the number of brownfields across the United States upwards of one million (Heberle and Wernstedt 2006, 480).

Critical brownfield scholarship has raised important questions about whether and how to balance environmental protection and human health with economic development—if this is even possible. A primary concern is whether the EPA-administered program—and its various state and local manifestations—may in fact redistribute land and resources to bolster private sector businesses at the expense of other goals and with unknown future effects. Critical questions that waste studies might ask of cases like those described in this chapter include: Is it appropriate to emphasize economic development over determining the extent of contamination in federal/state/local brownfield programs? How should public and private benefits of brownfields be weighed? Is the best "public use" its most "productive use"—and who determines who counts as the public and what constitutes productive or the best use? To what extent do the economic benefits of redeveloped brownfield property transfer wealth and economic activity from one part of a community or region to another? (Wernstedt et al. 2004, 21). In what ways do brownfield programs empower communities to address environmental injustices and economic stresses, and/or how do they facilitate violent cycles of displacement, exploitation, uneven development, and unequal environmental health? How might vulnerable populations—who often have the fewest tools and means to make brownfield development work—"keep the benefits of brownfield regeneration in their communities?" (Wernstedt et al. 2004, 21). What other ways of conceiving brownfields would this necessitate, such as collective property, land trusts, reparations, etc.?

Because brownfield policy largely relies on the property system framework to focus on individual parcels of land, it inhibits us from seeing brownfields as part of larger geographical relationships of race, space, and waste; brownfield policies "scale" problems to a discrete area without considering the interlinked spaces of concentrated wealth/health and poverty/illness. What is needed, instead, is to consider how decisions about cleanup and development at any given site shift "costs and benefits of economic revitalization and environmental remediation across space to other jurisdictions and across time to future generations" (Wernstedt and Hersh 1998, 172). This chapter has considered some of the ways remediation strategies at brownfield sites may further entrench race in space in such a way as to ensure exposure, predatory targeting, and disinvestment of nonwhite and poor communities. By implementing brownfields as an *antiracist* heuristic device, we can critically reflect on the racialized social production of property and the ways that notions of productivity, blight, improvement, and public health may intensify the racial disparities of property and land use, and unfairly, unwisely, unjustly transfer risk to future generations.

There are a number of ways to advance this antiracist land research and re/use further within waste studies and policy: For example, more wide-scale public waste scholarship and countermapping of the practices and geographies that have made and maintained articulations of waste and race.[5] The difficult work of researching brownfield waste and other environmental hazards requires not only technical skills, to investigate historic archives, real estate records, regulatory files, and hazard-tracking system reports, but also critical social awareness of legacies of native land appropriation, segregation, poverty, and racialized environments (Litt and Burke 2002, 464). The strong quantitative and qualitative research methods of waste studies scholars are essential to this task: Methodologies are needed that track the actual material stuff of waste and address the broader geographies of material-physical wastes that drift beyond property boundaries (downwind, downstream, in underground plumes, and so forth). National hazard-tracking systems do not capture a lot of the hazardous wastes and contamination that stem from over a century of heavy industrial operations; waste studies research contributes critical information on hazardous materials and potential physical and chemical risks to human health and the environment (Litt and Burke 2002, 466; Colten 1990).[6]

Moreover, particular futures are embedded in the soil composition and end use of land remediation projects and necessitate careful attention to the materiality and physical characteristics of brownfields. These facts on the ground are part of relationships between places and geographies of race, and thus wastes *in situ* can be investigated as deposits of racial histories. However, brownfield projects typically hide toxic foundations and material hazards from view via a fabricated geology and the creation of a transferable surface, annulling linkages between a particular site's material history and everyday lives, health, and ecology. In response, waste studies might work toward an *antiracist physical geography of brownfields* that combines histories and theories of race/racism and "critical attention to power relations with deep knowledge of biophysical science or technology in the service of social and environmental transformation" (Lave, Wilson, and Barron 2014, 2).

Specifically, *antiracist soil exegesis*—as one potential method of this critical physical geography of brownfields—would take soil seriously as something other than the simple truth or passive medium (Krupar 2018; Liboiron 2019). Drawing from sciences, social sciences, and humanities, such a transdisciplinary collaborative method would analyze the treatment of the soil as free disposal and waste container in terms of *sedimented* power relations—as an arrangement of human relations with biophysical nature; it would also consider the partitioning of surface and subsurface as a "color line" that allows for value creation through racialized transfer and financialization/devaluation. While claiming to be neutral, technocratic, and merely about the "bottom line", the administered division of surface and subsurface that allows for remediated property conversion and reuse would be critically reinterpreted and assessed in terms of how it perpetuates racist legacies and/or instigates racist effects, such as erasure, segregation, neglect, and qualified life. In doing so, antiracist soil exegesis would support claims to land use futures that do not simply entail technocratic systems of control, because such approaches obscure how they facilitate ongoing racial capitalism and close down what communities, ecologies, and entities can enter politics. Instead, the method facilitates projects to reclaim wastes and contaminated land in order to design reclamation and reparations into end uses, cleanup, and stewardship. Antiracist soil exegesis could also be enlisted in efforts to rethink relations between humans and materials and assemble alternative economies and political ecologies to the racial capitalist property system: For example, waste "communing" and forms of public provisioning that resist capitalist valuation and enclosure. Such critical physical geographies and methods would serve as "antidote field guides" that encourage alternative versions of sustainability and more equitable landscapes and modes of living that transform waste-race relations (Gibson-Graham 2011).

Antiracist brownfield research and policy would ultimately seek to *decolonize* land redevelopment by differentiating land use and ontology from *property* as financial instrument, to redress racial regimes of ownership, improvement, and blight.[7] This transformative work requires upending the settler-colonial maps, geopolitical divisions, and silences about race that often haunt waste studies. Emerging comparative work on the epistemologies and practices of brownfields across transnational contexts reveals the urgent need to examine waste and racial capitalism across the development divide too—to inaugurate waste studies of racial differentiation and inequality that mark land reuse along *global pedological* color lines.

Notes

1 Remediation may also include bioremediation strategies (i.e., petroleum-eating microbes) or thermal desorption (i.e., heating up the soil) to remove or burn off contaminants.
2 Seven hundred bodies were previously taken by archeologists to the University of California-Berkeley.
3 "Jim Crow tax shelters" refers to the tax-exempt/non-profit shadow of the state (humanitarian philanthro-capitalist organizations/services) that operates in response to the financial-environmental color line.
4 The "Eds & Meds" development strategy fails to address Camden's toxic industries or connect with the city's longstanding environmental justice activism (Jurand 2003).
5 Countermapping typically seeks to subvert official maps and/or render alternative spatial knowledges (Krupar 2015).
6 Two major national hazard-tracking systems include the Environmental Protection Agency's (EPA) Superfund and Resource Conservation and Recovery Act (RCRA) programs.

7   In particular, the transnational brownfield frontier of former U.S. military bases must be addressed, including settler-colonial land seizures and military occupation of vast tracts of "public lands" for U.S. national projects and war-making.

References

Albright, C. 2017. "Gentrification Is Sweeping Through America." *The Guardian*, November 10, 2017. https://www.theguardian.com/us-news/2017/nov/10/atlanta-super-gentrification-eminent-domain.

Arias, R. 2005. "'Shellmound' Documentary Exposes the Truth Behind, and Under, Bay Street Development." January 15, 2014. https://evilleeye.com/history/2005-shellmound-documentary-exposes-the-truth-behind-and-under-bay-street-developement/.

Ballogg, M., U.S. Environmental Protection Agency, and City of Tampa FL. 2013. "'Healthfields': Improving Access to Healthcare Through Brownfields Redevelopment." *Annual Georgia Environmental Conference*, August 22, 2013. http://www.georgiaenet.com/wp-content/uploads/2015/01/23MilesBallogg.pdf. Accessed January 21, 2021.

Berger, A. 2006. *Drosscape: Wasting Land in Urban America*. Princeton Architectural Press.

Bhandar, Brenna. 2018. *Colonial Lives of Property: Law, Land, and Racial Regimes of Ownership*. Duke University Press.

Cardno. n.d. "Who We Are." http://www.cardno.com (accessed January21, 2021).

Colten, C. E. 1990. "Historical Hazards: The Geography of Relict Industrial Wastes." *Professional Geographer* 42, no. 2: 143–156.

Conroy, W. 2018. "Studying Brownfields: Governmentality, the Post-Political, or Non-Essentialist Materialism?" *Fennia* 196, no. 2: 204–214.

Crysler, C. G. 2017. "Groundwork: (De)Touring Treasure Island's Toxic History". In L. Horiuchi and T. Sankalia (Eds.) *Urban Reinventions: San Francisco's Treasure Island*. University of Hawai'i Press. 175–186.

DelVecchio, R. 2002. "Urban Renewal Atop Sacred Past/Ohlone Protest Emeryville Project." *SFGate*, November 20, 2002. https://www.sfgate.com/bayarea/article/Urban-renewal-atop-sacred-past-Ohlone-protest-2752176.php.

Diamond, D. 2017. "How the Cleveland Clinic Grows Healthier While its Neighbors Stay Sick." *Politico*, July 17, 2017. https://www.politico.com/interactives/2017/obamacare-cleveland-clinic-non-profit-hospital-taxes/.

Dillon, L. 2014. "Race, Waste, and Space: Brownfield Redevelopment and Environmental Justice at the Hunters Point Shipyard." *Antipode* 46, no. 5: 1205–1221.

Dillon, L. and J. Sze. 2019. "Equality in the Air We Breathe: Police Violence, Pollution, and the Politics of Sustainability." In J. Sze (Ed.) *Sustainability: Approaches to Environmental Justice and Social Power*. New York University Press. 246–270.

Dizikes, P. 2020. "In Health Care, Does 'Hotspotting' Make Patients Better?" *MIT News*, January 8, 2020. https://news.mit.edu/2020/health-care-hotspotting-no-effect-0108.

Du Bois, W. E. B. 1996. *The Philadelphia Negro*. University of Pennsylvania Press.

Dull, M. and K. Wernstedt. 2010. "Land Recycling, Community Revitalization, and Distributive Politics: An Analysis of EPA Brownfields Program Support." *The Policy Studies Journal* 38, no. 1: 119–141.

Ehlers, N., and S. Krupar. 2019. *Deadly Biocultures: The Ethics of Life-Making*. University of Minnesota Press. https://manifold.umn.edu/projects/deadly-biocultures.

Freund, D. M. P. 2007. *Colored Property: State policy and White Racial Politics in Suburban America*. University of Chicago Press.

Gibson-Graham, J. K. 2011. "A Feminist Project of Belonging for the Anthropocene." *Gender, Place, and Culture* 18, no. 1: 1–21.

Gidwani, V. 2008. *Capital, Interrupted: Agrarian Development and the Politics of Work in India*. University Minnesota Press.

Gille, G. 2007. *From the Cult of Waste to the Trash Heap of History: The Politics of Waste in Socialist and Postsocialist Hungary*. University of Indiana Press.

Goldstein, J. 2013. "*Terra economica*: Waste and the Production of Enclosed Nature." *Antipode* 45, no. 2: 357–375.

Greenberg, M. R., and J. Hollander. 2006. "The Environmental Protection Agency's Brownfields Pilot Program". *American Journal of Public Health* 96, no. 2: 277–281.

Griner, A. 2019. "'On My Ancestors' Remains': The Fight for Sacred Lands." *Aljazeera*, December 16, 2019. https://www.aljazeera.com/features/2019/12/16/on-my-ancestors-remains-the-fight-for-sacred-lands.

Harris, C. 1993. "Whiteness as Property." *Harvard Law Review* 106, no. 8: 1707–1717.

Heberle L., and K. Wernstedt. 2006. "Understanding Brownfields Regeneration in the U.S." *Local Environment* 11, no. 5: 479–497.

Hula, R. C. 2001. "Changing Priorities and Programs in Toxic Waste Policy: The Emergence of Economic Development as a Policy Goal." *Economic Development Quarterly* 15, no. 2: 181–199.

Jurand, S. H. 2003. "Environmental Justice" *Movement Looks to Pivotal New Jersey Cases*. Trial. 12–17.

Kaysen, R. 2012. "Health Centers Find Opportunity in Brownfields." *The New York Times*, December 11, 2012. http://www.nytimes.com/2012/12/12/realestate/commercial/health-centers-find-opportunity-in-brownfields.html?_r=1&.

Krupar, S. 2015. "Map Power and Map Methodologies for Social Justice." *Georgetown Journal of International Affairs* 16, no. 2: 91–101.

Krupar, S. 2018. "Sustainable World Expo? The Governing Function of Spectacle in Shanghai and Beyond." *Theory, Culture & Society* 35, no. 2: 91–113.

Krupar, S. 2020. "Folklore of Operational Banality: Medical Administration and Everyday Violence." *Environmental Humanities* 12, no. 2: 431–453.

Lave, R., M. W. Wilson, and E. S. Barron, et al. 2014. "Intervention: Critical Physical Geography." *The Canadian Geographer* 58, no. 1: 1–10.

Li, T. M. 2014. "What Is Land? Assembling a Resource for Global Investment." *Transactions of the Institute of British Geographers* 39, no. 4: 589–602.

Liboiron, M. "Waste Is Not 'Matter Out of Place.'" *Discard Studies* September 9, 2019. https://discardstudies.com/2019/09/09/waste-is-not-matter-out-of-place/.

Liboiron, M., M. Tironi, and N. Calvillo. 2018. "Toxic Politics: Acting in a Permanently Polluted World." *Social Studies of Science* 48, no. 3: 331–349.

Litt J. S. and T. A. Burke. 2002. "Uncovering the Historic Environmental Hazards of Urban Brownfields." *Journal of Urban Health* 79, no. 4: 464–481.

Litt, S. 2018. "Opportunity Corridor Is Back on Track for 2021 Completion After Delay Caused by Taxpayer Lawsuit." Cleveland.com, February 14, 2018. https://www.cleveland.com/architecture/2018/02/opportunity_corridor_on_track.html.

Loyd, J. 2014. *Health Rights Are Civil Rights: Peace and Justice Activism in Los Angeles, 1963–1978*. University of Minnesota Press.

McNiece, C. M. 2006. "A Public Use for the Dirty Side of Economic Development: Finding Common Ground Between Kelo and Hathcock for Collateral Takings in Brownfield Redevelopment." *Roger Williams University Law Review* 12, no. 1: 229–255. https://docs.rwu.edu/rwu_LR/vol12/iss1/6.

Moore, J. W. 2011. "Ecology, Capital, and the Nature of Our Times: Accumulation and Crisis in the Capitalist World-Ecology." *Journal of World-Systems Research* 17, no. 1: 107–146.

Poindexter, G. C. 1996. "Separate and Unequal: A Comment on the Urban Development Aspect of Brownfields Programs." *Fordham Urban Law Journal* 24, no. 1: 1–16. https://ir.lawnet.fordham.edu/ulj/vol24/iss1/1/.

Pulido, L. 2016. "Flint, Environmental Racism, and Racial Capitalism." *Capitalism Nature Socialism* 27, no. 3: 1–16.

Sicotte, D. 2016. *From Workshop to Waste Magnet: Environmental Inequality in the Philadelphia Region*. Rutgers University Press.

Solomon N. and J. Pillets. 2019. "How Companies and Allies of One Powerful Democrat got $1.1 Billion in Tax Breaks." *ProPublica*, May 1, 2019. http://www.propublica.org/article/george-norcross-democratic-donor-tax-breaks.

Ranganathan, M. 2016. "Thinking with Flint: Racial liberalism and the roots of an American water tragedy." *Capitalism Nature Socialism* 27, no. 3: 17–33.

U.S. Environmental Protection Agency. 2008. "Improving Public Health in Brownfields Communities." EPA-560-F-07-253. http://www.epa.gov/brownfields/success/public_health08.pdf.

U.S. Environmental Protection Agency. 2019. "From Brownfields to Healthfields." https://epa.maps.arcgis.com/apps/Cascade/index.html?appid=fa7b68b3075a4340970b1e5c00c76cf4.

U.S. Environmental Protection Agency. n.d. "EPA Brownfields to Healthfields." https://www.arcgis.com/apps/MapSeries/index.html?appid=76cd82c0c167480799ab9e6f6f144e36 (accessed September 19, 2020).

U.S. Environmental Protection Agency, Office of Brownfields and Land Revitalization. 2018. "Incorporating Health Monitoring Activities into an EPA Brownfields Grant." EPA 560-F-18-187. https://www.epa.gov/sites/production/files/2015-09/documents/finalphandbffact.pdf.

Wernstedt, K., L. Heberle, A. Alberini, and P. Meyer. 2004. "The Brownfields Phenomenon: Much Ado About Something or the Timing of the Shrewd?" Discussion Paper 04-46, Resources for the Future. https://media.rff.org/archive/files/sharepoint/WorkImages/Download/RFF-DP-04-46.pdf (accessed January 21, 2021).

Wernstedt, K. and R. Hersh. 1998. "'Through a Lens Darkly'—Superfund Spectacles on Public Participation at Brownfield Sites." *Risk: Health, Society & Environment (1990–2002)* 9, no. 2: 153–173.

17

OF SHIPS OF DOOM AND ICEBERGS: EARLY PERSPECTIVES ON THE GLOBAL HAZARDOUS WASTE TRADE

Kate O'Neill

Introduction

This chapter takes an historical perspective on the global political economy of wastes and the waste trade. It focuses specifically on changing and often conflicting patterns and perceptions of the hazardous waste trade in the 1980s and 1990s, the years when the trade first became a matter of global concern, and when the major global governance arrangements to manage or prevent the trade were set up. This story predates the complexities of a truly global used electronics trade (Lepawsky 2018), controversial plastic and paper scrap imports into China (O'Neill 2017a), and the "global garage sale" of second-hand goods (Minter 2020). It does, however, show the shifting theoretical frames and literatures that grapple with the complexities of risks versus resources in the shipment of waste, used items or discards across national borders, and what this means for governing these impossible-to-ban shipments in a globalized world.

In other words, this chapter goes back to the "old school" literature in the field, on the hazardous industrial waste trade—toxic ash, oils, and sludge, contextualizing it in debates of that time, but also revisiting it in the context of today's debates over an ongoing, and hard-to-ban waste trade that has existed for centuries. Early recorded cases of industrial wastes crossing borders include shipments of brick dust from London to Russia in the first half of the 19th century, to aid in the rebuilding of Moscow after the Napoleonic Wars (Wilson 2007, 199). One of the most recent cases was the port explosion in Beirut in August 2020 caused by abandoned and badly stored ammonium nitrate, seized while in transit. The explosion killed over a hundred people and left hundreds of thousands homeless, and destroying Lebanon's main port. The cargo left the Republic of Georgia for its journey to Mozambique for use in explosives manufacture in 2013. While presumably never intended as hazardous waste, it became that when offloaded from the Moldovan-flagged Rhosus, which subsequently sank, never leaving Lebanese waters (BBC News 2020).

What follows is based on early analyses, mostly from the academic literature, especially Global Environmental Politics (GEP; for which the waste trade was a minor, though still important, case, even before climate change loomed so large on political and research agendas). It also draws on Science and Technology Studies (STS) and other critical perspectives on

DOI: 10.4324/9781003019077-17

wastes, risk, and globality and brings in some of the most important sources of data on the waste trade that existed at the time.

The next part of the paper reviews some of this early scholarship, before delving into how the waste trade emerged in the 1980s, and how global governance emerged in response to it. Part 3 examines how more critical literature challenged conventional global environmental justice narratives around the waste trade. One of those themes is that the "ships of doom" mentioned in the title of this chapter often went to countries like the United Kingdom (O'Neill 2000a). Another questions whether what we saw of the waste trade in the 1980s was the tip of the iceberg or the entire iceberg itself (Montgomery 1995). Finally, I connect this literature and its findings with work on the dominant forms of the global waste trade, and efforts to govern them, today.

Studying the hazardous waste trade in the 1980s and 1990s

Scholarship around the global politics of hazardous waste management systems, the Basel Convention, and the hazardous waste trade in the 1970s and 1980s, as they first emerged as global issues, remains surprisingly scarce. However, scholars who were or became preeminent in the critical (interpretive) social sciences, including GEP and STS did significant early work on globalizing hazardous waste. Important authors include Brian Wynne (e.g., 1987, 1987), Jennifer Clapp (e.g., 1994 and 2001), K.A. (Kenneth) Gourlay (1992), Jonathan Krueger (1998 and 1999), Laura Strohm (1993), Christoph Hilz (Hilz and Ehrenfeld 1991), Mark Montgomery (1994, 1995), Kate O'Neill (e.g., 1998, 2000, 2001), Katherine Kummer (1995), and others.[1]

These scholarly works share a number of themes. First, nearly all deal with industrial wastes and discards, not household, consumer or municipal solid waste. Although many studies of recycling, and now, of the international waste trade/trade in used goods, focus on consumer waste (like cardboard and plastic packaging, clothes, and personal electronics), Samantha MacBride (2012) and others have pointed out that most wastes are industrial, around 90 to 97%, depending on the data used (Gille 2016, 183). These studies emphasize this fact. Further, the wastes under discussion in these data compendia and related literature are primarily sludge, ash, or other residue that bears little resemblance to actual items that are thrown away. This is in strong contrast with electronic wastes, plastic packaging, end-of-life ships and other after-life items that have stronger potential for recycling and re-use, and which have through their very recognizability become symbols of wasteful consumerism (O'Neill 2019, Chapter 2).[2]

Second, most authors highlight the contextuality of hazardous wastes as a broad category, especially in terms of how they are handled at their destinations, and whether adequate infrastructure exists to manage them safely (O'Neill 2000a, 26–29; Gourlay 1992). The comparability of hazard(-ous wastes) is also contextual: to paraphrase Gourlay 1992, 20), what do a blob of mustard on a dinner plate, alternatively, rotten fish or meat have in common with a decommissioned nuclear power plant or oil rig as a case of hazardous waste?

Further, even though there is widespread acceptance of certain characteristics of hazardous wastes (toxic, reactive, corrosive, flammable, explosive, for example, as in the U.S. Environmental Agency's definition, see US EPA, n.d.), the absence of a coordinated cross-national definition of and harmonized data on hazardous waste is both an obstacle and an object of study for many works in the field (O'Neill 2000a, 71–72; Wynne). It is also an obstacle to effective global governance.

Third, this scholarship was among the earliest to situate global environmental justice within the wider fields of global environmental governance and politics (Clapp 2001; Pellow 2007; Agarwal et al. 2001). The emergence of the waste trade as a transboundary justice issue co-incided roughly with the first effort at the global level to define and institutionalize the concept of sustainable development, 1987s seminal report Our Common Future, also known as the Brundtland Report (World Commission on Environment and Development 1987). At that point in time, global climate governance had not formally begun. Nor was climate change treated as much beyond a northern concern that would not materialize for several generations, although one study at the time did connect climate policy—especially attribution of greenhouse gas emissions responsibilities—to colonialism (Agarwal and Narain 1991). Global climate justice activism only emerged at the 2000 UN Framework Convention on Climate Change meeting in the Hague. It built on the energy of the Alt-Globalization protest movement that had burst onto the stage at the WTO Ministerial Conference in Seattle in 1999 (Cockburn and St. Clair 2000). Therefore, studies of the hazardous waste trade at this time connected more easily with narratives of colonialism, economic inequalities and globalization and with the growing body of scholarship on the U.S. environmental justice movement (Pellow 2007) than with the (at that time) mainstream International Relations–dominated field of global environmental politics (O'Neill 2017b).

Fourth, many of the authors cited in this section address what have become some of the fundamentals of waste studies and discard studies (Liboiron 2018). They put front and center the complexities and contradictions of discards as a resource but also as a hazard (Gourlay 1992) and the perverse market incentives generated by reverse supply chains (Wynne 1987, 1989; Laurence and Wynne 1989), at home and across borders. These incentives impact flows of money and information as well as the wastes themselves, and the mostly informal sector workers who deal directly with them. Early work on ship-breaking highlights the disregard of human lives and bodies—the workers tiny in relation to the huge vessels they are disassembling—through visual media, reports, and articles (Crang 2010).

Adding the transboundary element enables the elaboration of the concept of distancing—literal and figurative displacement of hazardous waste from sight and mind (Clapp 1994; Princen 2002). To extend Brian Wynne's formulation: "the badly structured and, indeed, indeterminate behavioral-technical-risk-generating system" that "jumble[s] natural processes together in complex and widely variable ways" (Wynne 1987, 1) has been stretched thinly in places, thickly in others across the entire planet, and the political borders that crisscross its surface. According to Wynne et al. in the 1987 volume, hazardous waste management (or, the "hazardous management of wastes", 16), challenges scientific precepts in policy making and implementation, as well as public understandings of risk.

Fifth, many of these authors drew on new compendia of data—the most significant being *The International Trade in Wastes: A Greenpeace Inventory, Fifth Edition* (Vallette and Spalding 1990), weighing in at over 400 pages, and with cover art by U.S. artist Keith Haring (see Figure 17.1; the artwork shows a red foot bearing down on the figure below, with a yellow background and "Greenpeace" spelled in green letters).[3] It documents actual and attempted shipments of hazardous wastes from rich to poor countries (see also Third World Network 1989). The *Inventory* is hard to find these days, but it contains a treasure trove of early data that is one of the most comprehensive of its type.[4] The Organization for Economic Cooperation and Development, or OECD, also published data at this time on transfrontier movements of hazardous wastes (e.g., OECD 1985, 1994, 1997). The OECD's data—tables and numbers—is a contrasting format compared with the Greenpeace Inventory's case-by-case

Figure 17.1 Keith Haring artwork © Keith Haring Foundation, depicted on the cover of Vallette and Spalding (1990).

listing, each with a short narrative explanation where available. The *Inventory*, based on on-the-ground data collection and investigative reporting is a quite remarkable work, although not without inconsistencies, listing transactions that were not, technically or even in actual fact, toxic.[5]

Finally, many of these works questioned prevailing narratives, about "Southern country victimhood" in the face of considerable evidence of resistance by the Global South, not to mention the large hazardous waste trade going on among countries in the global north. By bringing insights in from other theoretical fields, like STS and GEP, these works brought a deeper understanding of the dynamics and global political economy global toxic wastes that can greatly inform today's debates about this trade, or trades. Cases like the 2006 Abidjan dumping case reveal, in the same way that critical studies of the e-waste trade do (e.g., Lepawsky 2018), the complexities of global shipping routes, ownership, and responsibility).

Illegal waste trading in the 1980s: setting the narrative frame

The illegal trade, or traffic, in hazardous wastes was thrust into the global spotlight by activists and the media in the 1970s and 1980s.[6] Several high-profile cases of waste dumping from the wealthy North to communities in the Global South galvanized action at the international level that led to the first global treaty directly addressing the global reach of toxics, the 1989 Basel Convention on Transboundary Movements of Hazardous Waste and Their Disposal. This section turns to the events and scandals that drove how the waste trade emerged onto the global scene and subsequent governance decisions.

According to the OECD (1985, 1994, 1997), around 300–500 million tons of hazardous wastes were generated annually across the world at this time, with 80–90% of that produced within the OECD, and around 10% shipped across national borders. Of that ten percent, 80–90% was legally shipped among OECD countries (Krueger 1998, 116; O'Neill 2000a, 35–39). Between 1986 and 1994, these same sources estimate that approximately 7.7 million tons were shipped to non-OECD countries in Eastern Europe and developing countries, most of these shipments occurring between 1986 and 1990 (Krueger 1998, 116). Now these, of course, are data that the OECD countries reported, and the actual numbers were likely to have been higher, although, as we see below, some analysts contest this claim.

The most notorious instances involved the dumping of toxic sludge, ash, and other illegal hazardous wastes containing substances such as mercury and dioxins. These shipments were driven by the desire of waste generators to pay to have those wastes disappear. Higher costs associated with regulatory demands and lower waste disposal capacity in developed countries (hand in hand with NIMBYism), along with cheaper global transportation made exporting hazardous waste even more attractive (Strohm 1993). What was trafficked was not always dumped. Some of it contained recyclable elements—such as batteries and metals. Others were technically saleable or usable products but of dubious or damaging properties (such as a shipment of purportedly radioactive cardboard from Florida to Honduras in 1999, reported in Vallette and Spalding 1990, 155). Tales of tangled and untraceable ownership of the wastes and of the vessels carrying them abound, enabled by flags of convenience regulation under international shipping law, and murky domestic controls in many countries. These factors added to the practical impossibility in most cases of assigning liability.

Many of these shipments, particularly to Africa and the Caribbean, inflicted lasting harm and even deaths directly attributable to waste exposure. Greenpeace cites Thor Chemicals, a British company that imported mercury waste from the United States and Europe into South Africa for incineration in a smelter located just outside what was then a "Homeland", the apartheid government's sanitized term for a segregated settlement for Black South Africans. The port town of Koko in Nigeria received several shipments of toxic waste form Europe between 1987 and 1989, in a deal between local officials and Italian broker company Jelly Wax (which appears in several of the cases described in Vallette and Spalding 1990).

Toxics-laden "ships of doom" (a headline in Britain's *Daily Mirror* in August 1994, referring to a "fleet" of trash laden barges headed from Germany to the United Kingdom), also termed "ships of death" (Clapp 1994) and "leper ships" (Laurence and Wynne 1989) plied the seas looking for ports (and willing local brokers) to offload their cargo, with no questions asked. In one of the most famous instances, the Liberian-flagged Khian Sea set sail from the United States in September 1986, carrying nearly 15,000 tons of toxic ash from the city of Philadelphia labeled as fertilizer. Port authorities refused it permission to offload its cargo in Bermuda, the Dominican Republic, Honduras, Guinea-Bissau, and the Antilles. In December 1987 it arrived in Haiti, and offloaded nearly a third of its cargo before ordered by the government to depart. In 1988, it traveled from Delaware to Yugoslavia and Singapore, undergoing three name changes, and at least one ownership transfer, in the process. By the time it reached Singapore, its holds were empty. The captain refused to say whether the ash had been dumped into the Atlantic and Indian Oceans, or left on-shore somewhere in Southeast Asia with an illegal broker (Vallette and Spalding 1990; Wynne 1989, 125). The ash dumped in Haiti stayed in barrels on that same beach for nearly 13 years, even visible in photos taken during the U.S. military incursion into Haiti in 1994. In 2000, it was repatriated to Florida, and in 2002 the final load of ash was buried at a landfill in south-central Pennsylvania, home again after 16 years.[7] Wynne (1989) documents others of the international hazardous waste incidents that fired up concern in those years, from planned asbestos exporting from New York City to Guatemala, to other ships—such as the Lynx, carrying toxic cargo from Italy, traveling from port to port for over a year (Wynne 1989; 125–128, drawn from Vallette and Spalding 1990).

Transboundary dumping was also common practice from Western to Eastern Europe during the era of the Iron Curtain. West Germany routinely dumped its waste—hazardous and otherwise—across the border to communist (and independent country at the time) East Germany. In 1972, East Germany agreed to import garbage from West Berlin—a sign of the thawing of the cold war, according to a Berlin official (reported in the *New York Times*, October 28 1972). Over subsequent decades, until reunification of the two Germanies, wastes, including toxic wastes, were dumped over the border from all over West Germany, with the most notorious being the Schönberg site (Lange 2020; Vallette and Spalding 1990, 217–219). Other frequently listed destinations included Poland, Lebanon, and Mexico (Vallette and Spalding 1990, 228–259, 202–204, 157–162, respectively).

The shaping of global waste trade governance

Collectively, all the cases and evidence of hazardous waste dumping from rich to poor or poorer countries shaped the emergence of a global hazardous waste trade governance regime. In its original form in 1989, the Basel Convention's goals were to manage and monitor hazardous waste trading, to ensure that any imports happened only with prior informed consent from the importing country, and then only if the wastes would be disposed of in an "environmentally sound manner". Greenpeace termed this outcome as "legalizing toxic terrorism worldwide" (Vallette and Spalding 1990, 12). The problems that NGOs and activist coalitions such as the International Toxic Waste Action Group highlighted included the lack of a legal definition of "environmentally sound manner", and the problematic nature of Prior Informed Consent (PIC) as defined under the Convention. Critics of PIC as a tool for global governance argue it is unclear who is giving that consent, under what influence, and with what capacity to enforce outcomes, especially when those giving the consent are poorer countries—sometimes simply (badly paid) customs officials (Krueger 1998).

One of the major problems many had with the Basel Convention in its original form was the so-called "recycling loophole". Essentially, pre-ban, anything not actually labeled as a "waste" was fair game for shipping if it could be designated as recyclable. This was a problem with the cases mentioned earlier. While now we have a different "visual image of something shipped across borders for recycling—something recognizably a cell phone, plastic bag, or item of clothing, more often than not these were barrels of contaminated ash or other substances labeled as fertilizer or for use in road building. Labeling—or designation—remained a problem after

Basel in its original form entered into force. High-profile examples include end-of-life ships, sailed to their final destination under their own steam and driven up onto beaches in South Asia to avoid being labeled as "wastes" (Greenpeace International and International Federation for Human Rights 2005). Other examples are the electronic devices—old computers and TVs, and now, increasingly, phones, tablets, iPads, and other devices—designated as charity or for recycling, in order to close the "digital divide" between North and South. These examples, as they unfolded, illustrated the accelerating collision between the global trade in scrap and the global trade in waste (O'Neill 2019). In the years immediately following the signing of the Basel Convention, developing countries drew up their own waste trade–related treaties. In 1991, 12 members of the African Union negotiated the Bamako Convention, which banned waste imports into Africa.[8] The 1995 Waigani Convention, among South Pacific island states, also banned hazardous and radioactive waste imports.[9]

Parties to the Basel Convention responded in 1995. Decision III.I, adopted at the Third Conference of the Parties, sought to strengthen the Convention to ban the shipment of wastes listed as hazardous (under Article 1.1a; essentially wastes with characteristics listed under Annex I) in the Convention from OECD countries to non-OECD countries (and non-Parties) for both disposal and recycling.[10] Yet, the Ban Amendment only entered into force in 2019, after a lengthy process of ratification in the signatory states, and the integration of a new procedure in 2011, designed to increase its flexibility (the "Country-Led Initiative"). A Liability Protocol, adopted in 1999, has not yet entered into force.

In the course of this journey, the Convention, despite its weaknesses and the Ban's long, slow journey to entry into force, exerted normative influence. North-South waste dumping became more universally condemned by the global community. Local and transnational NGOs kept an eye on governments (of importing and exporting countries), and other actors. High-profile cases, such as those detailed above, diminished in numbers, even with the increased scrutiny, and exceptions were all the more remarkable for their rarity. Perhaps the most high profile of these dumping incidents occurred in 2006. Truckloads of toxic sludge originally shipped from Amsterdam by the Singapore-based metals and energy trading corporation Trafigura were dumped (in the middle of the night) onto open dumpsites in Abidjan, the capital of Ivory Coast, immediately causing several fatalities and injuring thousands. The tragedy and subsequent outcry spurred guilty verdicts in 2010 in Dutch courts for Trafigura and for the Russian captain of the Probo Koala, the Panamanian cargo ship that carried the waste. The Nigerian national who owned the local company, Tommy, contracted to handle the waste, was sentenced to 20 years' jailtime in the Ivory Coast for his role (for more on this case, see Polgreen and Simons 2006 and BBC News Africa 2010).

However, the Ban Amendment opened the door to more conflict. First, ongoing deliberations on listing hazardous wastes and their characteristics in the Annexes proved more contentious than initially thought.[11] The global scrap industry (representatives including the Bureau for International Recycling) started to weigh in and was able to exert more influence than smaller groups (Clapp 1999; O'Neill 2001). Second, there was opposition from countries not listed in Annex VII at that time. These countries—India in the lead—disagreed with a blanket ban on imports destined for recycling. They objected to this "fixed list" approach, rather than finding ways to assess countries by their management and recycling systems. This potentially left (leaves) the Convention open to a dispute under World Trade Organization rules, once the Amendment enters into force (O'Neill and Burns 2005; Wirth 1998).

Yet, it can be argued—then and more recently—that Basel's flaws are related precisely to the prevailing framing of the hazardous waste trade at the time. It certainly responded to an immediate problem, symbolized by the "ships of doom" and the "tip of the iceberg" hypothesis.

However, it did not develop mechanisms and processes to learn from and respond to new information and knowledge about the trade—and the value, or lack thereof—of what was traded, and who could trade it.

Hence, at least at the level of rules, negotiations, and basic principles, the Convention has lacked the facility to learn, especially learning to reframe a problem and ideas on how to cope with it, over and above learning different ways to manage the same problem or implement unchanging goals (Young 1999, 262; roughly equivalent to "single" and "double loop" learning as used in STS and related scholarship (Pahl-Wostl 2009). The Ban Amendment has entered into force, but in its framing (banning shipments between OECD and non-OECD members, and between Parties and Non-Parties), it has not kept up with the more complex realities of today's waste trade. Nor has it been able to effectively regulate the e-waste trade (Lepawsky et al. 2017). This inability to adapt the structures of the regime are also likely to hamper efforts to control plastic waste shipments—except to the extent its normative force can shore up the decisions made by non-OECD member states (such as Malaysia, Thailand, or Vietnam) to stop waste imports into their territory (to the extent they have the capacity to do so).

Complicating the narrative

Even during the 1980s, as the above discussion demonstrates, the hazardous waste trade story was not as straightforward as maintained at the time. Competing narratives around the extent of the waste trade, the routes and directions of the trade all speak to the complexities of agency, risk, resources, and markets in current literature (O'Neill 2019; Lepawsky 2018; Grant and Oteng-Ababio 2016; Minter 2016), and demonstrate that the EJ lens on the global waste trade is necessary but not sufficient to explain why the trade happened. The approach taken by Basel Convention negotiators was not wrong, and certainly reflective of the political times, but it is far from complete, evidenced by later failures to ban the North-South trade, or effectively regulate the trade in discarded electronics.

The narrative most often challenged even at this time was that of the Southern victim/Northern perpetrator. Certainly, when it comes to the type of wastes that were at the forefront of people's minds at that time, the damage done to human health and local ecosystems from dumping the highly toxic ashes, sludge, and so on was considerable. It looked very different from the visual representations of the waste trade today. Given the costs of hazardous waste disposal in OECD countries at the time, what was shipped was often the most toxic materials. However, a different narrative emerges from a close examination of the data available at the time, especially in the Greenpeace Inventory. Many, maybe most, of the transactions listed did not in fact go ahead, especially those targeted in Africa and the Caribbean, and are listed as "refused". In other words, rather like Malaysia and imported plastic waste in 2019, countries like Guinea-Bissau, in the 1980s, were able to stand up and say no to some fairly egregious shipments (Montgomery 1994). These refused shipments challenge conventional wisdom about the extent and nature of the hazardous waste trade during these years.

In a 1995 article, Mark Montgomery challenged the "tip of the iceberg" theory that underpinned anti-waste trade activism at the time. He also used the Greenpeace Inventory, to demonstrate how often it appeared that many countries refused imports. In fact, imports were more likely to be successful when shipped to the Eastern Bloc or to other middle-income countries, rather than the poorest Asian, African, or Caribbean nation in the list. Why this was the case is not clear in every instance. However, local and international activists take some of the credit for whistle-blowing, and in others, governments themselves appear to push back. Or the deal just was not worth it for the exporters in the end. Given the necessary incompleteness of the data on hazardous wastes shipped to poorer countries, we do not know if we are seeing just the tip of the iceberg or the whole thing, but in tandem with the OECD data, it does not appear a tsunami of hazardous wastes was shipped to the Global South in the 1970s and 1980s, not least because governments were able to stand up and say no. Perhaps this has something to do with plenty of willing importers, not least some of the OECD countries themselves.

Also, while North–South waste trading (and dumping) dominated global policymaking in the 1980s and 1990s, it was less well known that much of the hazardous waste trade was legal, and accepted, under national law, by authorities in the importing country. The OECD data cited earlier suggests that back then, legal trade among its members accounted for around 80% of hazardous waste traded globally. This did not, however, mean that the most dangerous wastes went to places that could store, reprocess, or dispose of them most effectively. My first book (O'Neill 2000a; see also O'Neill 1997) addressed the question of why Great Britain became one of the world's largest net importers of hazardous waste (the other largest net importer was France).

Britain was not in a situation to store hazardous wastes safely. Its disposal capacity was low and even for those times, badly outdated. Most wastes were co-mingled and sent to unlined landfills, both highly risky operations. Germany, on the other hand, which had the best disposal facilities and excess capacity at the time, was a large net exporter, even after unification with the former East Germany. This study raised the broader question of why some countries appeared willing to take on more risk than others, and who benefited from that. The answers that I found lay in regulatory systems, industry structures, and state-society relations. The United Kingdom had a fragmented regulatory system but a highly concentrated and politically influential waste industry, and a political system that excluded environmental interests. Germany, on the other hand, had a more hierarchical, precautionary regulatory system, and direct Green Party in-volvement in its federal government. In fact, it seemed paradoxical, perhaps ironic, that one way it dealt with its waste problem was through export.

Either way, those cases and others took understanding the waste trade and import/export decisions into the realm of regulatory systems, political power, and state-society relations. The book also edged into understanding the waste trade in terms of demand rather than supply but did not take this on directly. Hazardous wastes were still framed very much as risks, not po-tential resources.

EU Directives on wastes, waste disposal, and hazardous chemicals transformed waste trade politics and practices among the member states. Germany has taken over as one of the world's largest waste importers, due to its advanced, high-capacity incineration facilities (Ludwig and Schmid 2007), although it is likely not importing toxic waste. Nonetheless, my work and others in this vein (e.g., Wynne 1987; Brickman, Jasanoff and Ilgen 1985; Vogel 1986) situate en-vironmental regulation and toxics/wastes governance deep in national styles and traditions of dealing with risk, and mediating science, policy, and state-society relations. Far from being technocratic, these practices need to be seen through these deeper political frameworks, lessons that can still be applied to today's waste (and used goods) trades across rich and poorer countries alike (Probst and Beierle 1999).

Informing later narratives: a changing waste trade world

Some of the trends that have preoccupied scholars of more recent forms of the waste trade (in the broadest sense) were pre-figured in this early work. These trends include ever more complex trade routes and a booming South-South trade (Lepawsky 2018). A South-South hazardous waste trade in the 1980s and even into the 1990s was rumored, but rarely documented. Wynne cites a case of hazardous waste trade from Singapore to Thailand (Wynne 1989, 126), while the 1999 discovery in Cambodia of barrels of toxic waste from Taiwan unsettled the tenth anniversary of the Basel Convention (O'Neill 2000b).

The global waste trade has been under scrutiny for decades. However, the early work failed by not looking directly at global scrap trades during that time frame, the existence of recycling as well as disposal, and of demand in recipient communities. Perhaps that is in part a function of the types of wastes we focused on, and which the Basel Convention focused on. These original hazardous wastes—the barrels of sludge, piles of incinerator ash, and so on—have been replaced in the public eye by consumer-driven discards—used electronics, post-consumer plastic and paper packaging materials, cars, clothes, and other goods all shipped across borders, often in defiance of the spirit and sometimes the letter of the Basel Convention.

The review and analysis in this chapter point towards several conclusions relevant today. First, there has always been rigorous contestation of prevailing understandings of the routes and directions of waste trading and its impacts. We can still learn from what went before, and apply new data and understanding to pressing transboundary problems of illegal waste dumping (Yang 2020). Understanding these continuities has the potential to open new courses of political action and intervention, at different points of the supply chain and as hazardous waste generation is embedded in domestic and local production and consumption.

Second, we are reminded that despite changing international norms, instances of hazardous waste dumping in this sense have not gone away. From Abidjan (2006) to Beirut (2020), carelessness, malice, and avarice still mean that some of the most toxic substances are still dumped on communities ill-equipped to handle them (Stoett and Omrow 2021).

Third, the "ships of doom" cases outlined in this chapter—the Khian Sea, the Probo Koala, and Rhosus—demonstrate yet another iceberg tip: the complex and unaccountable chains of ownership that can characterize global shipping lines and companies, an understudied component of the global political economy of the waste trade (van Wingerde and Bisschop 2019).

Finally, this narrative points towards the need for further work on phases of the global waste trade that overlap and co-exist with each other, but wax and wane over time (especially in terms of academic and public attention). These include scrap, hazardous, e-waste, plastics and paper, clothes, cars, and yet more: their continued existence, impacts (bad and good), and the governance challenges they pose representing the final iceberg this chapter identifies.

Notes

1 There is also an environmental economics literature on the waste trade, that connects with that discipline's debates around the Pollution Haven Hypothesis, e.g., Copeland (1991), Baggs (2009) and others, and early works on comparative politics of hazardous waste management include Forester and Skinner (1987) and Munton (1996).

2 It should also be noted that few of these studies engage directly with nuclear waste—either high or low level—shipped across borders, even though some of this still occurs. This is in part because nuclear wastes, even from nuclear power, fall into very different realms of global governance, specifically, under the purview of the governments of the nuclear powers, and the International Atomic Energy Agency (see O'Neill 1999; Morrison 2008).

3 Keith Haring (1958–1990) was an American artist who deployed street-style art as advocacy in the 1980s, particularly to advocate safe sex and AIDS awareness, but also broader causes.

4 This was an oddly open time for getting data on sensitive topics like the waste trade. In 1995 or so, I was able to download a whole cache of quite granular data, on waste imports into the United Kingdom, from the Department of the Environment's nascent website. I can only assume it

was up there because they did not think anyone would find it. A few years later, this was no longer the case: the webpage had become the sort of carefully controlled "user-friendly" format we still experience today.

5 One example being damaged tights and pantyhose from West Germany being offered to Poland (Vallette and Spalding 1990, 233). See also Montgomery 1995.

6 There is some question about whether we should use terms "shipment", "trade", or "traffic" to describe transboundary movement of hazardous waste. "Traffic" signals illegal dumping. "Trade" is more general, but still captures the fact that money is changing hands and goods are being moved across borders. "Shipments", approximating the language used in international law, is more general yet, but does not capture the economic transactions that characterize trade and traffic.

7 From a collection of articles about the Khian Sea from www.philly.com, downloaded in 2013 and archived by author.

8 The Bamako Convention went into force in 1998. It had, by the end of 2020, 29 parties and an additional 25 signatories (South Africa is a notable non-signatory).

9 The Waigani Convention's full title is The Convention to Ban the Importation into Forum Island Countries of Hazardous and Radioactive Wastes and to Control the Transboundary Movement of Hazardous wastes within the South Pacific Region. It entered into force in 2001. Radioactive wastes are not covered under Basel.

10 The OECD consists primarily of wealthier countries in the Global North; countries need to apply to become members. A full list of members can be found at www.oecd.org. Note that it includes non-traditional "northern" countries, including Turkey, Colombia, and Mexico.

11 This work was done in a Technical Working Group set up by the Convention secretariat, a process based on expert deliberation and multi-stakeholder involvement, but, as with all such deliberations, vulnerable to influence (Clapp 1999).

References

Agarwal, Anil, and Sunita Narain. 1991. *Global Warming in an Unequal World: A Case of Environmental Colonialism*. Centre for Science and Environment.

Agarwal, Anil, Sunita Narain, Anju Sharma, and Achila Imchen (Eds.). 2001. *Poles Apart: Global Environmental Negotiations 2*. Center for Science and Environment.

Baggs, Jen. 2009. "International Trade in Hazardous Waste." *Review of International Economics* 17, no. 1: 1–16.

BBC News Africa. 2010. "Trafigura Found Guilty of Exporting Toxic Waste." July 22, 2010. https://www.bbc.com/news/world-africa-10735255

BBC News. 2020. "Beirut Explosion: How Ship's Deadly Cargo Ended up at Port." August 7, 2020. https://www.bbc.com/news/world-middle-east-53683082

Brickman, R., S. Jasanoff, and T. Ilgen. 1985. *Controlling Chemicals: The Politics of Regulation in Europe and the United States*. Cornell University Press.

Clapp, Jennifer. 1994. "Africa, NGOs, and the International Toxic Waste Trade." *Journal of Environment and Development* 3, no. 2: 17–46.

Clapp, Jennifer. 1999. *The Global Recycling Industry and Hazardous Waste Trade Facilities*. Paper presented at the Annual Meeting of the International Studies Association, Washington DC.

Clapp, Jennifer. 2001. *Toxic Exports: The Transfer of Hazardous Wastes from Rich to Poor Countries*. Cornell University Press.

Cockburn, Alexander, and Jeffrey St. Clair. 2000. *Five Days That Shook the World: Seattle and Beyond*. Verso.

Copeland, Brian. 1991. "International Trade in Waste Products in the Presence of Illegal Disposal." *Journal of Environmental Economics and Management* 20: 143–162.

Crang, Mike. 2010. "The Death of Great Ships: Photography, Politics, and Waste in the Global Imaginary." *Environment and Planning A* 42: 1084–1102.

Forester, William S., and John H. Skinner (Eds.). 1987. *International Perspectives on Hazardous Waste Management*. Academic Press Inc.

Gille, Zsuzsa. 2016 "Ecological Modernization or Waste-Dependent Development? Hungary's 2010 Red Mud Disaster." In Helmuth Trischler and Odenziel Ruth (Eds.). *Cycling and Recycling*. Berghahn Press.

Gourlay, K.A. 1992. *World of Waste: Dilemmas of Industrial Development*. Zed Books.

Grant, Richard J., and Martin Oteng-Ababio. 2016. "The Global Transformation of Materials and the Emergence of Urban Mining in Accra, Ghana." *Africa Today* 62, no. 4: 2–20.

Greenpeace International and International Federation for Human Rights. 2005. *End of Life Ships: The Human Cost of Breaking Ships*. https://www.fidh.org/IMG/pdf/shipbreaking2005a.pdf.

Hilz, Christoph, and J. R. Ehrenfeld. 1991. "Transboundary Movements of Hazardous Wastes: A Comparative Analysis of Policy Options to Control the International Waste Trade." *International Environmental Affairs* 3, no. 1, 26–63.

Krueger, Jonathan. 1998. "Prior Informed Consent and the Basel Convention: The Hazards of What Isn't Known." *Journal of Environment and Development* 7, no. 2: 115–137.

Krueger, Jonathan. 1999. "What's to Become of Trade in Hazardous Wastes? The Basel Convention One Decade Later." *Environment* 41, no. 9: 10–21.

Kummer, Katharina. 1995. *International Management of Hazardous Wastes: The Basel Convention and Related Legal Rules*. Oxford University Press.

Lange, Sophie. 2020. "A Deal over Dirt: From a German-German Bargain to the Creation of an Environmental Problem in the 1980s." *Worldwide Waste: Journal of Interdisciplinary Studies* 3, no. 1: 1–10.

Laurence, D., and B. Wynne. 1989. "Transporting Waste in the European Community: A Free Market?" *Environment* 31, no. 6: 12–17, 34–35.

Lepawsky, Josh. 2018. *Reassembling Rubbish: Worlding Electronic Waste*. MIT Press.

Lepawsky, Josh, Erin Araujo, John-Michael Davis, and Ramzy Kahhat. 2017. "Best of Two Worlds? Towards Ethical Electronics Repair, Reuse, Repurposing and Recycling." *Geoforum* 81: 87–99.

Liboiron, Max. 2018. "The What and the Why of Discard Studies." *Discard Studies*. https://discardstudies.com/2018/09/01/the-what-and-the-why-of-discard-studies/

Ludwig, Udo, and Barbara Schmid. 2007. "Burning the World's Waste: Germany's Booming Incineration Industry." *Der Spiegel,* February 21.

MacBride, Samantha. 2012. *Recycling Reconsidered: The Present Failure and Future Promise of Environmental Action in the United States*. MIT Press.

Minter, Adam. 2016. "How We Think of E-Wastes Is in Need of Repair." *Anthropocene Magazine*. August.

Minter, Adam. 2020. *Secondhand: Travels in the New Global Garage Sale*. Bloomsbury Publishing.

Montgomery, Mark A. 1994. *Politics and Sustainable Development: Guinea-Bissau and Hazardous Waste Imports*. Presented at the Annual Meeting of the American Political Science Association, New York City.

Montgomery, Mark A. 1995. "Reassessing the Waste Trade Crisis: What Do We Really Know?" *Journal of Environment and Development* 4, no. 1: 1–28.

Morrison, Kerri. 2018. "National and Multinational Strategies for Radioactive Waste Disposal." *Environmental Law Reporter News & Analysis* 47, no. 4: 10300–10311.

Munton, Don, ed. 1996. *Hazardous Waste Siting and Democratic Choice*. Georgetown University Press.

O'Neill, Kate. 1997. "Regulations as Arbiters of Risk: Great Britain, Germany, and the Hazardous Waste Trade in Western Europe." *International Studies Quarterly* 41, no. 4: 687–718.

O'Neill, Kate. 1998 "Out of the Backyard: The Problems of Hazardous Waste Management at a Global Level." *Journal of Environment and Development* 7, no. 2: 138–163.

O'Neill, Kate. 1999. "International Nuclear Waste Transportation: Flashpoints, Lessons and Controversies." *Environment: Science and Policy for Sustainable Development* 41, no. 4: 12–15, 34–39.

O'Neill, Kate. 2000a. *Waste Trading among Rich Nations: Building a New Theory of Environmental Regulation*. MIT Press.

O'Neill, Kate. 2000b. "Managing Hazardous Waste: The Global Challenge." *Environment* 42, no. 3: 43–44.

O'Neill, Kate. 2001. "The Changing Nature of Global Waste Management for the 21st Century: A Mixed Blessing?" *Global Environmental Politics* 1, no. 1: 77–98.

O'Neill, Kate. 2017a. "Will China's Crackdown on 'Foreign Garbage' Force Wealthy Countries to Recycle More of Their Own Waste?" *The Conversation,* December 13 2017.

O'Neill, Kate. 2017b. *The Environment and International Relations*. 2nd Edition. Cambridge University Press.

O'Neill, Kate. 2019. *Waste*. Polity Press.

O'Neill, Kate, and William C. G. Burns. 2005. "Trade Liberalization and Global Environmental Governance: The Potential for Conflict". In Dauvergne Peter (Ed.) *Handbook of Global Environmental Governance*. Edward Elgar.

Organization for Economic Cooperation and Development (OECD). 1985. *Transfrontier Movements of Hazardous Wastes*. OECD Study 82. OECD.

Organization for Economic Cooperation and Development (OECD). 1993 *Transfrontier Movements of Hazardous Wastes, 1989–90 Statistics*. OECD.

Organization for Economic Cooperation and Development (OECD). 1994. *Transfrontier Movements of Hazardous Wastes, 1991 Statistics*. OECD.

Organization for Economic Cooperation and Development (OECD). 1997. *Transfrontier Movements of Hazardous Wastes: 1992–93 Statistics*. OECD.

Pahl-Wostl, Claudia. 2009. "A Conceptual Framework for Analyzing Adaptive Capacity and Multi-Level Learning Processes in Resource Governance Regimes." *Global Environmental Change* 19: 354–365.

Pellow, David Naguib. 2007. *Resisting Global Toxics: Transnational Movements for Environmental Justice*. MIT Press.

Polgreen, Lydia, and Marlise Simons. 2006. "Global sludge ends in tragedy for Ivory Coast." *The New York Times*, September 28.

Princen, Thomas. 2002. "Distancing: Consumption and the Severing of Feedback." In Thomas Princen, Michael F. Maniates, and Ken Conca (Eds.) *Confronting Consumption*. MIT Press.

Probst, Katherine N., and Thomas C. Beierle. 1999. "Hazardous Waste Management: Lessons from Eight Countries." *Environment* 41, no. 9: 22–30.

Stoett, Peter, and Delon Alain Omrow. 2021. "The Transnationalization of Hazardous Waste." In Stoett Peter and Delon Alain Omrow (Eds.) *Spheres of Transnational Ecoviolence: Environmental Crime, Human Security, and Justice*. Springer International Publishing.

Strohm, Laura. 1993. "The Environmental Politics of the International Waste Trade." *Journal of Environment and Development* 2, no. 2: 129–153.

Third World Network. 1989. *Toxic Terror: The Dumping of Hazardous Wastes in the Third World*. Third World Network.

US EPA (n.d.), "Defining Hazardous Waste: Listed, Characteristic and Mixed Radiological Wastes." https://www.epa.gov/hw/defining-hazardous-waste-listed-characteristic-and-mixed-radiological-wastes

Vallette, Jim, and Heather Spalding. 1990. *The International Trade in Hazardous Wastes: A Greenpeace Inventory*. 5th Edition. Greenpeace International Waste Trade Project.

van Wingerde, Karin, and Lieselot Bisschop. 2019. "Waste Away. Examining Systemic Drivers of Global Waste Trafficking Based on a Comparative Analysis of Two Dutch Cases." *Erasmus Law Review* 4: n/a

Vogel, David. 1986. *National Styles of Regulation: Environmental Policy in Great Britain and the United States*. Cornell University Press.

Wilson, David C. 2007. "Development Drivers for Waste Management." *Waste Management & Research* 25: 198–207.

Wirth, David A. 1998. "Trade Implications of the Basel Convention Amendment Banning North-South Trade in Hazardous Wastes." *RECIEL* 7, no. 3: 237–248.

World Commission on Environment and Development (WCED). 1987. *Our Common Future*. Oxford University Press.

Wynne, Brian (Ed.). 1987. *Risk Management and Hazardous Waste: Implementation and the Dialectics of Credibility*. Springer-Verlag.

Wynne, Brian (Ed.). 1989. "The Toxic Waste Trade: International Regulatory Issues and Options." *Third World Quarterly* 11, no. 3: 120–146.

Yang, Shiming. 2020. "Trade for the Environment: Transboundary Hazardous Waste Movements after the Basel Convention." *Review of Policy Research* 37, no. 5: 713–738.

Young, Oran R. (Ed.). 1999. *The Effectiveness of International Environmental Regimes: Causal Connections and Behavioral Mechanisms*. MIT Press.

18

OIL WASTING: THE NECROAESTHETICS OF ENERGY EXPENDITURE

Amanda Boetzkes

If waste has become a primary access into the study of environment and society it is because it has emerged as the seat of the global economy. Which is to say that the global economy is bound to processes of energy harvesting, accumulation, and expenditure that destine the planet for ecological disaster while profiting from that disaster. Waste drives energy use in ways that consolidate globalization as a determining planetary force. In this chapter, I will consider the wasting of a key resource by which this operation becomes explicit: oil wasting. While I will discuss the global energy economy in order to give some precision to oil as a specific resource commodity that has been vigorously pursued since the mid-20th century to the point of protracted periods of active warfare, it is the operation of oil wasting specifically that throws the logic of this economy into view. I therefore address oil wasting by shifting a perspective from the economy's need for oil consumption toward its practices of oil expenditure.

I attune my analysis of oil wasting to the conditions of visibility furnished by the aesthetics of contemporary art. Through an examination of *Behind the Sun*, a video by Kuwaiti-artist Monira Al Qadiri, I will chart a distinction between consuming and expending energy—economy and ecology—that allows us to rethink the concept of waste itself. Through George Bataille's theorization of energy economies, I chart a fundamental distinction between consuming and expending energy that invites us to think critically about the in-strumentalization of oil waste. I argue that since the postwar era, the capitalist drive to waste oil has transformed energy consumption into a more broad, but wholly destructive mode of energy expenditure. The difference between the consumption of an energy resource for profit and the expenditure of energy as waste arises precisely in the gap between human economy and planetary ecology. This gap is, however made visible and materially relevant when we study oil wasting at the site of its toxic environmental effects. Oil waste thus throws our understanding of economic stability into question entirely. It also makes visible the coextensiveness of human emiseration and planetary destruction that founds the substructure of the global economy.

Political theorist, Achille Mbembe suggests that sites of destructive warfare are an integral part of the exertion of power in such a way that individual subjects are besieged in the service of a broader project of population management. Elaborating Michel Foucault's concept of bio-power, Mbembe's concept of *necropower* posits that death itself has been put to work in the

DOI: 10.4324/9781003019077-18

pursuit of a paradoxical conception of sovereignty by which freedom is achieved in and through death. Where for Foucault, populations are set towards the thriving of life through the affirmation of reproductive subjectivities, for Mbembe, régimes of global necropower identify populations of people who are disposable, and whose besieged bodies or whose death becomes the primary ground against which life is defined as such. Thus, he draws from Bataille to explain that "…although it destroys what was to be, obliterates what was supposed to continue being, and reduces to nothing the individual who takes it, death does not come down to the pure annihilation of being. Rather, it is essentially self-consciousness.…"[1] Death is an absolute that lies beyond the horizon of meaning. It is the ultimate excess, an anti-economy that binds the political to a promise of freedom in and through death. Necropolitics therefore appears as a "spiral transgression", a difference that disorients the very idea of a limit.[2] Such paradoxes shape the conspicuous superfluity of waste in the age of global oil wars and the aesthetics of art that attempt to capture this specific visibility.

Behind the Sun: the burning of the Kuwait oil fields

In her ten-minute video, *Behind the Sun*, Kuwaiti artist Monira Al Qadiri screens her filmed VHS footage of the burning of the Kuwait oil fields in 1991 (Figure 18.1). This momentous act of willful destruction—the burning of the limited resource over which the United States went to war with Iraq, intervening when the latter country occupied Kuwait—points to the contradictory process of wasting that underpins the global oil economy. When Iraqi troops were forced by American forces to retreat from Kuwait, they ignited over 700 Kuwaiti oil wells as well as the adjacent open oil lakes. The Iraqi army's scorched-earth strategy of withdrawal was a last ditch effort to display its command over the oil economy by wasting that resource. It was also a tactical decision to pollute the air in order to obscure further American air strike targets. The burn lasted for ten months, at a rate of over 6 million barrels of oil a day. The spoilage of land and air was therefore instrumentalized to preserve Iraq's market power and control of the territory. The burning of oil secured the transformation of oil from a material resource into a virtual form of profit while it quite literally obstructed the U.S. army's view of the area.

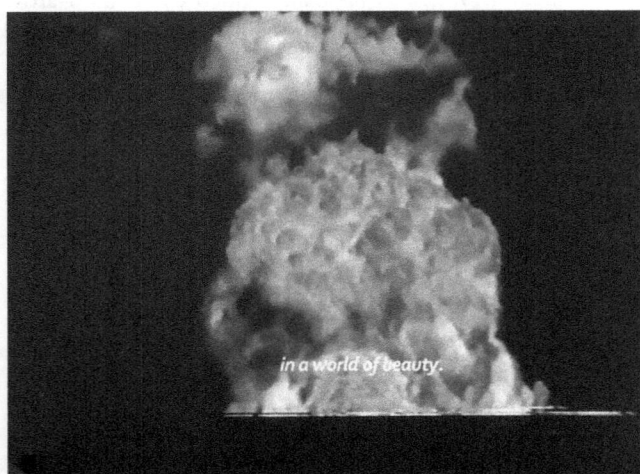

Figure 18.1 Monira Al Qadiri, behind the sun, single-channel video, 2013.

Al Qadiri's video captures how this event appeared to be nothing short of an apocalyptic vision. She captured scenes both from a distance as a glowing light akin to a sunset on the horizon, and more immediately footage from the road near the area, filling the camera frame with the seemingly endless clouds of thick, black smoke that billowed from the fiery land. From these different perspectives, the sight of the burning fields oscillates between the terrifying appearance of a volcanic eruption or the mushroom cloud of a nuclear bomb, and a series of beautiful abstractions—shimmering orange light on the water of a nearby lake or the bio-morphic form of a rose with intermingling red, orange and pink flame rising up against a velvety black backdrop. It is at the edge of this contradiction between earthly catastrophe and natural beauty that Al Qadiri overlaid the footage with a deep male voice reading Arabic poetry. The poetry therefore draws out a mutation at play within the video's imagery. Using the enchanting tones of the voice to conjure the genre of religious media in which imagery of waterfalls and forests are paired with recitations from the Koran to effect spiritual mysticism and religious authority, the video pairs popular natural spiritualism with a harrowing inversion: a wholly destructive oil war.

Behind the Sun points to two mutually corroborating contentions regarding the burning of the Kuwait oil fields: first, that acts of willful destruction have a visual efficacy in the strategizing of warfare, and second that they can be read as aesthetic expressions of the necropolitical machinations that fuel the oil economy and its imaginary. The video is a document of both the *use* of apocalyptic visions of oil wasting for war and the *exchange* of oil wasting from an act of war into a cultural sensibility. War, environmental destruction, economy, and art therefore converge in this scene of glorious expenditure. Further, as I will suggest, the video corroborates the integration of necropolitcs into cultural life.

Energy expenditure in the age of oil

I invoke the concept of *glorious expenditure* in reference to the writings of Georges Bataille, the mid-20th-century poet and theorist. In the 1940s, Bataille published *The Accursed Share*, a transhistorical account of economy from the lens of energy expenditure.[3] Bataille suggests that organized societies inevitably produce a surplus of resources; excess energy which it gathers in order to expand. Inevitably, that expanding society is driven to use this excess energy in rituals of conscious and openly wasting which he calls glorious expenditure or *la dépense*. But if that society does not burn that excess, the energy will burst forth through the unpredictable disasters that result from the suppression of its inherent drive to waste. In pre-capitalist economies, such as those of the Aztecs or Northwest Coast First Nations, societies invoked strategies of energy expenditure such as sacrificial offerings, the ritualistic destruction of objects of value, and even the production and consumption of art. In restricted economies, such as those that appeared under the competing political régimes of the 20th century, this excess energy is expended in the form of destructive spoilage, such as war and the atom bomb. His critique of capitalism is that its refusal to expend energy—its reingestion of energy in the form of profit—generated a cycle of consumption and stockpiling of energy that was driven towards ever more totalizing forms of destruction as the economy expands into a global form.

Bataille's theory of economy has inspired a number of recapitulations that draw out the relevance of energy economies to the politics of oil and ecological subject positions. Allan Stoekl, for example, suggests that Bataille's understanding of energy expenditure has a bearing on our concepts of sustainability and how these are foreclosed by the oil economy.[4] Noting that oil is both a scarce resource and yet a primary fuel of the world market, he notes from a Bataillean perspective that the oil industry is in the perilous situation of relying on the consumption of a scarce resource while consistently waging war over it, driving towards its own destruction. This is the hallmark of a restricted economy: the consumption of energy with no avenue for expenditure except in the form of anarchical destruction.

Stoekl foregrounds how Bataille's analysis of energy economies locates an ethics of systemic balance in and through transgressive operations of energy expenditure. For example, the ritualized sacrifice of people and objects in non-Western societies was geared towards the violent crossing of the limits of the individual subject and its private property through the generous donation of energy to stimulate social mobility. This would retroactively effect a location and preservation of the limits of the social organization as a totality but allow the overturn of hierarchy within that organization. But the rise of capitalism put all energy to productive use with no avenue to expend surplus energy. It therefore hypostatized the individual, class hierarchy, and energy slavery in the service of the accumulation of profit with no culture of energy expenditure. Capitalism eliminated the possibility of a truly generous burn-off of energy and instead drove itself toward an infinitely expanding form that has breached the systemic carrying capacities and led to more problematic forms of destruction. The rise of nuclear energy and the Cold War was a symptom of this continual need to expend energy precisely because of the fortification of the economic prohibition to do so. Further, nuclear energy was not merely a marker of the globalization of this energy economy, it threatened catastrophe on a planetary scale.

This dynamic now animates the oil economy, the primary energy resource driving our current era. Consider the colossal expenditures of oil at stake in the two Iraq Wars, or the Deepwater Horizon Oil Spill in the Gulf of Mexico in 2010, when a leak in the seafloor pipeline of the massive offshore drilling project caused an explosion that killed 11 people, and then a three-month long gush of hundreds of millions of gallons of crude oil into the Gulf. As Matthew Huber has readily shown, oil is integral to the sociotechnological character of a global capitalist energy regime.[5] But expenditure is not merely a matter of the spoliation of abstract energy; the oil economy expends lives and ecological life itself as a means to perpetuate itself and expand. Further, the expenditure of oil is neither accidental nor unconscious. Indeed, oil expenditure has been put to work in such a way that economic force can be gleaned from the destruction of lives and ecologies.

Necropolitics and the wasting of lives

Achille Mbembe recovers Bataille's reflections on waste economies to examine the current age of necropolitics, a predicament in which war becomes the primary objective of power since it is through war that lives and populations can be enslaved, tortured, and killed for profit. Through Mbembe's account of necropolitics not only does the intimate tie between race, slavery, and the disposability of lives become clear, but so also does the logic of necropolitical sites such as plantations, concentration camps, genocides, and terrorism begin to take shape. For example, he argues that the plantation is both a political-juridical structure and a space in which death itself is wielded on the body of the slave as a continuous threat and presence of the power of the sovereign and the economy of the master-slave relation. The slave is alive but is always kept in a state of injury, in a "phantom-life world of horrors and intense cruelty and profanity… Violence, here becomes an element in manners, like whipping or taking of the slave's life itself … Slave life, in many ways, is a form of death-in-life.…"[6] The concatenation of power, terror, siege, and race extended from the plantation to the colonies, and ultimately to what Mbembe describes as the late-modern colonial state.

In this form, necropower generates mass expulsions, the resettlement of stateless people in refugee camps and infrastructural warfare on those very sites—the demolition of homes, resources, communications systems, and roadways in order to reterritorialize the area. As the colonial power dominates the occupied territory, its protracted state of siege obscures the difference between an internal and external enemy. Keeping with the logic of the colony, occupation marks the colonized as "savages" with whom the colonizers cohabit; where they resist, they are criminalized rather than identified as external combatants. Under the governance of the new state, the population is starved of resources and income and eventually it is yielded to a condition of disposability, or what Giorgio Agamben calls, bare life. In this regard, Mbembe's argument resounds with Michael Mcintyre and Heidi Nast's concept of (bio)necropolitics, which details the way in which capitalism cultivates and profits from impoverished labor power through techniques of emiserating and torturing surplus populations of racialized peoples such as migrant workers, Indigenous peoples, and refugees.[7] The production of a necropolis, a domain of expropriated, alienated, and socially dead people is the outcome of the biopolis of sovereign European land and subjects. The instrumentalization of killing in the service of the reproduction of European life is cast out of the colonial consciousness. Yet death itself pervades the besieged bodies and territories of the late-modern colonial states outside of Europe.

Death as the wasting of life becomes an engine by which land is besieged, marked, and in the case of the Kuwaiti oil fields, scorched so that even profitability itself is sacrificed for the abstraction of a necropolitical sovereign, discovered in and through the wielding of death. Death is both mobile and distributed; as Mbembe points out, necropolitical wars are "hit and run" affairs rather than strategies of acquisition and conquest.[8] This paradoxical form of headless sovereignty—death as sovereignty in and of itself—achieved new technological heights during the Gulf War when the United States initiated a program of precision targeting of the land from the air, so called "surgical strikes" using bombs covered with depleted uranium, laser-guided missiles, asphyxiation bombs, and unmanned aerial vehicles. The language of precision striking was touted as a way to obscure the extent of the damage and deaths as well as the generalized brutalization waged against the civilian population. It was not the case that the enemy could be yielded as an isolated target, but rather that an entire population had been concentrated and neutralized for disposability.

Here is where Mbembe arrives at the important reflection that the logic of martyrdom and the logic of survival confront one another paradoxically but nevertheless productively within the paradigm of necropolitics.[9] Once the spatialization of death-in-life has been established in besieged areas—he takes Gaza as a prime example—bodies can be primed for a double coding as both homicidal weapon and a suicidal claiming of death as freedom. The terrorization of the population marks the body in such a way that within the imaginary, the only possibility of subjective survival is in and through death, in and through an acceptance of a struggle with the enemy that will culminate in the killing of both simultaneously. This "new semiosis of killing" constellates in the suicide bomber who takes hold of the siege by cloaking the body in the very materials of war and transforming the body into its very necrotic topography in which ballistic weaponry and flesh, occupied and occupier, ordinary scene, and catastrophic event of death combine and scatter into indistinguishable debris.

Necroaesthetics and the glorious expenditure of oil waste

Mbembe's exposition on late-modern colonization via Bataille's conception of death and the expenditure of life as avenue to sovereignty allows us to consider the contradictory position of land and sacrifice at play in the burning of the Kuwait oil fields. Further, it invites a consideration of the aesthetics of oil wasting in more specific terms. For what we see in Al Qadiri's video is not just a documentary, it is an aesthetic formulation of the logic of death-in-life that pervaded the event. Mbembe argues that the necropolitical logic intertwines with aesthetic form in and through the besieged body of the slave. Once the slave has been sealed off within the spatial paradigm of the plantation, they have entered a state of exemption by which their body exists as a pure instrument of production, yet is nevertheless able to stylize almost any object, language or gesture into a performance. "The slave is able to demonstrate the protean capabilities of the human bond through music and the very body that was supposedly possessed by another".[10] Thus, the state of subjugation is also an aesthetic capability and avenue into a paradoxical form of plasticity.

However, that very plasticity is an extension of a cultural sensibility in which cruelty toward the enslaved has become a form of play conditioned for the pleasure of the master. Necroaesthetics thus appear as a superfluous visibility, a remarkable burst of beauty and life in a field of misery and exploitation. This flourish points to the spatial interpenetration of the necropolis and the biopolis and the sadistic relation that subtends them. It also marks a divergence from the historic spatial paradigm by which, as anthropologists following Mary Douglas have duly noted, societies eliminate their waste to zones that become charged with the danger of taboo. Yet, as Thomas Eriksen and Elisabeth Schober chart, within the totalizing regime of neoliberal globalization and in the context of widespread ecological crisis, there is no space for waste, rather spheres of social coexistence with it.[11] The new visibility of necroaesthetics is therefore a symptom of the collapse of boundaries between the necropolis and the biopolis at stake in the ecological crisis of oil wasting.

In this respect we can reflect on the entanglement of the imagery of burning oil and the metaphors of natural beauty at play in Al Qadiri's video. As the rich voice embellishes the apocalyptic scenes with its homage to nature—sparkling stars, the burst of spring, the quivering of new life—nature as such and its violent destruction are fused. While the flames undulate against the black sky and the voice summons the image of roses, nature is not only exposed as an abstract ideal but converted into its opposite, an exuberant display of a totalizing end of life. The possession of land in and through its destruction is revealed as an aesthetic expression and corroboration of the global imaginary. The ruinous topography of burning oil seemingly celebrates ecological death as the pinnacle of energy expenditure.

Al Qadiri's video invites the viewer to take consciousness of glorious expenditure as a necroaesthetic form. For the burning of the Kuwait oil fields was not an accidental or unconscious expenditure. It was as much a military tactic as it was an economic maneuver by which to seize control over a natural resource. Paradoxically, power was exerted through a willful planetary destruction, in a logic born of global pressures and intimidation. Expenditure, profit, victory, and beauty are bound together through the rhetorical unfolding of the imagery. Yet she invites us to recuperate an ethical reflection at the site of transgression. While the scenes are strangely compelling, Al Qadiri lays bare the perversity of destructive oil expenditure. Her video enacts before our eyes the transformation from an appreciation of natural beauty to an appreciation of its death. While the logic of necropower might be subsumed into the language of war operations, Al Qadiri draws out its fundamental underpinning in the sensations of glory and ecstasy in the event of sacrifice.

Beneath the Sun was exhibited at MoMA PS1 in the exhibition *Theater of Operations, The Gulf Wars 1991–2011*. The exhibition convened works by Iraqi, Kuwaiti, and Palestinian artists alongside works by American and Europeans. The exhibition captured the destruction of civilian lives and land, both from the perspective of those civilians under siege and Americans who sought to contest and expose the deceitful visual régime of the American government over the course of the Gulf War and the Iraq War. The American necroaesthetic sensibility registers through video, paint, photography, film, installation, and sound. Over the 20-year period, artists parlayed the blood, carnage, fire, besieged, and dead bodies into a shared resistance to the American government's concealment of their attacks on civilian populations and their land under the guise of "surgical strikes" and covert strategic operations. As the artists all show through their respective media, the destruction in Iraq and Kuwait was a bloodbath—a total wasting of land and life. *Behind the Sun* encapsulates this willful destruction in the event of the Kuwait oil fields, but it is through the context of the exhibition, which displays the necropolitics of America's oil wars on Iraq, that we understand a subtle reversal of that event. While it seems a scandal that Iraq would ignite this coveted resource in an effort to scorch the earth, at the same time there can be no doubt that this waste was the outcome of the U.S. siege. The beauty and quasi-mystical danger that Al Qadiri visualizes is the aesthetic outburst characteristic of the sadistic sensibility of the prevailing necropolitical power, the administrations of George Bush Sr. and George W. Bush.

I have briefly charted a movement from the concept of energy expenditure as a form of systemic balance, to the tragic negligence of the need to waste well that results from the global oil economy's instrumentalization of oil wasting. The case of the Kuwait oil fires is a privileged instance of the usefulness of oil expenditure in the context of a necropolitical régime, whereby power is gained through destroying lives and ecological life. If, as Al Qadiri's video shows us, this regime is vindicated through its pleasure in planetary death, then we might wonder about other forms of ecological wasting at the hands of the oil economy, such as those created by carbon emissions and global warming. Might we take our cue from her visualization of a necroaesthetics to seek out the logic of oil expenditure in the beauty and violence of ecological destruction?

Notes

1 Achille Mbembe, *Necropolitics*, trans. Steven Corcoran (Durham, NC: Duke University Press, 2019).
2 Ibid., 16.
3 Georges Bataille, *The Accursed Share: An Essay on General Economy. Volume 1 Consumption*, trans. Robert Hurley (New York: Zone Books, 1991).
4 Allan Stoekl, *Bataille's Peak: Energy, Religion, and Postsustainability* (Minneapolis: University of Minnesota Press, 2007).
5 Matthew T. Huber, *Lifeblood: Oil, Freedom, and the Forces of Capital* (Minneapolis: University of Minnesota Press, 2013).
6 Mbembe, 21.
7 Michael McIntyre and Heidi Nast, "Bio(necro)polis: Marx, Surplus Populations and the Spatial Dialectics of Reproduction and Race", *Antipode* 43.5 (2011): 1465–1488.
8 Mbembe, 30.
9 Ibid., 39.
10 Ibid., 22.
11 Thomas Hylland Eriksen and Heidi Schober, "Waste and the Superfluous: an Introduction", *Social Anthropology* 25.3 (2017): 282–287.

References

Bataille, Georges. 1991. *The Accursed Share: An Essay on General Economy. Volume 1 Consumption.* Translated by Robert Hurley. Zone Books.
Eriksen, Thomas Hylland, and Heidi Schober. 2017. "Waste and the Superfluous: An Introduction." *Social Anthropology* 25, no. 3: 282–287.

Huber, Matthew T. 2013. *Lifeblood: Oil, Freedom, and the Forces of Capital*. University of Minnesota Press.

Mbembe, Achille. 2003. "Necropolitics." *Public Culture* 15, no. 1: 11–40.

Mbembe, Achille. 2019. *Necropolitics*. Translated by Steven Corcoran. Duke University Press.

Mcintyre, Michael, and Heidi Nast. 2011. "Bio(necro)polis: Marx, Surplus Populations and the Spatial Dialectics of Reproduction and Race." *Antipode* 43, no. 5: 1465–1488.

Stoekl, Allan. 2007. *Bataille's Peak: Energy, Religion, and Postsustainability*. University of Minnesota Press.

19

WASTEPICKER ORGANIZATIONS AND URBAN SUSTAINABILITY

Jutta Gutberlet

Introduction

Cities are core drivers of waste problems, and are also places, where significant change and innovation can happen. The problematic realities that emerge from handling or mishandling waste obviously require robust solutions. Local governments have important long-term decisions to make on how to best deal with complex waste issues. Major impacts in the Global South cities are related to the lack of waste collection in informal settlements and the deficiencies or absence of separate waste collection to recover the organic fraction for composting or biogas production and the inorganic fraction for material recycling. Given these deficiencies, neoliberal approaches have surfaced with privatizing municipal waste management and developing Public Private Partnerships (PPPs), usually to finance high-tech solutions in many parts of the world (Latendresse and Bornstein 2017). These waste management regimes frequently dismiss local waste-pickers who are already working with recycling. Sophisticated solutions for municipal solid waste management, often developed in industrialized countries (e.g., highly automated material separation, incinerators, compactor trucks), have limited applications in the social, economic, institutional, and legal context of the Global South (Guerrero, Maas and Hogland 2013). In this chapter, I will argue that locally developed solutions, grounded in social and economic inclusion, are more effective in addressing local and regional waste management, while also attaining social and environmental aims.

In many parts of the world, waste-pickers are already the major recycling force, particularly when formal recycling programs are lacking (Chaturvedi and Gidwani 2010; Gutberlet 2008, 2016; Linzner and Lange 2013; Medina 2000; Wilson, Velis and Cheeseman 2006, 2012; Samson 2009; Scheinberg 2011). An estimated 15–20 million people are operating globally as waste-pickers, collecting recyclables informally, by filtering them out at curbside collection, at dumpsters or landfills (ILO 2013). Just in Latin America and the Caribbean there are up to 3.8 million waste-pickers, most of them working independently (Terraza and Sturzenegger 2010). While prejudice and stigma affect the livelihoods of these workers and their families, waste-pickers worldwide are building networks, partnerships, and coalitions, gaining more visibility. Particularly, in view of waste-related impacts on climate change, resource recovery schemes involving waste-pickers have received more attention by academia, international organizations, NGOs, and some governments, as greenhouse saving waste management options (Lau et al. 2016; Gutberlet, Besen, and Moraes 2020).

There is a potential to improve selective waste collection and recycling in the city, to increase material recovery rates, levels of reuse, and material diversion into the circular economy with the inclusion of organized waste-pickers (Aparcana 2017; Ezeah, Fazakerley, and Roberts 2013). Yet, integration and formalization of waste-pickers does not necessarily imply a transition from precarity and poverty to formal privilege (O'Hare 2020). Millar (2014) questions whether waste-pickers were actually better off when working at dumps than after a neoliberal shutdown of landfills with the inclusion of only a few of these workers into a recycling plant. Increased exclusion as a result of landfill closure is a very common situation, reported also in other locations (Almeida et al. 2016). Policies, funding, and programs that secure the livelihoods of all waste-pickers when transitioning from the landfill to recycling plants need to be secured. Formalization, unfortunately, can also mean falling through the cracks, particularly when waste-pickers no longer have access to waste and see their livelihoods threatened. The literature also shows that there are significant hurdles to the integration of waste-pickers in formal waste management, often involving tedious bureaucratic requirements and expensive fees (e.g., to regularize and maintain the status of waste-picker organizations) (Colombijn and Morbidini 2017). Adequate supportive, inclusive public policies, programs, and structures have to be in place to create opportunities for waste-pickers to improve their livelihoods and to enhance selective waste collection and expand the circular economy.

Innovations from waste-pickers have enabled better selective waste collection programs in many cities. These programs have generated more social and economic inclusion, and have contributed to building better communities and a cleaner environment (Dias 2016; Gutberlet 2016, 2018). While hurdles and shortcomings still remain, there is a noticeable momentum of rising policy urgency related to waste issues, which calls for extended producer responsibility, more efficient waste management practices, and which pushes for the integration of waste-pickers in municipal waste management. Innovative environmental waste governance builds on governments partnering with waste-picker organizations and co-producing collection and diversion services for recyclable resources. These initiatives clearly also have the potential to address some of the Sustainable Development Goals (SDGs) (UNDP 2020), making cities more livable and communities more inclusive (Gupta and Vegelin 2016).

This chapter starts by contextualizing waste-picker organizations in Brazil and by introducing public policies that support the collective organization of waste-pickers and their integration into municipal waste management targeting household waste. Then, the concept of grassroots social innovations is presented, focusing specifically on the experiences of waste-pickers, informed by the literature and by empirical ethnographic and community-based research, consisting of extensive and embedded qualitative fieldwork (Fox 2003; Studdert and Walkerdine 2016; Wallerstein et al. 2019) conducted over the past decade, mostly in Brazil. This chapter continues with a discussion about the transformative effects of social innovations on cities, communities, and environments and the diverse practices and policy engagements involving waste-pickers through their organizations. These positive impacts are also translated into SDGs (Gutberlet 2021). The final section concludes with the proposal of inclusive waste governance for more sustainable cities.

Member-based organizations (MBOs) of waste-pickers

Informal waste-pickers are defined as individuals or enterprises who are involved in private-sector recycling and waste management activities (primarily geared towards household waste and

sometimes also industrial waste) that are not sponsored, financed, recognized, supported, organized, or acknowledged by the formal solid waste authorities, or which sometimes operate in
violation of or in competition with formal authorities (IWWG-Task Group Informal Recycling).
Waste-pickers in the informal sector work in diverse ways. They may operate independently;
collect and separate recyclable materials on their own or with their families (children and elderly
included); collect and separate at landfills, dumps, or in streets; or they might have established
routes and specific clients (households, businesses, offices) to regularly collect materials from.
Some waste-pickers also work for scrap dealers, who provide them with the space for separation,
besides variable other benefits but also dependencies. All informal waste-pickers sell their daily
collection to so-called intermediaries or middlemen (scrap dealers, recycling shops).

In some countries (e.g., Argentina, Brazil, Colombia, Nicaragua) informal waste-pickers
have organized in MBOs, specifically cooperatives, associations, federations, and networks. In
other parts of the world (e.g., India), they have created unions and are integrated as informal
workers in the formal collection, separation, and recycling of waste (Linzner and Lange 2013).

In Brazil, the number of waste-pickers is estimated between 400,000 and 600,000 (IPEA
2013). The 2010 census (IBGE 2012) counts 387,910 self-declared waste-pickers. There could
be significant underreporting, given the stigma attached to this activity. The census data sheds
some light on the socio-economic characteristics of this category. In 2010, the average age of
waste-pickers in Brazil was 39 years, the majority were Afro-descendants (66%), only 25% had
completed their basic education, 20% were considered illiterate and 31% of all informal waste-
pickers were female (IBGE 2012). According to IBGE, 38.6% of the waste-pickers in Brazil
were already organized in MBOs (mostly cooperatives and associations). See Figure 19.1 for an
example of a waste-picker cooperative in Brazil. Among organized waste-pickers, the number
of women is usually slightly higher (between 50 and 60%) compared to men.

Figure 19.1 Avemare, a waste-picker cooperative in Santana de Parnaiba. Photo by Jutta Gutberlet.

The goal of MBOs is to work collectively, to establish democratic deliberations, and to thrive for emancipation and inclusion. However, not all groups follow the economic practice of self-management, where the workers are the owners of production, and where decisions are made collectively. MBOs are complex and can also be contradictory. In her research on grassroots organizations in Latin America. Gago (2017), for example, has observed what she called "neoliberalism from below" induced by historic paternalistic government control and implicit oppression. "Neoliberalism is articulated with communitarian forms, with popular tactics for making a living, enterprises that drive informal networks, and modes of negotiating rights that rely on the workers' economic strategies to negotiate the expansion of those rights" (Gago 2017, 19). In his work, Paulo Freire provides an explanation for these challenges. He explains that oppression is domesticating and "[o]ne of the gravest obstacles to the achievement of liberation is that oppressive reality absorbs those within it and thereby acts to submerge human beings' consciousness" (Freire 2005, 52). Not all organized waste-pickers have been exposed to the "liberating" concept presented in pedagogy of the oppressed (Freire 2005), which asserts that the world can only be transformed by means of praxis: the reflection and action upon ones' everyday experiences.

Formal waste-picker organizations are relatively easily traced and supervised by the state. Taxation and legal obligations (e.g., formalizing the cooperative and maintaining the documentation up-to-date) as well as regulatory requirements (e.g., abiding to occupational health regulations and safety norms), are often bureaucratic, time consuming and expensive duties, difficult for cooperatives to comply with (Do Carmo and de Oliveira 2010). The relatively high taxes waste-picker cooperatives have to pay also constitute an impediment for their survival; a tax reform taking into consideration the popular economy is needed.

There are many benefits from being part of a MBO, these can include social and health benefits, access to educational programs for themselves and their children, access to municipal and state funding for infrastructure projects, safer working conditions, as well as networking opportunities with NGOs and other social actors. These organizations also usually bring an improved social status to waste-pickers. They mostly work in uniforms, access workplace health protective equipment, as cooperative member they hold an ID card, and have access to basic sanitation and kitchen facilities allowing them to have a warm meal at work. They take pride in their work and present themselves as professionals in the waste management sector. The benefits and challenges of becoming integrated into formal waste management have been discussed by many authors and for different geographic contexts (Wilson, Velis, and Cheeseman 2006; Asim, Batool, and Chaudhry 2012; Velis et al. 2012, 2017; Scheinberg 2012; Dias and Samson 2016; Gutberlet and Carenzo 2020).

Worldwide, different waste-picker organizations and networks provide innovative formulations on how to manage waste in a better way, how to promote participatory waste management, and how to improve the livelihoods of waste-pickers, disrupting existing assumptions and preconceived ideas about them. Many of the leaders of these local or regional networks come together forming a social movement. In Brazil, waste-pickers congregate in a national movement, the *Movimento Nacional de Catadores(as) de Materiais Recicláveis*—MNCR (MNCR, 2012). The MNCR currently lists 58 affiliated cooperatives in the city of São Paulo and 95 in the metropolitan region of São Paulo (a region that spans 39 municipalities and hosts 22.2 million inhabitants). There are currently six networks and one Federation (Fepacore) established in this metropolitan region (MNCR 2020). The movement is also part of the Latin American recyclers' movement (RedLacre—*Rede Recicladores*) and of the virtual global network of waste-picker organizations (Global Alliance of Waste Pickers—*GlobalRec*), which is supported by WIEGO (Women in Informal Employment: Globalizing and Organizing) and brings together 17 different countries, mainly in Latin America, Asia, and Africa. The Brazilian waste-pickers' movement MNCR was created in 2004, with the goal to expand inclusive solid waste management programs throughout the country and to integrate the struggle of waste-pickers for self-determination and inclusion in the praxis of handling and diverting solid waste, which also meant for them better access to funding and credit lines. The mission of the MNCR is to increase the recognition of waste-pickers and to contribute to the construction of fair and sustainable societies based on the socio-productive inclusion of waste-pickers. Socio-productive inclusion represents more than just access to the market, but rather includes access to economic and social rights, which consists of funding and appropriate financing, continuous technical assistance, and social security coverage. The concept implies an intersectoral articulation and an integrated perspective of livelihoods and basic needs (Figueiredo and Bastos 2019).

The MNCR has been able to actively influence government legislation and institutional structures for social inclusion (Campos 2014). The principles that guide this social movement are direct actions, class independence, solidarity, direct democracy, and mutual support (MNCR 2020). Waste-pickers in Brazil, as in many other countries, demand remuneration for the public services they provide. With the COVID-19 pandemic, waste-pickers are also demanding to be recognized as "essential service provider" (Gutberlet 2020).

The MNCR has played an important role in the elaboration of federal Law 12,305/2010, the National Solid Waste Policy (*Política Nacional de Resíduos Sólidos PNRS*) (Brazil 2010), which among other aspects establishes waste-picker associations and cooperatives as key actors in selective waste collection. Since 2002, the Ministry of Labor recognizes the profession "waste-picker" (*catador*) in the Brazilian Code of Occupations (Besen and Dias 2011); which represents a first step of recognizing this category. By means of the Federal Policy on Public Sanitation Services (Law 11,445), launched in 2007, governments are allowed to contract the services of waste-picker cooperatives without the need of going through a bidding process. To stimulate reverse logistics, the national legislation supports the establishment of sector-wide agreements to achieve set recycling targets (Campos 2014). Over the past decade, extended producer responsibility (EPR) projects began to involve waste-picker organizations. Since 2007, the National Secretariat of Solidarity Economy (*Secretaria Nacional de Economia Solidária*—SENAES) implemented specific funding programs for waste-pickers, such as the federal program *Cataforte* until 2016, when President Dilma Rousseff was impeached. The national bank foundation *Fundação Banco do Brasil*, the state-owned oil company *Petrobras* and other institutions have also funded specific projects with waste-pickers. The social and solidarity economy (SSE) policies in Brazil have helped reduce poverty at the national level (Caruana and Srnec 2013). These policies were vital also to the consolidation of the MNCR and their member organizations, enhancing their visibility and stimulating the creation of networks.

Regionally, waste-picker MBOs have started to organize as networks. Motta (2017) who studied three waste-picker networks in the state of São Paulo concluded that networks are generally born with a commercial focus and then evolve different additional tasks. While cooperation, limited to instrumental and economic purposes, is not sufficient, collective commercialization certainly facilitates network consolidation. Networks evolve into important organizational arrangements in the face of the complex challenges presented by the recycling market and the local and regional political landscapes (Aquino, Castilho, and Pires 2009; Boeira, Campos, and Ferreira 2007).

Grassroots social innovations in waste management

Seyfang and Smith have coined the term "grassroots innovations" to "describe networks of activists and organizations generating novel bottom-up solutions for sustainable development; solutions that respond to the local situation and the interests and values of the communities involved" (2007, 585). Grassroots social innovations are bottom-up, reflect ordinary peoples' everyday life experiences and struggles and involve democratic processes, actively engaging community members in the design, development or production of an innovation, to benefit their community or the public at large, by bringing social change which is developed, approved, and owned by the grassroots. Social innovations can be linked to a technology, a strategy, and a governance practice. This involves "any type of collaborative social undertaking that is organized at the local community level, has a high degree of participatory decision-making and flat hierarchies" (Grabs et al. 2016, 100).

Grassroots social innovations are common among waste-picker organizations, and involve the exploration of alternative models in waste management, low-cost technologies, adapted infrastructure to improve working conditions, changed administrative procedures or alternative internal governance structures. Often these innovations have an impact on wider local governance outcomes when it comes to waste management. Since the late 1990s, social and technological innovations have brought important changes to municipal waste governance in many countries, particularly creating alternative forms of MBOs, allowing for new models in waste management to emerge (Aquino, Castilho and Pires 2009; Boeira, Campos and Ferreira 2007).

In Brazil, the collective organization of waste-pickers through networks is a social innovation that has helped increase the income, allowing waste-picker cooperatives to sell directly to the recycling industry, avoiding middlemen. By compiling specific materials from different groups and then selling them collectively, the network *Rede Solidária CataVida,* e.g., was able to sell white High-Density Polyethylene (PEHD) in 2019 at R$1.80 to the industry, while middlemen were paying only R$0.50 to R$0.80 (Gutberlet 2021).

Other innovative responsibilities of waste-picker networks include capacity-building activities for their members (e.g., related to cooperative values, work safety, waste management policy, waste minimization), negotiation with local governments to establish formal contracts, project development, fundraising, environmental education involving the wider community, helping with internal conflict resolution or inspections, and certifications at the cooperative level. Sometimes the networks also provide consultancy on waste management to governments, formal recycling centers, or businesses and commerce. Network representatives participate in municipal council meetings, public hearings, or municipal and regional working groups to contribute to the redesign of public policy in waste management. Networks have greater bargaining power, often allowing them to negotiate municipal contracts for individual-associated cooperatives (Gutberlet 2021).

Successful networks can also develop innovative technologies. *Rede Solidária CataVida,* in the metropolitan region of São Paulo, e.g., started a polymer factory, recovering different types of plastics and producing pellets for the recycling industry. They also have adapted a process to collect and transform cooking oil into soap and animal feed. Waste-pickers can do more than just collect and separate different materials; they also develop appropriate technologies to transform resources and make products of higher value.

Smith, Fressoli, and Thomas (2014) highlight the fact that grassroots movements seek socially inclusive knowledge, processes, and outcomes geared towards local communities. Waste-picker cooperatives and associations are autonomous worker-run organizations, with an elected administrative board, following democratic principles in decision making, through regular and extraordinary general assemblies, where all participants have a vote. Waste-picker cooperatives and associations provide services to their communities, by doing door-to-door household collection and running volunteer drop-off points for recyclables in the community and often also through participating in environmental education activities.

Grabs et al. (2016) describe the different levels (individual and societal) that influence successful grassroots innovations. At the individual level, those involved in the innovation process act as role models, providing the social stimulus for the development of the innovation. We can find numerous natural leaders among organized waste-pickers, many of whom are women, leaders such as Aline Souza from *Centercoop* in Brasilia, Maria Monica da Silva, from *CooperFenix* in Diadema, or Ionara Pereira dos Santos from *Avemare* in Santana de Parnaíba, who act as outstanding role models for members of this social movement of waste-pickers in Brazil. Each one of them is a leader in their cooperative as well as an activist raising awareness for the social and environmental contributions made by their group. Such leaders show resilience and perseverance in the struggle for recognition and redistribution, standing up for their category as well as for broader environmental issues, such as the current fight against waste incineration.

MBOs promote community actions, engagement, and support, which increases the capacity to expand the innovation. These leaders are typically well connected to their community, and are recognized not only within their cooperative as more than just leaders and co-workers. The backing of an innovation by the general public is essential for change to become sustained. For example, the regular performance of separate household collection of recyclables by a local cooperative, needs to gain the trust and acceptance of the community in order for this service to be expanded and disseminate environmental awareness about recycling and zero waste issues.

In Mogi das Cruzes, e.g., *Rede Catasampa* provides services to the municipal recycling program *Reciclamogi*, a resilient program that has survived several political transitions. The waste-pickers receive a fixed value for the collection and sorting based on operation costs. They further benefit from subsidized transport, food, health coverage and social benefits. Their monthly income is around one minimum salary (R$1,163.55, approximately US$300, based on 2019 exchange rate) plus benefits (which is above the average earning of other cooperatives in the region). A similar case in the region is *Rede Verde Sustentável*, that has supported the contract negotiations for *Avemare,* in Santana de Paraiba, who is now providing selective waste collection services to the city. Another cooperative, *CMR* in Itapevi and part of *Rede Verde Sustentável* has established a partnership with the *Instituto Eurofarma* for environmental education projects, they started out with 5 schools and now support 44 schools, giving talks to students and collecting recyclable material at the schools, who have created voluntary drop-off centers for their neighborhoods. Not always are waste-pickers remunerated for these educational services. Sometimes municipalities make a fixed contract with the local waste-picker cooperative which includes collection and diversion as well as community outreach for education.

There are also limitations to the process of bottom-up social innovations, primarily linked to the vulnerable situations of grassroots members in terms of unstable employment, low income, racial inequality, social exclusion, low education levels, lack of access to social services, and difficult access or insufficient health care. Other barriers to successful grassroots innovations are the overall precarious and vulnerable living conditions and locations of their members, the lack of recognition, and even stigmatization as well as the little support provided by community and local governments reinforcing prejudice. Widespread stigma against waste-pickers is one of the key causes preventing these individuals from making even larger contributions to resource diversion at municipal but also industrial levels. Another reason is described as "lack of political will" to integrate waste-pickers into municipal waste management despite existing inclusive policy frameworks. Community support is crucial for an innovation to develop. Those waste-pickers who were able to build relationships with the local community, e.g., through door-to-door collection; also have stronger local voices. It is also about continuity, persistence, and engagement, involving long-term learning and adaptation to change. Organized waste-pickers are remarkably resilient, given the lack of resources, the volatile economic and financial environment, and the oppressive political circumstances under which they often have to operate. In light of the SDGs that can be tackled with the socio-productive inclusion of waste-pickers, the essentiality of the service provided by these workers has to be recognized (Gutberlet 2020).

Waste-pickers tackling sustainable development goals

Similar to other international organizations (UN 2019; 2015; WBGU 2016), the 2016 UN Habitat World Cities Report stresses that "the new urban agenda should promote sustainable cities and human settlements that are environmentally sustainable and resilient, socially inclusive, safe and violence-free, economically productive; and better connected to and contributing towards sustained rural transformation" (UN Habitat, 2016, 2). The 2030 Agenda for Sustainable Development focuses on 17 SDGs to address sustainability transition (Gupta and Vegelin 2016; Le Blanc 2015) and inclusive waste governance creates opportunities to tackle some of these goals. For example, by establishing fair service contracts with waste-picker organizations, remunerating the environmental and community services they provide on a daily basis, goal number one, to reduce poverty, can be addressed. Those groups that have established contracts with the city are in better economic shape than groups whose income only depends on the resale value of the recyclables, which is subject to market fluctuations and exploitation by middlemen.

Despite the influence of Social and Solidarity Economy (SSE) policies in Brazil (Singer 2009), there is not enough political will in most Brazilian cities to socially innovate waste management; particularly now that many of the social policies and institutions have been dismantled under the current government of Jair Bolsonaro. Municipal contracts are a pathway to providing income security. Without the support of local governments, and without paying a fair price for the work of waste-pickers the aim of eradicating poverty will not be reached. In addition to income security, the International Labor Organization ILO (2002) underlines the necessity for representation security for decent working conditions. The better the representation, e.g., by the national waste-pickers movement (MNCR), or waste-picker networks, the more powerful waste-pickers are in negotiating contracts with the government.

Many international organizations recognize that "the cooperative enterprise is the type of organization that best meets all dimensions of reducing poverty and exclusion" (Wanyama 2014, 59). This has to do with the collective and value-based approach cooperatives take, that can empower the disadvantaged to fight for their rights and interests, and that provides job security. The mandate of waste-picker networks includes improving the working conditions and outcomes (work flow organization, work efficiency and effectiveness, health and safety at work, etc.), following labor laws and regulations, helping implement "decent working" conditions, thus tackling SDG number eight (UNDP 2020).

The organized work of waste-pickers also contributes to increasing gender equality. The cooperative space in Brazil usually attracts more women than men (Dias 2002; INSEA 2007).

It is an environment that affirms culturally shared female identities, allowing them to value themselves, develop self-esteem and provide opportunities for personal growth, empowerment, and leadership development (Dias and Ogando 2015). Self-management (*autogestão*) of the cooperative changes work relations that are particularly empowering for women. The cooperative is a space that generates collective consciousness; it enables women to become leaders and to develop their skills. Women value the social relations within the working space, thus helping to avoid alienation (Gambina and Roffinelli 2013). Cooperatives contribute towards gender equality, not just by increasing female membership and providing them with an income, but also by expanding the opportunities for women to empower themselves, to engage in capacity development and education, as well as expanding their leadership skills and helping other women. However, many of these women still remain in vulnerable situations (single parent, abusive relationships, poverty, illiteracy, ill health, etc.). Being part of an MBO facilitates the implementation of specific educational and skill building programs by the government or by non-governmental organizations to reach the vulnerable population.

SDG Goal number 11 is about making cities and human settlements inclusive, safe, resilient, and sustainable. Adequate waste management is a pre-condition for environmental and human health, particularly in densely populated areas. The goal specifically targets the reduction of the adverse environmental impacts of cities, by transitioning to circular economy. This goal puts emphasis on the removal and the adequate destination of waste, reduction, reuse, and recycling. Waste-pickers have the knowledge of making recommendations to the public, e.g., on materials that are not recyclable and are contaminating the recycling stream, thus further congesting landfills, or new packaging trends they observe over time. Some multi-layered packaging, such as multi-layered PET bottles used as milk containers, e.g., currently have no market, causing an environmental hazard and a nuisance for the waste-pickers who find them during material separation. The following figures (Figures 19.2 and 19.3) exemplify the problem posed by non-recyclable materials at the cooperative level. Every day, they separate these bottles from the rest of the material, without being able to sell them and having to stock them on their already limited space in the cooperative. Waste-picker organizations can play an

Figure 19.2 Some of the brands that use aseptic PET for the packaging of milk drinks. Photo by Jutta Gutberlet.

Figure 19.3   Bales of milk bottles at Cooper Viva Bem, in São Paulo that have no customer because of the composite materials. Photo by Jutta Gutberlet.

important role in educating and making consumers more aware about problematic materials and their recyclability, which is a direct contribution to waste reduction.

Selective waste collection and environmental education services provided by organized waste-pickers bring great benefits to local communities, by helping increase resource recovery, reduce the waste of materials, and diminish the amount of waste deposited in landfills. Particularly during door-to-door collection, waste-pickers have the opportunity to speak to household members and can make them aware of ways to reduce the waste of resources, e.g., with clean source separation or the avoidance of buying products that come in certain packaging. Waste-pickers clean up cities and thus their work reduces the risk of water logging and flooding as well as the spread of diseases and consequent risks to human and ecological health. Neighborhoods that are regularly serviced by waste-pickers tend to have less recyclables littering the curbside (such as PET plastic bottles), which could potentially cause water logging. Trained waste-pickers also pick up those materials with value which have been randomly thrown away, ending up in the drainage system.

Waste-picker networks participate in different city forums and contribute to municipal waste management by sharing new ideas and their experiences. Waste-pickers have been adamant in questioning waste incineration, speaking up in public audiences and seminars. Waste-pickers already engage in community work and environmental awareness campaigns for zero waste and responsible consumption, thus also addressing SDG number 12 (UNDP 2020). They give presentations at schools, universities, or businesses; participate in conferences, workshops, and seminars; make videos; participate in art projects; and disseminate environmental education materials on 3Rs and how to best separate for the circular economy. Particularly when performing door-to-door collection of recyclables, they create opportunities to engage with neighbors explaining the need for resource recovery. They are active in discussions on responsible production and consumption and are a central piece in the current reverse logistics programs with industry sectors (Ferri, Chaves, and Ribeiro 2015). As exemplified earlier, waste-pickers know the recycling market and they are able to identify materials and packaging with no market value that are not recycled. Currently many products, and the packages that contain them, are sold to consumers without enough knowledge as to how, or even if, they can be recycled. Organized waste-pickers could be the whistle-blowers, helping society hold businesses accountable for the waste they create. Unrecyclable materials, from Styrofoam to multi-layered plastics, often pile up in waste-picker cooperatives, carefully bundled, while they receive no reward for collecting and separating these products, that occupy their space.

Final considerations

For cities to transition into more sustainable urban agglomerations, the inclusion of grassroots actors, such as waste-pickers, is essential (Lemos and Agrawal 2006). This means shifting the practice towards inclusive and environmental governance, changing rules, regulations and public policies, decision making, and behaviors. The growing amounts of urban waste generated worldwide and the increased socioeconomic disparity due to prevailing waste management systems within cities and across regions are consistent challenges of global concern. While waste-pickers are already key protagonists, working in the circular economy, current development trends in waste governance further exclude them from waste management solutions. Waste-to-Energy, promoted by transnational companies, and implemented via public-private partnerships, are currently competing with small-scale, grassroots, and community-based recycling, as exemplified for Brazil (Gutberlet, Bramryd, and Johansson 2020).

Waste management is a critical public service and waste that is not collected or littered creates serious environmental and human health problems, also affecting the global community (leaking into lakes and oceans; contaminating soils, animals, and consequently our food chains; and generating GHG emissions that affect the climate). Waste governance delimitates strategic goals and guiding principles, applies legal and financial instruments, institutional arrangements, procedures and modes of operation, and involves different actors to address waste reduction and resource recovery (Kooiman 2003). Particularly polycentric (Andersson and Ostrom 2008) and multilevel governance (Benz 2007; Kübler and Pagano 2012) are non-hierarchical and complementary approaches involving separately constituted bodies with overlapping jurisdictions, with the potential to enhance innovations and stimulate cooperation for better outcomes and innovative learning (Skelcher 2005). By tailoring governance to the specific local circumstances, involving local actors (e.g., waste-picker representatives) and by enhancing accountability and transparency, more trust is developed within the community, creating the ability to address big challenges, such as waste and poverty.

The process of governing waste involves the articulation of different local and regional institutions, practices, and actors that collaborate with national government partners and other value chain actors, in order to collectively constitute *local governance actor* groups. In the city of São Paulo, for example, such a group advises on the design and implementation of projects and is consulted on changes in policies and waste systems (Besen et al. 2014; Besen and Fracalanza 2016;Fracalanza and Besen 2016). Waste-pickers are represented in this group through network leaders and representatives from the MNCR, specific forums (e.g., *Forum Lixo e Cidadania*), and committees (local waste management committees). Participation in these groups and forums is voluntary and the participants are not compensated for their consulting activity. The resources required to participate have to come from the collective of waste-pickers (e.g., networks are funded through monthly membership fees of the co-operatives, or the National Movement is funded through projects and partnerships). Resources are always scarce, which poses a limit to the actions of waste-pickers. Funding should be in place for waste-picker representatives to fully participate in those important decision-making spaces.

Partnerships established between waste-picker cooperatives (often via the networks) and the city or specific industries, to deliver waste services and many other activities related to the value chain of waste, are another example for multilevel and polycentric governance structures at the municipal level, allowing for waste reduction, collection, and diversion as well as socio-productive inclusion. In Argentina, e.g., the waste-picker cooperative *Reciclando Sueños* has signed a contract with a chemical industry, to collect and divert their recyclable waste into the circular economy. They were able to significantly reduce the amount of industrial waste dumped on landfills, increasing waste diversion, and thus saving the industry money and contributing to a better environmental standard (Gutberlet and Carenzo 2020). An example from Brazil shows how the waste-picker network *Coopcent-ABC* has established a contract with the Brazilian Association of Cosmetics (ABIHPEC), and is responsible for collecting and separating the packaging produced by that industry, helping the industry to comply with their reverse logistics requirements, established by Brazil's reverse logistics policy (Gutberlet and Carenzo 2020).

Innovative forms of governance are decentralized, participatory, and inclusive, focused on waste reduction, resource recovery, and the circular economy. It also addresses poverty reduction, builds community resilience, and increases environmental sustainability. Optimizing municipal selective waste collection and recycling reduces final disposal in landfills (Rada, Ragazzi, and Fredizzi 2013), which is crucial for large urbanized areas where landfill space is scarce. The integration of waste-pickers in municipal selective waste collection programs not only increases recycling rates and drives the circular economy but also reduces poverty, which is in line with achieving SDGs. The diversity of materials in solid waste and the varying degree of recyclability and compostability requires specific forms of management and regulations, making governance more complex, insofar as it also involves different segments and actors. Greater synergies between governments and civil society organizations are needed, since the lack of respect and profound prejudice against waste-pickers still debilitates the dialogue between the multiple actors, preventing effective collaboration in waste management.

Research shows that social movements are experts and they are a key ingredient of ecological transformation of society. In addition to social movements' confrontational and resistance activities (particularly related to a shift towards waste-to-energy), these grassroots have become expert consultants for government and industry. However, with the advent of hyper-neoliberal governments whose policies heavily reduce the state's investment and involvement in public service provision, the proposed change in their roles is severely held back.

Further research with waste-pickers is needed to provide evidence and demonstrate the long-term and broad benefits from integrating waste-pickers in waste management, where they can fulfill different roles as diverters and waste ambassadors. Many waste-pickers would like to explore different tasks, involving responsibilities that go beyond collecting, separating, and selling materials towards transforming, adding value (for recycling, refurbishing, and reuse), and consulting (educating the public, government, and industry). Exploratory research could focus on finding these economic niches for waste-pickers. Furthermore, MBOs play a crucial social role in welcoming and supporting informal workers. There are very few other low-barrier job opportunities available for socially and economically excluded and often highly vulnerable individuals. The quality and extent of the livelihoods provided by waste-picker groups is still under-researched. Particularly women are often taking the lead in accommodating individuals with health and addiction conditions, living in poverty, or are difficult to employ (e.g., due to criminal records, migrant background). The social and economic benefits for cities from these contributions of waste-picker cooperatives are to be better assessed and made visible. The more socioeconomic data available on waste-pickers, the better their valuable services can be supported with adequate policy, technology, and governance measures.

References

Almeida, E. T. V. de, A. C. Balthazar, A. P. Echkardt, F. A. Cordeiro, S. Figueiredo, N. Donato, M. P. R. Paes, M. L. Gomes, and dos Santos. 2016. "Protagonismo e esfera pública em campos dos Goytacazes: a trajetória recente dos catadores do lixão da Codin." In B. C. Jaquetto Pereira and F. Lira Goes (Eds.) *Catadores de materiais recicláveis: um encontro nacional*. Instituto de Pesquisa Econômica Aplicada (Ipea). 201–216. ISBN 978-85-7811-267-7.

Andersson, K. P., and E. Ostrom. 2008. "Analyzing Decentralized Resource Regimes From a Polycentric Perspective." *Policy Sciences* 41: 71–93.

Aparcana, S. 2017. "Approaches to Formalization of the Informal Waste Sector Into Municipal Solid Waste Management Systems in Low- and Middle-Income Countries: Review of Barriers and Success Factors." *Waste Management*, 61: 593–607.

Aquino, I. F., A. B. Castilho Jr., and T. S. D. L. A. Pires. 2009. "Organização em rede dos catadores de materiais recicláveis na cadeia produtiva reversa de pós-consumo da região da grande florianópolis: uma alternativa de agregação de valor." *Gestão and Produção* 16, no. 1: 15–24.

Asim, M., S. A. Batool, and M. N. Chaudhry. 2012. "Scavengers and their role in the recycling of waste in Southwestern Lahore." *Resources, Conservation and Recycling*, 58: 152–162.

Benz, A. 2007. "Multilevel Governance." In A. Benz et al. (Eds.) *Handbuch Governance. Theoretische Grundlagen und empirische Anwendungsfelder*. Verlag für Sozialwissenschaften. 297–310.

Besen, G. R. and S. M. Dias. 2011. "Gestão Pública Sustentável de Resíduos Sólidos—uso de bases de dados oficiais e de indicadores de sustentabilidade." *Revista Pegada*, Special Issue, July 2011.

Besen, G. R., H. Ribeiro, W. M. R. Gunther, and P. R. Jacobi. 2014. "Selective Waste Collection in the São Paulo Metropolitan Region: Impacts of the National Solid Waste Policy." *Ambiente and Sociedade* XVII, no. 3: 253–272. doi:10.1590/S1414-753X2014000300015.

Besen, G. R., and A. P. Fracalanza. 2016. "Challenges for the Sustainable Management of Municipal Solid Waste in Brazil." *Journal disP – The Planning Review* 52, no. 2: 45–52. doi:10.1080/02513625.2016. 1195583.

Boeira, S. L., L. M. D. S. Campos, and E. Ferreira. 2007. "Redes de catadores- recicladores de resíduos em contextos nacional e local: do gerencialismo instrumental à gestão da complexidade?" *Organizações and Sociedade* 14, no. 43: 37–55.

Brazil. 2010. Política Nacional de Resíduos Sólidos Lei n° 12.305, 02.08.2010. http://www.planalto.gov. br/ccivil_03/_ato20072010/2010/lei/l12305.html. Accessed February 17, 2021.

Bulkeley, H., M. Watson, R. Hudson, and P. Weaver. 2005. "Governing Municipal Waste: Towards a New Analytical Framework." *Journal of Environmental Policy and Planning* 7, no. 1: 1–23. doi:10.1080/15239080500251700.

Campos, H. K. T. 2014. "Recycling in Brazil: Challenges and Prospects." *Resources, Conservation and Recycling* 85: 130–138.

Caruana, M., and C. Srnec. 2013. "Public Policies Addressed to the Social and Solidarity Economy in South America. Toward a New Model?" *Voluntas* 24, no. 3: 713–732.

Chaturvedi, B., and V. Gidwani. 2010. "The Right to Waste: Informal Sector Recyclers and Struggles for Social Justice in Post-Reform Urban India". *India's New Economic Policy: A Critical Analysis* 15: 125–153.

Colombijn, F., and M. Morbidini. 2017. "Pros and Cons of the Formation of Waste-Pickers' Cooperatives: A Comparison Between Brazil and Indonesia." *Decision* 44, no. 2: 91–101.

Dias, S., and M. Samson. 2016. *Informal Economy Monitoring Study Sector Report: Waste Pickers.* WIEGO. https://www.wiego.org/sites/default/files/publications/files/Dias-Samson-IEMS-Waste-Picker-Sector-Report.pdf. Accessed February 2, 2021.

Dias, S. M. 2002. *Construindo a cidadania: avanços e limites do Projeto de Coleta Seletiva em parceria com a ASMARE.* Master Thesis. Instituto de Geociências, Universidade Federal de Minas Gerais, Belo Horizonte.

Dias, S. M. 2016. "Waste Pickers and Cities." *Environment and Urbanization* 28, no. 2:375–390. doi:10.1177/0956247816657302.

Dias, S. M., and A. C. Ogando. 2015. *Engendering Waste Pickers Cooperatives in Brazil.* ILO Research Conference on Cooperatives and the World of Work, Antalya, Turkey, 9–10 November 2015. http://www.wiego.org/sites/default/files/publications/files/Dias-Engendering-Wastepicker-Cooperatives-Brazil.pdf. Accessed February 17, 2021).

Do Carmo M. S., and J. A. P. de Oliveira. 2010. "The Semantics of Garbage and the Organization of the Recyclers: Implementation Challenges for Establishing Recycling Co-Operatives in the City of Rio de Janeiro, Brazil." *Resources, Conservation and Recycling* 54: 1261–1268.

Ezeah, C., J. A. Fazakerley, and C. L. Roberts. 2013. "Emerging Trends in Informal Sector Recycling in Developing and Transition Countries." *Waste Management* 33: 2509–2519.

Figueiredo, F. F., and V. P. Bastos. 2019. "Política Nacional de Resíduos Sólidos, indústria da reciclagem e a inclusão sócio-produtiva de catadores: um debate socioambiental." https://registro.alasperu2019.pe/Static/archivos/0454befd8c787566182361eca8fa88968c526821.pdf. Accessed February17, 2021.

Ferri, G. L., G. L. D. Chaves, and G. M. Ribeiro. 2015. "Reverse Logistics Network for Municipal Solid Waste Management: The Inclusion of Waste Pickers as a Brazilian Legal Requirement." *Waste Management* 40: 173–191.

Fox, N. J. 2003. "Practice-based Evidence: Towards Collaborative and Transgressive Research." *Sociology* 37, no. 1:81–102. doi:10.1177/0038038503037001388.

Fracalanza, A. P., and G. R. Besen. 2016. "Challenges for the Sustainable Management of Municipal Solid Waste in Brazil." *Journal disP—The Planning Review* 52, no. 2: 49–56.

Freire, P. 2005. *Pedagogy of the Oppressed* (30th anniversary edition). The Continuum International Publishing Group Inc.

Gago, V. 2017. *Neoliberalism from Below: Popular Pragmatics and Baroque Economies.* Duke University Press.

Gambina, J., and G. Roffinelli. 2013. "Building Alternatives Beyond Capitalism". In C. Piñeiro Harnecker (Ed.) *Cooperatives and Socialism, A View from Cuba.* Palgrave Macmillan. 46–59.

Gerometta, J., H. Haussermann, and G. Longo. 2005. "Social Innovation and Civil Society in Urban Governance: Strategies for an Inclusive City." *Urban Studies* 42, no. 11: 2007–2021.

Grabs, J., N. Langen, G. Maschkowski, and N. Schapke. 2016. "Understanding Role Models for Change: A Multilevel Analysis of Success Factors of Grassroots Initiatives for Sustainable Consumption." *Journal for Cleaner Products* 134: 98–111. doi:10.1016/j.jclepro.2015.10.061.

Guerrero, L. A., G. Maas, and W. Hogland. 2013. "Solid Waste Management Challenges for Cities in Developing Countries." *Waste Management* 33: 220–232.

Gupta, J., and C. Vegelin. 2016. "Sustainable Development Goals and Inclusive Development." *International Environmental Agreements* 16: 433–448.

Gutberlet, J. 2021. "Grassroots Waste Picker Organizations Addressing the UN Sustainable Development Goals." *World Development* 138: 105195. doi:10.1016/j.worlddev.2020.105195.

Gutberlet, J. 2020. "Transforming Cities Globally: Essential Public and Environmental Health Services Provided by Informal Sector Workers." *One Earth* 3, no. 3: 287–289. doi:10.1016/j.oneear.2020.08.018.

Gutberlet, J. 2018. "Waste in the City: Challenges and Opportunities for Urban Agglomerations." In M. Ergen (Ed.) *InTech Urban Agglomeration* (Edited Volume). doi:10.5772/intechopen.72047.

Gutberlet, J. 2016. *Urban Recycling Cooperatives: Building Resilient Communities.* Routledge Taylor and Francis Group.

Gutberlet, J. 2015. "Cooperative Urban Mining in Brazil: Collective Practice in Selective Household Waste Collection and Recycling." *Waste Management* 45:22–33.

Gutberlet, J. 2008. *Recycling Citizenship, Recovering Resources: Urban Poverty Reduction in Latin America.* Ashgate.

Gutberlet, J., G. R. Besen, and L. Moraes. 2020. "Participatory Urban Solid Waste Governance in the Global South." *Detritus* 13: 167–180. 10.31025/2611-4135/2020.14024.

Gutberlet, J., T. Bramryd, and M. Johansson. 2020. "Expansion of the Waste-Based Commodity Frontier: Insights from Sweden and Brazil." *Sustainability* 12, no. 7: 2628. doi:10.3390/su12072628.

Gutberlet, J., and S. Carenzo. 2020. "Waste Pickers at the Heart of the Circular Economy: A Perspective of Inclusive Recycling from the Global South." *Worldwide Waste: Journal of Interdisciplinary Studies* 3, no. 1: 6, 1–14. doi:10.5334/wwwj.50.

IBGE (Instituto Brasileiro de Geografia e Estatistica). 2012. *Pesquisa Nacional por Amostra de Domicílio 2012.* IBGE. https://www.ibge.gov.br/estatisticas/sociais/trabalho/17270-pnad-continua.html. Accessed February 17, 2021.

ILO (International Labour Office). 2002. *Decent Work and the Informal Economy—Report VI presented for the General Discussion at the International Labour Conference 2002, Geneva.* International Labour Office. ISBN 92-2-112429-0. www.ilo.org/public/english/standards/relm/ilc/ilc90/pdf/rep-vi.pdf. Accessed February 17, 2021.

ILO (International Labour Organization). 2013. *The Informal Economy and Decent Work: A Policy Resource Guide Supporting Transitions to Formality.* International Labour Office, Employment Policy Department.

INSEA (Instituto Nenuca de Desenvolvimento Sustentável). 2007. *Perfil sócio-econômico dos catadores da rede CATAUNIDOS—2007.* INSEA/UFMG/FELC.

IPEA (Instituto de Pesquisa Economica Aplicada). 2013. *Situação Social das Catadoras e dos Catadores de Material Reciclável e Reutilizável.* IPEA.

International Waste | Working Group (IWWG)—Task Group Informal Recycling. https://www.tuhh.de/iue/iwwg/task-groups/informal-recycling.html. Accessed February 17, 2021.

Kooiman, J. 2003. *Governing as Governance.* Sage Publication.

Kübler, D., and M. A. Pagano. 2012. "Urban Politics as Multilevel Analysis." In K. Moosberger et al. (Eds.) *The Oxford Handbook of Urban Politics.* Oxford University Press. 114–129.

Latendresse, A., and L. Bornstein. 2017. "Urban Development: Cities and Slums in the Global South." In P. A. Haslam et al. (Eds.) *Introduction to International Development,* Chapter 19, 2nd Edition. Oxford University Press. 357–372.

Lau, A. W. W. Y., Y. Shiran, R. M. Bailey, E. Cook, R. Martin, J. Koskella, ... J. E. Palardy. 2016. An integrated global model and scenario analysis of environmental plastic pollution , 2016–2040.

Le Blanc, D. 2015. "Towards Integration at Last? The Sustainable Development Goals as a Network of Targets." *Sustainable Development* 23, no. 3. doi:10.1002/sd.1582.

Lemos, M. C., and A. Agrawal. 2006. "Environmental Governance." *Annual Review of Environment and Resources* 31: 297–325.

Linzner, R., and U. Lange. 2013. "Role and Size of Informal Sector in Waste Management—a Review." *Resources, Conservation and Recycling* 166, no. 2: 69–83.

Medina, M. 2000. "Scavenger Cooperatives in Asia and Latin America." *Resources, Conservation and Recycling* 31, no. 1:51–69.

Millar, K. 2014. "The Precarious Present: Wageless Labor and Disrupted Life in Rio de Janeiro, Brazil." *Cultural Anthropology* 29, no. 1:32–53.

MNCR (Movimento Nacional dos Catadores de Materiais Recicláveis). 2020. Declaração de princípios e objetivos do MNCR. www.mncr.org.br/ (accessed 17.2.2021).

Motta, V. P. 2017. Dinâmicas de cooperação e a sustentabilidade das redes de cooperativas de catadores de materiais recicláveis: estudo de casos múltiplos.174 f. Dissertação (Mestrado em Administração)—Centro Universitário FEI.

O'Hare, P. 2020. "'We Looked after People Better when We Were Informal': The 'QuasiFormalisation' of Montevideo's Waste-Pickers." *Bulletin of Latin American Research* 39: 53–68.

Rada, E. C., M. Ragazzi, and P. Fredizzi. 2013. "Web-GIS Oriented System Viability for Municipal Solid Waste Selective Collection Optimization in Developed and Transient Economies." *Waste Management* 33: 785–792. doi: 10.1016/j.wasman.2013.01.002.

Samson, M. (Ed.). 2009. *Refusing to be Cast Aside: Waste Pickers Organizing Around the World.* WIEGO.

Scheinberg, A. 2011. *Value Added, Modes of Sustainable Recycling in the Modernisation of Waste Management Systems.* PhD thesis. Wageningen University, 120 pp. https://library.wur.nl/WebQuery/wurpubs/411095.

Scheinberg, A. 2012. *Informal Sector Integration and High Performance Recycling: Evidence from 20 Cities.* WIEGO Working Paper (Urban Policies) N 23 March 2012.

Seyfang, G., and A. Smith. 2007. "Grassroots Innovations for Sustainable Development: Towards a New Research and Policy Agenda." *Environmental Politics* 16, no. 4: 584–603.

Singer, P. 2009. "Relaciones entre sociedad y Estado en la economía solidaria." *Íconos: Revista de Ciencias Sociales* 33: 51–65.

Skelcher, C. 2005. "Jurisdictional Integrity, Polycentrism, and the Design of Democratic Governance." *Governance: An International Journal of Policy, Administration and Institutions* 18, no. 1: 89–110. doi:10.1111/j.1468-0491.2004.00267.x.

Smith, A., M. Fressoli, and H. Thomas. 2014. "Grassroots Innovation Movements: Challenges and Contributions." *Journal of Cleaner Production* 63: 114–124. doi:10.1016/j.jclepro.2012.12.025.

Studdert, D., and V. Walkerdine. 2016. *Rethinking Community Research: Inter-Relationality, Communal Being and Commonality.* Palgrave Macmillian.

Terraza, H., and G. Sturzenegger. 2010. "Dinámicas de organización de los Recicladores informales: três casos de estúdio en América Latina." *Banco Interamericano de Desarrollo, Sector de Infraestructura y Medio Ambiente.* Nota Técnica.

Tirado-Soto, M. M., and Zamberlan, F. L. 2013. "Networks of Recyclable Material Waste-Picker's Cooperatives: An Alternative for the Solid Waste Management in the City of Rio de Janeiro." *Waste Management* 33: 1004–1012.

UNEP (United Nations Environment Programme). 2019. *Global Environment Outlook GEO-6, Healthy Planet, Healthy people.* UNEP. doi:10.1017/9781108627146.

UN (United Nations, Department of Economic and Social Affairs, Population Division). 2015. *World Urbanization Prospects: The 2014 Revision,* (ST/ESA/SER.A/366). https://population.un.org/wup/Publications/Files/WUP2014-Report.pdf, Accessed February17, 2021.

UN (United Nations, Department of Economic and Social Affairs, Population Division). 2019. *World Population Prospects 2019: Highlights* (ST/ESA/SER.A/423). https://population.un.org/wpp/Publications/Files/WPP2019_Highlights.pdf, Accessed February 17, 2021.

UNDP (United Nations Development Program). 2020. "Sustainable Development Goals." https://www.undp.org/content/undp/en/home/sustainable-development-goals/.

UN-HABITAT. 2016. World Cities Report 2016. Urbanization and Development. Emerging Futures. UN-HABITAT. http://wcr.unhabitat.org/, Accessed February17, 2021.

UN-HABITAT (United Nations Human Settlements Programme). 2016. World Cities Report 2016 (HS/038/16E), United Nations Human Settlements Programme (UN-HABITAT), Nairobi, Kenya, ISBN: 978-92-1-132708-3, https://unhabitat.org/sites/default/files/download-manager-files/WCR-2 016-WEB.pdf, Accessed September 27, 2021.

Velis, C. A. 2017. "Waste Pickers in Global South: Informal Recycling Sector in a Circular Economy Era." *Waste Management and Research* 35, no. 4, 329–331.

Velis, C. A., D. C. Wilson, O. Rocca, S. Smith, A. Mavropoulos, and C. R. Cheeseman. 2012. "An Analytical Framework and Tool ('InteRA') for Integrating the Informal Recycling Sector in Waste and Resource Management Systems in Developing Countries." *Waste Management and Research* 30, no. 43: 43–66.

Wallerstein, N., K. Calhoun, M. Eder, J. Kaplow, and C. H. Wilkins. 2019. "Engaging the Community: Community-Based Participatory Research and Team Science." In K. Hall, A. Vogel, and R. Croyle (Eds.) *Strategies for Team Science Success.* Springer. 123–134. doi:10.1007/978-3-030-20992-6_9.

Wanyama, F. O. 2014. *Cooperatives and the Sustainable Development Goals: A contribution to the post-2015 development debate.* ILO.

WBGU (German Advisory Council on Global Change). 2016. *Humanity on the Move: Unlocking the Transformative Power of Cities.* WBGU.

Wilson, D., L. Rodic, A. Scheinberg, C. Velis, and G. Alabaster. 2012. "Comparative Analysis of Solid Waste Management in 20 Cities." *Waste Management and Research* 30, no. 3: 237–254.

Wilson, D., C. Velis, and C. Cheeseman. 2006. "Role of Informal Sector Recycling in Waste Management in Developing Countries." *Habitat International* 30: 797–808.

20

WASTE, LABOR, AND LIVELIHOODS IN SOUTH AFRICA

Mary Lawhon, Nate Millington, and Kathleen Stokes

Introduction

Waste flows, like so many material flows in South Africa, are shaped by socio-political processes and deeply entwined with apartheid histories and ongoing inequality. South African waste studies draw on ideas from across the Global Northern and Southern literatures, often situating South Africa as a middle-income country where wealth and poverty coexist. High rates of (often conspicuous) consumption contribute to the production of valuable discarded material. High unemployment and a relatively strong labor movement contribute to diverging views over labor, including disagreements over who works and how work is organized and compensated. Apartheid spatial policies created sprawling cities with many un- and under-serviced areas, and post-apartheid municipalities initially sought to expand services in the midst of global trends towards the privatization of services. A long-standing environmental movement increasingly incorporates environmental justice demands and connects with global social movements, but in practice often struggles to fuse old and new problem frames. The extremes of South Africa thus produce spaces where waste processes and politics are both fraught and full of possibility.

There is growing interest in rethinking and reworking waste flows and waste politics in South Africa, making it an important site for waste research. Such changes are happening in conjunction with inquiries into the appropriate role of the state and its relationship to the economy. The South African state is often considered developmental: it seeks to guide and shape the market to produce growth and, at times, economic justice, although what this means in practice is inconsistent (Freund 2007; Turok 2010; Parnell and Robinson 2012). Waste is increasingly framed as part of a green economy, an underexploited resource waiting to be rendered productive. Various state-organized and state-supported recycling efforts emphasize economic growth, waste minimization, skills development, and/or increased employment. Ongoing interventions are neither conventional expansions of the state nor a straightforward story of neoliberal privatization; instead, the primary ambition is for the state to support private actors as they grow an economically viable, multi-racial recycling sector. These changes at times reinforce and at times challenge different imaginaries of what (good) infrastructure is and what types of infrastructure are (financially, materially, politically) possible.

DOI: 10.4324/9781003019077-20

Much waste scholarship in South Africa, as elsewhere in the Global South, has focused on understanding and emphasizing the politics of informal waste reclaiming and critiqued the privatization of municipal services. Ongoing efforts to reshape waste flows do not only exclude already marginalized reclaimers and outsource labor: they also seek to implement different understandings of the appropriate relationship between technology, labor, and profit across the waste sector. Recent scholarly investigations into waste governance, waste economies, technological processes, and labor dynamics have helped to blur the artificial dichotomies of formal/informal and public/private waste management. In doing so, they have shifted analytical focus to the heterogeneous spaces, processes, actors, and relations that (re)produce waste management infrastructures within and across landscapes (Millington and Lawhon 2019). Here, we review this literature and emphasize the diverse forms and conditions of frontline work throughout the waste sector. We show how questions of who will benefit from this "underexploited resource"—who owns waste, who labors[1] to turn it into a resource, who pays for waste services—are reworked in waste experiments and contested by various actors throughout the waste sector.

Waste flows, actors, and institutions

As is true globally, municipal household solid waste is both a small portion of the wider waste stream (4%) and the primary focus of most South African social science waste scholarship. Roughly 90% of the waste produced in South Africa ends up in landfills (Department of Environmental Affairs 2018). This figure obscures a high degree of variability, however, given the preponderance of hazardous waste produced by industries such as mining. Statistics SA reports that 66.9% of South African households received weekly waste removal services in 2016, rising to 91% of households in metropolitan areas (Statistics South Africa 2018, 13–14). These rates are high for the Southern African region, but the quality and nature of such services (including who provides them) varies across municipalities and neighborhoods. Further, these numbers obscure interruptions of service as pick-ups are frequently delayed, cancelled, or inconsistent. Some residents pay private companies or individuals for supplementary collections and additional services (Oyekale 2015). For instance, some households pay for private recycling services if they are outside a municipal pilot area, while others pay individuals to dispose of excess waste when municipal refuse bins have been stolen or are too small for the household. Uneven waste services are not exceptional: such patterns can be found elsewhere (Myers 2005; Gidwani 2015; Cornea, Véron, and Zimmer 2017; Alene 2018; Fredericks 2018) and for other basic services in South Africa.

While official reports suggest low rates of recycling, these representations are contested and some claim that rates of recycling for specific materials are quite high. For instance, the plastics industry suggests that South Africa's recycling rates for PET plastics are amongst the highest in the world (Plastics SA 2019; although see also Carnie, 2020). Much of this recycling is predicated on a significant informal sector, and it is estimated that 60–90,000 people collect discarded materials from streets and landfills (Godfrey and Oelofse 2017). As has been demonstrated elsewhere, South African reclaimers have long been disregarded or blatantly excluded from broader infrastructural investment, policies, and strategies (e.g., Samson 2009; Schenck and Blaauw 2011; Viljoen, Blaauw, and Schenck 2015; Peres 2016). While some efforts have been made to consult and include reclaimer organizations in public participation processes, reclaimers have largely struggled to gain legitimacy in the eyes of the state (Samson 2015; Samson et al. 2020; Sekhwela and Samson 2020).

Scholarship about waste reclaimers has described their collection practices and conditions (Schenck and Blaauw 2011; Schenck, Blaauw, and Viljoen 2016; Sentime 2011; Viljoen,

Blaauw, and Schenck 2015; Blaauw et al. 2020; Yu et al. 2020), their relationship to state practices particularly around questions of stigma (Peres 2016), and their political activity and connection to state practices of formalization (Blaauw et al. 2015; Samson 2019, 2020; Schenck, Blaauw, and Viljoen 2019; Samson et al. 2020; Sekhwela and Samson 2020). A nascent body of scholarship has focused on the entire waste value chain (Langenhoven and Dyssel 2007; Hoffman and Schenck 2020). Once materials have been picked, reclaimers sell recyclables to intermediaries, often at buy-back centers. These intermediaries then sell the materials either locally or globally, to mills, processing facilities, recyclers, and other sites where materials are converted (Viljoen, Blaauw, and Schenck 2012). Waste reclaimers typically work individually, although some largely unsuccessful efforts have been made to develop cooperatives (Godfrey and Oelofse 2017; Godfrey et al. 2017; Muswema et al. 2018). As is true in most contexts, handling of waste is not legal without authorization from the state but enforcement is irregular. While reclaiming is frequently precarious and dangerous, it is also argued to be a response and alternative to the oppressive and exploitative realities of formal employment (Makina 2020; see Barchiesi 2008).

Most South African landfills were created during the apartheid era and are nearing capacity. Intensified national environmental and waste management regulation has required many existing landfills to be upgraded or closed altogether (Godfrey and Oelofse 2017). Costly upgrades and public resistance to new landfill sites (Bond and Dada 2007; Bond and Sharife 2012; Leonard 2012) have made it difficult to continue landfilling high volumes of waste. This pressure is part of ongoing justifications for reducing waste volumes and increasing recycling and sits uneasily with efforts to increase the collection of materials from underserviced areas.

Parallel to concerns about landfill capacity and the financial and political difficulties with creating new ones, a second set of motivations for altering waste flows has emerged. The recycling sector is old, long underpinned by the work of reclaimers, but has received renewed attention as waste is framed no longer as a problem but as an opportunity (Samson 2015; see also Dias 2016; Gutberlet et al. 2016). While the state seeks to grow the recycling sector through various experiments and interventions to increase the capture of waste volumes, researchers have emphasized that this growth is not simply additive: state efforts also *rework* existing flows. As we discuss in Section 4, this reworking formalizes permissions and diverts resources from some who have long benefited from *de facto* access to municipal waste.

State oversight and finance

The state has historically played two roles in South African waste management: as a service provider for households and as a regulator of waste flows. These roles, originally conceived through the lens of environmental protection, are blurring and expanding: following the 2008 Waste Act, an array of new institutional arrangements, regulatory standards, and market mechanisms have emerged (Godfrey and Oelofse 2017). In this context, the state is increasingly embracing a developmental role, emphasizing the economic possibilities alongside ecological ones. This includes complementing long-standing programs on waste awareness and education with projects promoting waste work and entrepreneurship (Higgs and Hill 2019; Stokes 2020).

Regulating waste

Waste regulations exist at multiple scales and according to cooperative governance principles, but ultimately municipal waste in South Africa is the responsibility of municipalities. After the end of apartheid in 1994, basic services (including waste collection) were promised to all citizens (Republic of South Africa, 1996, schedule 5). This commitment led to the implementation and updating of waste legislation and accompanying policy strategies and plans to ensure all citizens have access to waste collection and that such services and infrastructures are regulated and operate to a decent standard. Following the first National Waste Summit, the South African government published the Polokwane Declaration on Waste Management in 2001. With the vision of working towards "zero waste" by 2020, the declaration called upon all South Africans to participate in, contribute to, and promote waste minimization and recycling efforts across the country (Department of Environmental Affairs and Tourism 2001; Matete and Trois 2008). Although the goal of zero waste is far from being achieved, it has nevertheless informed efforts to widen participation and imagine alternative waste flows (e.g., Snyman and Vorster 2011; Trois and Jagath 2011).

The national government passed the Waste Act in 2008 and the associated National Waste Management Strategy (NWMS) in 2011. National legislation largely casts a wide net of responsibility, suggesting a role for the private sector. Civil society is encouraged to build its capacity and become empowered in order to contribute to advocacy and lobbying, uphold regulatory measures, and participate in recovery and waste management programs. A South African Waste Information System (SAWIS) has been maintained by the Department of Environmental Affairs (DEA)'s South African Waste Information Centre (SAWIC) since 2005 to capture government and industry data for the waste sector, such as tonnage. SAWIC also assists with the requirement for each provincial and municipal government to produce an Integrated Waste Management Plan (Godfrey 2008; Godfrey et al. 2012). Provincial governments are considered the "primary regulatory authority for waste services" (Department of Environmental Affairs 2011, 55), and must collect, aggregate, and publish waste information (such as the volume and weight of different waste materials) while also overseeing municipal plans. These efforts are supported by a Waste Research, Development, and Innovation Roadmap initiative, jointly run by the Department of Science and Innovation and the Council for Scientific and Industrial Research (CSIR).

The NWMS promotes waste as a sector for job creation, setting targets for the waste sector. These include the proposed creation of 69,000 new jobs in the waste sector by 2016, and an additional 2,600 small and medium-scale enterprises (SMMEs) and cooperatives (Department of Environmental Affairs 2011 6–7). Impressions from many studying the waste sector suggest that reclaimers are primarily black African (some South African, some not) and established waste businesses (including buy-back centers) are primarily white-owned; we are unaware of data at the national or urban scale that provide representative data (see Samson 2010 who makes a similar observation in her literature review). State efforts to promoting socioeconomic development through waste are underpinned by broader political trajectories addressing post-apartheid socioeconomic inequalities and racialized labor market exclusions, and specifically Broad-Based Black Economic Empowerment (BBBEE). Many new waste businesses have been supported through BBBEE initiatives, and established businesses sought to obtain and increase their level of BBBEE certification (this includes hiring, training, promoting, and increasing ownership of black workers in the businesses). While the overall efficacy of BBBEE is the subject of much scrutiny, there is a clear sense in the waste sector of the importance of BBBEE certification for obtaining state support (e.g., Theron and Visser 2010; Muswema 2012; Higgs and Hill 2019).

While the NWMS has been revised and updated twice, with the most recent version published for public consultation in December 2019, the promises of job growth and "waste as resource" remain prevailing governmental discourses (Department of Environmental Affairs

2019). Provincial governments also earmark waste management as a priority area within broader job creation strategies. For instance, waste applies to several of Western Cape government's strategic priorities: job creation and growth, improved environmental conditions, integrated service delivery, and tackling social ills (Western Cape Government 2015, 1).

There is substantial pressure on the state to implement extended producer responsibility (EPR) schemes; while at the municipal level there is little clarity about what this would mean, efforts have been made to develop EPR for particular waste streams and there are plans to expand these projects. The DEA has been tasked with creating a national Waste Bureau to oversee the successful execution of EPR alongside general waste sector development. Following global initiatives, a plastic bag levy was developed in 2004, and initiatives for tires and electronic waste followed. These efforts have largely been fraught, however, as conflicts of interest and disagreements over who exactly pays for this responsibility have marked the process (Lawhon 2012a; 2012b; Dikgang, Leiman, and Visser 2012; Nkosi et al. 2019). Many have raised concerns about the ability of the state to adequately oversee such processes (Nahman 2010; Nahman and Godfrey 2010; Pillay 2020). Producer Responsibility Organizations, speaking largely for industrial producers and packagers, have called for industry-led voluntary mechanisms tied to specific materials, and have highlighted their success in increasing recycling rates through material-specific levies. Efforts to develop EPR projects in South Africa are ongoing, as of June 2020, but there is little clarity about what form they will take in the future.

This overarching national framework ultimately leaves much scope for municipalities to determine actual waste practices, including what kind of service is needed, how to fund it, and who will provide it. Local policies and plans focus on execution, compliance, and the practical delivery aspects of municipal waste management. In a wider context of uncertainty over how to provide and finance services, municipal practices vary and across South Africa there are ongoing efforts to experiment with different configurations of labor, technology, and finance.

Funding and financing waste management

Across all spheres of government, waste is increasingly and repeatedly framed as both a funding source and a burden: it holds potential value and is a source of economic growth and job creation, but equally is a risk for human health, the environment, and community pride and well-being. For instance, the Waste Act's preamble identifies the disproportionate impact of current waste management service levels on the poor and the value and economic opportunities of waste (Republic of South Africa 2008, preamble; See too Miraftab 2004; Langenhoven and Dyssel 2007; Godfrey and Oelofse 2017; GreenCape 2020).

At all levels of government, cost recovery is considered necessary for sustainable waste management services. Provinces and municipalities collect their own revenue to supplement national transfers (Republic of South Africa 1996, sect. 227), but often struggle to sufficiently fund all nationally mandated activities. Expanding service provision as well as promoting waste minimization, re-use, and recycling requires new infrastructures and services (Department of Environmental Affairs 2011). Yet, as is true in many other countries and for other basic services in South Africa, waste management is both a substantial portion of municipal budgets and under-resourced (Department of Environmental Affairs 2011; Karani and Jewasikiewitz 2007; Makgae 2011). Funding and financing the waste sector varies across municipalities and many have experimented with a range of options. Drawing on international trends and influenced by international development agencies, national and local states are experimenting with public-private partnerships, privatization, and the outsourcing of responsibility for waste to citizens.

Changing dynamics of waste work

State efforts to rework waste are intended to increase the volume of waste captured and create a waste system that is legible to the state, one with established rules and responsibilities and measurable volumes and economic impacts. In this context, informality is framed as inefficient, insufficiently capitalizing on available resources, criminal, and dangerous. What precisely a legible/formal/modern new system will look like, however, is both unclear and contested. Ongoing questions include what waste labor should look like and who will garner the benefits from the waste economy (see Theron and Visser 2010; Godfrey and Oelofse 2017). Further, there is uncertainty as to whether a real material waste economy can live up to the discursive promise of waste as a resource (see Gille 2007).

Recycling in the Global North often depends on mechanized work to capture, transport, and sort waste; the small margins associated with the industry compel the technological advancement of productivity through conveyors, sophisticated trucks, new modes of compression, and increasingly complex global supply chains. The acceleration of mechanization in recycling and waste management in the Global North has put pressures on countries of the Global South to keep up with trends, pushing the prices of materials downwards. Technological transfer subsequently marks the waste sector throughout Africa and the Global South more broadly, but its application is uneven and geographically situated (Gregson and Crang 2015). Adoption of mechanized processes are often pre-dicated on assumed gains in productivity but are also shaped by global aspirations towards modern service provision (see Cooper 2010; Redfield and Robins 2016). Yet this is not simply a matter of increased value generation, for capturing more waste does not necessarily mean more profit. The inputs that capture this extra waste are also burdens: the cost of managerial labor and use of trucks as well as the inclusion of less profitable discarded materials means that it is not a foregone conclusion that there is much more profit to go around.

In South Africa, there is skepticism that the mechanized model of recycling is appropriate and considerable doubts as to the financial viability of expanded household waste services. Despite widespread dissatisfaction with current practices and northern models, there is sub-stantial uncertainty as to what might be a better system (Lawhon, Millington, and Stokes 2018). Job creation has powerful significance in South Africa, in part due to its high rates of un-employment (Stats SA 2019) and long-standing valorization of full-time waged labor (Barchiesi 2008; Lawhon, Millington, and Stokes 2018). This means that capital-intensive approaches to waste are largely understood to be both economically and politically unviable. Public discourse frequently lauds green development's potential to be labor intensive, and picking up litter, collecting waste, and recycling are frequently cited examples (Department of Environmental Affairs 2011; Theron and Visser 2010). Waste work is framed as low-skilled labor and therefore a particularly appropriate opportunity for unemployed citizens. *Making Waste Work*—as says a 2017 slogan for the Polyolefin Recycling Company—is a phrase that helps to explain current priorities in the waste sector and mechanisms through which the private sector justifies its role in the eyes of the state (see Lawhon, Millington, and Stokes 2018). Waste management de-cisions are predicated, therefore, on complex interplays between manual and technical processes and are informed not just by economic efficiency but political priorities and cultural values. For example, investments in technology at materials recycling facilities may increase recovery rates, but there is evidence of a hesitation to engage in capital investments in a social context of high unemployment (Lawhon, Millington, and Stokes 2018).

Efforts to establish full-time, state-employed waste workers *and* collect waste across whole cities have largely been stymied as municipalities have proven unable or unwilling to finance such efforts and generate resistance from reclaimers who have long generated livelihoods from waste. While some urge that increased recycling can reduce the cost of waste services, on the whole the state and private sector have largely promoted experimental alternatives that separate waste and recycling collection and funding. As has been well-documented for waste studies and beyond, private companies operating services on behalf of (or in addition to) municipalities often work with informal laborers or employ workers through casual subcontracting, affording fewer rights or protections to workers than their municipal counterparts (Miraftab 2004, 2005; Samson 2007; Millstein and Jordhus-Lier 2012). Yet these are not the only ways in which waste work is changing as the state pursues a more formal, legible waste sector that captures more waste but does not depend on continuous subsidization. Here, we describe additional strategies that have emerged that rework waste labor and material access: responsibilizing waste management, establishing and enforcing permissions, encouraging waste entrepreneurialism, and retaining waste work without providing wages.

Municipalities have, for example, developed curbside recycling initiatives in an effort to increase recycling. These have typically been piloted in higher income areas, but the results of such experiments have demonstrated that vehicles and labor make this practice expensive and controversial. Collected materials often only justify the costs in high-income areas, and only if materials are not diverted first. Efforts to develop separation at source programs in Johannesburg have generated pushback from reclaimers contesting what Samson (2020) frames as a form of enclosure linked to colonialist logics. Reclaimers argue that they play a critical municipal service, defining themselves as "those who save the city millions of rands by mining recyclables in landfills" (African Reclaimers Organization 2019). The new separation at source programs resulted in considerable loss of income for reclaimers working both in the street and at the landfill; expanding formal employment to areas where informal labor is active (such as curbside recycling) does not guarantee that individuals informally operating will gain similar roles within the formal economy. In 2018, the African Reclaimers Organization was established and organized protests in response to the alleged failure of integration efforts on behalf of the municipality (see Samson 2020; Samson et al. 2020). Such controversies emphasize that these "new" collections are not capturing an entirely untapped resource, but instead rework who is able to legitimately claim materials. Conflicts over incineration and waste-to-energy projects have also been notable: while many are concerned with the associated potential pollution, such projects would reduce overall possibilities for reclaimers and other waste workers (Chamane 2016; Ernstson and Swyngedouw 2018).

Provincial and municipal governments alike have sought to formalize reclaimers into waste cooperatives and enterprises through dedicated consultations, procurement schemes, and broader material, financial, and regulatory support mechanisms. Yet efforts to develop cooperative waste collection companies, inspired by successes elsewhere, have largely failed. Waste sector professionals speak of the difficulties of drawing informal collectors into the formal sphere (Simatele, Dlamini and Kubanza 2017; Samson 2019; Samson et al. 2020). Formal jobs are often structured in ways that do not work with the logics and desires of reclaimers (Sekhwela and Samson 2020; Samson 2020) or implicitly exclude them (Stokes 2020). For example, many reclaimers are not able to work legally, while others do not want to work for an employer or for set hours (Makina 2020).

While reclaimers have been the subject of much of the Global South waste literature, other parts of the waste sector are also being reworked. Informational campaigns as well as national and local cleanups have proliferated in recent decades, often as part of a broader array of campaigns and initiatives which simultaneously cast particular forms of waste work as a civic duty, an issue of community pride, and economic opportunity. Anti-litter campaigns are longstanding and ongoing in many of South Africa's metropolitan regions, and as with their counterparts globally, emphasize environmental protection, cleanliness and beautification (Steyn 2002). The City of Cape Town has run the WasteWise campaign in various forms since 2002 (Armien-Ally 2014), including recently instructing residents in informal settlements how to use litter bins in 2018 ("Green Litter Bin Education and Awareness Project for Informal Settlements", 2018). Similarly, the City of Johannesburg has launched numerous informational campaigns related to effective recycling and waste management behaviors over the last two decades, including #WasteStopsWithME and A Re Sebetseng (A Re Sebetseng Background 2018). Public cleaning and recyclable collections have also been promoted for "clean and green" community volunteerism and entrepreneurship through various initiatives. These strategies have often proved limited, increasingly conceived as part of an environmentalism that is out of touch with the priorities of urban residents; involvement in such schemes remains limited in many cases (Kubanza and Simatele 2020).

Other state initiatives, at times running in parallel and through some of the same agencies and organizations, work away from older notions of cleanliness, beautification, and volunteerism. Often adopting the discourse of the green economy, they seek non-waged ways to incentivize waste work. This includes pilot programs led by stipend-paid community facilitators as per Green Zones in Cape Town's WasteWise campaign. Entrepreneurialism is also promoted, and waste programs also promote community cooperatives and enterprises to deliver local services such as Jozi@Work in Johannesburg. Other programs provide technological gifts for waste cooperatives and financial prizes for community "clean and green" projects as seen with Gauteng Province's *Bontle ke Botho* campaign. In contrast to a purely civic approach, these latter efforts also frame waste work as (potentially) remunerative: some programs offer a stipend or low-paid, temporary wages, or emphasize the possibility that participation will result in future jobs or economic benefit. Remuneration from such programs tend to be piecemeal and temporary, however, usually relegated to short-term contracts with little benefit beyond a temporary wage.

Efforts to enroll the work of citizens without providing jobs is not limited to the state. The private sector is also experimenting with new modes of labor relationships that can legalize and legitimize the collection of recyclables without paying full-time employees or even contracting labor. Pack-a-Ching, for example, is a mobile buyback center designed to encourage collection in the township of Langa in Cape Town which has subsequently expanded to sites in other cities. The intention of this project is to open waste recycling to a wider population, encouraging more people to participate in the recycling economy. Doing so puts residents in direct competition with reclaimers but is imagined as a more legitimate operation that increases rates of material capture.

Understanding efforts to rework waste labor, including creating a greater role for more residents as citizens and economic actors, are crucial to understanding ongoing changes in the waste sector beyond a simple dichotomy of formal/informal. Ongoing experiments with waste management in South Africa not only often exclude reclaimers, they also seek to enroll labor without ensuring secure livelihoods. In the context of a limited developmental state, state and private-sector actors seek to address socioeconomic obligations while devolving obligations, costs, and risk (Stokes 2020).

Emerging questions about waste and work in
South Africa and beyond

Across South Africa, there are increasing efforts to rework waste flows. Much of the literature on waste in South Africa and elsewhere has understood these changes through the lenses of privatization and dispossession, and particularly emphasized the politics and practices of informal waste reclaiming. Yet ongoing changes to waste governance in South Africa tell a more expansive story, including interplays between the state and various non-state actors responding to local as well as global pressures. These changes weave together state priorities around job creation with (at times questionable) environmental ambitions in a context of municipal cost recovery, competition over resources, existing labor practices, and the materialities of discarded objects. Thinking about waste governance and waste economies from South Africa helps us to think through what kinds of waste practices are possible and plausible not only in a neoliberal political economy, but also with the support of a developmental state. We suggest that ongoing changes to waste governance in South Africa speak not just to other places in the global south where formal and informal co-exist intimately; they also have significance for thinking through the possibilities and limitations of the waste economy more broadly.

Critical questions facing waste scholarship, particularly in the Global South, concern the role of waste work and waste workers, and the relationship between the value of waste materials and the rights of those who collect and process them. Converting waste into resource is not an easy task. Ongoing conflicts over waste initiatives demonstrate the complex politics of tying livelihoods to waste and the capacity of the sector to create (decent) jobs. Yet much of the contemporary discourse around the green economy underestimates the work (and the cost of decent compensation) required to collect material as well as the interests of those already working with waste. Ongoing waste initiatives reshape existing patterns of labor and value, but are struggling to create the green economy, including growth and jobs, that many have hoped for. As such, attention to the mechanics of not just informal waste work, but also the formal, semi-formal, and formalizing labor configurations within the sector, is critical to ongoing efforts to figure out how waste work should be remunerated and under what conditions.

Further, the relationship between waste materials, waste work, value, and distribution underpins the "success" of waste economies and efforts to reconfigure them. Much work in waste scholarship has understood and sought to defend waste as an informal "open" commons, where those who have done the necessary labor of collecting and reprocessing reap the benefits (in divergent ways). Yet exclusion and competition for scarce resources, even amongst reclaimers, and a history of racial dispossession reminds us that there are no easy answers to who ought have rights to waste and how the existing and potential profits from waste should be distributed (Makina 2020).

Asking who has rights to waste and its benefits are empirical and ethical questions, ones with implications for the waste economy currently being developed in South Africa (and potentially elsewhere). While the labor that underpins the waste sector is undeniably crucial to recovery rates, municipal waste itself is collectively produced through the everyday habits of residents who use and discard materials in a number of different ways. If waste workers are deserving of the profits from waste due to their labor, are there more just or more egalitarian mechanisms for deciding who gets to do this work and how it is compensated for? And if all workers across the waste sector are paid a just wage, are there actually profits to be made or are subsidies needed from the state and households to support this work? More broadly, what are the ultimate implications of developing a waste economy given the limited role played by recycling in responding to the planetary waste crisis? Does framing waste as an economy supported by a developmentalist state ultimately undermine environmental efforts to reduce the production of waste?

Despite being a middle-income country, inequality means that many South Africans consume at rates that are on par with Global North rates. This, in combination high unemployment, means that many have imagined South Africa to be well-positioned to build a thriving waste economy. In some ways, it is. Yet more attention is needed to the processes through which waste is rendered into value, and the different understandings and meanings of waste that are presently circulating. Attention to the specificities of the waste economy demonstrates how conflicts over waste flows are, at times, based on differing ideas of what waste is and who should pay for and benefit from it. In South Africa, the common refrain that waste is or can be rendered into value obscures the complex social and political dynamics that underpin the industry, with implications for ongoing experimentation in the sector. Attention to the waste economy in South Africa subsequently opens up a series of questions that are critical to global waste studies but also to broader ongoing conversations about the contemporary green economy, infrastructural politics, and contemporary labor.

Note

1   We use the term "work" in its broadest sense, as effort that can be human or non-human, labor; to mean paid work and "job" to mean full-time regular employment.

References

A Re Sebetseng Background. 2018. "A Re Sebetseng." Accessed 13 August 2020.

African Reclaimers Organization. 2019. *Letter to City*, March 27. https://www.wiego.org/sites/default/files/resources/files/ARO%20Reclaimers%20March27_2019%20letter%20to%20city.pdf.

Alene, N. B. 2018. "The Everyday Politics of Waste Collection Practice in Addis Ababa (2003–2009)." *Environment and Planning C: Politics and Space* 36: 1195–1213.

Armien-Ally, J. 2014. *An Evaluation of the City of Cape Town Municipality's Waste Wise Sustainable Education and School Recycling Programme.* MA Thesis. University of Cape Town.

Barchiesi, F. 2008. "Hybrid Social Citizenship and the Normative Centrality of Wage Labor in Post-Apartheid South Africa." *Mediations* 24, no. 1: 52–67.

Blaauw, P. F., J. M. M. Viljoen, C. J. Schenck, and E. C. Swart. 2015. "To 'Spot' and 'Point': Managing Waste Pickers' Access to Landfill Waste in the North-West Province." *AfricaGrowth Agenda* 4: 18–21.

Blaauw, P., A. Pretorius, K. Viljoen, and R. Schenck. 2020. "Adaptive Expectations and Subjective Well-Being of Landfill Waste Pickers in South Africa's Free State province." *Urban Forum* 31, no. 1: 135–155.

Bond, P. and R. Dada. 2007. "A Death in Durban: Capitalist Patriarchy, Global Warming Gimmickry and Our Responsibility for Rubbish." *Agenda* 21, no. 73: 46–56.

Bond, P. and K. Sharife. 2012. "Africa's Biggest Landfill Site: The Case of Bisasar Road." *Le Monde Diplomatique*, https://mondediplo.com/outsidein/africa-s-biggest-landfill-site-the-case-of.

Carnie, T. 2020. *Recycling Myths Debunked.* New Frame. https://www.newframe.com/recycling-myths-debunked/.

Chamane, M. 2016. "An Influx of Waste-to-Energy Projects in South Africa." https://globalrec.org/2016/04/13/an-influx-of-waste-to-energy-projects-in-south-africa/.

Cooper, T. 2010. "Recycling Modernity: Waste and Environmental History." *History Compass* 8, no. 9: 1114–1125.

Cornea, N., R. Véron, and A. Zimmer. (1 April 2017). "Clean City Politics: An Urban Political Ecology of Solid Waste in West Bengal, India." *Environment and Planning A: Economy and Space* 49, no. 4: 728–744.

Department of Environmental Affairs. 2018. *South Africa State of Waste Report. Second Draft Report, Pretoria.* Department of Environmental Affairs.

Department of Environmental Affairs. 2011. "National Waste Management Strategy." Republic of South Africa.

Department of Environmental Affairs. 3 December 2019. National Environmental Management: Waste Act, 2008 (Act No. 59 of 2008) Consultation on the Draft Revised and Updated National Waste Management Strategy." *Government Gazette*, No. 42879, Government Notices NO. 1561. https://www.gov.za/sites/default/files/gcis_document/201912/42879gon1561.pdf.

Department of Environmental Affairs and Tourism. 2001. The Polokwane Declaration on Waste Management. Republic of South Africa.

Dias, S. M. 2016. "Waste Pickers and Cities." *Environment and Urbanization* 28, no. 2: 375–390.

Dikgang, J., A. Leiman, and M. Visser. 2012. "Analysis of the Plastic-Bag Levy in South Africa". *Resources, Conservation and Recycling* 66: 59–65.

Dlamini, S., M. D. Simatele, and N. Serge Kubanza. 2019. "Municipal Solid Waste Management in South Africa: From Waste to Energy Recovery Through Waste-to-Energy Technologies in Johannesburg." *Local Environment* 24, no. 3: 249–257.

Ernstson, H. and E. Swyngedouw. 2018. "Wasting CO2: The Remarkable Success of a Climate Failure." In *Heterogeneous Infrastructures in African Cities Workshop*, 1–32. Working Paper TLR Project. The Situated UPE Collective, Kampala. https://bit.ly/3c9nXOZ.

Fredericks, R. 2018. *Garbage Citizenship: Vital Infrastructures of Labor in Dakar, Senegal.* Duke University Press.

Freund, B. 2007. "South Africa as Developmental State?: Part Two: Policy and Political Choices." *Africanus* 37, no. 2: 191–197.

Gidwani, V. 2015. "The Work of Waste: Inside India's Infra-Economy." *Transactions of the Institute of British Geographers* 40, no. 4: 575–595.

Gille, Z. 2007. *From the Cult of Waste to the Trash Heap of History: The Politics of Waste in Socialist and Postsocialist Hungary.* Indiana University Press.

Global Alliance of Waste Pickers. n.d. "South African Waste Pickers Association (SAWPA)". https://globalrec.org/organization/south-african-waste-pickers-association-sawpa/. Accessed August 12, 2020.

Godfrey, L., and S. Oelofse. 2017. "Historical Review of Waste Management and Recycling in South Africa." *Resources* 6, no. 4: 57.

Godfrey, L., A. Muswema, W. Strydom, T. Mamafa, and M. Mapako. 2017. "Co-Operatives as a Development Mechanism to Support Job Creation and Sustainable Waste Management in South Africa." *Sustainability Science* 12, no. 5: 799–812.

Godfrey, L. 2008. "Facilitating the Improved Management of Waste in South Africa Through a National Waste Information System." *Waste Management* 28, no. 9: 1660–1671.

Godfrey, L., Scott, D., Difford, M., and Trois, C. 2012. "Part 1: The Role of Waste Data in Building Knowledge: The South African Waste Information System." *Waste Management* 32, no. 11: 2154–2162.

Godfrey, L., M. T. Ahmed, K. G. Gebremedhin, J. H. Y. Katima, S. Oelofse, O. Osibanjo, U. H. Richter, and A. H. Yonli. 2019. "Solid Waste Management in Africa: Governance Failure or Development Opportunity?" In N. Edomah (Ed.) *Regional Development in Africa.* IntechOpen. doi:10.5772/intechopen.86974.

Gregson, N., and Crang, M. 2015. "From Waste to Resource: The Trade in Wastes and Global Recycling Economies." *Annual Review of Environment and Resources* 40: 151–176.

GreenCape. (2020). "Waste 2020 Market Intelligence Report." https://www.greencape.co.za/assets/WASTE_MIR_20200331-v2.pdf.

Gutberlet, J. J.H. Kain, B. Nyakinya, D. H. Ochieng, N. Odhiambo, M. Oloko, J. Omolo, E. Omondi, S. Otieno, P. Zapata and M. J. Zapata Campos. 2016. "Socio-Environmental Entrepreneurship and the Provision of Critical Services in Informal Settlements." *Environment and Urbanization* 28, no. 1: 205–222.

Higgs, C. J., and T. Hill. 2019. "The Role That Small and Medium-Sized Enterprises Play in Sustainable Development and the Green Economy in the Waste Sector, South Africa." *Business Strategy and Development* 2, no. 1: 25–31.

Hoffman, M., and C. Schenck. 2020. "The Value Chain and Activities of Polyethylene Terephthalate Plastics in the South African Waste Economy". *Local Economy* 35, no. 5: 523–535. 10.1177/0269094220931697.

Karani, P., and Jewasikiewitz, S. M. 2007. "Waste Management and Sustainable Development in South Africa." *Environment, Development and Sustainability* 9, no. 2: 163–185.

Kubanza, N. S. and M. D. Simatele. 2020. "Sustainable Solid Waste Management in Developing Countries: A Study of Institutional Strengthening for Solid Waste Management in Johannesburg, South Africa." *Journal of Environmental Planning and Management* 63, no. 2: 175–188.

Kubanza, N. S. In press. "The Role of Community Participation in Solid Waste Management in Sub-Saharan Africa: A Study of Orlando East, Johannesburg, South Africa." *South African Geographical Journal* 103, no. 2: 223–236.

Langenhoven, B., and Dyssel, M. 2007. "The Recycling Industry and Subsistence Waste Collectors: A case study of Mitchell's Plain." *Urban Forum* 18, no. 1: 114–132.

Leonard, L. 2012. "Another Political Ecology of Civil Society Reflexiveness against Urban Industrial Risks for Environmental Justice: The Case of the Bisasar Landfill, Durban, South Africa." *Singapore Journal of Tropical Geography* 33, no. 1: 77–92.

Lawhon, M. 2012a. "Contesting Power, Trust and Legitimacy in the South African e-waste Transition." *Policy Sciences* 45, no. 1: L69–L86.

Lawhon, M. 2012b. "Relational Power in the Governance of a South African e-waste Transition." *Environment and Planning A* 44, no. 4: 954–971.

Lawhon, M., Millington, N., and Stokes, K. 2018. "A Labor Question for the 21st Century: Perpetuating the Work Ethic in the Absence of Jobs in South Africa's Waste Sector." *Journal of Southern African Studies* 44, no. 6: 1115–1131.

Makgae, M. 2011. "Key Areas in Waste Management: A South African Perspective." *Integrated Waste Management* 2, 1169.

Makina, A. 2020. *Logics Used to Justify Urban Appropriations: An Examination of Waste Picking in Tshwane.* PhD Diss. University of Oklahoma.

Matete, N., and Trois, C. 2008. "Towards Zero Waste in Emerging Countries—A South African Experience." *Waste Management* 28, no. 8: 1480–1492.

Millington, N., and Lawhon, M. 2019. "Geographies of Waste: Conceptual Vectors From the Global South". *Progress in Human Geography* 43, no. 6: 1044–1063.

Millstein, M., and Jordhus-Lier, D. 2012. "Making Communities Work? Casual Labour Practices and Local Civil Society Dynamics in Delft, Cape Town." *Journal of Southern African Studies* 38, no. 1: 183–201.

Miraftab, F. 2004. "Neoliberalism and Casualization of Public Sector Services: The Case of Waste Collection Services in Cape Town, South Africa." *International Journal of Urban and Regional Research* 28, no. 4: 874–892.

Miraftab, F. 2005. "Informalizing and Privatizing Social Reproduction: The Case of Waste Collection Services in Cape Town, South Africa". In N. Kudva and L. Bernia (Eds.) *Rethinking Informalization: Poverty, Precarious Jobs and Social Protection.* Cornell University Press. 148–162.

Muswema, A. P., A. Okem, H. Blottnitz, and S.Oelofse. 2018. "Cooperatives in Waste and Recycling: A Recipe for Failed Waste Hierarchy Implementation?'Johannesburg: CSIR Researchspace." https://researchspace.csir.co.za/dspace/handle/10204/10508.

Muswema, A. P. 2012. Wrestling with Empowerment: A BBBEE (Enterprise Development) Case Study and Model for the Waste Sector from Durban. Waste Con 2012, East London International Convention Centre, East London, 9-12 October 2012.

Myers, G. A. 2005. *Disposable Cities: Garbage, Governance and Sustainable Development in Urban Africa.* Routledge.

Nahman, A. 2010. "Extended Producer Responsibility for Packaging Waste in South Africa: Current Approaches and Lessons Learned." *Resources, Conservation and Recycling* 54, no. 3: 155–162.

Nahman, A., and L. Godfrey. 2010. "Economic Instruments for Solid Waste Management in South Africa: Opportunities and Constraints." *Resources, Conservation and Recycling* 54, no. 8: 521–531.

Nkosi, N., E. Muzenda, M. Belaid, C. Mateescu, and P. Bilal. 2019. "A Review of the Recycling and Economic Development Initiative of South Africa (REDISA) Waste Tyre Management Plan: Successes and Failure". In *2019 7th International Renewable and Sustainable Energy Conference (IRSEC).* IEEE. 1–8.

Oyekale, A. S. 2015. "Factors Explaining Households' Cash Payment for Solid Waste Disposal and Recycling Behaviors in South Africa." *Sustainability* 7, no. 12: 15882–15899. https://www.mdpi.com/2071-1050/7/12/15795.

Palmer, I., Moodley, N., and Parnell, S. 2017. *Building a Capable State: Service Delivery in Post-Apartheid, South Africa.* Zed Books Ltd.

Parnell, S., and Robinson, J. 2012. "(Re) Theorizing Cities From the Global South: Looking Beyond Neoliberalism." *Urban Geography* 33, no. 4: 593–617.

Peres, T. S. 2016. *Stigma Management in Waste Management: An Investigation into the Interactions of "Waste Pickers" on the Streets of Cape Town and the Consequences for Agency.* PhD Diss. University of Cape Town.

Pillay, A. S. 2020. "The Development of an Extended Producer Responsibility (EPR) Mechanism for Waste Management in South Africa." In Sadhan Kumar Ghosh (Ed.) *Sustainable Waste Management: Policies and Case Studies.* Springer. 43–52.

Plastics SA Admin. (22 August 2019) "South African PET Recycling Rates Amongst the Highest in the World." *Plastics SA.* https://www.plasticsinfo.co.za/2019/08/22/south-african-pet-recycling-rates-amongst-the-highest-in-the-world/ Accessed August12, 2020.

Redfield, P., and S.Robins. 2016. "An Index of Waste: Humanitarian Design, "Dignified Living" and the Politics of Infrastructure in Cape Town." *Anthropology Southern Africa* 39, no. 2: 145–162.

Republic of South Africa. 1996. "Constitution of the Republic of South Africa, Act 108."

Republic of South Africa. 2008. "National Environmental Management: Waste Act, Act 59."

Rodina, L. 2016. "Human Right to Water in Khayelitsha, South Africa–Lessons from a 'Lived Experiences' Perspective." *Geoforum* 72: 58–66.

Samson, M. 2007. "When Public Works Programmes Create 'Second Economy' Conditions: Part Two: Policy and Political Choices." *Africanus* 37, no. 2: 244–256.

Samson, M. 2009. "Wasted Citizenship? Reclaimers and the Privatised Expansion of the PublicSphere". *Africa Development* 34, 3-4.

Samson, M. 2010. *Reclaiming Reusable and Recyclable Materials in Africa: A Critical Review of English Language Literature.* WEIGO. Accessed August 12, 2020. https://www.wiego.org/sites/default/files/publications/files/Samson_WIEGO_WP16.pdf.

Samson, M. 2015. "Accumulation by Dispossession and the Informal Economy–Struggles Over Knowledge, Being and Waste at a Soweto Garbage Dump." *Environment and Planning D: Society and Space* 33, no. 5: 813–830.

Samson, M. 2017. "Not Just Recycling the Crisis: Producing Value at a Soweto Garbage Dump." *Historical Materialism* 25, no. 1: 36–62.

Samson, M. 2020. "Whose Frontier is it Anyway? Reclaimer "Integration" and the Battle Over Johannesburg's Waste-based Commodity Frontier." *Capitalism Nature Socialism* 31, no. 4: 60–75.

Samson M., S. Timm, T. Chidzungu, N. Dladla, G. Kadyamadare, K. Maeka, M. Mahlase, A. Mokobane, K. Molefe and L. Ndlovu. 2020. "Lessons from Waste Picker Integration Initiatives: Development of Evidence Based Guidelines to Integrate Waste Pickers into South African Municipal Waste Management Systems. Final Technical Report: Johannesburg Case Study." *Waste Research Development and Innovation Roadmap Research Report.* https://wasteroadmap.co.za/wp-content/uploads/2020/03/1-Wits_Final_Report_Johannesburg-Technical-Report.pdf. Accessed August 12, 2020.

Schenck, C. J., P. F. Blaauw, and J. M. M. Viljoen. 2016. "The Socio-Economic Differences between Landfill and Street Waste Pickers in the Free State Province of South Africa." *Development Southern Africa* 33, no. 4: 532–547.

Schenck, C. J., P. F. Blaauw, E. C. Swart, J. M. M. Viljoen, and N. Mudavanhu. 2019. "The Management of South Africa's Landfills and Waste Pickers on Them: Impacting Lives and Livelihoods." *Development Southern Africa* 36, no. 1: 80–98.

Schenck, R. and P. F. Blaauw. 2011. "The Work and Lives of Street Waste Pickers in Pretoria-A Case Study of Recycling in South Africa's Urban Informal Economy." *Urban Forum* 22, no. 4: 411–430.

Sekhwela, M. M., and Samson, M. 2020. "Contested Understandings of Reclaimer Integration—Insights from a Failed Johannesburg Pilot Project." *Urban Forum* 31, no. 1: 21–39.

Sentime, K. 2011. "Profiling Solid Waste Pickers: A Case Study of Braamfontein—Greater Johannesburg." *Africanus* 41, no. 2: 96–111.

Simatele, D. M., S. Dlamini, and N. S. Kubanza. 2017. "From Informality to Formality: Perspectives on the Challenges of Integrating Solid Waste Management into the Urban Development and Planning Policy in Johannesburg, South Africa." *Habitat International* 63: 122–130.

Snyman, J., and Vorster, K. 2011. "Towards zero waste: a case study in the City of Tshwane". *Waste Management and Research* 29, no. 5: 512–520.

Statistics South Africa. 2018. Environment, in-depth analysis of the General Household Survey 2002–2016. GHS Series Report Volume IX. Statistics South Africa, Pretori.

Statistics S. A. 2019. Government Statistics, Statistical Release, Pretoria.

Steyn, P. 2002. "Popular Environmental Struggles in South Africa, 1972–1992." *Historia* 47, no. 1: 125–158.

Stokes, K. 2020. *Waste Labour and Infrastructural Citizenship: Promises and Perils of State-led Community Waste Initiatives in South African Cities.* PhD Thesis. University of Manchester.

Theron, J., and M. Visser. 2010. "Waste Management and the Workplace." *Law, Democracy and Development* 14. https://www.ajol.info//index.php/ldd/article/view/68290.

Timm, S. 2015. *Modalities of Regulation In the Informal Economy: A Study of Waste Collectors in Cape Town.* PhD Diss. University of Cape Town.

Trois, C., and Jagath, R. 2011. "Sustained Carbon Emissions Reductions Through Zero Waste Strategies for South African Municipalities." *Integrated Waste Management* 2: 442–460.

Turok, I. 2010. "Towards a Developmental State? Provincial Economic Policy in South Africa." *Development Southern Africa* 27, no. 4: 497–515.

Viljoen, J. M. M., P. F. Blaauw, and C. J. Schenck. 2012. "The Role and Linkages of Buy-Back Centres in the Recycling Industry: Pretoria and Bloemfontein (South Africa)." *Acta Commercii* 12, no. 1: 1–12.

Viljoen, K., P. Blaauw, and R. Schenck. "'I Would Rather Have a Decent Job': Barriers Preventing Street Waste Pickers from Improving Their Socioeconomic Conditions." *Economic Research Southern Africa*, ERSA working paper, no. Working paper 498 (5 February 2015). http://www.econrsa.org/system/files/publications/working_papers/working_paper_498.pdf.

Western Cape Government. 2015. Provincial Strategic Plan 2014 – 2019. Cape Town: Western Cape Government Department of the Premier.

Yu, D., D. Blaauw, and R. Schenck. 2020. "Waste Pickers in Informal Self-Employment: Over-Worked and on the Breadline." *Development Southern Africa* 37, no. 6: 971–996. 10.1080/0376835X.2020.1770578.

21

PREPPING FOR THE [INSERT HERE] APOCALYPSE AND WASTING THE FUTURE

Myra J. Hird and Jacob Riha

Introduction

One of us (Myra) clicks on the local online news and the annoying advertisement pop-up is about toilet paper.[1] Before, I would have muted the sound, but now I watch the advertisement. Charmin, the advertiser, shows a family of bears sitting together watching television. The voice-over claims that Charmin is working hard to ensure that we all have enough toilet paper to make it through the COVID-19 pandemic. And I find myself asking: *do* I have enough toilet paper?

Before the global COVID-19 pandemic, environmental tipping points, global climate change, political instability, dwindling primary resources, the ever-widening gap between wealthy and poor, mass migration, and exponential capitalist growth—amongst other critical realities—were enough to furnish myriad imagined apocalyptic futures. Now that we are experiencing successive waves of the COVID-19 pandemic, an apocalyptic future has mainstreamed to millions of people whose economic and political privilege has hitherto cocooned us from the lived experiences of environmental collapse. And while the international COVID-19 response by-line might be "We're All in This Together", our daily news feed suggests otherwise.

As the WHO urges a coordinated global effort to prevent the further spread of COVID-19 and the Intergovernmental Panel on Climate Change and other supra-governmental bodies remain focused on developing systems of resilience and adaptation for populations vulnerable to the effects of climate change, a disparate group of people self-defined as "preppers" have already been mobilizing themselves and their loved ones for myriad imagined apocalyptic scenarios. Once represented by social media as individuals on the margins of mainstream society if not mental stability, the COVID-19 pandemic is positioning prepping—if not preppers themselves—as a rational and, indeed, responsible answer to the pandemic. Preppers distinguish themselves from "hoarders" based on their purposeful stock-piling of things anticipated to be vital to surviving short- or long-term disaster. Whereas hoarders are characterized by the volume of disorganized clutter of a jumble of things with no direct survival value, preppers systematically purchase and organize things for survival under anticipated immanent and/or future conditions of extreme societal collapse. But nor are preppers part of a growing "preventer" identity and movement: preppers aim to protect themselves and their loved ones *after*

DOI: 10.4324/9781003019077-21

disaster strikes; preventers focus their actions on preventing disaster from occurring in the first place. This includes small everyday actions designed to lower environmental footprints at the individual level, such as reusing and repairing clothing and small appliances, dumpster-diving, and scavenging from other people's discards to much more community-engaged actions such as civil protest, government lobbying, and so on.

This chapter focuses on prepping as a particular response to the uncertainty of our species' survival. Drawing on a range of theoretical traditions and empirical observations, we critically examine the various discourses and practices that preppers deploy in preparing themselves and their loved ones for what they believe is the probability, if not certainty, of a survivalist future. Far from the experiences of millions of people who have already been forced into relentless adaptation due to unremitting poverty, inequality, and global changes in climate, preppers largely plan for their imagined future by accumulating survivalist skills and *things*. That is, what most characterizes preppers is their heightened consumption: preppers spend many thousands of dollars—sometimes millions of dollars—stockpiling stuff: weapons, "bug-out" gear, non-perishable food, water, clothing, generators, and more, sometimes in secret bunkers or other hidden places. WalMart, Costco, and other mainstream consumer havens offer emergency food storage kits, as publishing and distributing companies sell prepping guides for adults and children alike. Not only, then, do they eschew initiatives that seek to prevent an imagined future apocalypse, as preventers do, but as we will show, preppers influence the very conditions that they then feel compelled to respond to as they intensify the hegemony of over-consumption (and its rather more silent twin, overproduction). As such, we argue that the increasingly popular phenomenon of prepping is a contemporary reiteration of western consumer/trashing culture, which feeds the global neoliberal capitalist system responsible for the very apocalyptic conditions to which preppers believe they must respond.

Anchoring our analysis are two interrelated questions familiar to waste studies scholars that concern the dialectical relationship between the individual and society. Firstly, how do we avoid simply characterizing prepping as an individual behavior, and the phenomenon of prepping as the aggregate of these individual actions? That is, how do we conceptualize prepping less as a psychological coping mechanism and more as a *socius logy*, or logic of society? And, secondly, how do we consider the related problem of scale (Liboiron 2013, 2014; Hird 2021) whereby preppers are held more accountable for the environmental costs of their comparatively small-scale accumulation of consumer products and the waste it produces than environmental costs (including waste) accrued through the extraction, production, and distribution of products that are—by orders of magnitude—greater.

Si vis pacem, para bellum (If you want peace, prepare for war)

While demonstrating to his sons how to shoot guns, self-proclaimed "prepping entrepreneur" Tim Ralston shot himself in the thumb. Appearing on National Geographic's Doomsday Preppers, a reality television series that aired from 2012 to 2014, in many ways Ralston reifies the prepper stereotype: he is American, white, middle-aged, male, and appears fond of reciting NRA mantras such as "every American household should have a gun. Never have enough guns, never enough ammo" (Doomsday Preppers 2012, S1, E3). This particular reality show participant's apocalyptic scenario of choice is a nuclear disaster, which causes an electro-magnetic pulse that at least temporarily (and perhaps permanently) disables America's energy grid and knocks out its transportation and communications systems. Also corresponding well with the prepper stereotype, Ralston began his prepper lifestyle in secret, hiding from his family the approximate US$30,000 he purchased in weapons, ammunition, food, and other supplies.

Commenting, with relief, about his firearm accident, Ralston told the interviewers "thank God it wasn't my son" (Ibid.).

During its run, the reality TV show *Doomsday Preppers* featured weekly snapshots of individuals preparing for either cataclysmic natural or human-made disasters. The show may have provided insights into the lived experiences of people preparing for disaster, but its focus is nothing new. Numerous films, some based on novels, depict either human-made environmental disasters (The Day After Tomorrow, Children of Men, Mad Max, The Road, On the Beach, Threads and our personal favorites Miracle Mile (Jacob) and Dr. Strangelove (Myra)) or natural disasters (Sunshine, Melancholia) or zombie apocalypse (World War Z) or global alien invasion (War of the Worlds, Edge of Tomorrow, Signs) or robot ascension (The Terminator, I, Robot) or—yes, global viral infection (The Seventh Seal, The Omega Man, 12 Monkeys, Outbreak, Contagion) or some combination of these. Of course, the mainstreaming of disaster and apocalypse stretches much further back in time than these films. The so-called "first doom boom" occurred during the Cold War, as Americans in particular (but also Soviets—see Brown 2013) worried about nuclear war and its aftermath. Further back still, the Christian *New Testament*'s Book of Revelation, also known as the Apocalypse of John, details the "apokalypsis" or unveiling—the reckoning of all humanity with the forces of good and evil. We find variations on the end-of-days theme in other major religions such as Islam (The Hour), Judaism (Day of the Lord, War of Gog and Magog), and Hinduism (Vishnu's return to earth in the form of Kulki to destroy the forces of evil). Max Brooks, author of *The Zombie Survival Guide: Complete Protection from the Living Dead,* nicely enjoins the contemporary agnostic with the ancient religious: "If you believe you can accomplish everything by 'cramming' at the eleventh hour, by all means, don't lift a finger now. But you may think twice about beginning to build your ark once it has already started raining" (2003, 159).

As Indigenous scholars and activists (for instance Watt-Cloutier 2015; Hoover 2017), environmental racism researchers (for example Nixon 2011; Adeola 2012) and, more recently, supranational organizations such as the IPCC (2020) have pointed out, real apocalyptic conditions arrived hundreds of years ago for Indigenous peoples violently subjugated by colonization; Black and other racialized people enchained through slavery and subsequent discriminatory laws, policies, and practices; people forced to migrate due to the effects of global warming, and poor people devastated by inequitable labor systems, the World Bank, the IMF, and so on. The UNHCR, for instance, "estimates that there are over 65.6 million forcibly displaced people worldwide, of whom approximately two-thirds are internally displaced and therefore unprotected by international law...[with another] 10 million stateless individuals" (UNHCR 2016 in Alexander and Sanchez 2018, 11). These are people who live in a state of unrelenting emergency. One-third of the global human population have no electricity. Moreover, it is clear that those most responsible for climate change are not living with its already-here effects: the poorest 45% of human beings are responsible for just 7% of anthropogenic carbon emissions while the richest 7% create 50% of the world's carbon emissions (Malm and Hornborg 2014).

But these past and present unnamed millions are not preppers; they are *actually* living with chronic disasters that have shifted into permanently degraded living conditions, in the here and now. Media and research studies most frequently depict preppers (for instance Clearwater 2010; Foster 2014; Klein 2008; Mills 2019; Robbins and Moore 2012; Foster 2014) as motivated by the *anticipation* of future disaster, or an apocalypse yet to come. Preppers fear a diverse range of impending disasters, from solar flares that wipe out electrical grids to terrorist attacks to peak-oil to cataclysmic floods, that more closely resemble something out of a movie rather than the already-here apocalyptic living conditions of the world's abject (Kristeva 1982; Giorgio and

Daniel 1998; Scheider-Mayerson 2013). A series of polls—taken pre-COVID-19, and summarized by Michael Mills—shows that U.S. residents, in particular, have a host of anxieties: "over 40% of Americans fear losing a loved one to terrorism…international conflict (47.5%), economic collapse (44.4%), cyber-attacks (39.1%), a collapse of the electrical grid (35.7%) and biological warfare (41.8%)" (2018, 3). Research estimates that (pre-COVID-19) there were between 3 to 5 million Americans who identified with the prepping movement (Ibid). Of course, these are vague estimates because a critical feature of the prepper is that s/he prepares for disaster in secret in order to prevent other people from stealing her/his things, a point we return to later in the chapter.

And while the media and researchers report on prepping as a phenomenon found in a range of countries and cultures in the globalized north, the most popular representation of the prepper is the American man, woman, and/or family because the United States is most strongly associated with the quintessential "personality trait" of preppers: consumerism with an emphasis on buying guns and ammunition. Indeed, prepping is a consumer bonanza. There are the lazy-person's ready-made non-perishable prepping staples: for instance, TheReadyStore.com sells the READY pre-2000 Food Storage Supply Kit for a 27-year life span for US$3,683.25. Filtration and cleaning supplies may be purchased from the Berkey Light System for US$200–300. Wise Food Storage sells a one-year supply of food for two people for US$2,595 and Costco's Chef's Banquet All-Purpose Readiness Kit sells a 20-year supply of 600 meals for one person for US$149.99. Caro shipping containers may also be purchased for US$1,000 each in order to store all of the products we buy. For the more committed prepper, there is a cornucopia of "essential" products to purchase. Bugout gear includes predictable things like generators, candles and hand-crank emergency radios, knives, guns, and ammunition but also less obvious things like gold and silver. The list is, by design, endless. As Gwendolyn Foster highlights, "There is a lot of money to be made out [of] prepping" (2014, 16). To wit, the revenue for some dehydrated and preserved food companies increased by as much as 708% in 2007/8 (Murphy 2013). There are also books to be bought: *A Complete Beginners Guide to Prepping* (Tactical.com, n.d.), *The Prepping Guide: SHTF Plans* (Brown n.d.), *The Prepper's Water Survival Guide* (Luther 2015), *When Technology Fails: A Manual for Self-Reliance, Sustainability, and Surviving the Long Emergency* (Stein 2008), *The Prepper's Blueprint* (Pennington and Luther 2014), *When All Hell Breaks Loose: Stuff You Need to Survive When Disaster Strikes* (Lundin 2007), and many more. All in all, and before the COVID-19 pandemic, prepping was a US$500 million per year consumer industry (Kelly 2016). This sum doesn't include Larry Hall's Survival Condo bunker, which alone cost some US$20 million (Garrett 2020).

And since preppers live in a constant state of preparation (there are never enough guns and never enough ammo), they lament that it is only their finances that constrain further consumption: it's "all I can afford" says Meegan Hurwitt (Doomsday Preppers 2012 S1, E1), "right now [fuel is] too expensive" says Dennis Evers (Ibid., S1, E2) and there is simply "not enough space in the apartment" says Jason Charles (Ibid., S1, E3), leading preppers to use whatever space is available until eventually, those who can afford it move to bigger houses, purchase bigger secret bunkers, and/or more storage containers. Indeed, Charmaine Eddy's analysis suggests that "ideals of…property ownership and proper object consumption" centrally occupy the prepper's ambition (2014, 2). More stuff needs more space and, as Eddy points out, in this regard it can be a fine line between prepping and hoarding. Guns and ammunition are particularly popular prepper purchases. Pat and Lynette Brabble, for instance, a retired American couple in their 60s and featured on *Doomsday Preppers*, strongly tether their religion, gun ownership, and the American Constitution together. As Pat remarks from their secret stock-piled room, they thank God "for all the provision[s] that he's [God] made [for them]", which includes over 100 guns (2012 S1, E3). Indeed, the experts on the show who assess preppers' provisions often counsel participants to purchase and store more products, especially guns, and to carry their guns with them at all times, "for your protection".

And all of this consumption takes place within the context of a global neoliberal capitalist system dependent on ever-increasing extraction, manufacturing, distribution, consumption, and, inevitably, waste. To wit, Paul Crutzen, the scientist who revitalized the term Anthropocene, originally stated that the Anthropocene began with the large-scale extraction of fossil fuels, but he recently changed his mind. He now places the start of the Anthropocene on 16 July 1945—the Trinity detonation, the first test of a nuclear device, and its radioactive waste fallout (Zalasiewicz and Mark 2015). The World Bank (n.d.) conservatively estimates that 2.01 billion tons of municipal solid waste is generated annually, and that most of this waste is disposed of in open dumps. And municipal solid waste—the waste that individuals and households produce post-consumption—accounts for a small fraction of the waste that the extractive and manufacturing industries produce. In Canada, where we write this chapter, over 95% of our waste is generated at the extraction and production stages rather than at the consumer stage (Hird 2021). For example, Josh Lepawsky (2018) details that, in the United States, some 489,840 U.S. tons of e-waste is recycled through Dell's Reconnect program. But this already enormous amount of waste is far smaller than the 902,792 U.S. tons of sulfuric acid waste produced by a single smelter operation in Mexico that produces copper and other metals used in electronics production. Indeed, Lepawsky observes that the waste produced in extracting some of the materials used in electronics *from one smelter* is 1.8 times larger than e-waste exports *from the entire* United States of America. And managing all of this waste is big business: not only is haulage and disposal a multi-billion-dollar industry, but so is the remediation of Superfund and other contaminated waste legacy sites (Hird 2021; Beckett and Keeling 2019).

The global production of waste and its contamination of human bodies, wildlife, land, and water, itself constitutes one of the key dystopian futures that preppers seek to shield themselves and their loved ones from. And thus, prepping through hyper-consumption accelerates the very environmental condition that they are preparing themselves for: environmental degradation through accelerating industrial and military production, and its corollary, increasingly toxic and contaminating waste polluting the environment. Like the *Terminator,* who creates the conditions of its own existence, preppers invoke their own environmental apocalypse. Or as Franklin Ginn astutely observes, "Fantasies of apocalypse are both a product and a producer of the Anthropocene" (2015, 352). And now, with the global COVID-19 pandemic, prepping has escaped its containment as an identity on the margins of normal society. Prepping is mainstreaming.

Properly masked and drenched in Purell[2]

It is a common children's fable: one little pig who put pleasure before responsibility, a second little pig who simply did not do enough, and a third little pig who was sufficiently prepared to not only save himself but magnanimously save his two unprepared brothers as well. As one of us (Jacob) ran out of toilet paper during the first COVID-19 wave, and was left going from store to store, I wondered—am I one of the lazy brothers? As COVID-19 began sweeping the globe and preppers everywhere settled into a feeling of vindication, many of us hitherto non-preppers were left stockpiling *stuff* in hopes that consumption might translate into resilience.

At the time of writing, the novel coronavirus (SARS-Cov-2), the agent of COVID-19, has infected more than 47 million people across 190 countries, and caused over 1.2 million deaths

(Fauci, Lane, and Redfield 2020; John Hopkins University & Medicine 2020). The world's most privileged economies, health care systems, and, indeed, governments, are on the verge of collapse. Extraordinary measures designed to stop the spread of COVID-19 (social distancing, border closures, curfews, PPE regulations) have been variably enforced by governments around the world, leaving some 7 billion people scrambling to meet uncertain and shifting economic, social, behavioral, and political conditions (Sohrabi et al. 2020). By March 2020, much of the world was forced into lockdown as a result of the outbreak, meaning most consumers were following stay-at-home orders and restricted to only leaving their homes for "essential items". Like animals preparing for winter, we were told to "stockpile food and medication in [our] homes" (Canadian Health Minister Patty Hadju, *National Post* Feb. 26, 2020 (Bharti 2020)). At the advice of our governments and media alike, we were urged to purchase "extra stores of things like toilet paper, pet food and feminine hygiene products" (Collie 2020). Panic buying set in as retailers and business owners around the world were forced to limit the number of product purchases like hand sanitizer, bleach, and cold medicine. Suddenly, once exclusively the obsession of fringe survivalists, disaster preparedness has become a national pastime: toilet paper is now protected by security guards (Toh 2020) and we are "drowning in Purell" (Hesse and Zak 2020, n.p.). While before COVID-19, a modest 3–5 million Americans and fewer Canadians were concertedly prepping for future disasters; since the COVID-19 pandemic began, some 52 million Americans and 14 million Canadians have begun stockpiling food and other "necessities" (Laycock and Binstead 2020).

Disasters typically influence both production and consumption patterns (Larson and Shin 2018; Pantano et al. 2020; Sheu and Kuo 2020). For instance, during WWII, industrial productivity increased by 96% (Nelson 1991). Since the onset of COVID-19, people across the world have displayed stockpiling behaviors that differ in amount and product types from their usual shopping habits—incited by a fear of being like the unprepared and lazy pig brothers who simply did not do enough (Ansalam et al. 2020). Sales of household cleaning and disinfectant products have dramatically increased since the WHO declared a global health emergency. According to Statistics Canada, for instance, year-over-year sales growth of hand sanitizer, masks, and various cleaning supplies reached as high as 639%, 404%, and 180%, respectively; while paper towels and toilet paper increased by 288% during the same period (2020; Bedford 2020). In the United States, the same pattern occurred: panic-induced purchasing increased sales of aerosol disinfectants (385%), multipurpose cleaners (148%), and tissue products (60%) (Conway 2020). Similar stockpiling of grocery items saw household spending in Canada increase by roughly 46% in spring of 2020—led most notably by rice (239%), dried pasta (205%), canned vegetables (180%), and flour (179%) (Statistics Canada 2020). Alongside this, fresh produce sales decreased by 15% while pre-packaged foods and frozen items—wrapped in plastic packaging—surged by 31% (Roe, Bender, and Qi 2020).

As we continue to see more retailers offering emergency kits and survivalist materials (Costco, Walmart, etc.), the disaster prep industry has been experiencing a massive sales surge from customers concerned about the risks associated with COVID-19. Hazmat suits, tactical gear, water purification tablets, and air filters are just a few of the survival goods sold on TheEpicenter—an emergency gear and food supply website that has stopped picking up the phone because it cannot keep up with the sudden influx of orders (Popken 2020). A similar company, LHB Industries, boasts about sales of emergency supplies soaring by between 200% and 1,700% during the first few months of the pandemic (Ibid.). The coronavirus has also caused a surge in firearm sales: just 12 days after former President Trump announced a national emergency (March 12, 2020), gun sales in the United States rose from 80,000 per day to more than 176,000 per day—initiating what would grow to become the three highest single months of gun purchases in U.S. history and surpassing 9/11 and the Columbine massacre by more than 12% (Collins and Yaffe-Bellany 2020). In addition, many of these are first-time gun purchases.

Years from now, in what we all hope will be post-pandemic, the most defining images of the coronavirus will certainly include exhausted health care workers, body bags piled in makeshift mortuaries, and single-use masks washed up on beaches. The COVID-19 crisis has propelled a rapid expansion in the production of plastic products (masks, gloves, body bags, etc.), with governments, hospitals, residential care facilities, schools, and so on competing to boost their private stockpiles. Meanwhile, everyday consumers are left fighting for their share of supplies and overwhelmed nations like the United States cease exports of PPE products to Canada and Latin America despite "significant humanitarian implications" (BBC 2020, n.p.). Globally, we are on pace to waste more than 129 billion face masks and 65 billion gloves (Silva et al. 2020); undoing years of work attempting to address the global issue of plastics pollution and demand for fossil fuel derivatives (Adyel 2020). According to reports from Wuhan, the epicenter of the COVID-19 outbreak, the increased demand for medical supplies produced more than 240 tons of single-use plastic waste per day (Eroglu 2020). If this projection is accurate, the United States will produce an entire year's worth of medical waste in just 2 months (Silva et al. 2020). Adding to this, individual choices during lockdown have seen take-out meals and pre-packaged groceries grow the plastics packaging market from US$909.2 billion in 2019 to US$1012.6 billion by 2021 (Ibid.). Collectively, the global health crisis has put immense pressure on waste management systems, posing major environmental challenges to municipal solid waste and hazardous biomedical waste management.

Before the pandemic, an estimated 2 billion people worldwide lacked access to waste collection, while 3 billion people lacked access to proper waste disposal (Sarkodie and Owusu 2020). Now, the impact on this already burdened industry has significantly amplified. In some areas, like New York City, commercial and industrial waste has decreased by roughly 50% (Kulkarni and Anantharama 2020). However, those same areas have seen residential solid waste generation increase by 5–30%, with the total volume of waste in the United States peaking nationally at 20% higher than normal in April 2020 (Ibid.). As social distancing has led to increases in online shopping and takeout services, a plethora of plastic-wrapped products being delivered to homes has only added to the enormous influx of plastic waste. For instance, the small island city-state of Singapore discarded an additional 1,470 tons of plastic waste from takeout packaging alone during just an eight-week period of lockdown measures (UNCTAD 2020). Add to this an overwhelming production of medical waste and those parts of the world hit hardest by COVID-19 might see an unprecedented increase in waste by upwards of 445% (Kulkarni and Anantharama 2020; Sarkodie and Owusu 2020). Collectively, the unrestrainable production of waste as a result of COVID-19 has caused illegal dumping to increase by an estimated 300% during lockdown (Sarkodie and Owusu 2020). Historical data tells us that more than 75% of COVID-related waste will end up mismanaged, filling landfills and city streets and floating in oceans for decades to come (UNCTAD 2020).

I've already been in this movie[3]

In *Shopping Our Way to Safety: How We Changed from Protecting the Environment to Protecting Ourselves*, Andrew Szasz persuasively argues that the relationship between individualism, capitalism, and consumerism effects a pronounced and insidious consequence—the decline of collective action, or what he calls "political anaesthesia" (2007, 195). Szasz's starting point is the well-traversed risk theory literature (Beck 1992) that argues that modernity is characterized by a significant increase in people's understanding that they live (and die) with not simply an increase in both the numbers and diversity of potentially harmful hazards, but also an appreciation that risks are inherent to our technologically dependent and driven society:

> ... indoor air is more toxic than outdoor air. That is because many household cleaning products and many contemporary home furnishings - carpets, drapes, the fabrics that cover sofas and easy chairs, furniture made of particle board - outgas toxic volatile organic chemicals. Ok, we will go outside - only to inhale diesel exhaust, particulates suspended in the air, molecules of toxic chemicals wafting from factory smokestacks. (2007, 1)

With little difficulty, we may add a growing number of risks to this list: COVID-19 and other viruses and bacterial infections and diseases (for instance, salmonella-infected onions recently occupied our local news), heavy metals, endocrine disrupting chemicals, phthalates, herbicides, pesticides, and various gases including methane, carbon dioxide, carbon monoxide, hydrogen, oxygen, nitrogen, and hydrogen sulphide. There are over 7 million known chemicals—80,000 of which are in commercial circulation with another 1,000 new chemicals being introduced each year (Wynne 1987, 48). Add to this the approximate 14,000 food additives and con-taminants added to landfills from discarded food. As Brian Wynne points out, the toxicity of household waste—batteries, cleaning and polishing fluids, cosmetics, medicines, etc.—is often higher than that of some industrial wastes (Ibid.). The greatest source of environmental cadmium is thought to be from batteries thrown out in domestic waste. Municipal waste incinerators are also known to emit dioxins and furans with their aerial discharges, possibly to worse levels than toxic waste incinerators (Ibid.; see also Hird 2021).

For Beck and other risk scholars, risk is perceived, filtered, and made sense of through neoliberal capitalism, which emphasizes a market economy, enhanced privatization, an overall decrease in government control in favor of industry control, and a general entrepreneurial approach to profit maximization (Crooks 1993; Foote and Mazzolini 2012; for general dis-cussions of neoliberal capitalist governmentality, see Burchell, Gordon, and Miller 1991; Foucault 1984, 1988). Neoliberal capitalism has successfully transformed citizens into in-dividuals, and individuals into consumers. Thus, it is as consumers that we are supposed to make sense of environmental risk:

> ... the brutal fact of ontological insecurity always has an ultimate addressee: the recipient of the residual risk of the world risk society is the *individual*. Whatever propels risk and makes it incalculable, whatever provokes the institutional crisis at the level of the governing regime and the markets, shifts the ultimate decision-making responsibility onto the individuals, who are ultimately left to their own devices with their partial and biased knowledge, with undecidability and multiple layers of uncertainty. This is undoubtedly a powerful source of right-wing radicalism and fundamentalism.... (Beck 2007, 195, italics in original)

What Beck identifies as right-wing radicalism and fundamentalism takes the much-publicized form of the U.S. Republican party's explicit climate-change denial (DePryck and Francois 2017), but also much more mundane everyday individual responses found in preparedness behaviors. Of course, survivalism in America has long been associated with conservatism, racism, sexism (women can goods while men shoot guns), and white supremacy movements (Lamy 1996). Sean Hannity, for instance, recommends prepping to *Fox News* viewers

(Kelly 2016). Of the myriad forms that individual responses to environmental risk might take, it is primarily in the "modality of consumer" within neoliberal capitalist societies (Szasz 2007, 4). As such, much of preparedness takes the form of consuming (and ultimately wasting) products; some products, as we have seen, that are specifically marketed as disaster preparation, such as high-end bunkers for billionaires, while far more products—indeed everything that *can* be purchased—now marketed as "general preparedness" for an unknowable future. The COVID-19 crisis has transformed such mundane items as toilet paper, tampons, bottled water, and toothpaste into consumer stockpile must-haves. During COVID-19's first wave, as social media overran with posts about which stores still had toilet paper in stock, companies such as Charmin reassured consumers that increased consumption was the most responsible response to the pandemic rather than as a citizen within a polity, which would implicate individuals within their community and society.

Thus, within neoliberal capitalist systems, attenuating environmental risk is to be achieved through relentless individual consumption. Each purchase is directed at protecting the individual consumer and her family rather than her community or society as a whole. We witness the enthusiasm, passion and time, attention, and energy that consumers devote to determining the "best" bottled water or the "best" organic underarm deodorant in the numerous posts on Facebook, Reddit, and other online forums devoted to consumer lifestyles. And having purchased the phthalate-free shampoo and Dasani bottled water, the individual demonstrates her care of herself (Foucault 1976), care for her family, and care for her environment. Millions of privileged individual acts of consumption that attempt to insulate the individual and her loved ones against environmental risk leave behind an increasing number of individuals, households, and communities that struggle with issues of safe drinking water, air quality, non-toxic food, and so on:

> An affluent minority—a savvy and influential minority whose political influence is disproportionately greater than their numbers—buys out of the toxic environment, believes it has taken care of the problem for themselves, and loses further interest in that particular toxic issue. Support for more substantive reform weakens. At best, as with organic foods, the situation will tend toward the creation of a permanent dual market, the larger of which consists of products manufactured in a toxic work environment and contain toxic ingredients, and whose production and consumption continues to discharge these substances into the environment. (Szasz 2007, 208)

Companies such as Airinum have rushed in to profit from the new COVID-19 "mask mania" by selling designer masks for $99 or more. From Airinum's website we learn that:

> We take around 20,000 breaths every day and air is essential for our survival. When we breathe poor quality air, it can severely damage our health and contribute to asthma, respiratory diseases, cancer, strokes and even death. In fact, 7 million people die annually as a result of poor air quality, a number far too high. Whether you need protection from toxic air pollution, itchy pollen, to stay away from bacteria or simply want to be at your very best, the Urban Air Mask will empower you to breathe cleaner and healthy air.

Airinum explicitly acknowledges (indeed highlights) the effect of air pollution in killing 7million people per year. But the solution is not to improve air quality—to take the lead in producing less, lobbying governments, supporting affected communities—but to increase consumption—what industry calls "empowering" the individual privileged enough to afford their designer masks. And, familiarly, the privileged few respond accordingly within the neoliberal capitalist script, as these testimonials on Airinum's website attest:

> I was so pleased with these respirators, they fit perfectly, look great, and actually do the job. After wearing it for about an hour, I decided to take it off, thinking that the bushfire smoke had dissipated. However, as I began to remove the mask, I quickly realised this was not the case, rather *the mask had filtered out the smoke so well that it made me think the air was clear.* (David M., our emphasis)

> I wear the mask when walking through the centre of London for half an hour, two times a day. The mask fits well and keeps my nose and mouth covered. It does it's job, I can't smell the pollution or cigarette smoke—which I was really happy with as I have a strong sense of smell! The filter feels substantial, *I feel protected from the polluted environment in built-up areas.* (Emma M., our emphasis)

David need not question the association between climate change and Australia or California's recent forest fires and Emma need not question why London's air is polluted; they opt out of these concerns by insulating themselves from risk by buying designer masks. Indeed, Gwenyth Paltrow, whose company Goop is valued at $250 million, sports an Urban Air Mask from the comfort of her private jet in her Instagram post to which she claims to have "already been in this movie", referring to her role in the virus pandemic film *Contagion*. How luxurious to experience the COVID-19 pandemic as a movie role. Paltrow, the Kardashians, other celebrities and elites like the Walton family—who own Walmart, which sells prepping kits—each endorse prepping.

Designer masks and designer hand sanitizer are just the tip of a vast, expensive iceberg of products that the privileged are taking advantage of during this global pandemic. Those wealthy enough are opting to use their private planes or charter jets for "evacuation flights" out of infected areas and into vacations: "Avoid coronavirus by flying private...Request a quote today!" advertises Southern Jet. Others are opting to spend time on their yachts or rented islands (feat. Kim Kardashian's private island birthday party in October 2020) to isolate themselves from infected shoreline communities. "It totally makes sense", the president of B&B Yacht Charter Jennifer Saia exclaimed, whose family spent its spring vacation on a yacht in the Bahamas instead of their regular villa in Italy:

> You're keeping your family contained in a very small, should-be-clean environment. And going from your car to your F.B.O. [private jet terminal] to your private jet right onto the tarmac. And from there, right onto your yacht, and not having to deal with the public. (quoted in Williams and Bromwich 2020, n.p.)

The environmental gain of fewer middle-class people flying overseas has been offset by the dramatic increase in wealthier people using private jets (Sullivan 2020; Wagner 2020). And now, according to *The New York Times* (2020) investigation, U.S. health care providers are offering what amounts to private VIP emergency room memberships so that wealthy people may further avoid contact with "the public" should they require hospital care. Luxury bunker sales are on the rise again, as they were at the outset of the Cold War.

A burgeoning literature illuminates the relationship between environmental risk and inequity (see for instance Adeola 2012; Nixon 2011) and consumption and environmental harm more specifically (see for instance Princen, Maniates, and Conca 2002; Dauvergne 2010). In this chapter, our focus is on consumption as the primary mode of environmental risk preparedness, and its consequences for waste generation and management. Structural inequality is built into risk:

> There is no ontology of risk. Risks do not exist independently, like things. Risks are risk conflicts in which there is a world of difference between the decision-makers, who could ultimately avoid the risks, and the involuntary consumers of dangers, who do not have a say in these decisions and onto whom the dangers are shifted as "unintentional, unseen side effects". (Beck 2007, 195)

Risk and privilege are relative: middle-class suburbanites may not control the economic system that, say, defines and decides acceptable levels of DDT administered on crop lands in Somalia (although they may), but they do have the economic power to move themselves and their families to more environmentally safe enclaves, and to further insulate themselves through product purchases that are prohibitively expensive for millions of people. As the *Doomsday Preppers* television series recounts in episode after episode, many preppers assume the risk of financial ruin in order to stockpile US$100,000 or more on food stocks (see for instance, S1E2).

As we have demonstrated, prepping constitutes an increase—often exponentially—in consumption and what Peter Dauvergne (2010) terms its "shadow": increasing extraction and manufacturing producing environmental degradation, which significantly includes waste. As Szasz points out with regard to the consumption of bottled water:

> To make a bottle full of (more or less) clean water, mountain springs are turned into industrial sites, plastic polymers have to be produced, energy used, hazardous wastes and postconsumer, "solid" wastes generated. The production process transforms nature, pollutes nature. Clean consumption is compatible with, even requires, dirty production. (2007, 197)

Just as we know that most waste is produced during products' production phase (see for example Liboiron 2013, 2014; Hird 2021) rather than its post-consumption phase, we also know that factors such as the "extremely large scale of modern industrial life", economic globalization, and economic inequality all contribute to what Jennifer Clapp terms "waste distancing" or the geographical, economic, and political distance placed between consumers and waste (2002, 155). This distancing serves to manage waste in ways that do not disturb—and indeed tends to increase—circuits of mass production, mass consumption, and, *ergo*, industry profit (Hawkins 2006; Kollikkathara et al. 2009; Lynas 2011).

Global climate change, soil depletion and degradation, biodiversity loss, contaminated water, and air pollution are all exactly this: global in their ultimate reach and effects. Environmental racism and inequality research demonstrate that, so far, the most acute and profound effects of environmental degradation are experienced by already disenfranchised millions. In other words, people who prepare for environmental disaster are those who are sufficiently privileged to be able to prepare; everyone else is already living (and dying) with environmental disaster. Our contemporary neoliberal capitalist characterization of prepping during the COVID-19 pandemic—defending ourselves and loved ones, accumulating survivalist skills and things—obscures the complex web of interdependence that *produces* the products and services that we consume. That is, ever-increasing consumption encourages individuals to believe that they are protected while masking (as double-entendre) the intended and unintended consequences of profit-driven production and consumption in terms of rights-diminished and impoverished labor, community economic, political and cultural devastation, (increasingly toxic) waste generation, and so on. As such, prepping to protect ourselves and our loved ones contributes to the very disasters that this behavior seeks to avoid.

An ounce of prevention is worth a pound of cure

The end of the Cold War may have abated anxieties about nuclear conflict and its radioactive fallout (although even this now feels more fragile thanks to Trump's apocalyptic bromance with North Korea) or we may be simply living with additional apocalyptic scenarios. The COVID-19 pandemic presents an opportunity to revisit the phenomenon of prepping from the perspective of living in present disaster rather than the more abstract disaster-to-come anticipated by the traditional prepper. Where once represented as an extreme response to general anxieties by a relatively small and generally white, conservative group of people living at the margins of society, COVID-19 joins climate change, biodiversity loss, and other already-here realities that are making preppers out of everyone with the means to purchase and stockpile consumer goods. In short, the COVID-19 pandemic has mainstreamed prepping.

Since the Cold War, the neoliberal capitalist leitmotif to environmental disaster and degradation is *blame and shop, shop and blame*. Patriarch Filaret, head of the Ukrainian Orthodox Church of the Kyiv Patriarchate, publicly stated that the COVID-19 pandemic is "God's punishment for the sins of men and sinfulness of humanity...First of all, I mean same-sex marriage. This is the cause of the coronavirus" (Villareal 2020, n.p.). After Filaret tested positive for the coronavirus, his church defended him, stating, "As the head of the church and as a man, the Patriarch has the freedom to express his views, which are based on morality" and asked the public to "pray for His Holiness Patriarch Filaret, so that the All-Merciful and Almighty Lord God will heal the Patriarch" (Ibid). Former President Donald Trump and his supporters blame China and the WHO for the "China virus" while the Chinese government blames the U.S. army for the virus (Winter 2020). And as we have shown in this chapter, along with the familiar blame-game, we are following the same, and in many cases wildly increased, patterns of consumption: "eco-doom as consumerist spectacle" as Ginn puts it (2015, 353). By marketing design, we consume to insulate ourselves from each other:

> Community, empathy and logic are not easy to merchandise; they are not profitable to corporations who like to keep us divided and conquered. There is *a lot* of money to be made out [of] prepping. Paranoia sells. Guns and ammo and other prepping gear such as underground bunkers and security, food and water, and so on is an extremely lucrative marketplace. (Foster 2014, 16, emphasis in original)

Insidiously, these individual acts of consumption (multiplied by millions of people, repeated billions of times), are integral to how neoliberal capitalism defines and structures morality, or our relationship to both our immediate and distant communities:

> In combination with neoliberalism, the individual becomes his own "moral entrepreneur" and thus holds the fate of civilization in [her] hands. The result is a new "categorical imperative": act as though the fate of the world depends on your action. Separate your waste, ride a bicycle, use solar energy, etc. The key contradiction which is both obscured and revealed here is that the individual is

condemned to individualization and self-responsibility, even vis-a-vis global threats, despite the fact that [she] is severed from the decision contexts which escape [her] influence. (Beck 2007, 169)

Recall that immediately after the 9/11 attacks on September 11, 2001, President George W. Bush advised citizens to go out and shop (Leonard and Conrad 2010). After the first COVID-19 wave, France's government told its citizens to "go out and start spending" (The Local 2020, n.p.). Not only are we encouraged to believe that increased consumption will resolve global problems such as disease and terrorist attacks, but as Max Liboiron observes, this individualization of responsibility produces a "constant misalignment of scale" (2014, n.p.) whereby the extractive and manufacturing industries produce far more waste in the production of the products we buy (or that remain stock-piled in warehouses if we do not buy) and individuals voluntarily shoulder the burden of responsibility for the ensuing environmental degradation, including waste (Hird 2021). Increases in extraction, production, and distribution (far more than consuming) brings peak oil, toxic waste, and other global environmental crises closer. As such, what characterizes both pre- and post-COVID prepping is increased industrial production—including weaponry—and increasing waste that is increasingly toxic, increasingly finding its way into land and lakes, rivers, seas and oceans, and into our bodies via waste, such as microplastics. One of prepping's significant ironies is that instead of leading to independence from the neoliberal capitalist industrial complex, prepping depends upon it. Indeed, neoliberal capitalism and the industrial manufacturing complex *created* prepping, and thus cannot be called upon to resolve ever-increasing production and consumption.

Former U.S. President Trump's 2020 election rhetoric rehashed the neoliberal capitalist accusation familiar since America's participation in the Cold War: that any emphasis on community is "socialist and anti-American" (Foster 2014, 16). Yet, we also see a lively counter-narrative in grassroots movements, such as Black Lives Matter and Idle No More, that are clearly demonstrating that millions of people do not have access to the (privileged) individual response of self-insulation from harm, and are instead building and supporting community-level solutions. Various alternative futures, from economic steady-state theories, de-growth initiatives, transition towns, and citizen libraries to anti-capitalist youth movements challenge individual-level neoliberal capitalist responses that are accelerating environmental degradation and disaster (Kallis and March 2015; Liboiron 2014). Even some preppers resist their portrayal in the media as hyper-consumers. Indeed, as Mills found, prepper concerns are "overwhelmingly non-apocalyptic" and "that any worst-case disaster was…going to be temporary" (2019, 6). Out of the 39 preppers in his ethnographic study, all of them stockpiled food for only a few months, and only one couple had a bunker. Still, others identify themselves as preventers rather than preppers. For example, preventers put their energies into adopting renewable energy systems such as solar panels and windmills, as well as building environmental resiliency through permaculture and other Indigenous practices. Preventers reuse, refurbish, and repurpose materials, frequenting junkyards and yard sales rather than Costco and Amazon. For some, the COVID-19 crisis has provided an opportunity to build stronger connections with their communities; adapting to limitations of the global supply chain and transitioning to a more sustainable way of living (Bodenheimer and Leidenberger 2020; Forster et al. 2020). So, whereas preppers largely practice what Andrew Szasz terms "inverted quarantine", aiming to *isolate themselves* from environmental disaster through heightened consumption (and thus more waste production)—a practice that has significantly increased during the COVID-19 pandemic through both voluntary and forced confinement—preventers set their intentions on challenging the conditions that are leading to disaster.

Perhaps the third little pig has it right: not only does he save himself, but he works hard to create a robust shelter for his brothers. We might interpret this fable within the traditional prepper context: through hard work and preparation, save yourself and your loved ones from whatever wild calamity might befall you (a wolf that can destroy homes with his breath). But we choose a more generous reading: it might be a hard lesson learned (two pigs lose their homes), but the best way to thrive in the world is to prevent disaster *collectively*.

Notes

1   Myra Hird thanks Samm Medeiros for her early assistance in reviewing the literature on preppers.
2   Quote by Nellie Bowles, *The New York Times*, April 24, 2020.
3   Quote by Gweneth Paltrow via Instagram. 26 February 2020.

References

Adeola, F. O. 2012. *Industrial Disasters, Toxic Waste, and Community Impact: Health Effects and Environmental Justice Struggles Around the Globe.* Lexington Books.

Adyel, T. M. 2020. "Accumulation of Plastic Waste During COVID-19." *Science* 369: 1314–1315.

Alexander, C., and A. Sanchez. 2018. *Indeterminacy: Waste, Value, and the Imagination.* Berghahn Books.

BBC News. 2020. "Coronavirus: US 'wants 3M to end mask exports to Canada and Latin America'", [online], 3 April. Available from: https://www.bbc.com/news/world-us-canada-52161032. Accessed May 21, 2021.

Beck, U. 1992. *Risk Society: Towards a New Modernity.* SAGE.

Beck, U. 2007. *World at Risk.* Polity Press.

Beckett, C., and A. Keeling. 2019. "Rethinking Remediation: Mine Reclamation, Environmental Justice, and Relations of Care." *Local Environment* 24: 216–230.

Bedford, E. 2020. "Canadian Sales Growth of Personal Care and Cleaning Products Due to Coronavirus 2020." Statista. https://www.statista.com/statistics/1110802/coronavirus-growth-in-personal-care-and-cleaning-product-sales-canada/

Bharti, B. 2020. "Coronavirus Updates: Stockpile Food and Meds in Case of Infection, Canada's Health Minister Says." *National Post.* https://nationalpost.com/news/world/coronavirus-live-updates-who-covid19-covid-19-italy-china-canada-wuhan-deaths. Accessed November 6, 2020.

Bodenheimer, M., and J. Leidenberger. 2020. "COVID-19 as a Window of Opportunity for Sustainability Transitions? Narratives and Communication Strategies Beyond the Pandemic." *Sustainability: Science, Practice and Policy* 16: 61–66.

Bowles, N. 2020. "I Used to Make Fun of Silicon Preppers. And Then I Became One." *The New York Times.* https://www.nytimes.com/2020/04/24/technology/coronavirus-preppers.html.

Brown, B. nd. "The Prepping Guide, SHTF Plans." https://theppreppingguide.com/about/. Accessed October 26, 2020.

Brown, K. 2013. *Plutopia: Nuclear Families, Atomic Cities, and the Great Soviet and American Plutonium Disasters.* Oxford University Press.

Burchell, G., C. Gordon, and P. Miller. 1991. *The Foucault Effect: Studies in Governmentality.* University of Chicago Press.

Carr, M. 2012. "The Doomsday 'Preppers' Portfolio." Investment U. https://www.investmentu.com/article/detail/30710/the-doomsday-preppers-portfolio. Accessed November 10, 2020.

Clapp, J. 2002. "The Distancing of Waste: Overconsumption in a Global Economy." In T. Princen, M. Maniates, and K. Conca (Eds.) *Confronting Consumption.* The MIT Press. 155–176.

Clearwater, D. A. 2010. "Living in a Militarized Culture: War, Games and the Experience of U.S. Empire." *Topia (Montreal)* 23–24: 260–285.

Collie, M. 2020. "Yes, You Should Have a Coronavirus Emergency Kit. Here's What to Include." *Global News.* https://globalnews.ca/news/6665520/coronavirus-emergency-kit/. Accessed November 10, 2020.

Collins, K., and D. Yaffe-Bellany. 2020. "About 2 Million Guns Were Sold in the U.S. as Virus Fears Spread." *The New York Times.* https://www.nytimes.com/interactive/2020/04/01/business/coronavirus-gun-sales.html. Aaccessed November 10, 2020.

Conway, J. 2020. "Cleaning Products Sales Growth From the Coronavirus in the U.S. in March 2020." *Statista*. https://www.statista.com/statistics/1104333/cleaning-product-sales-growth-from-coronavirus-us/.

Crooks, H. 1993. *Giants of Garbage: The Rise of the Global Waste Industry and the Politics of Pollution Control*. Lorimer.

Dauvergne, P. 2010. *The Shadows of Consumption: Consequences for the Global Environment*. The MIT Press.

Davis, H. 2015. "Life & Death in the Anthropocene: A Short History of Plastic." In H. Davis and E. Turpin (Eds.) *Art in the Anthropocene: Encounters Among Aesthetics, Politics, Environments and Epistemologies*. Open Humanities Press. 346–358.

DePryck, K., and G. Francois. 2017. "The Denier-in-Chief: Climate Change, Science and the Election of Donald J. Trump." *Law and Critique* 28: 119–126.

Doomsday Preppers, Season 1. 2012. *Produced by National Geographic & Sharp Entertainment*. National Geographic.

Eddy, C. 2014. "The Art of Consumption: Capitalist Excess and Individual Psychosis in Hoarders." *Canadian Review of American Studies* 44: 1–24.

Edwards, F., and D. Mercer. 2007. "Gleaning From Gluttony: An Australian Youth Subculture Confronts the Ethics of Waste." *Australian Geographer* 38: 279–296.

Eroglu, H. 2020. "Effects of COVID-19 Outbreak on Environment and Renewable Energy Sector." *Environment, Development and Sustainability* 23, no. 1: 4782–479.

Fauci, A. S., H. C. Lane, and R. R. Redfield. 2020. "COVID-19—Navigating the Uncharted." *The New England Journal of Medicine* 382: 1268–1289.

Foote, S., and E. Mazzolini. 2012. *Histories of the Dustheap: Waste, Material Cultures, Social Justice*. The MIT Press.

Foster, G. A. 2014. *Hoarders, Doomsday Preppers, and the Culture of Apocalypse*. Palgrave Macmillan.

Forster, P. M., H. I. Forster, M. J. Evans et al. 2020. "Current and Future Global Climate Impacts Resulting from COVID-19." *Nature Climate Change* 10: 913–919.

Foucault, M. 1976. *The History of Sexuality: The Care of the Self*. Editions Gallimard.

Foucault, M. 1984. "The Politics of Health in the Eighteenth Century." In P. Rabinow (Ed.) *The Foucault Reader*. Pantheon Books. 273–289.

Foucault, M. 1988. "Technologies of the Self." In L. H. Martin, H. Gutman, and P. H. Hutton (Eds.) *Technologies of the self: A seminar with Michel Foucault*. Tavistock Publications. 16–49.

Freud, S. [1920] 2020. *A General Introduction to Psychoanalysis*. Outlook Verlag GmbH.

Garrett, B. 2020. "Doomsday Preppers and the Architecture of Dread." *Geoforum*. doi:10.1016/j.geoforum.2020.03.014.

Ginn, F. 2015. "When Horses Won't Eat: Apocalypse and the Anthropocene." *Annals of the Association of American Geographers* 105: 351–359.

Giorgio, A., and H. R. Daniel. 1998. *Homo Sacer: Sovereign Power and Bare Life*. Stanford University Press.

Gordillo, G. R. 2014. *Rubble: The Afterlife of Destruction*. Duke University Press.

Hannan, J. 2018. "Trolling Ourselves to Death? Social Media and Post-Truth Politics." *European Journal of Communication* 33: 214–226.

Hawkins, G. 2006. *The Ethics of Waste: How We Relate to Rubbish*. Rowman & Littlefield.

Heath, R. 2020. "Who's to Blame?" *Politico Nightly*. https://www.politico.com/newsletters/politico-nightly-coronavirus-special-edition/2020/04/03/whos-to-blame-488802 (accessed January 21, 2021).

Hesse, M., and D. Zak. 2020. "The History and Mystery of Purell, the Most Sacred Goo of Our New Era." *The Washington Post*. https://www.washingtonpost.com/lifestyle/style/the-power-of-purell-compels-you/2020/03/26/41243960-6dde-11ea-b148-e4ce3fbd85b5_story.html. Accessed November 6, 2020.

Hird, M. J. 2021. *Canada's Waste Flows*. McGill-Queen's University Press.

Hoover, E. 2017. *The river is in us: Fighting toxics in a Mohawk community*. University of Minnesota Press.

James, A. 2016. "10 Multi-Billion Dollar Niches that Are Prime for the Taking." *Brand Builders*. https://www.brandbuilders.io/10-great-niches-for-marketers/. Accessed November 10, 2020).

John Hopkins University & Medicine. 2020. "COVID-19 Dashboard by the Centre for Systems Science and Engineering (CSSE) at John Hopkins University." https://coronavirus.jhu.edu/map.html

Kallis, G., and H. March. 2015. "Imaginaries of Hope: The Utopianism of Degrowth." *Annals of the Association of American Geographers* 105: 360–368.

Kanygin, J. 2020. "Gun and Ammunition Sales up in Alberta Amid COVID-19 Pandemic." *CTV News Calgary*. https://calgary.ctvnews.ca/gun-and-ammunition-sales-up-in-alberta-amid-covid-19-pandemic-1.4867969. Accessed November 6, 2020.

Kelly, C. R. 2016. "The Man-Pocalypse: Doomsday Preppers and the Rituals of Apocalyptic Manhood." *Text and Performance Quarterly* 36: 95–114.

Klein, N. 2008. *The Shock Doctrine: The Rise of Disaster Capitalism*. Vintage Canada.

Kristeva, J. 1982. *Powers of Horror: An Essay on Abjection*. Columbia University Press.

Kollikkathara, N., H. Feng, and E. Stern. 2009. "A Purview of Waste Management Evolution: Special Emphasis on USA". *Waste Management* 29: 974–985.

Kulkarni, B. N., and V. Anantharama. 2020. "Repercussions of COVID-19 Pandemic on Municipal Solid Waste Management: Challenges and Opportunities." *Science and the Total Environment* 743: 140693.

Lamy, P. 1996. *Millennium Rage: Survivalists, White Supremacists, and the Doomsday Prophecy*. Springer.

Larson, L. R. L., and H. Shin. 2018. "Fear During Natural Disaster: Its Impact on Perceptions of Shopping Convenience and Shopping Behaviour." *Services Marketing Quarterly* 39: 293–309.

Laycock, R., and S. Binstead. 2020. "Doomsday Prepper Statistics." *Finder*. https://www.finder.com/ca/doomsday-prepper-statistics.

Leonard, A., and A. Conrad. 2010. *The Story of Stuff: The Impact of Overconsumption on the Planet, Our Communities, and Our Health—and How We Can Make it Better*. Free Press.

Lepawsky, J. 2018. *Reassembling Rubbish: Worlding Electronic Waste*. MIT Press.

Liboiron, M. 2013. "Modern waste as strategy." *Lo Squaderno: Explorations in Space and Society* 29: 9–12. http://www.losquaderno.professionaldreamers.net/?p=1480.

Liboiron, M. 2014. "Solutions to Waste and the Problem of Scalar Mismatches." *Discard Studies*. https://discardstudies.com/2014/02/10/solutions-to-waste-and-the-problem-of-scalar-mismatches/. Accessed October 20, 2020.

The Local. 2020. "French Told to 'Go Out and Start Shopping' to Relaunch Economy." https://www.thelocal.fr/20200529/french-told-go-out-and-start-spending-to-relaunch-the-economoy?fbclid=IwAR1H1SNc227Sm3LdJ65L4-sPbhkhYBWN8h53TPuC9LuUChB-K3tvSHUmvbQ. Accessed November 10, 2020.

Lundin, C. 2007. *When All Hell Breaks Loose: Stuff You Need to Survive When Disaster Strikes*. Gibbs Smith.

Luther, D. 2015. *The Prepper's Water Survival Guide*. Ulysses Press.

Lynas, M. 2011. *The God Species: How Humans Really Can Save the Planet*. Fourth Estate.

Malm, A., and A. Hornborg. 2014. "The Geology of Mankind? A Critique of the Anthropocene Narrative." *The Anthropocene Review* 1: 62–69.

Mills, M. F. 2019. "Preparing for the Unknown…Unknowns: 'Doomsday' Prepping and Disaster Risk Anxiety in the United States." *Journal of Risk Research* 22: 1–13.

Murphy, T. 2013. "Preppers Are Getting Ready for the Barackalypse." *Mother Jones News*. https://www.motherjones.com/politics/2012/12/preppers-survivalist-doomsday-obama/.

Nelson, R. R. 1991. "Diffusion of Development: Post-World War II Convergence Among Advanced Industrial Nations." *The American Economic Review* 81: 271–275.

Nixon, R. 2011. *Slow Violence and the Environmentalism of the Poor*. Harvard University Press.

Ozimek, A. 2017. "How to Doomsday Prep Like an Economist." *Forbes*. https://www.forbes.com/sites/modeledbehavior/2017/09/30/how-to-doomsday-prep-like-an-economist/#690f62584d47. Accessed November 11, 2020.

Pantano, E., G. Pizzi, D. Scarpi, and C. Dennis. 2020. "Competing During a Pandemic? Retailers' Ups and Downs During the COVID-19 Outbreak." *Journal of Business Research* 116: 209–213.

Paratus Business News. 2017. "How Survival Businesses Can Profit From New Markets (a.k.a. Liberal Preppers)." http://paratusnews.com/forbes-how-survival-businesses-can-profit-from-new-markets-a-k-a-liberal-preppers/. Accessed November 11, 2020.

Paterson, S. 2016. "Japan's Tsunami Debris: Five Remarkable Stories." *BBC News Asia*. https://www.bbc.com/news/world-asia-35638091. Accessed October 26, 2020.

Pennington, T., and D. Luther. 2014. *The Prepper's Blueprint: The Step-By-Step Guide to Help You Prepare for Any Disaster*. CreateSpace Independent Publishing Platform.

Popken, B. 2020. *From Disaster Bunkers to Dried Food, Survival Supply Sales Are Spiking*. NBC News. Accessed November 10, 2020.

Princen, T., M. Maniates, and K. Conca. 2002. *Confronting Consumption*. The MIT Press.

Robbins, P., and S. A. Moore 2012. "Ecological Anxiety Disorder: Diagnosing the Politics of the Anthropocene." *Cultural Geographies* 20: 3–19.

Roe, B. E., K. Bender, and D. Qi. 2020. "The Impact of COVID-19 on Consumer Food Waste." *Applied Economic Perspectives and Policy* 3, 1: 401–411. 10.1002/aepp.13079

Sarkodie, S. A., and P. A. Owusu. 2020. "Impact of Meteorological Factors on COVID-19 Pandemic: Evidence From Top 20 Countries With Confirmed Cases." *Environmental Research* 191: 110101.

Scheider-Mayerson, M. 2013. "Disaster Movies and the 'Peak Oil' Movement: Does Popular Culture Encourage Eco-Apocalyptic Beliefs in the United States?" *Journal for the Study of Religion* 7: 289–314.

Sheu, J., and H. Kuo. 2020. "Dual Speculative Hoarding: A Wholesaler-Retailer Channel Behavioural Phenomenon Behind Potential Natural Hazard Threats." *International Journal of Disaster Risk Reduction* 44: 101430.

Silva, A. L. P., J. C. Prata, T. R. Walker et al. 2020. "Increased Plastic Pollution Due to COVID-19 Pandemic: Challenges and Recommendations." *Chemical Engineering Journal* 126683.10.1016/j.cej. 2020.126683

Sohrabi, C., Z. Alsafi, N. O'Neill et al. 2020. "World Health Organization Declares Global Emergency: A Review of the 2019 Novel Coronavirus (COVID-19)." *International Journal of Surgery* 76: 71–76.

Statistics Canada. 2020. "Canadian Consumers Prepare for COVID-19." https://www150.statcan.gc.ca/n1/en/pub/62f0014m/62f0014m2020004-eng.pdf?st=tk2_Wyh_.

Stein, M. 2008. *When Technology Fails: A Manual for Self-Reliance, Sustainability and Surviving the Long Emergency.* Chelsea Green Publishing.

Strasser, S. 1999. *Waste and Want: A Social History of Trash.* Metropolitan Books.

Sullivan, P. 2020. "Wealthy Fliers Worried About Coronavirus Turn to Private Jet Service." *New York Times.* https://www.nytimes.com/2020/05/30/your-money/coronavirus-private-jets.html. Accessed October 26, 2020.

Szasz, A. 2007. *Shopping Our Way to Safety: How We Changed from Protecting the Environment to Protecting Ourselves.* University of Minnesota Press.

Tactical.com. nd. "A Complete Beginners Guide to Prepping." https://www.tactical.com/complete-beginners-guide-to-prepping/. Accessed October 26, 2020.

Toh, M. 2020. "'It's Crazy': Panic Buying Forces Stores to Limit Purchases of Toilet Paper and Masks." *CNN.* https://www.cnn.com/2020/03/06/business/coronavirus-global-panic-buying-toilet-paper/index.html. Accessed November 6, 2020.

UNHCR. 2016. "UNHCR: Global Trends, Forced Displacement in 2016". Retrieved 17 May 2018 from http://unhcr.org/5943e8a34.pdf

UNCTAD. 2020. "Growing Plastic Pollution in Wake of COVID-19: How Trade Policy Can Help." https://unctad.org/news/growing-plastic-pollution-wake-covid-19-how-trade-policy-can-help

Vanham, P. 2020. "World Economic Forum launches COVID-19 Action Platform to fight coronavirus." *World Economic Forum.* https://www.weforum.org/agenda/2020/03/world-economic-forum-launches-covid-action-platform/. Accessed November 11, 2020.

Villareal, D. 2020. "Ukrainian Church Leader Who Blamed COVID-19 on Gay Marriage Tests Positive for Virus." *NBC News.* https://www.nbcnews.com/feature/nbc-out/ukrainian-church-leader-who-blamed-covid-19-gay-marriage-tests-n1239528. Accessed October 15, 2020.

Wagner, S. 2020. "Can You Own a Private Jet If You Care About Climate Change?" *BNN Bloomberg.* https://www.bnnbloomberg.ca/can-you-own-a-private-jet-if-you-care-about-climate-change-1.1365986. Accessed October 20, 2020.

Watt-Cloutier, S. 2015. *The Right to Be Cold: One Woman's Story of Protecting Her Culture, the Arctic and the Whole Planet.* Penguin Canada.

Williams, A., and J. E. Bromwhich. 2020. "The Rich Are Preparing for the Coronavirus Differently." *The New York Times.* https://www.nytimes.com/2020/03/05/style/the-rich-are-preparing-for-coronavirus-differently.html. Accessed October 7, 2020.

Winter, L. 2020. "Chinese Officials Blame US Army for Coronavirus." *The Scientist.* https://www.the-scientist.com/news-opinion/chinese-officials-blame-us-army-for-coronavirus-67267. Accessed October 26, 2020.

Wood, E. M. 2002. "The Agrarian Origin of Capitalism." In *The Origin of Capitalism: A Longer View.* Verso. 95–121.

World Bank. "Trends in Solid Waste Management." https://datatopics.worldbank.org/what-a-waste/trends_in_solid_waste_management.html. Accessed October 24, 2020.

Wynne, B. 1987. *Risk Management and Hazardous Waste: Implementation and Dialectics of Credibility.* Springer-Verlag.

Zalasiewicz, J., and W. Mark. 2015. "First Atomic Bomb Test May Mark the Beginning of the Anthropocene." *The Conversation.* https://theconversation.com/first-atomic-bomb-test-may-mark-the-beginning-of-the-anthropocene-36912. Accessed August 5, 2020.

INDEX

A

ABC News 227
Aboriginal Land Rights Act (1974, Australia)
    226, 230
Abu-Lughod, L. 85
Abu-Rish, Z. 78
The Accursed Share: An Essay on General Economy
    (Bataille) 269
Actor Network Theory 6
actor-network-theory (ANT) 105
Adam, Barbara 90, 100, 233
Adeola, F. O. 307, 314
Adyel, T. M. 311
African Reclaimers Organization 297
Agamben, Giorgio 271
Agarwal, Anil 256
Agency for Toxic Substances and Disease Registry
    (US) 245
agentive materialities 105
Age of Humans 90
Agrawal, A. 285
Aguiar, L. 48
Aguiar-Castillo, L. 201
Agyeman, J. 41
Ahmann, Chloe 45
Akese, G. AS. 144
Alabi, Okunola 129
Alaimo, S. 234
Albertarelli, S. 198
Albright, C. 242
Alene, N. B. 292
Alexander, Catherine 6, 11, 20, 21, 22, 23, 25, 26,
    27, 63, 65, 106, 142, 213, 215, 307
Alexander, I. 144
Alexievitch, Svetlana 216, 217

Alexis-Martin, Becky 216, 217, 230
Allred, N. 228, 229
Almeida, E. T. V. de 276
Al-Qadiri, Monira 14, 16, 263, 268, 269, 272, 273
Al-Shuruq newspaper 77
Alt-Globalization protest movement 256
American Depression of 1930s 98
anaerobic digesters 6
Anand, N. 97
Anantharama, V. 311
Anders, Günther 216
Anderson, E. N. 105, 108, 109
Andrady, Anthony 63
animism as relational ontology 105
Anthropocene epoch 63, 90, 91, 100, 215, 309
anti-litter campaigns 69
Aparcana, S. 276
apartheid spatial policies (South Africa) 291
apocalypse 269; environmental 309; preppers and
    307–309
Appadurai, Arjun 98, 212
Appel, H. 97
Aquino, I. F. 279, 280
archaeology 91
Arefin, Mohammed Rafi 10
Arenas-Huertero, F. J. 129
Argyrou, V. 74
Arias, R. 244
Armien-Ally 298
art and aesthetics in landfill sites 64–65
Associated Press 216
The Association of Plastic Recyclers 182
Aston, J. 231
Atomic Energy Authority, UK 225
atomic fission 215

atomic tests 215; in Australia by UK 224, 225, 226, 227, 229–230
Aurisicchio, M. 142
Austin, T. 228
Australian Institute for Aboriginal Torres Strait Islander Studies (AIATSIS) 228
avoidance to keep matter in place 36–37
Ayalon, O. 205

B
Badges, Levels/Leaders, Achievements, Points (BLAP) gamification system 197
Baines, Kristina 108, 111
Bakker, K. 145
Balayannis, Angeliki 35, 36, 142
Ballogg, Miles 245, 246
Bamako Convention, 1991 (African Union) 260
Bandhauer, Karen 182
Barchiesi, F. 293
Barham, Brad 189
Barles, S. 91, 93, 94
Barnard, Alex V. 7
Barnett, M. 49
Barron, E. S. 250
Barthe, Yannick 215
Bartlett, H. C. 32
Bartlett, Lesley 125
Basel Convention on the Control of Transboundary Movements of Hazardous Wastes and Their Disposal, 1992 7, 8, 14, 255, 258, 259, 263; recycling loophole in 259–260
Bastos, V. P. 279
Bataille, George 14, 267, 269, 270, 271
Baum, K. 36
Bauman, Zygmunt 9, 41
Baxter, W. 142
BBC News 254
BBC News Africa 260
Beck, Ulrich 216, 311, 312, 315, 317
Beckett, C. 36
Bedford, E. 310
Behind the Sun (video, Al-Qadiri) 14, 267, 268–269, 273
Beierle, Thomas C. 262
Being Nuclear: Africans and the Global Uranium Trade (Hecht) 21
Beirut port explosion, August 2020 254
Bell, Lucy 106, 232
Bell, Shannon 38
Bender, K. 310
Bennett, Carys E. 90
Bennett, J. 41, 105
Bennett, Jane 213
Bentham, Jeremy 9
Benton, Adia 49
Benz, A. 285

Berger, A. 240, 242, 243
Berkes, Fikret 105
Berlant, L. 45
Berthier, Castillo 127
Bertoni, Filippo 62
Besen, G. R. 276, 279, 285
Bess, Michael K. 128
Bhandar, Brenna 240
BinCam system 201, 205
BinLeague online platform 201
Binstead, S. 310
biography of things 98
bio-metaphors 98
bionecropolitics 271
biopolitics 16
biopower 46, 267
bioscience waste and whiteness 42
Bird Rose, Deborah 229, 232
Blaauw, P. F. 292, 293
black and indigenous people of color (BIPOC) 41, 45, 46, 47, 48, 49, 239, 240
Blainey, G. 224
Blight, Geoffrey 185
Bloom, Jonathan 6
Blowers, Andrew 211
Bodenheimer, M. 317
Boeira, S. L. 279, 280
Boetzkes, Amanda 14, 16, 64
Bohm, Robert A. 175
Boisvert, E. 227
Bol, Lucia 111
Bolsinaro, Jair 282
Bonatti, Valeria 71
Bond, P. 293
Bornstein, E. 49
Bornstein, L. 275
Boswell, John 134
Boudia, S. 99
Bowker, Geoffrey C. 178
Boyer, Paul 216
Brabble, Lynette 308
Brabble, Pat 308
Bradford method of controlled tipping 57, 62
Bramryd, T. 285
Brand, James H. 56
Brewer, John D. 131
Brickman, R. 262
Bridge, G. 145
Bringing Them, Home Report (final AIATSIS report, 1977) 231
Brinkley, C. 37
British sanitary science 56–57
Broad-Based Black Economic Empowerment (BBBEE) 294
Brook, Daniel 5
Brooks, Marianne Su-Ling 171

Brooks, Max 307
Broom, A. 46
Brown, B. 307
Brown, Kate 215, 216
Brown, Steve 103
Brownfields Program (1995, EPA) 241
Brugh, Mercedes 186
Brulle, Robert J. 5
Brunet, P. 215
Bryant, B. 41
Bubandt, Nils 62
Buch, Elana 48
Bulkeley, Harriet 8, 169
Bull, Richard 129
Bullard, Robert D. 5, 36, 41, 58
Bunker, Stephen G. 189
Burchell, G. 312
Burke, A. 231
Burke, T. A. 249
Burns, William C. G. 260
Bush, George W. 273, 317
Bush, George W., Sr. 273
Butt, Waqas 37, 38, 43, 44

C
Cairns, S. 97
The Cairo Cleanliness and Beautification
    Authority 74
Callon, Michel 105
Calvillo, N. 41, 45, 240
Campbell, Hugh 106
Campos, D. S. 279, 280
Campos, H. K. T. 279
Canadian settler colonialism 6
Candel, Sébastien 186
Cannon, C. 146
capitalism 10; culture/nature and subject/object
    dualisms in 103
Cappellini, B. 106
capture rate 181–182, 184
Carenzo, S. 145, 286
care workers 48–49; low-wage 49; use of polluting
    substances 49; whiteness and waste
    interrelationship and 48–49
Carmin, J. 41
Carnie, T. 292
Carpenter, Ele 216
Carson, Willa 245, 246
Carsten, J. 71
Caruna, M. 279
Castillo, A. B., Jr. 279, 280
cataclysmic natural or human-made disasters 307
catadores (waste-pickers) 60
Caverley, Nicholas 45
Chadwick, Edwin 9
Chale, Pastor Valdez 105, 108, 109

"Character and Eroticism" (Freud) 32
Chattergee, Syantani 38
Chaturvedi, B. 275
Chaves, G. L. D. 285
Cheeseman, C. R. 275
"The Chemistry of Dirt" (Freud) 32
Chernobyl nuclear disaster, April 1986 215,
    216–217, 234
Chikowore, N. R. 144
Children of Men (movie) 307
Childs, P. 142
Chin, Mel 64, 65
Choong, Weng-Wai 196, 198
Christensen, Lance 186
Cirelli, Claudia 133
citizenship and responsibility (as part of waste
    discourse) 76–78
city sanitation 9
"Civil Laboratory for Environmental Action
    Research (CLEAR)" 47
Clapp, Jennifer 7, 8, 255, 256, 258, 315
Clark, Corrie 171
Clark, J. F. M. 56
"Classify plastic waste as hazardous" (Rochman) 36
cleanliness (as part of waste discourse) 75–76
cleanliness as part of faith 78–79
Clearwater, D. A. 307
climate change 90, 256, 275, 315
Cloke, J. 141
Cloudylabs 213
Cockerill, K. 145
Coh, Thomas 111
Cole, C. 142
Cole, T. 49
Colen, S. 49
Coles, B. 142
Collins, John 74
Collins, K. 311
Colombijn, F. 276
colonialism 6, 10, 43, 44; waste as matter of
    230–233
colonial rule and sanitation 10
colonization of Indigenous peoples 307
Colten, C. E. 249
Comber, R. 201, 202
commercial content moderation (CCM) 15
commodity foods 112–113
Commonwealth of Australia Bureau of Resource
    Sciences 227
comparative history 128
comparative study method 123–124; across time
    and spatial scales 125; case selection 126–127;
    framework for systematic comparisons 127–129;
    horizontal or vertical comparisons 125; narrative
    approach 129–130; number of cases required for
    124; for policy analysis 123–124; productive

systems across countries 129; reasons for 124–125

A Complete Beginner's Guide to Prepping (Preissman) 308

Comprehensive Environmental Response, Compensation, and Liability Act (CERCLA) 241

computer graveyard 95

Conca, K. 315

Conrad, A. 317

Conroy, W. 239

conspicuous consumption and waste 5, 255; systematic and cultural factors in 141

consumer waste studies 5–6, 11, 89

Contagion (movie) 307

contaminated brownfield 14, 238–244; antiracist research into 250; critical scholarship on reclamation of 249–250; as heuristic of racial justice 240; public participation in reclamation of 243–244; public use of reclaimed 242–243; racial segregation in US and 240; redevelopment of in early 1990s in US 241–242

contamination rate 182–183, 184; drivers of 182–183

continuous environmental monitoring (CEM) systems 186

Conway, J. 310

Cook, G. 108

Coole, Diana 105

Cooper, Tim 8, 57, 296

cooperative enterprises 282; self-esteem amongst females in 283; self-management in 283

Corbett, Jack 134

Cornea, N. 292

Corteel, D. 95

Corvellec, H. 148

Cossu, Raffaello 185

Costa, Luiz 105

Cousins, Joshua J. 184

COVID pandemic 23, 82, 279, 305–306, 308, 309–310, 311, 312; living in present rather than abstract future apocalypse 316–317; prepping after 317; prepping during 315–316; stockpiling and 313; waste collection and 311

Cox, S. J. B. 37

Cram, Shannon 214, 215

Crang, Mike 7, 93, 145, 169, 256, 296

Crawford, Alan 183

Crewe, Louise 20, 105

criminalization of land protectors 37

critical waste studies 141–149

Crooks, H. 312

cross-cultural view of humanity 69

Crutzen, Paul 309

Crysler, C. G. 238

cultural notions of waste 41

cultural particularism of commodification 114

cultural studies 10; drawback of waste studies 11; of waste 10–11

Culture and Waste: The Creation and Destruction of Value (Hawkins & Myecke) 11

Curie, Marie 234

curses 73–74

D

Dada, R. 293

Dahlén, Lisa 171

Dalakoglou, Dimitris 61

Daniel, H. R. 308

Danzinger, Eve 105, 108, 109

DART Container Corporation 35

Darwish, S. 73, 79

da Silva, Maria Monica 281

Dauvergne, Peter 315

Davies, A. 148

Davies, T. 36

Davies, Thom 217, 230

Davis, A. 48

The Day After Tomorrow (movie) 307

Deci, E. 197, 198

Deci, E. L. 205

de Coverly, Edd 36

deflation 169

degrowth paradigm 149

de Hoop, Evelien 10

De León, Jason 47

Delgado, Otoniel Buenrostro 129

Delhi, Mughal 10

Deloitte Sustainability 183

DelVecchioi, R. 244

Demaria, F. 146

Denes, Agnes 64, 65

denigrated labor, devalued objects and denied humanity 43, 44

Denison, Richard A. 186

Denton, C. 93, 94

DePryck, K. 312

Descola, Philippe 105

Deterding, S. 196, 197

Dewey, John 212, 213, 215

Dewitt, Marc 173

Dhawan, A. 93

Dhillon, C. M. 143

Dhillon, J. 37

Diamond, D. 247

Dias, S. M. 276, 279, 282, 283

digital sorting game 201

Dikgang, J. 295

Dillon, Lindsey 142, 185, 239, 240, 248

Dinler, D. S. 7

dirt: conditions of 31; definition 4, 15, 32; as distinct from waste 12, 15; as matter out of place 32; as part of system 31, 33; power and 33–34

"Dirty hands: The toxic politics of denunciation" (Fiske) 37
discards 169–170; composition of 179–181; contamination and 182–183; possession and dispossession of 187–189
discard studies 15, 256; anti-racist and anti-colonial literature as part of same history of 51; compared to waste studies 15; as normative field 31
Discard Studies 3
Discard Studies blog 142, 143
Discard Studies Twitter conference (2020) 140
disposables as litter 34
disposal 172, 173–174; compared to diversion 174–175; as disposition outcome 173; diversion as 20th century term for 174
disposition 169, 173
dispossession 169
diversion 174–176; disposal compared to 174–175; rate 175, 176, 184
Dizikes, P. 247
Dlamini, S. 297
Docquier, Nicolas 186
Doherty, J. 46
Dong, Jun 172
don't litter slogans 68, 72
Doomsday Preppers (National Geographic) 306–307, 308–309
Doron, A. 46
double engaged ethnography 132–133
Douglas, Mary 4, 11, 12, 15, 31, 32–33, 34, 35, 36, 37, 38, 55, 69, 70
Dr. Strangelove (movie) 307
Duan, Ning 125
Du Bois, W. E. B. 43, 46, 51, 241
Dull, M. 239, 241, 244
dumpster diving (freeganism) 7
Dunaway, F. 34
Dunlap, Riley E. 216
Dyer, R. 197
Dyssel, M. 293, 295

E

eco-doom as consumerist spectacle 316
ecologically-oriented art works 65
economic geography 25
Economics of Recycling: The Global Transformation of Materials, Values and Social Relations (Alexander & Reno) 11, 20
economy *vs.* ecology 267
Eddy, Charmaine 308
Edelman, J. 233
Edge of Tomorrow (movie) 307
Edmonds, R. L. 154
Ehlers, N. 247
Ehrenfeld, J. R. 255
Ehrig, Hans-Jürgen 185
Eichstaedt, Peter H. 215

Ekström, K. M. 142
Eleazars, Abraham 96
electronic waste (e-waste) 7, 146–147, 260, 309
Elimelech, E. 205
Ellen MacArthur Foundation 142
Elliot, Susan J. 12
Ellis, E. C. 90
energy consumption *vs.* energy expenditure 267, 269–271
energy economics theory 14, 267, 269–270
energy expenditure 14
Engelke, P. 90
Enlightenment 8–9
environment (as part of waste discourse) 80–83
environmental apartheid 240
environmental governance 154
environmental inequality formation theory 143
environmental injustice 51
environmentalism 31; critiques of waste in capitalism 46; trope of 71
environmental justice 5, 185; critique and reform 41
environmental legislation 154
environmental management 154
Environmental Police (Tunisia) 81
environmental pollution 238–239
Environmental Protection Agency (EPA) 96, 146, 241, 244, 245, 255
environmental racism researchers 307
Environment and Climate Change Canada 36
environment as value sphere 71–72
Eriksen, Thomas 272
Eroglu, H. 311
Ert, E. 205
essentialism 80
Estes, N. 37
ethics 11, 71, 141, 299; lifeboat 37; of responsibility 63; waste 212
The Ethics of Waste: How We Relate to Rubbish (Hawkins) 69
ethnic hierarchization 44
ethnography 13, 131; double engaged 132–133
Eurostat 155
Evans, David 105, 106, 141
Evans, D. M. 197
Evans, James 129
Evans, Peter 5, 6
Evard, Olivier 215
excess of modernity 63
experience of radioactive waste 212–213
extended producer responsibility (EPR) 279, 295
externality as discursive reality 233
extractivism 225; Australia as site for 227–228; imaginaries 228; imaginary in postcolonial Australia 231
Ezeah, C. 276

F

Fadlalla, A. 49
Fanon, Frantz 10
Fardon, Richard 31, 32, 33
Fassin, D. 49
Fauci, A. S. 310
Faust, B. 105, 109
Fausto, Carlos 105
Fawcett-Atkinson, M. 36
Fazakerley, J. AS. 276
Federal Taxpayer Relief Act (1997, US) 242
Fei, F. 157
fei zhenggui (informal waste recycling) 158–159
Fels, D. I. 198, 205
feminist inhuman epistemology 63
feminist materialism 6
feminist/performative theories of the body 6
Ferreira, Devair Alves 216
Ferreira, E. 269, 280
Ferri, G. L. 285
Fetherston, J. T. 173
field science, technology and society (STS) 6
fieldwork 12–13
fieldwork-based scholarly research on waste 127
Figueiredo, F. F. 279
financialization of waste 25, 26
Fink, Ann 111
Finney, C. 46
First Nations peoples 144, 228, 269; accounts of atomic tests in Australia by 229–230; alternative ways of living with waste 234; colonial violence against Australian 228; disenfranchisement of Australian 228–229; waste's different status for 232
first peopling fantasy 229
Fiske, Amelia 37
Fitzgerald, Jenrose 38
Flannery, Tim 232
Flyvbjerg, Bent 125
Fontaine, Guillaume 123
food: commodification of 114; as important focus of respect 109–110; as object in cost-benefit analyses 104; overprovisioning of 141; as part of relational system 103; production-consumption cycle 103
food disposal 104; as central aspect of everyday human existence 104–105
food gifting 142
food use 105; alternative paradigms for 106
food waste 5–6, 12; cultural connotations of 103–104; efficiency as dominant paradigm governing 115–116; morally fraught disposal practices 105; as social and environmental concern 106; as social construct of capitalist cosmologies 103
food waste management 148
Foote, S. 312

Foreman, Peter J. 8
formality-informality continuum 127
Fortier, C. 37
Foster, Gwendolyn A. 307, 308, 317
Foucault, Michel 9, 267, 268, 312, 313
Fox, Louise 144
Fox, N. J. 276
Fox News 312–313
Fracalanza, A. P. 285
Fragkou, Maria Christina 184, 185
framings of waste 68–69, 70–72; public signage 72–74
Francois, G. 312
Fraser, C. 106, 145
Fredericks, Rosalind 6, 44, 45, 78, 292
Fredizzi, P. 286
Freire, Pablo 139, 278
Fresno sanitary landfill 57
Fressoli, M. 280
Freud, Sigmund 32, 33
Freudenburg, William R. 216
Freund, B. 291
Freund, D. M. P. 239
Friedmann, Harriet 105
Friends of the Earth 227
Frost, Samantha 105
Fukushima nuclear disaster, March 2011 211
Fundação Banco do Brasil 279
Furniss, Jamie 7, 12, 69

G

Gabrys, J. 93
Gago, V. 278
Gambina, J. 283
games with a purpose (GWAP) 198
gamification approaches 13, 196–206; as buzzword 197; case studies 199–204; challenges to 204–206; educational perspective 198; games *vs.* play 197; to motivate behavior changes 196–197; pro-environmental behaviors and 198–199; role of in distancing 199; as synonymous with rewards 197–198
Gange, D. 89
garbage as anchor for blackness 43
garbology project 91
García-Madariaga, J. 196, 205
Garcier, Romain J. 11, 13, 211, 213, 214
Garnett, E. 142
Garrett, B. 308
Geddes, Barbara 125, 134, 135
Geertz, C. 106
Geha, C. 78
gender equality 283
gentrification 45
The Geographies of Garbage Governance: Interventions, Interactions, and Outcomes (A. Davies) 148
Georgescu-Roegen, N. 95

Gidwani, V. 6, 9, 43, 51, 142, 240, 275, 292
Giles, David B. 7, 49
Gille, Zsuzsa 6, 58, 69, 93, 105, 106, 145, 148, 169, 180, 187, 240, 255, 296
Gilligan, Keith 186
Gilliland, J. A. 196, 205
Ginn, Franklin 309, 316
Ginsburg, F. D. 49
Giordano, C. 205
Giorgio, A. 307
Girel, Mathias 213
Giroux, H. A. 140
Gleaners (movie, Varda) 25
Glenn, E. N. 48
Global Alliance of Waste Pickers 278–279
global environmental governance 256, 259–261
global environmental justice 256
global environmental politics (GEP) 254, 255, 256, 257
global garage sale of second-hand goods 254
global governance of waste trading 14, 259–261
Global North 6; electronic waste and 7; mobility as generator of new waste streams 23; scholarship on food waste focused on 106; waste labor in 7
global plastic waste generation 63
Global South 6; electronic waste and 7; e-waste reclamation in 93; food waste research in 106; informal economies in 14; redirection of waste to 58; waste labor in 7
global trade in pharmaceuticals 50
global waste trade 24; as novel toxic form of colonialism 89
glorious expenditure 269
Godfrey, L. 292, 293, 295, 296
Gohil, Nisarg 171
Goiania radioactive contamination incident, 1987 216
Goldstein, Jesse 9, 37, 155, 240
Good, C. 145
Goorhuis, M. 155
Gordon, C. 312
Gourlay, Kenneth A. 55, 255
governance of waste 7–8; issue of scale in 8; within the UK 8
Grabs, J. 280, 281
Graham, Stephen 6, 97
grassroots social innovations 280–282; among waste-picker organizations 280–281; to help cities transition into more sustainable ones 285; individual and social levels that influence 281; limitation to process to bottom-up 281–282
Gray-Cosgrove, C. 36
Graziano, Valeria 6
The Great Charter of the Forest of 1225 9
Green, Jim 215, 227
Greenberg, M. R. 241

GreenCape 295
Greene, Krista L. 177
green economy 14; waste as part of 291
green health projects 240–241
greenhouse gas (GHG) emissions 185, 256, 276, 285
Greenpeace 147, 256–257, 258, 259, 260, 261
greenwashing 245–248
Greer, Germaine 231, 232
Gregson, Nicky 6, 7, 8, 16, 20, 21, 22, 23, 24, 25, 26, 89, 105, 106, 141, 145, 169, 177, 296
Griner, A. 244
Gu, B. 156
Gu, Y. 155
Guerrero, L. A. 275
Guillaumet, A. 81
Gulf War (1991) 14
Gullett, Brian K. 177
Gupta, A. 97
Gupta, J. 276, 282
Gupta, Joyeeta 124
Gutberlet, Jutta 7, 14, 15, 127, 130, 144, 187, 275, 276, 279, 280, 282, 285, 286, 293

H
Haberl, Helmut 184
Hackett, E. J. 140
Hadju, Patty 310
Haebich, A. 228
Halbmayer, E. 105
Hall, Larry 308
Hallett, L. 142
Halvorson, Britt 12, 16, 49, 50
Hamari, J. 197, 205
Hamblin, Jacob Darwin 214
Hannity, Sean 312
Hannon, J. 142
Hansen, K. T. 49
Haraway, Donna 105
Hardin, Garrett 37
Haring, Keith 256
Harris, C. 239
Harris, C. I. 47, 48
Harris, M. 226
Harrison, Kathryn 129
Harvey, David 59, 187
Hawkins, Gay 11, 69, 71, 141, 205, 212, 231, 234, 315
Haynes, Murrae 215
hazardous waste 7, 125, 128, 144–145, 171–172; broad category of 255; laborious and hazardous conditions for working with 95; lack of coordinated cross-national definition of 255; new compendia of data 256–257
hazardous waste facilities 5
hazardous waste trade 254–261; in 1970s and 1980s

258–259; in 1980s and 1990s 254, 255–258; global governance arrangements for 254; legal side of 262; Southern victim/Northern perpetrator narrative 261; UK as world's largest net importer 262

healthfields 240, 245–248; negative outcome of 246–247

Heberle, L. 248

Hecht, Gabrielle 6, 20, 21, 22, 24, 25, 26, 27, 90, 214, 215, 225, 226

Helwege, Ann 127

"Here's where you donated clothing really ends up" (*CBC News*) 146

Herod, A. 48

Hersh, R. 240, 241, 244, 249

Hershkowitz, Allen 186

Herzfeld, M. 73

Hesse, M. 310

Hetherington, Kevin 105

"Hidden Mountain: The Social Avoidance of Waste" (de Coverly, McDonagh, O'Malley & Patterson) 36

hierarchies of care 48–49

Higgs, C. J. 293, 294

Hill, T. 293, 294

The Hill God and the Hunter (Coh & Bol) 111–112, 114

Hilz, Christopher 255

Hird, Myra J. 6, 10, 14, 15, 41, 62, 63, 187, 231, 232, 234, 306, 307, 309, 312, 315, 317

Hiroshima nuclear bombing 216, 234

historicity of waste 90, 91

histories of garbage and sanitation 8–9

history of enclosure 9–10

history of landfills 56–58

history of matter out of place 32–33

history of waste studies 4–11

Hoag, Colin 62

Hodson, Mike 184

Hofacker, C. F. 198

Hogland, W. 275

Hollander, J. 241

Hommels, A. 93

Hoover, E. 307

Hopkins, N. S. 81

Hornborg, Alf 105, 307

Howard, John 231

Huang, P. C. C. 160

Huber, Matthew 185, 270

Huggan, G. 225

Hula, R. C. 244

Hultman, J. 148

human-computer interaction (HCI) 198, 201

human-environment relationships 103

Huotari, K. 197

hygiene management 9

hyperobject (Morton) 234

I

I, Robot (movie) 307

Ialenti, Vincent 36

identity co-optation 38

identity formation 141–142

Ilgen, T. 262

immigrants 44

inchoate stream of experience 212

incinerators 56

Indeterminacy: Waste, Value and The Imagination (Alexander & Sanchez) 20

"Indian Givers" (documentary, I. Alexander & Mallon) 144

Indigenous activists 307

Indigenous animistic worldviews 105

industrial capitalism 42, 43; conjoined with race 43–44

industrialization 56, 90

industrial pollution 56

industrial waste 171, 255; agricultural wastes 171; construction and demolition (C&D) debris 171; early cases of 254; excavation fill 171; medical discards 171; mining wastes 171

informal laborers 43, 122

informal recycling 154–156

informal recycling sector (IRS) 155–156; institutional involvement in PRC in 165–166; size, performance and mechanisms in PRC 158–160; survey activities in PRC 161–165

informal waste: comparative politics of 130; reclamation of 292

informal waste pickers 15

informal waste-picking processes 131; comparative ethnography of 131–132

information-based policy instruments 129

infrastructural discontent 10

innovation cycle theories 98

inside-out Earth 24

Institute of Scrap Recycling Industries (ISRI) 183

institutionalized U.S. medicine 50

Instituto Brasileiro de Geografia e Estatistica (IBGE) 277

Instituto de Pesquisa Economica Aplicada (IPEA) 277

Instituto Nenuca de Desenvolvimento Sustentável (INSEA) 282

intellectual property 48

interdisciplinary research 27

Intergovernmental Panel on Climate Change (IPCC) 305, 307

International Atomic Energy Agency (IAEA) 214, 215

International Commission on Radiological Production (ICRP) 214

International Federation for Human Rights 260

international governance of waste 7–8

International Journal of Labor and Working-Class History 7
International Labor Organization (ILO) 275, 282
International Monetary Fund (IMF) 307
international non-governmental organizations (INGOs) 107
International Toxic Waste Action Group 259
international trade and traffic of waste 14
The International Trade in Wastes: A Creative Inventory, Fifth Edition (Vallette & Spalding) 256
intimate dirty work 48–49
Iraq Wars 270
Isenberg, Nancy 41, 46

J
Jackson, Steve J. 97
Jacobs, J. M. 97
Jagath, R. 294
Jambeck, Jenna 171, 174
James, C. L. R. 43
Jasanoff, S. 262
Jay, P. 146
Jewasikiewitz, S. M. 295
Johansson, M. 285
Johnson, D. 198
Jolé, Michèle 78
Jordhus-Lier, D. 297
junk DNAs 16, 42, 47–48
jurisdiction, operational 170–171

K
Kallianos, Yannis 61
Kallis, G. 317
Karani, P. 295
Kardashian, Kim 314
Karidis, Arlene 185
Katz, C. 146
Kaviraj, Sudipta 74
Kaza, Silpa 172, 173, 178
Keane, W. 71
Kelly, C. R. 308
Kemberling, M. 143
Kerr, J. M. 144
Khoo, S. M. 145
Kim, Trogal 6
Kim, T. W. 197
Kimmerer, Robin Wall 189
kin concept 16
King, J. P. 140
Kinsella, W. J. 145
Kirsch, Scott 215, 216
Klein, N. 307
Knowles, Caroline 6
Koestner, R. 205
Kohn, A. 198
Kohn, Tamara 51
Koivisto, J. 205

Kollikkathara, N. 315
Kooiman, J. 285
Kraft, Michael E. 216
Krausmann, F. 90
Kristeva, J. 307
Krones, Jonathan Seth 177
Krueger, Jonathan 255, 258, 259
Krupar, Shiloh 14, 16, 215, 247, 250
Kubanza, N. S. 297, 298
Kübler, D. 285
Kuchenbuch, L. 91
Kuletz, Valerie L. 6, 10
Kulkarni, B. N. 311
Kummer, Katherine 255
Kummu, Matti 106
Kuo, H. 310
Kütting, Gabriela 125, 126
Kuwaiti oil fields on fire, 1991 268, 269, 271

L
labeling and categorizing waste 35; demarcation creation of control 35–36
Labib (friend of environment) mascot (Tunisia) 82–83
labor theory of value 24
Laceby, Patrick 215
Ladner, Sam 131
Lagerkvist, Anders 171
Lambek, M. 71
Lamond, J. 229
Lamy, P. 312
Landecker, Hannah 105
landfill management 58; human and non-human temporalities in 62; spacebugs in 62; undomestication process 62
landfills 6, 55, 94; blocked 61; capacity of 293; chemicals in 62–63; criticisms of 63; as economic entities 58; evolution of 65; future of 63–64; geographic location of 58; historic 62; historical background 56–58; incinerators and 56–57; as life-less places 55; litter in 34; metrics and 184–185; narrow view of 56; out of sight out of mind 65; ownership and control 59; in Palestine 58–59; as place for rubbish but little else 55; in PRC 155; social science of 12; socio-material life of 55–56; in South Africa 59, 293; as spatial undertakings 58; waste from across borders 58–59
Landman, Todd 124, 125
landraising 56
land reuse 233; of brownfield sites 242–243
landscape decontamination 215
Lane, H. C. 310
Lang, J. M. 139
Lange, U. 275
Langenhoven, B. 293, 295
Langton, Marcia 230

Laplantine, François 70
Larson, L. R. L. 310
late-modern colonization 270, 271
Latendresse, A. 275
Latour, Bruno 105
Lau, A. W. W. Y. 276
Laurence, D. 258
Lave, R. 250
Law, J. 105
Lawhon, Mary 14, 292, 295, 296
Lawson, B. E. 41
Laycock, R. 310
Lea, T. 228, 229
Le Blanc, D. 282
Lebreton, Laurent 63
Lee-Geiller, Seulki 125, 126
Leidenberger, J. 317
Leiman, A. 295
Le Lay, S. 95
Lemos, M. C. 285
Leonard, A. 317
Leonard, L. 293
Lepage, Emmanuel 216
Lepawsky, Josh 6, 7, 15, 36, 58, 89, 93, 95, 133, 142, 146, 147, 169, 170, 180, 188, 212, 254, 261, 263, 309
leper ships 258–259
Levictus 36
Li, Belinda 15
Li, S. 155, 157, 158, 159, 163
Li, T. M. 244
Liboiron, Max 4, 6, 12, 13, 15, 32, 35, 36, 37, 41, 42, 45, 47, 50, 51, 55, 142, 143, 145, 149, 170, 187, 189, 197, 240, 250, 306, 315, 317
Licbbach, Mark Irving 124
life cycle analyses (LCAs) 186
lifespans 98
Lim, Timothy C. 124, 125
Lima, Mairon G. Bastos 124
Linbaugh, Peter 25
Linzner, R. 155, 160, 275
Litt, J. S. 249
Litt, S. 248
litter 34; as out of place 34; as Western dualist system of thought 232
Littercoins 199, 201
Little, P. C. 144
Liu, T. 157
Living with Things: Ridding, Accommodation, Dwelling (Gregson) 20
Locke, John 9, 228, 229
London Hanged (Linbaugh) 25
Longstreth, Frank 124
Los Alamos National Laboratory 36
Loukil-Tlili, B. 81, 83
Love vs. Commonwealth of Australia (2020) 233
Low, Sheau-Ting 196

low caste workers 43
Lowe, Edward D. 131
Loyd, J. 239
Luckin, Bill 56, 57
Ludwig, Udo 262
Lundin, C. 308
Luo, Y. 201, 204, 205
Lynas, M. 315
Lynch, S. 199, 205
Lyons, Donald 175

M
Maas, G. 275
MacBride, Samantha 13, 16, 63, 71, 142, 146, 158, 171, 175, 177, 185, 188, 255
Macfarlane, Alan 33
MacLaren, Virginia 15
macroeconomics 36
Mad Max (movie) 307
Maennling, Nicolas 129
Magna Carta 9
Mahoney, James 125, 134
Makgae, M. 295
Makina, A. 293, 297, 299
Making Waste Pay slogan 296
Malin, Stephanie A. 215
Malkki, L. 41
Mallon, S. 144
Malm, A. 307
Mangan, K. 50
Maniates, M. 315
M. Anjos, R. 216
Mao Tse-tung 155
Marable, Manning 45, 51
Maralinga Rehabilitation Project (Australia) 232–233
March, H. 317
Marello, Marta 127
marine microplastics 142–143
marine plastics 6
Mark, W. 309
Maroney, K. 197
Márquez-Benavides, Liliana 129
Marsh, Jillian K. 215
Marshall, Cody 182
Martin, B. 226
Masco, Joseph 215, 216
mass production and consumption 63, 90; resource extraction, production and distribution for 93
materialist foundation for social difference and inequality 43
materiality of waste 144–145
materials flow analysis (MFA) 184
Matete, N. 294
Mather, Charles 147, 188, 212
matter in place 38; avoidance to keep 36–37; combination of techniques for 38; to create

order again 35; demarcation creation of control 35–36; if you can't beat them ceremonially join them 37–38; labeling of anomalous events as dangerous 37; physical control of anomaly 36; techniques for keeping it in place 35–38
matter out of place 31; dirt *vs.* rubbish 31; keeping matter in place 35–38; litter as 34; rubbish definition correlated to 55
Maya animist cosmology 103, 108, 112
Maya Leaders Alliance (MLA) 108, 111
Mayan loss of cultural identity 111–112
Mayan Old Ways 112
Maya respect-based food disposal 106
Maya traditional environmental knowledge (TEK) 105
Mazzolini, E. 312
Mbembe, Achille 14, 16, 267–268, 270, 271
McDonagh, Pierre 36
McDowall, W. 154
Mcintyre, Michael 271
McKay, A. 225, 226
McKenzie-Mohr, D. 196
McMichael, Philip 105
McNeill, J. R. 90
McNiece, C. M. 242
Meah, A. 145
Mebratu-Tsegaye, Tehtena 129
mediations on waste 69, 70–72
medical waste 311
Medina, M. 275
megacities 6
Melancholia (movie) 307
Melosi, Martin V. 9, 63, 89, 91, 99
member-based organizations (MBOs) of waste-pickers 276–279; benefits 278; goal of 278; grassroots social innovation s and 280; as networks 279; promotion of community actions, engagement and support 281
Menzies, Robert 225
Merill, T. 107
Mersky, Ronald L. 172
Messick, B. 73
metabolism 145–147
Metcalfe, A. 105
Meuser, M. 5
miasmatic theory of disease 9
Miezitis, Y. 225, 226
Mignolo, W. 50
migrants 41, 47
Mildenberger, Matto 37
Millar, Kathleen 6, 43, 44, 45, 46, 49, 60, 276
Millar, S. 145
Miller, Chaz 182
Miller, P. 312
Millington, N. 292, 296
Mills, C. 44
Mills, M. F. 307

Mills, Michael 308
Millstein, M. 297
Milward-Hopkins, Joel 178
Minerals Council of Australia 225, 228
mining waste 6
Ministry of Commerce (MOC) of PRC 159, 160
Ministry of Housing and Urban Rural Construction (MOHURD) of PRC 158
Minter, Adam 254
Miracle Mile (movie) 307
Miraftab, Faranak 7, 295, 297
Mitchell, T. 82
Mittmann, J. D. 226, 234
model spaces concept 82
modernist histories of waste 9
modernity 43
modernity/coloniality 50
Mohai, P. 41
Mohammed, A. H. 196
mononaturalism 105
Montgomery, Mark 255, 261
Moore, J. W. 239
Moore, S. A. 307
Moore, Sarah 5, 63, 142
Mora, A. 197, 198
Moraes, L. 276
Morbidini, M. 276
Morganti, L. 199, 206
Morlino, Leonardo 122, 125, 126
Morris, Julian 186
Morrow, O. 148
Morton, Timothy 234
Möser, K. 98
Motta, S. C. 140, 279
Mourad, M. 148
Movimento National de Catadores(as) de Materiais Recicláveis (MNCR) 278, 279, 282, 285
Mu, Z. 154
Muecke, Stephen 6, 11
Mukherjee, C. 186
Mulpiddie, Edie 229
municipal solid waste (MSW) 155, 292–293; generators of 171; parallel systems in PRC 157; in PRC 154–157; records on data in PRC 156–157
municipal solid waste (MSW) governance 157–158
municipal waste 15
Muniz, Vik 60
Muñoz-Cadena, C. E. 129
Murcott, Anne 106
Murphy, Cullen 64, 91
Murphy, M. 48
Murphy, T. 308
The Mushroom at the End of the World (Tsing) 69
Muswema, A. P. 293, 294
Myers, G. A. 292

N

Nagasaki nuclear bombing 216, 234
Nagle, Robin 42, 144, 173
Nahman, A. 295
Nakao, Atsushi 215
Napoleonic Wars 254
Narain, Sunita 256
Nast, Heidi 271
National Parks and Wildlife Conservation Act (1975, Australia) 226
National Programme for Cleanliness and Support for the Environment (Tunisia) 73
National programme for the cleanliness of the environs and the beauty of the environment (Tunisia) 74
National Radioactive Waste Agency (ANDRA) 218
National Resources Defense Council (NRDC) 144
National Secretariat of Solidarity Economy (SENAES) 279
National Solid Waste Policy (Brazil) 279
National Waste Management Strategy (NWMS) (2011, South Africa) 294–295
Nature 36
necroaesthetics 272
necropolitics 14, 16
necropower 267–268, 270–272
Negrusa, A. L. 198
Nelson, A. 47
Nelson, R. R. 310
neoliberal capitalism 312–313; morality structured by 316–317
neoliberal globalization 272
Newell, Joshua P. 184
new genetics 12
Newman, Richard 11
new materialism 6, 41, 105
new social movement (US) 5
New York City Department of Sanitation 177, 183
The New York Times 259, 314
Nguyen, M. T. N. 49
Nicholson, S. 197, 198
Nitivattananon, V. 196
Nixon, R. 41, 45, 100, 307, 314
Nkosi, N. 295
no dumping slogans 72, 77–78
non-caste workers 43
non-governmental organizations (NGOs) 8, 50, 74, 75, 85, 107, 187, 188, 259, 276
not in my backyard (NIMBY-ism) 258
Novak, J. 198
Nucho, Joanne Randa 61
nuclear colonialism 225
nuclear dump 227
nuclear fallout 215
nuclearity 214, 215

nuclear nation 224–227; Australia as 224–227
nuclear power plant removal (Germany, 2020) 88
nuclear waste 6, 13–14, 227; in Australia 224; cultural tropes in management of 11; as presence or absence experience 217; as risk and danger experience 216–217; sociocultural framing and construction of 216; storage of in Australia 233; as technological condition experience 216; uranium tailings 214–215
nuclear waste storage 88

O

O'Brien, Martin 5, 105, 211
O'Brien, Mary H. 11
obsolescence in consumer goods 5, 97–99, 147
obsolescence of mankind 216
occularcentric approach to waste 71
Oelofse, S. 292, 293, 295, 296
Offenhuber, Dietmar 169, 180, 188
Ogando, A. C. 283
Ogle, V. 89
O'Hare, Patrick 6, 7, 12, 49, 59, 213, 276
oil economy 14, 270
oil wasting 267–273; aesthetics of 272
Ojeda-Benítez, Sara 129
Oloko, Patrick 9, 142
O'Malley, Lisa 36
The Omega Man (movie) 307
Omrow, Delon Alain 263
O'Neill, Kate 8, 14, 128, 141, 142, 147, 254, 255, 258, 260, 262, 263
On our Selection (Lamond) 229
On the Beach (movie) 307
OpenLitterMap app 199
operant conditioning 197
Oppong-Tawiah, D. 198, 199
Organization for Economic Cooperation and Development (OECD) 256–257, 258, 260, 261, 262
Our Common Future (a.k.a. Brundtland Report) 256
Ouroboros symbol 96
Outbreak (movie) 307
out of sight, out of mind 7, 99; landfills as 65
Owojori, Oluwatobi 177
Owusu, P. A. 311
Oyekale, A. S. 292

P

Pace, N. 144
Pacheco-Vega, Raul 13, 16, 126, 127, 129, 130, 131, 133, 148
Packard, Vance 5
Pagano, M. A. 285
Pahl-Wostl, Claudia 261
Pajo, Judi 216
Paltrow, Gwenyth 314

Pantano, E. 310
Parizeau, K. 130, 131, 144, 145
Parizeau, Kate 13, 37, 106, 127, 129, 133
Parkinson, A. 233
Parnell, S. 291
Parreñas, Rachel 49
Parsons, E. 106
Partial Test Ban Treaty (PTBT), 1963 215, 226
Pasternak, Judy 215
Patterson, Maurice 36
pedagogy 13, 139–140
Pellow, David Naguib 5, 7, 41, 58, 143, 144, 256
People's Republic of China (PRC) 154–155
Pepinsky, Thomas B. 125
Peredo, Ana Maria 130
Pereira da Silva, T. M. 145
Peres, T. S. 292, 293
Pérez-Chiqués, Elizabeth 124
Persaud, Donny 6
Personal States: Making Connections between People and Bureaucracy in Turkey (Alexander) 21
Peters, B. Guy 123
Peterson, Kristin 50
Petryna, A. 215
Petrzelka, Peggy 215
Petts, Judith 129
Pezzullo, Phaedra C. 11
Phillips, C. 145
physical control of anomaly 36
Piepzna-Samarsinha, L. L. 48
Pillay, A. S. 295
Pillets, J. 247
Pires, T. S. D. L. A. 279, 280
plastic and paper imports into China 254
plasticity 272
plastic packaging 255
plastics industry 292
"Plastic Wars" (documentary, NPR & PBS Frontline) 146
plastic waste 7, 36, 146
Pockley, P. 226, 227
Poindexter, G. C. 240
Polgreen, Lydia 260
political anaesthesia (Szasz) 311–312
political-economic nexus 25
pollution 6; as central to maintaining power structures 51; as class of danger to power 33; environmental 238–239; feminist and queer analyses 142; uneven distribution of globally 41
Pollution and Colonialism (Liboiron) 143
pollution release and transfer registries (PRTRs) 8
Polokwane Declaration on Waste Management (South Africa) 294
polychlorinatedpiphenyls (PCBs) 7
Popken, B. 310
post-colonialism 44
post-humanism 41

post-mining tourism 21
poststructuralism 6
Potter, Emily 14, 231
Povinelli, E. 48
power relations 6, 31, 143–144; gender binaries as central to power 34; theory of in waste relations 33–34
Prashad, Vijay 10
The Prepper's Blueprint (Pennington & Luther) 308
The Prepper's Water Survival Guide (Luther) 308
prepping (stockpiling) 14–15; in anticipation of future disaster 307–308; money to be made out of 308; popular representation of 308
The Prepping Guide: SHTF Plans (Brown) 308
primordial waste (Locke) 228
Prince, R. J. 49
Princen, T. 199, 315
prior informed consent (PIC) 259
Probst, Katherine M. 262
Project Ploughshare (US) 216
"Protecting the power to pollute: Identity co-optation, gender and the public relations strategies of fossil fuel industries in the United States" (Bell, Fitzgerald & York) 38
Protestantism 71
Puar, J. 45, 48
Public Health Act of 1875 (Britain) 56
public private partnerships (PPPs) 275, 295
public signage (Tunisia) 68–69, 72–74; pure injunctions in 83–84
Pulido, L. 239
Purity and Danger: An Analysis of Concepts of Pollution and Taboo (Douglas) 31, 33, 35, 69
Purnell, Phil 178

Q
Qi, D. 310
qualitative approaches 12, 131
quantitative approaches 12, 13, 124; defilement 13; depossession 13; measurement categories 13
Quested, T. E. 205

R
Rabaka, R. 43, 46, 48
Raban, D. 197
race: conjoined with capitalism 43–44; constitutive role of in waste and whiteness 41; as devaluer of people and waste devaluer of matter 46; forgetability or deniability of like waste 42; profound and constitutive impact on world 42; relationship to waste 41–43; in terms of whiteness 42–43
racial capitalism 14, 45
racial geography 45
racialization 12; as set of sociopolitical relations 43
racialized hierarchies of care 49
Rada, E. C. 286

The Radiance of France: Nuclear Power and National Identity After World War II (Hecht) 21
radioactive waste 36, 211; cleanup 232–233; Indigenous modalities 232–233; as object of experienced 212–213; properties 213–214; specifications 214
radioactive waste contamination 216–217
Rafaeli, S. 197
Ragazzi, M. 286
Ralston, Tim 306
Ramírez-García, Robert 171
Ramón-Gallegos, E. 129
Rams, Dagna 58
Ramusch, Roland 155, 160, 164, 166
Ranganathan, M. 239
Ranger Uranium Environmental Inquiry, 1976 (Australia) 227
Rapp, R. 49
Rathje, William 64, 91
Rathoure, A. K. 142
Ratti, Carlo 169, 180, 188
Rau, H. 145
Rawski, T. G. 160
Ray, Daisy 215
Reassembling Rubbish: Worlding Electronic Waste (Lepawsky) 147
Recycle Me project (Tunisia) 75–76
recyclers 122
recycling 63, 255; closed-loop vision of 96; composition of 179–181; designated for 183; ethical imperatives and implications of 11; litter 34; mechanization of 296–297; multi-racial sector in South Africa 291; of plastics 292; polystyrene 35; in PRC 155; as waste management 92–93
Recycling Reconsidered: The Present Failure and Future Promise of Environmental Action in the United States (MacBride) 146
Reddy, R. N. 43, 51, 142
Redfield, P. 49, 296
Redfield, R. R. 310
RedLacre (Rede Recliladores) recyclers movement 278
reducing, reusing and recycling (Three Rs) 146
refugees 41
refuse revolution 56
"Re-Imagining Land Ownership in Australia" (Verran) 228
religion (as part of waste discourse) 78–80
remediation and aftercare engineering 100
Reno, Joshua Ozias 4, 5, 6, 11, 12, 16, 20, 41, 42, 48, 57, 58, 62, 145, 211
residuals 25
residues 22
Resnick, Elana 44
Resource Conservation and Recovery Act (RCRA) 241

resource conserving policy approaches 154–155
Resource Recycling 183
reuse practices 94
reverse logistics of waste business 93–94; of waste disposal 94
Revival Field (art installation, Chin) 64
Reynolds, Christian 177
Rhodes, R. A. W. 134
Ribeiro, G. M. 285
Rice, Murray 175
Richter, G. 197, 198
ridding work 141–142
Right to Repair movement 147
Riha, Jacob 14, 15, 307
Ro, M. 205, 206
The Road (movie) 307
Robbins, P. 307
Roberts, C. L. 276
Roberts, J. GT. 143
Robins, S. 296
Robinson, Cedric 43, 51
Robinson, Howard 185
Robinson, J. 291
Robinson, Susan 182
Rochat, David 184
Rochman, Chelsea 36
Roe, B. E. 310
Roediger, David 51
Roff, S. R. 232
Roffinelli, G. 283
Rosa, B. 142
Rosa, Elisabetta 133
Rosa, Eugene A. 216
Rose, Richard 124
Rosenblatt, W. H. 50
Royal Commission into British Nuclear Tests in Australia, 1985 226, 227, 229, 232
rubbish definition correlated to matter out of place 55
Rubbish Theory: The Creation and Destruction of Value (Thompson) 69
Rudd, Steele 229
Ruiz-Alba, J. L. 196, 205
Rundle, G. 231
Runyan, A. S. 142
Rutherford, Ernest 234
Ryan, R. 197, 198
Ryan, R. M. 205

S
Saia, Jennifer 314
Salehabadi, D. B. 88
Salhofer, Stefan Petrus 155, 160, 164, 166
salvage markets 96
Samson, Melanie 59, 144, 187, 275, 292, 293, 297
Sanchez, Andrew 20, 142, 307

sanitary landfill 57, 59, 64, 65, 89, 90, 91, 99, 185;
science behind 57
Santayana, V. 225
Sarkodie, S. A. 311
Satyro, W. C. 147
Saurabh, Arora 10
Scanlan, John 8
Scarlett, Lynne 174
scavenging as social movement 7
Scheider-Mayerson, M. 308
Scheinberg, Anne 129, 275
Schenk, C. J. 292, 293
Schiele, K. 197, 205
Schindler, S. 146
Schmid, Barbara 262
Schmid, M. 100
Schmidt, Michelle 12
Schnailberg, Allan 7
Schnegg, Michel 131
Schober, Elisabeth 272
Schulz, Y. 155, 159, 161
science, technology, engineering and mathematics
(STEM) fields 140
science and technology studies (STS) 254, 255,
257, 261
Scopus 32
Scott, James 82
scrap business 94–95
Seaborn, K. 198, 205
Seabrook, J. A. 196, 205
second-hand cultures 22
Second-Hand Cultures (Gregson & Crewe) 20, 21
Second Treatise of Government (Locke) 228
*Seeing Like a State: How Certain Schemes to Improve
the Human Condition Have Failed* (Scott) 82
Sekhwela, M. M. 292, 293, 297
Selwood, Daniel 182
Senior, A. 225
sensitization 21
Sentime, K. 292
settler colonialism 225; focus on settlement 231
The Seventh Seal (movie) 307
Sewell, William H., Jr. 125
Seyfang, G. 280
shamification 205
Sharife, K. 293
Sheu, J. 310
Shi, Jiangwei 185
shipbreaking 93
ships of doom 255, 256, 258–259, 260, 263
Shiu, H. 310
*Shopping Our Way to Safety: How We Changed from
Protecting Ourselves* (Szasz) 311
Shove, E. 197
Shu, Shi 185
Sicotte, D. 247
Signs (movie) 307

Silva, A. L. P. 311
Silverman, D. G. 50
Simatele, D. M. 297, 298
Simmons, Erica S. 126
Simons, Marlise 260
Singh, Vijai 171
Sipper, Bill 183
situatedness of waste 4
Skelcher, C. 285
Skinner, B. F. 197
Skocpol, Theda 125
Slate, N. 43
slavery 43, 270, 307
sludge waste 255
small and medium-scale enterprises (SMMEs) 294
*Small Business Liability Relief and Brownfields
Revitalization Act* (2002, US) 241
Smith, A. 280
Smith, Adam 9
Smith, Nicholas Rush 126
smoothing mechanisms 36
Snyder, Richard 126
Snyman, J. 294
Sobolev, Sergei 216
social and solidarity economy (SSE) policies 282
social construction of waste 4–5
social innovations 280
social scientist framing 22
socio-economic metabolism of societies 90
Sohrabi, C. 310
solid waste governance 130
solid waste management 130
Solomon, M. 41
Solomon, Marisa 44, 45
Solomon, N. 247
Soma, Tammara 13, 15, 106, 142, 196, 202, 203,
204, 205
Somers, Margaret 125
Sontag, Susan 70
South African Waste Information Centre
(SAWIC) 294
South African Waste Information System
(SAWIS) 294
Souza, Aline 281
Spalding, Heather 256, 258, 259
Srnec, C. 279
Stamatopoulou-Robbins, Sophia 44, 58–59, 65
Star, Susan Leigh 178
State Council of the PRC 155
Statistics Canada 310
Stefanovich, O. 144
Stegmann, Rainer 185
Steinberg, Michael K. 107, 110
Steinmo, Sven 124
Steuer, Benjamin 13, 16, 155, 157, 158, 159, 161,
163, 164, 165, 166
Steyn, P. 298

stigmatization of people involved in waste 36–37
stockpiling 310
Stoekl, Allan 269, 270
Stoett, Peter 263
Stokes, K. 293, 296, 297, 298
Stolen Generations (Australia) 228, 231
"The Story of Bottled Water" (video, Fox) 144
Strach, Patricia 124
Strasser, Susan 8, 89, 94
Strohm, Laura 255, 258
structured comparisons 123
Stuart, Tristram 6, 106
Studdert, D. 276
Sturzenegger, G. 275
Styrofoam 35
Sullivan, Kathleen 124, 314
Sunder Rajan, K. 50
Sundqvist, Göran 214
Sunshine (movie) 307
supranational organizations 307
survivalist materials 310
survival of mankind 306
sustainability 175; cost recovery as necessity for
    295–296
sustainable development goals (SGDs) 276; goal
    number 11 283; goal number 12 284–285;
    waste-pickers and 282–285
Sutherland, Anne 107
Suttibak, S. 196
Šyc, Michal 186
Szasz, Andrew 5, 311, 313, 315, 317
Sze, J. 239

T
TallBear, Kim 47
Tao, D. 155, 158, 161, 163
Tarr, Joel 88
tax increment funding (TIF) 242
Taylor, Dorceta 187
teaching critical waste studies 139–149; broad and
    multi-disciplinary 140; classroom resources
    140–141; metabolism 145–147; pedagogical
    approach 139–140; power dynamics 143–144;
    waste flows 145–147; waste governance
    147–149; waste materiality 144–145; waste
    regimes 147–149
technofossils 90–93
Tedlock, Dennis 110
Temple, Henry John, 3rd Viscount Palmerston 32
temporal dimensions of waste 89–90
temporality 12; material culture's 98
The Terminator (movie) 307, 309
terra nullius (British legal framing of land claim) 228,
    230, 231
Terraza, H. 275
Theater of Operations, The Gulf Wars 1991–2011
    (MoMA exhibition) 272

Thelen, Kathleen 124
Theron, J. 294, 296
Thieme, A. 201, 202
Thomas, H. 280
Thomashausen, Sophie 129
Thompson, Gregory 105, 111
Thompson, Michael 5, 47, 69
Thrift, Nigel 6, 97
throwaway consumption 89
throwaway society 141
Tiffin, H. 225
The Times 32
tip of the iceberg hypothesis 260, 261
Tironi, M. 41, 45, 240
Tobon, S. 196, 205
to frame, the term 70–71; meanings 70–71
Toh, M. 310
Toledo Maya Cultural Council (TMCC) 110, 111
Tong, X. 155, 158, 161, 163
Tonjes, David J. 177
Townsend, Timothy 171
toxicity of household waste 312
toxic waste removal 35
tragedy of the commons 37
Trainer, T. 149
transnational NGOs 260
Tremblay, Crystal 130
Trois, C. 294
Trump, Donald 310, 316, 317
Tsing, Anna L. 63, 69
Tuçaltan, G. 148
Tufano, Linda 175, 184
Tuhiwai Smith, Linda 132
Turner, S. 41
Turok, I. 291
12 Monkeys (movie) 307
Tynan, E. 229, 230, 232, 233
Tzuc, Felix Medina 105, 108, 109

U
Uddin, Sayed Mohammad Nazim 129
ultimate sink 90, 93, 95–96, 98
UN Conference on Trade and Development
    (UNCTAD) 311
"Understanding Wastes" (O'Neill) 142
UN Development Program (UNDP) 276, 282
UN Framework Convention on Climate Change
    (2000, the Hague) 256
UN Habitat World Cities Report 282
United Nations High Commissioner for Refugees
    (UNHCR) 307
universal nature 46
unmaking the made 88–89, 93; complexities of 88;
    questionable temporal concepts 88–89; tedious
    and hazardous processes in 95–96; temporalities
    of 96
Uraltes chymisches Werk (Eleazars) 96

uranium: Australian histories of 228–230; exports of from Australia 225–226; extraction, production and use of 224; laws to limit harm done by 226; settler-colonial imaginaries about 224
urbanization 56, 90
urban sanitation 92
urban waste economies 14
used electronics trade 254
US-UK Combined Development Authority (CDA) 225
utilitarianism 9

V
Vallette, Jim 256, 258, 259
value or waste determinacy 142
Valverde, Mariana 170
Van Bemmel, A. 145
Van Der Linden 205
van der Straeten, J. 89, 90
van der Werf, P. 196, 205
Van Horen, Basil 7
van Wyck, Peter C. 215
Varda, Agnes 25
Vargas, Joao H. Costa 45
Varul, M. Z. 5
Vavrus, Frances 125
Veblen, Thornstein 5, 11
Vegelin, C. 276, 282
Velis, C. A. 275
Veracini, L. 225
Véron, R. 292
Verran, Helen 228
Vibrant Matter: A Political Ecology of Things (Bennett) 213
Viljoen, J. M. M. 292, 293
visceral perception of waste 22
Visser, M. 294, 295, 296
Viveiros de Castro, Eduardo 105
Vogt, Evon 111, 262
Vogtz, Franziska 131
Volk, T. 4
Vorster, K. 294
Voyles, Traci Brynne 215, 224, 225, 226, 234

W
Wachal, Robert 175
wage labor discipline 43
Wagner, S. 314
Waigani Convention, 1995 260
Wainwright, Joel 107
Waitt, G. 145
Walkerdine, V. 276
Wallerstein, N. 276
Walsh, Daniel C. 179
Walther, Daniel J. 10
Wang, F. 160

Waring, George Edwin 174
War of the Worlds (movie) 307
Warren, Hayley 183
waste: character of 4; circulation of leads to violence 41, 45–46; connection with modernity 8–9; as concept 104; as creation of capitalist modes of valuation 115; definition of as danger to nature 99; how it becomes 141–142; materiality of 5; as opposite of value 5; origins of word 4; as part of system 25; place nexus of 7; race relationship to 41–43; as reflection of cultural values 103; temporal dimensions of 89
Waste (O'Neill) 142
Waste Act (2008, South Africa) 294, 295
waste activism 148
Waste and Want: A Social History of Trash (Strasser) 8
WasteApp 201
waste as cultural exchange 111–112
waste as object *vs.* waste as streams 141
waste chains 7
waste characterization 178–179
waste collection 6–7
waste colonialism 143, 189
waste composition 177–178
waste disposal 94; comparison of treatment and disposal systems 122
waste distancing 315
waste dumping 258; of toxic sludge 258, 260; transboundary 259
waste economies 270
waste-engendering social patterns 23
waste engineering 93
waste ethnographies 212
waste flows 145–147; rethinking of in South Africa 291–300
waste generators 171; classification of 171; disposal and 172
waste governance 13, 147–149, 285–286; comparative studies in 124, 129
waste-inducing system breakdowns 23
waste labor 7; changing dynamics of 296–298; intersectionality of dirtiest aspects of 7
wasteland 10; Australia as 224, 226, 229
The Waste Makers (Packard) 5
waste management (WM) 93; in Britain 56–57; capture rate 181–182; contamination rate 182–183; as critical public service 285; as cultural performance 71; fragmented systems in PRC 155; grassroots social innovations in 280–282; processing costs 175; reasons for disposal over discard 175–176; social costs of 187; South African state roles in 293–295; spaces and technologies of 89
Waste Management and Sustainable Consumption: Reflections on Consumer Waste (Ekström) 142
waste management (WM) industry 24–25; forgetability or deniability of waste in 42

waste management (WM) measurements 169–178; capture category 169, 181–182; composition category 169, 177–181; contamination category 169, 182–183; disposal category 169, 173–174; diversion category 169, 174–177; generation category 169, 171–173
waste management (WM) regimes 275
waste management (WM) sites 55–56
waste management (WM) systems 154–155; continuous measurement approach 183–184; tip scale as measurement method 176–177
waste materiality 144–145
"The Waste of the World" (Gregson) 20, 22, 24
waste-picker cooperatives 286
waste-picker networks 284–285
waste-pickers 59, 127; global numbers of 275–276; innovations by 276; labor rhythms of 60–61; as major recycling force 275–276; member-based organizations (MBOs) of 276–279; organization of 276–278; reasons for working as 60; sustainable development goals (SGDs) and 282–287; worldwide organizations 278–279
waste politics (in South Africa) 291–300
waste reclaimers 292–293
waste regimes 93, 147–149
waste regulations 293–294
waste scholars 21; theory of power in 33–34
waste studies 15; awareness campaigns 196; comparative studies in social science of waste 121–122; compared to discard studies 15; comparison as methodological strategy 132; education to engage communities in 196; fundamentals of 256; gamification approach to 196–206; gendered analyses in 142; interdisciplinary nature of 27; land reuse as central case for 238–239; landscapes and people as disposable 142; as normative field 31; political economy of waste and value at core of 25; radioactive waste 211–212; on social construction of waste 104–105
waste systems three Rs as guide for 146
Waste Today 174
waste-to-energy 64, 65, 285
waste treatment 122
wastewater treatment 129
Waste Wise campaign (Cape Town, South Africa) 298
waste workers 37, 38
water access 129
water governance 8
water practices ethnographies 105
Waters, Colin N. 63, 215
water workers 144
Watson, Matt 145, 169
Watt-Cloutier, S. 307
Weart, Spencer R. 216
Weber, H. 88, 89, 90, 93, 94, 95, 96, 98, 100

Weber, Max 71
Weber-Waltraud, Heike 12
Wee, S. C. 198
Weheliye, A. 43
Weible, Christopher M. 124
Weinberg, Adam 7
Werbach, K. 197
Wernstedt, K. 239, 240, 241, 244, 248, 249
Westermann, A. 90
Westra, L. 41
"The What and Why of Discard Studies" (Liboiron & Lepawsky) 142
What a Waste 2.0: A Global Snapshot of Solid Waste Management to 2050 (World Bank) 172–173
Wheatfield (art installation, Denes) 64
When All Hell Breaks Loose: Stuff You Need to Survive When Disaster Strikes (Lundin) 308
When Technology Fails: A Manual for Self-Reliance, Sustainability, and Surviving the Long Emergency (Stein) 308
white privilege 46
white supremacy 41–42; in built environment 45; scientific objectivity and 47
White Trash: The 400 Year Untold History of Class in America (Isenberg) 46
Whitlam, Gough 227
Wilk, Richard R. 105, 108, 111
Wilson, D. 275
Wilson, David C. 129, 175, 254
Wilson, I. 230
Wilson, M. W. 250
Windfield, Elliott Steen 171
Winfield, Mark 129
Winiwarter, Verena M. 100
Winter, L. 316
Wirth, David A. 260
Wittmer, Josie 37, 130, 144
Wolfe, Patrick 225
Women in Informal Employment: Globalizing and Organizing (WIEGO) 279
Wonneberger, A. 205
World Bank 57, 64, 172, 307, 309
World Commission on Environment and Development 256
World Health Organization (WHO) 310, 316
World Trade Organization (WTO) 256
World War Z (movie) 307
Worldwide Waste: Journal of Interdisciplinary Studies 3
Wright, M. W. 142
Wynne, Brian 7, 8, 36, 62, 255, 256, 258, 259, 262, 263, 312

X
Xiao, S. 155, 158

Y
Yaffe-Bellamy, D. 311

Yang, Shiming 263
Yasunari, Teppei J. 215
Yates, M. 43, 51
Year Zero (Rose) 229
Yee, Amy 185
York, Richard 38
Young, Oran R. 261
Yu, D. 293

Z
Zahara, Alex 132
Zak, D. 310
Zalasiewicz, Jan 63, 90, 309
Zaman, A. U. 142
Zapata, Patrick 59
Zapata Campos, María José 59
Zelinka, I. 201, 204, 205

Zeller, Vanessa 184
Zero Waste movement 142, 187
zero waste vision 294
Zhang, A. 46
Zhang, M. 159
Zhao, J. 201, 204, 205
Zhu, J. 154
Zhu, Wei 185
Zimlich, Rachel 185
Zimmer, A. 292
Zimring, Carl 41, 43, 44, 46, 94
Zoellner, Tom 214
The Zombie Survival Guide: Complete Protection from the Living Dead (Brooks) 307
Zonabend, Françoise 211
Zuckerman, Alan S. 124

For Product Safety Concerns and Information please contact our EU
representative GPSR@taylorandfrancis.com
Taylor & Francis Verlag GmbH, Kaufingerstraße 24, 80331 München, Germany

www.ingramcontent.com/pod-product-compliance
Lightning Source LLC
Chambersburg PA
CBHW081047220326
41598CB00038B/7016